Advances in Science, Technology & Innovation

IEREK Interdisciplinary Series for Sustainable Development

Advances in Science, Technology & Innovation (ASTI) is a series of peer-reviewed books based on the best studies on emerging research that redefines existing disciplinary boundaries in science, technology and innovation (STI) in order to develop integrated concepts for sustainable development. The series is mainly based on the best research papers from various IEREK and other international conferences, and is intended to promote the creation and development of viable solutions for a sustainable future and a positive societal transformation with the help of integrated and innovative science-based approaches. Offering interdisciplinary coverage, the series presents innovative approaches and highlights how they can best support both the economic and sustainable development for the welfare of all societies. In particular, the series includes conceptual and empirical contributions from different interrelated fields of science, technology and innovation that focus on providing practical solutions to ensure food, water and energy security. It also presents new case studies offering concrete examples of how to resolve sustainable urbanization and environmental issues. The series is addressed to professionals in research and teaching, consultancies and industry, and government and international organizations. Published in collaboration with IEREK, the ASTI series will acquaint readers with essential new studies in STI for sustainable development.

More information about this series at http://www.springer.com/series/15883

Federico Rossetti • Ana Crespo Blanc
Federica Riguzzi • Estelle Leroux
Kosmas Pavlopoulos • Olivier Bellier
Vasilios Kapsimalis
Editors

The Structural Geology Contribution to the Africa-Eurasia Geology: Basement and Reservoir Structure, Ore Mineralisation and Tectonic Modelling

Proceedings of the 1st Springer Conference of the Arabian Journal of Geosciences (CAJG-1), Tunisia 2018

Springer

Editors
Federico Rossetti
Università Roma Tre
Roma, Italy

Ana Crespo Blanc
University of Granada
Granada, Spain

Federica Riguzzi
Istituto Nazionale di Geofisica e Vulcanologia
Roma, Italy

Estelle Leroux
Unité Géosciences Marines—IFREMER
Plouzané, France

Kosmas Pavlopoulos
Paris Sorbonne University Abu Dhabi
Abu Dhabi, United Arab Emirates

Olivier Bellier
Centre Européen de Recherche et d'Enseignement en Géosciences de l'Environnement (CEREGE)
Aix-en-Provence, France

Vasilios Kapsimalis
Institute of Oceanography
Anavyssos, Greece

ISSN 2522-8714 ISSN 2522-8722 (electronic)
Advances in Science, Technology & Innovation
IEREK Interdisciplinary Series for Sustainable Development
ISBN 978-3-030-01454-4 ISBN 978-3-030-01455-1 (eBook)
https://doi.org/10.1007/978-3-030-01455-1

Library of Congress Control Number: 2018960212

This Springer imprint is published by the registered company Springer Nature Switzerland AG
The registered company address is: Gewerbestrasse 11, 6330 Cham, Switzerland

Foreword

Deep seismic imagery and mantle tomography have considerably increased our understanding of the lithosphere and crustal architecture, as well as sedimentary basin formation and deformation processes. In the meantime, high-resolution seismic reflection profiles, as calibrated against exploration wells and outcrop studies, also provide a better control of the sedimentary infill and 3D reservoir architecture. Seemingly, source-to-sink studies, coupled with subsidence and kinematic modelling calibrated against paleothermometers, provide accurate reconstructions of the long-term tectono-stratigraphic evolution of sedimentary basins. Thermal and fluid flow modelling are now routinely used to better understand and predict the formation and distribution of ore deposits and petroleum potential, along with the permeability distribution of their hosting reservoirs.

These proceedings volume comprises 74 papers accepted for presentation during the 1st Springer Conference of the Arabian Journal of Geosciences (CAJG-1), Tunisia, 2018. The major topics and issues treated include: (i) Basement geology; (ii) Fluid–rocks interactions, hydrothermalism and ore deposits; (iii) Reservoir geology, structure and stratigraphy; (iv) Mediterranean tectonics; (v) The Alpine-Himalayan convergence zone; (vi) Tectonic modelling.

Individual papers presented in this respect range from local case studies, predominantly relating to the North African, Mediterranean and Mid-Eastern regions, to more global geodynamic syntheses dealing with continental break-ups, oceanic accretion and passive margins' development. The present document will provide an updated portfolio, evidencing the input of structural geology to the understanding of lithosphere and crustal-scale processes activating in the broader Afro-Eurasian realm. It is intended to stand as a highly interesting reference for both students and senior scientists and professionals.

August 2018 François Roure

François Roure is a graduate of the Ecole Normale Supérieure of St-Cloud, France, holder of a Doctorate in Sciences from the University of Paris VI. He joined the IFP's Geology-Geochemistry-Geophysics Division in 1984, following a 4-year career at the CNRS. His research is predominantly focused on the study of sedimentary basins (architecture and geodynamics, thermicity, oil-bearing systems, fluid/rock interactions and reservoir characterization). He was selected as an extraordinary Professor at the Free University of Amsterdam (VU). He also chaired the working group on sedimentary basins in the International Lithosphere Program (ILP). He has contributed more than a hundred articles, published in prestigious international journals. In 2010, François Roure was honoured by the EAGE (European Association of Geoscientists and Engineers), which selected him as the winner of the 2010 Wegener Award, as a recognition of his contribution to the geoscientific research in the area of petroleum exploration of frontier areas and the search for new reserves, particularly in mountain belts.

Acknowledgements

Our appreciation is extended to the authors of the papers for their hard and diligent work and producing high-quality contributions. We would like to thank the reviewers of the papers for their in-depth reviews and great efforts in improving the quality of the papers. Also, thanks are extended to Amjad Kallel who supervised and handled the evaluation process, to Sahbi Moalla who handled the submission and evaluation system for the ten conference proceedings volumes, and the publishing staff of Springer headed by Nabil Khélifi, Senior Editor for their efforts and contributions in completing this conference proceedings volume. All the above-mentioned efforts were very important in making this book a success.

About the 1st Springer Conference of the Arabian Journal of Geosciences (CAJG-1), Tunisia 2018

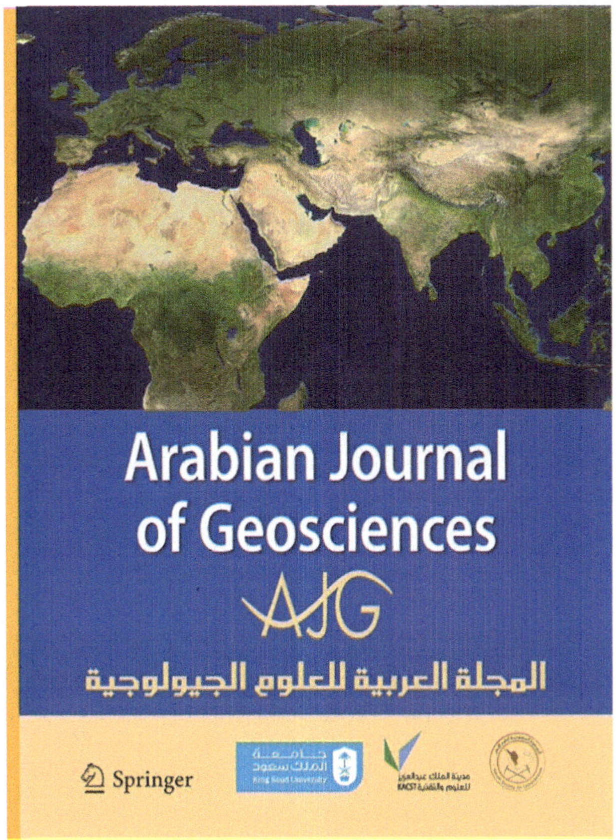

The *Arabian Journal of Geosciences* (AJG) is a Springer journal publishing original articles on the entire range of Earth sciences in partnership with the Saudi Society for Geosciences. The journal focuses on, but not limited to, research themes which have regional significance to the Middle East, the Euro-Mediterranean, Africa, and Asia. The journal receives on average 2000 submissions a year and accepts around 500 papers for publication in its 24 annual issues (acceptance rate 25%). It enjoys the participation of an editorial team of 100 international associate editors who generously help in evaluating and selecting the best papers.

In 2008, Prof. Abdullah Al-Amri, in close partnership with Springer, founded the Arabian Journal of Geosciences (AJGS). In this year, the journal celebrates its tenth anniversary. On this occasion and to mark this event, the Founder and Editor-in-Chief of the AJGS Prof. Al-Amri organized in close collaboration with Springer the 1st Conference of the Arabian Journal of Geosciences (1st CAJG) in Hammamet, Tunisia from November 12 to 15, 2018 (www.cajg.org).

The conference was an occasion to endorse the journal's long-held reputation for bringing together leading authors from the Middle East, the Euro-Mediterranean, Africa, and Asia who work at the wide-ranging fields of Earth sciences. The conference covered all crosscutting themes of Geosciences and focused principally on the following ten tracks:

- Track 1. Climate, paleoclimate, and paleoenvironmental changes
- Track 2. Geoinformatics, remote sensing, geodesy
- Track 3. Geo-environmental engineering, geomechanics and geotechnics, geohazards
- Track 4. Geography, geoecology, geoarcheology, geotourism
- Track 5. Geophysics, seismology
- Track 6. Hydrology, hydrogeology, hydrochemistry
- Track 7. Mineralogy, geochemistry, petrology, and volcanology
- Track 8. Petroleum engineering and petroleum geochemistry
- Track 9. Sedimentology, stratigraphy, paleontology, geomorphology, pedology
- Track 10. Structural/petroleum/mining geology, geodynamics, marine geology

The dynamic four-day conference provided more than 450 attendees with opportunities to share their latest unpublished findings and learn the newest geoscience studies. The event also allowed attendees to meet and discuss with the journal's editors and reviewers.

More than 950 short contributing papers to the conference were submitted by authors from more than 70 countries. After a pre-conference peer review process by more than 500 reviewers, 700 papers were accepted. These papers were published as chapters in the conference proceedings by Springer.

The conference proceedings consist of ten edited volumes, each edited by the following group of Arabian Journal of Geosciences (AJGS) editors and other guest editors:

Volume 1. Patterns and Mechanisms of Climate, Paleoclimate, and Paleoenvironmental Changes from Low-Latitude Regions

Zhihua Zhang (AJGS Editor): Beijing Normal University, Beijing, China
Nabil Khélifi (AJGS Editor): Earth Sciences Editorial Department, Springer, Heidelberg, Germany
Abdelkader Mezghani (Guest Editor): Norwegian Meteorological Institute, Norway
Essam Heggy (Guest Editor): University of Southern California and NASA Jet Propulsion Laboratory, Caltech, USA

Volume 2. Advances in Remote Sensing and Geo informatics Applications

Hesham M. El-Askary (Guest Editor): Schmid College of Science and Technology at Chapman University, USA
Saro Lee (AJGS Editor): Korea Institute of Geoscience and Mineral Resources, Daejeon, South Korea
Essam Heggy (Guest Editor): University of Southern California and NASA Jet Propulsion Laboratory, Caltech, USA
Biswajeet Pradhan (AJGS Editor): University of Technology Sydney, Sydney, Australia

Volume 3. Recent Advances in Geo-Environmental Engineering, Geomechanics and Geotechnics, and Geohazards

Amjad Kallel (AJGS Editor): ENIS, University of Sfax, Tunisia
Zeynal Abiddin Erguler (AJGS Editor): Dumlupinar University, Kutahya, Turkey
Zhen-Dong Cui (AJGS Editor): China University of Mining and Technology, Xuzhou, Jiangsu, China
Ali Karrech (AJGS Editor): The University of Western Australia, Australia
Murat Karakus (AJGS Editor): University of Adelaide, Australia

Pinnaduwa Kulatilake (AJGS Editor): Department of Materials Science and Engineering, The University of Arizona, USA
Sanjay Kumar Shukla (AJGS Editor): School of Engineering, Edith Cowan University, Perth, Australia

Volume 4. Exploring the Nexus of Geoecology, Geography, Geoarcheology, and Geotourism: Advances and Applications for Sustainable Development in Environmental Sciences and Agroforestry Research

Haroun Chenchouni (AJGS Editor): University of Tebessa, Algeria
Ezzoura Errami (Guest Editor): Chouaïb Doukkali University, El Jadida, Morocco
Fernando Rocha (Guest Editor): University of Aveiro, Portugal
Luisa Sabato (AJGS Editor): Università degli Studi di Bari 'Aldo Moro', Bari, Italy

Volume 5. On Significant Applications of Geophysical Methods

Narasimman Sundararajan (AJGS Editor): Sultan Qaboos University, Muscat, Oman
Mehdi Eshagh (AJGS Editor): University West, Trollhättan, Sweden
Hakim Saibi (AJGS Editor): United Arab Emirates University, Al-Ain, Abu Dhabi, UAE
Mustapha Meghraoui (AJGS Editor): Université de Strasbourg, Strasbourg, France
Mansour Al-Garni (AJGS Editor): King Abdulaziz University, Jeddah, Saudi Arabia
Bernard Giroux (AJGS Editor): Centre Eau Terre Environnement, Québec, Canada

Volume 6. Advances in Sustainable and Environmental Hydrology, Hydrogeology, Hydrochemistry, and Water Resources

Helder I. Chaminé (AJGS Editor): School of Engineering—ISEP, Polytechnic of Porto, Portugal
Maurizio Barbieri (AJGS Editor): University of Rome La Sapienza, Italy
Ozgur Kisi (AJGS Editor): Ilila State University, Tbilisi, Georgia
Mingjie Chen (AJGS Editor): Sultan Qaboos University, Muscat, Oman
Broder J. Merkel (AJGS Editor): TU Bergakademie Freiberg, Freiberg, Germany

Volume 7. Petrogenesis and Exploration of the Earth's Interior

Domenico Doronzo (AJGS Editor): Consejo Superior de Investigaciones Cientificas, Spain
Emanuela Schingaro (AJGS Editor): Università degli Studi di Bari Aldo Moro—UniBa, Italy
John S. Armstrong-Altrin (AJGS Editor): The National Autonomous University of Mexico, Mexico
Basem Zoheir (Guest Editor): Benha University, Egypt and University of Kiel, Germany

Volume 8. Advances in Petroleum Engineering and Petroleum Geochemistry

Santanu Banerjee (AJGS Editor): Indian Institute of Technology Bombay, Mumbai, India
Reza Barati (AJGS Editor): The University of Kansas, Lawrence, KS, USA
Shirish Patil (Guest Editor): Saudi Aramco and King Fahd University of Petroleum and Minerals, Dhahran, Saudi Arabia

Volume 9. Paleobiodiversity and Tectono-sedimentary Records in the Mediterranean Tethys and Related Eastern Areas

Mabrouk Boughdiri (AJGS Editor): University of Carthage, Amilcar, Tunisia
Beatriz Bádenas (AJGS Editor): University of Zaragoza, Zaragoza, Spain

Paul Selden (AJGS Editor): University of Kansas, Lawrence, Kansas, USA
Etienne Jaillard (Guest Editor): Université Grenoble Alpes, France
Peter Bengtson (AJGS Editor): Universität Heidelberg, Heidelberg, Germany
Bruno R. C. Granier (AJGS Editor): Université de Bretagne Occidentale, Brest, France

Volume 10. The Structural Geology Contribution to the Africa-Eurasia Geology: Basement and Reservoir Structure, Ore Mineralisation and Tectonic Modelling

Federico Rossetti (Guest Editor): Università Roma Tre, Roma, Italy
Ana Crespo Blanc (Guest Editor): University of Granada, Spain
Federica Riguzzi (Guest Editor): National Institute of Geophysics and Volcanology, Roma, Italy
Estelle Leroux (Guest Editor): IFREMER, Unité Géosciences Marines, Plouzané, France
Kosmas Pavlopoulos (Guest Editor): Paris Sorbonne University Abu Dhabi, Abu Dhabi, UAE
Olivier Bellier (Guest Editor): CEREGE, Aix-en-Provence, France
Vasilios Kapsimalis (Guest Editor): Institute of Oceanography, Hellenic Centre for Marine Research, Anavyssos, Greece

About the Conference Steering Committee

General Chair

Abdullah Al-Amri: Founder and Editor-in-Chief of AJGS, King Saud University, Saudi Arabia

Conference Supervisor

Nabil Khélifi: Senior Publishing Editor, Springer Middle East and North African Program Springer, a part of Springer Nature, Heidelberg, Germany

Scientific Committee Chair

François Roure: Guest of Editorial Board of AJGS, IFP—Energies Nouvelles, France

Walter D. Mooney: Guest of Editorial Board of AJGS, US Geological Survey Western Region, USA

Local Organization Chair

Mabrouk Boughdiri: Associate Editor of AJGS, University of Carthage, Bizerte, Tunisia

Evaluation Chair

Amjad Kallel: Assistant Editor of AJGS, ENIS, University of Sfax, Tunisia

Publication Chair

Biswajeet Pradhan: Associate Editor of AJGS, University of Technology Sydney, Sydney, Australia

Essam Heggy: Guest of Editorial Board of AJGS, University of Southern California and NASA Jet Propulsion Laboratory, Caltech, USA

Program Chair

Hakim Saibi: Associate Editor/Assistant Editor of AJGS, United Arab Emirates University, Al-Ain, Abu Dhabi, UAE

Domenico Doronzo: Associate Editor/Assistant Editor of AJGS, Consejo Superior de Investigaciones Cientificas, Spain

Communication Chair

Mohamed Ksibi: Guest of Editorial Board of AJGS, ISBS, University of Sfax, Tunisia

English Language Advisory Committee

Abdelmajid Dammak: ENIS, University of Sfax, Tunisia
Chokri Khalaf: FMS, University of Sfax, Tunisia
Dhouha Mabrouk: FLSHS, University of Sfax, Tunisia
Mohamed Elbahi: ENIS, University of Sfax, Tunisia
Sami Shami: ENIS, University of Sfax, Tunisia
Yasmine Basha: FLSHS, University of Sfax, Tunisia

Conference Manager

Mohamed Sahbi Moalla: Coordinator of AJGS, ISET, University of Sfax, Tunisia

Contents

Part III Fluid-Rock Interaction, Hydrothermalism and Ore Deposits

Part IV Reservoir Geology, Structure and Stratigraphy

About the Editors

Dr. Federico Rossetti is a holder of a degree in Geological Sciences (1995), and a Ph.D. in Earth Sciences (1999) from the Sapienza University of Rome, Rome (Italy). He is currently Professor of Structural Geology at the University of 'Roma Tre', Roma (Italy). His research is focused on understanding ductile and brittle continental tectonics in different geodynamic settings, as observed on different spatio-temporal scales. He has executed extensive fieldwork across the Alpine-Himalayan convergence zone (Betic-Rif chain, Apennine and Alpine chain, Iran), the Pan-African orogen in Tanzania and the Ross-Delamerian orogen in Antarctica. He was awarded the 2005 'Felice Ippolito Prize' for the category Earth Sciences, for the research he conducted in Antarctica. His approach is multidisciplinary, integrating geological field survey, structural geology, metamorphic petrology and geodynamic synthesis. He has been occupying the position of a Topical Editor for the EGU journal Solid Earth ever since 2010, Chief Editor since 2018. He is currently Associate Editor of the AGU journal Tectonics. He is the author of more than 100 research papers published in internationally indexed and refereed journals. In 2015, Dr. Rossetti joined the AJGS as an Associate Editor responsible for evaluating submissions pertaining to the areas of Structural Geology, Continental Tectonics and Metamorphic Petrology.

Ana Crespo Blanc born in Switzerland, is a Full Professor of Geodynamics at the University of Granada (Spain) ever since 1993. She occupied the position of President of the Geological Society of Spain, between 2008 and 2012. Her research has been focused mainly on structural and kinematic analysis, relationships between deformation and metamorphism, isotopic geochemistry, marine geology and analogue modelling. She applied these different methodologies to the knowledge of the tectonic evolution of key areas of the Hercynian chain of Iberia, the Alps, the Himalaya or the Nankay Trough. Since the 90s, her research has mainly been centred on the Mediterranean region, specifically, the Betic-Rif orogen around the Gibraltar Arc. She is particularly interested in the architecture and the strain partitioning of such an arcuate orogen. Presently, she is interested in the kinematic reconstructions of the Western Mediterranean, regarding which she characterized, collaboratively with her research group, the different deformation events affecting the whole area. She is simultaneously involved in significant outreach research activities closely associated with earth sciences.

Federica Riguzzi is a holder of a degree in Physics and a Ph.D. in Earth Sciences at the University of Rome La Sapienza. Senior Researcher at Instituto Nazionale di Geofisica e Vulcanologia, Rome, Italy. Her research activity is focused mainly on geodetic methods applied to geophysics. It concerns both methodological and applied studies dedicated to the realization and measurement of crustal deformation control and to fundamental Physics experiment networks; time series analyses of GPS observations, as related to the study of the deformation fields and seismotectonic processes, and the use of historical information and recent seismic data applied to seismic risk areas. The geodetic methodologies have been applied to geophysical studies relating to different regions and with various aims. Her contributions are predominantly centred on the definition of global-scale plate motion models in different reference frames; the study of kinematics and deformation fields of the central Mediterranean and Suez-Sinai areas, along the Betic Cordillera (Spain and Morocco) and the Eastern Himalaya Syntaxis (NE India) through GNSS observations; the analysis of co- and post-seismic deformations after the occurrence of significant seismic events; the estimation of deformation source of the Colli Albani volcano (Roma); the definition of a new methodology based on the joint analysis of deformation rates and seismicity data to assess the earthquake short-term occurrence probability. She has also conducted a number of macroseismic surveys, following the occurrence of significant seismic events, and historical seismicity- related studies. She collaborates with national and international research groups, the Universidad de Jaen (Spain), the National Research Institute of Astronomy and Geophysics (Egypt), as well as the Geological Survey and the Raman Center for

Applied and Interdisciplinary Sciences of Kolkata (India). She has been responsible for the Research Units, financed by the Gruppo Nazionale di Difesa dei Terremoti, and the Department of Civil Defense. Additionally, she took part in several projects supported by the Italian Space Agency. She is an Associate Editor of the Annals of Geophysics, and the author of various scientific papers published in reputably rated national and international journals.

Estelle Leroux obtained her marine technician degree at the National Institute of Science and Marine Technology (INTECHMER), France. She began her career in 2000 working for a few years in a Petroleum Seismic exploration for CGG International (General Company of Geophysics), especially in the Gulf of Mexico offshore. Subsequently, she spent several years reviewing data acquisition and processing on scientific oceanographic vessels, and in different French academic research institutes (IFREMER—French Institute for Marine Research, CNRS—French National Center for Scientific Research, UBO—University of Western Brittany, and IUEM—European University Institute for Marine Research).

She, then, decided to pursue a scientific career. After completing her Ph.D. at UBO/IFREMER on seismic stratigraphy in the Gulf of Lion, in 2012, she joined the UPMC-Sorbonne University, Paris, actively resuming work on seismic and sequence stratigraphy. Her key study areas were centred on the Cenozoic Western Mediterranean area.

In 2016, Estelle Leroux integrated the IFREMER Center in Brest as a marine geologist. Her research was predominantly focused on the combined application of stratigraphy and modelling, in the analysis of ancient and modern sedimentary basin fills. Her expertise is mainly centred on seismic and sequence stratigraphy, as well as numerical geological modelling, both in 'Source-to-sink' and 'Mud-to-Mantle' approaches.

Her aim is targeting to decode the sediment sequences, sediment architectures and deposit features in past sedimentary basins. It is also dedicated to provide key guides on both qualitative and quantitative paleo-variations of sediment fluxes, sea level, and subsidence in a bid to understand climate and geodynamic changes. She is the (co-)author of scientific publications on internationally recognized peer-reviewed journals, and is also involved in several multidisciplinary projects.

Kosmas Pavlopoulos (KP) was born in Athens, Greece. He received his B.Sc. in Geology, and Ph.D. in Environmental Geomorphology at the University of Athens, Greece in 1986 and 1992, respectively. Since 2013, KP has been a Professor of environmental geomorphology, in the Geography and Planning Department at the University of Paris-Sorbonne, Abu Dhabi. He had been previously appointed in the Geography Department at the Harokopio University, Athens, Greece ever since 1999. He was Chair of the working group on geoarcheology of IAG from 2009 to 2017, and member of the Scientific Committee of the French Archaeological School in Athens, from 2010 to 2017. Since 2017, he has been member of the Executive Committee of IAG (International Association of Geomorphologists), as a Publishing Editor.

He is a recognized specialist in the field of coastal geomorphology, sedimentology, Quaternary geology, seal level changes. His research is predominantly focused on the sediments depositional environments on arid and semi-arid areas, geomorphological evolution of marginal landscapes, sea level changes, hydrogeology and the Wadi dynamics, water resources' management as well as the natural hazards, as predominant in the Mediterranean and Mid-Eastern areas.

He has participated in 62 research projects, and has been the coordinator in 20 among them. He has more than 140 publications in peer-reviewed scientific journals (68 currently shown in Scopus), and his work received more than 677 citations (h-index = 16, co-authors 141, source Scopus) and (156 currently shown in Research Gate), 1181 citations (h-index = 30.88, source Research Gate). KP has been the supervisor of 9 Ph.D., and a member of the supervising committee of 12 Ph.D. students. As an expert in Geosciences and environment, he is member for the evaluation of research projects, launched by the Greek Ministry of Research since 2000, by the French institute of CNRS 2004–2007 (ECLIPSE program), by the French Ministry of Education and Research from 2012–2013 (ANR project), by the MIUR (the Italian Ministry for Education, University and Research) 2015–2016 (Reprise project), and by the Ford Motor Company in 2016 (Conservation and Environmental Grants).

Olivier Bellier is Full Professor of Tectonics at the Aix Marseille University, CNRS, IRD, INRA, Coll France, CEREGE, Aix-en-Provence, France since 2000. He had been a Joint Director of the Science Faculty of Aix-Marseille (up to 2013), then a Joint director (in charge of the Research) of OSU Pythéas Institute—Observatory of the Universe and Environment of Aix-Marseille University (2013–2018). He is presently a Director of the CEREGE Laboratory (AMU, CNRS, IRD, Collège de Fr., INRA). His research is focused mainly on Tectonics and Seismic hazard: Brittle and active tectonics, Tectonic geomorphology, Structural geology, which he applied to the study of Geodynamics and Geological risks. His research is partly centred on investigating the Mediterranean area: Tunisia, Morocco, SE France, Italy, etc. However, he developed other research collaborations: USA-USGS, China, Peru, Argentina, Venezuela, Indonesia, Turkey, Iran, Armenia, etc. He was a member of the scientific committee of the Sigma program (Seismic Ground Motion Assessment), Consortium EDF, CEA, Areva, ENEL, Italy, (2011–2015), and a member of the scientific committee of the RGF (Référentiel Géologique de la France), the BRGM program, 2014–2017. He was also a member of the Steering Committee of Geological Society of France (Société Géologique de France), 2012–2015, and a member of the expert committee of ECOS Nord (Committee of the scientific cooperation evaluation between France and South America from the Foreign Affairs Ministry (Hubert Curien Program) for Columbia, Venezuela, Mexico. He was leader of the Steering Committee of the International Scientific Congress «Provence 2009 : Seismic risk in moderate seismicity area»—Aix-en-Provence, July 2009, and a leader of the Steering and Scientific Committees of the First International workshop GeoriskMor «Geosciences and Risk assessment in Morocco: an integrated approach» Tétouan (Morocco) 28–30 April 2016, Faculty of Sciences, Abdelamalek-Essaâdi, the University of (Tanger-Tétouan), GeoriskMor project, http://georisk-morocco.cerege.fr/, the OT-med workshop. He has been the supervisor of 21 Ph.D. theses since 1991 (19 being presently defended, other in process) and took part in 75 Ph.D. thesis juries, 7 being defended in foreign universities (Iran, Tunisia, Italy, Swiss, etc.). He is the author of 120 international scientific publications (108 indexed in Isi-web of Sciences, in 2018). Besides, he is a co-writer of five scientific books' attached chapters and a co-writer of a chapter of a scientific culture specialized book. He also participated as a co-writer of a jointly scientific culture published book dealing with the Provence earthquake.

Dr. Vasilios Kapsimalis is a Research Director of the Institute of Oceanography of the Hellenic Centre for Marine Research. His scientific interests pertain to the field of marine geology, sediment dynamics and quality, submarine geomorphology, Late Quaternary and Holocene stratigraphy, relative sea level changes, underwater geoarchaeology and management of dredged material coastal erosion. He has published more than 35 papers in peer-reviewed journals, and presented more than 70 original contributions in international and Greek-based conferences. In addition, he has coordinated 25 research projects dealing with sediment quality assessment, monitoring and impact of sediment disposal operations on marine environment, geomorphological mapping of the coastal and offshore areas, paleo-geographical reconstructions of the Greek continental shelf. As a chief scientist, he has organized and participated, in several coastal (in estuarine, lagoonal, beach areas, etc.) and offshore (continental shelves, slopes and deep basins) multidisciplinary surveys, deploying the HCMR's fleet (R/V AIGAIO and S/V ALKYON). His educational work is related to the teaching of graduate and postgraduate courses at the Harokopio University of Athens (Oceanography), and the National Technical University (Marine Geology and Geophysics), along with the supervision of B.Sc., M.Sc. and Ph.D. theses. Additionally, he has organized a number of international and Greek congresses and workshops of various oceanographic and environmental subjects, and disseminated the results of his work in well-recognized magazines and newspapers. He was the representative of the HCMR (2010–2016) to the Licensing Committee on Marine Research of the Greek Foreign Ministry, which issues licenses for each marine research conducted within the Greek-owned waters.

Part I
Keynote

Mediterranean Sea Level and Bathymetry of the Deep Basins During the Salt Giant Deposition: Inference from Seismic and Litho-Facies

C. Gorini, L. Montadert, and B. Haq

Abstract

The partial sequestration of the Mediterranean Sea from adjacent oceans at the end of the Miocene caused an evaporation surfeit that increased the evaporite concentration on the seafloor and peripheral basins. As a result, an up to 2-3 km thick sequence of evaporites were deposited in the center of the deep basins. This coincided with the concomitantly intense subaerial erosion of the adjacent margins. The Messinian salinity crisis (MSC) ended with the catastrophic re-flooding of the Mediterranean. During the Zanclean, slopes and shelves prograded to reach the modern configurations. Neither the widespread erosion nor the regional extent of the very thick halite beds are compatible with a totally desiccated basin. The volume of evaporites deposited in the deep basin also implies a permanent or intermittent connection with the world oceans.

Keywords

Messinian salinity crisis • Deep mediterranean basins
Seismic stratigraphy • Paleoshores

The partial sequestration of the Mediterranean Sea from adjacent oceans at the end of the Miocene caused an evaporation that increased the evaporite concentration on the seafloor and peripheral basins. As a result, up to 2–3 km thick evaporites were deposited in the deep basins [1–3] (Fig. 1). Concomitantly intense subaerial erosion of the adjacent margins took place. The Messinian salinity crisis (MSC) ended with the catastrophic re-flooding of the Mediterranean [2] and it took a few Myr. to rebuild the slopes and shelves to their pre-MSC configurations [3–7].

C. Gorini (✉) · B. Haq
Institut des Sciences de la Terre Paris, Sorbonne Université, CNRS-INSU, ISTeP UMR, 4 place jussieu, 7193 Paris, France
e-mail: christian.gorini@upmc.fr

L. Montadert
Beicip-Franlab, 232 AV. N. Bonaparte BP 213, 92502 Rueil-Malmaison Cedex, France

Here we illustrate the seismic facies of the salt deposited in the deep Mediterranean basins during the MSC with several examples taken from our new observations and from previous works done by numerous researchers. "Deep basins" refers to their position in the deep central parts of the extant Messinian basins that are approximately delineated by the present-day feet of the Mediterranean slopes and abyssal plains around the Ligurian-Provencal basin (30 Myr old basement), the central basins (Ionian) and the eastern basins (older Tethyan basement, see Fig. 1) even if the configuration of these basins were different 6 Myr ago. The paleo-bathymetry of these deep basins has been estimated to be around 2500 m deep in the late Miocene. Given the original depth of the basins, their bathymetry was still deep even after the estimated 1500 m sea level fall. The distribution of evaporitic series and clastics in the eastern and western Mediterranean deep basins, identified by their seismic facies, clearly shows that following the sea-level fall and erosion of the margins, the evaporitic succession in the deep basins was deposited in hypersaline waters in a bathymetric setting of around 1000 m deep. The debate concerning the amplitude of the sea-level drop associated with this crisis and the depth of deposition of the halite continues because of the confusion in distinguishing between variations in sea level of the Mediterranean Messinian and the paleobathymetry of the Mediterranean sub-basins. Moreover the evaluation of paleogeography during the Messinian is complicated by rapid subsidence due to the overloading of the basins (vs. the uplift and tilting of the emerged platforms [4]) and the active tectonics that has rapidly remodeled the morphologies of Mediterranean basins in the last 6 Myr. We use sequence-stratigraphic terminology to facilitate the description and the interpretation we propose, but it should be kept in mind that the MSC depositional mega-sequence requires the definition of a model-independent methodology for the deep Messinian Mediterranean seafloor, which was partly connected to the global oceans during the MSC. MSC seismic units correspond to systems tracts deposited in three main stages that represent a complete sea level cycle: (1) A

F. Rossetti et al. (eds.), *The Structural Geology Contribution to the Africa-Eurasia Geology: Basement and Reservoir Structure, Ore Mineralisation and Tectonic Modelling*,
Advances in Science, Technology & Innovation, https://doi.org/10.1007/978-3-030-01455-1_1

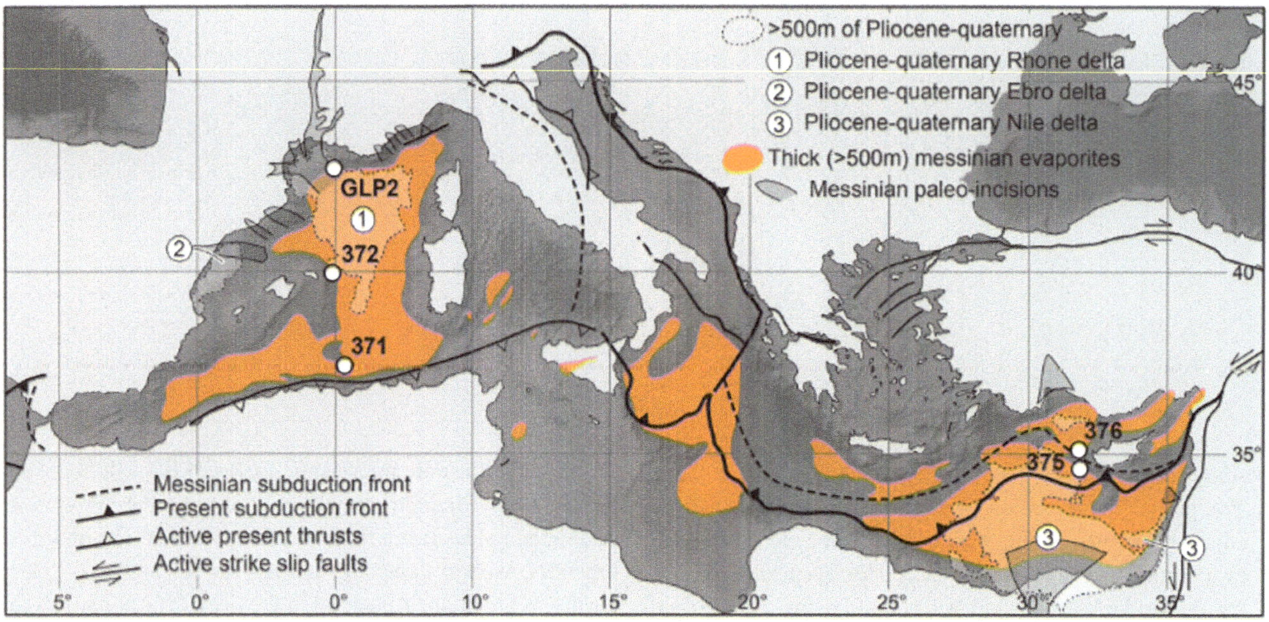

Fig. 1 Position of the thick Messinian salinity crisis evaporites in the western, central, and eastern present-day basins. The position of the extant major deltas is indicated

falling stage characterized by mass—transport deposits and/or forced regressive prograding units related to the increasing rate of relative sea level fall. This stage is characterized by a marked shift in the depocenters towards the deep basins. The falling stage is also characterized by massive clastic inputs from major Messinian rivers (the Rhone, Nile, and rivers in the Antalya Gulf) or from smaller river systems (offshore south Lebanon). These clastics were deposited in basins that had been oversaturated since the salinity crisis began, as evidenced by homogeneous non-transparent seismic facies displaying high seismic velocities at the base of the MSC depositional series (e.g., upper Miocene turbidites) and their inter-fingering with transparent halite facies; (2) Massive regionally-continuous halite deposition characterized by transparent facies with no clastic input follows the mixed lower clastics and evaporites deposition stage; (3) The upper part of the evaporites clearly onlaps the Messinian Erosional Surface of the western basin margins. During this last stage, the results of deep sea drilling projects point to shallow water deposition of the evaporites. Upslope the presence of smooth surfaces corresponds to a transgression surface that subsequently re-shaped the Messinian Erosional Surface which are good indicators of the position of Messinian paleoshorelines at the end of the MSC (sea level 600–800 m under the present sea level) [8, 9]. In the vicinity of the Eratosthenes seamount, the upper Miocene platforms are also a good indicator of the position of the Messinian paleoshores before the reflooding [6]. If we have good indicators of the MSC paleoshore positions at the end of the MSC (e.g. Mediterranean sea level), paleodepth of deposition of the top of the evaporitic succession is still controversed.

In conclusion, (1) Neither the wide dispersal of the clastics nor the regional extent of the very thick halite beds are compatible with a long term desiccated basin. (2) The volume of evaporites deposited in the deep basins also implies a permanent or intermittent connection with the world oceans. (3) The top of the clastic event is coeval with the lowest MSC sea level (but not the lowest deposition depth) and with the sharp transition to the Messinian transparent halite. (4) The thickness of the evaporites records the rapid precipitation of massive halite that filled the accommodation space in a relatively short time and also implies that it was accompanied by an increase in the accommodation (flexural subsidence) and continuous supply of brine.

Bibliography

1. Feng, Y.E., Ynkelzon, A., Steingberg, J., Reshef, M.: Lithology and characteristics of the Messinian evaporite sequence of the deep Levant Basin, eastern Mediterranean. Mar. Geol. **376**, 118–131 (2016)
2. Garcia-Castellanos, D., Estrada, F., Jiménez-M, I., Gorini, C., et al.: Catastrophic flood of the Mediterranean after the Messinian crisis. Nature **462**, 778–781 (2009)
3. Gorini, C., Montadert, L., Rabineau, M.: New imaging of the salinity crisis: dual Messinian low stand mega sequences recorded in the deep basin of both the eastern and western Mediterranean. Mar. Pet. Geol. (2015). https://doi.org/10.1016/j.marpetgeo.2015.01.009

4. Leroux, E., Rabineau, M., Aslanian, D., Granjeon, D., Droz, L., Gorini, C.: Strati-graphic simulations of the shelf of the Gulf of Lions: testing subsidence rates and sea-level curves during the Pliocene and Quaternary. Terra Nova **26**(3), 230–238 (2014). https://doi.org/10.1111/ter.12091 (insu-00948717)
5. Lofi, J., Gorini, C., Berné, S., Clauzon, G., Reis, A.T., Ryan, W.B. F., Steckler, M.S.: Erosional processes and paleo-environmental changes in the Western Gulf of Lions (SW France) during the Messinian salinity crisis. Mar. Geol. **217**(1–2), 1–30 (2005)
6. Papadimitriou, N., Gorini, C., Nader, F.H., Deschamps, R., Symeoub, V., Lecomte, J.C.: Tectono-stratigraphic evolution of the western margin of the Levant Basin (offshore Cyprus) (2017)
7. Urgeles, R., Camerlenghi, A., Garcia-Castellanos, D., De Mol, B., Garcés, M., Vergés, J., Haslam, I., Hardman, M.: New constraints on the Messinian sealevel drawdown from 3D seismic data of the Ebro Margin, western Mediterranean. Bas. Res. **23**, 123–145 (2011)
8. Rabineau, M., Leroux, E., Aslanian, D., Bache, F., Gorini, C., Moulin, M., Molliex, S., Droz, L., dos Reis, A. T., Rubino, J. L., Guillocheau, F., Olivet, J. L.: Quantifying subsidence and isostatic readjustment using sedimentary paleomarkers, example from the Gulf of Lion. Earth Planet. Sci. Lett. **388**, 353–366 (2014)
9. Bache, F., Popescu, S. M., Rabineau, M., Gorini, C., Suc, J. P., Clauzon, G., Olivet, J. L., Rubino, J. L., Melinte-Dobrinescu, M. C., Estrada, F., Londeix, L., Armijo, R., Meyer, B., Jolivet, L., Jouannic, G., Leroux, E., Aslanian, D., Reis, A. T. D., Mocochain, L., Dumurdžanov, N., Zagorchev, I., Lesić, V., Tomić, D., Namık Çağatay, M., Brun, J. P., Sokoutis, D., Csato, I., Ucarkus, G., Çakır, Z.: A two-step process for the reflooding of the Mediterranean after the Messinian Salinity Crisis. Basin Res. **24**(2), 125–153 (2012)

The Alps Revisited—4D-MB, the German Contribution to the Alp Array Mission

Michael Weber, Mark Handy, and Emanuel Kästle

Abstract

4D-MB is a multi- and interdisciplinary project that stands as an integral part of the international Alp Array mission. It helps test the advanced hypothesis that re-organizations of the Earth's mantle during the collision of tectonic plates have immediate as well as long-lasting effects on the earthquake distribution, crustal motion and landscape evolution in the mountain belts. The Alp-Array consortium involves 64 institutions distributed in 17 countries.

Keywords

Alps • Tectonics • Seismology • Seismic arrays

1 Introduction

The target is the European Alps (Fig. 1) and the neighboring mountain belts (Apennines, Carpathians, Dinarides), lying in the central Mediterranean area [1]. To this end, an interdisciplinary initiative funded by the German Science Foundation (DFG), involving Geophysics, Geology, Petrology, Modeling and other fields, was launched. During the first three years of the programs, a strong focus was laid on geophysics, mainly broadband seismology, applying the data derived from the long-term backbone deployment (ca. 600 stations with ca. 50 km spacing). A denser network of 150 BB stations, along with 150 stations based in the central Alps, was also deployed. On the German side of the Alp Array, more than twenty Institutions and Universities took part.

M. Weber (✉)
GFZ, Potsdam, Germany
e-mail: mhw@gfz-potsdam.de

M. Weber
University Potsdam, Potsdam, Germany

M. Handy · E. Kästle
Freie University Berlin, Berlin, Germany

2 Settings: Materials and Methods

The object consists in investigating links associating the Earth's surface and mantle through integration of the 3D-seismological imaging of the crust-mantle system, as persisting underneath the Alps, with geologic observations and modeling, to help us look both backwards and forwards in time: the fourth dimension.

This integrative 4D approach entails the treatment of four research related issues:

1. Reorganizations of the lithosphere during mountain building using high-resolution seismic imaging of the lithospheric slabs lying beneath the Alps. The aim is to determine the origin of switches lying in subduction polarity, particularly, to understand how lithospheric slabs persisting tears nucleate and propagate in time and space.
2. Surface response to mountain structure changes, on different time scales, helps test the controversial idea that time-integrated denudation and uplift rates could partly reflect slab-tearing and break-off events occurring in the mantle.
3. Deformation of the crust and mantle taking place during the mountain building process helps resolve the question of whether deep structures manifest early stages of mountain-building, or primarily preserve the imprint of later events (indentation, lateral escape). This process constrains rates of structural change affecting the crust and mantle.
4. Motion patterns and seismicity help identify the spatial and temporal patterns of faulting and seismicity, to draw an overall motion picture from the present back in time. The aim is to understand whether the presently active earthquakes have been established in response to a new tectonic regime.

F. Rossetti et al. (eds.), *The Structural Geology Contribution to the Africa-Eurasia Geology: Basement and Reservoir Structure, Ore Mineralisation and Tectonic Modelling*,
Advances in Science, Technology & Innovation, https://doi.org/10.1007/978-3-030-01455-1_2

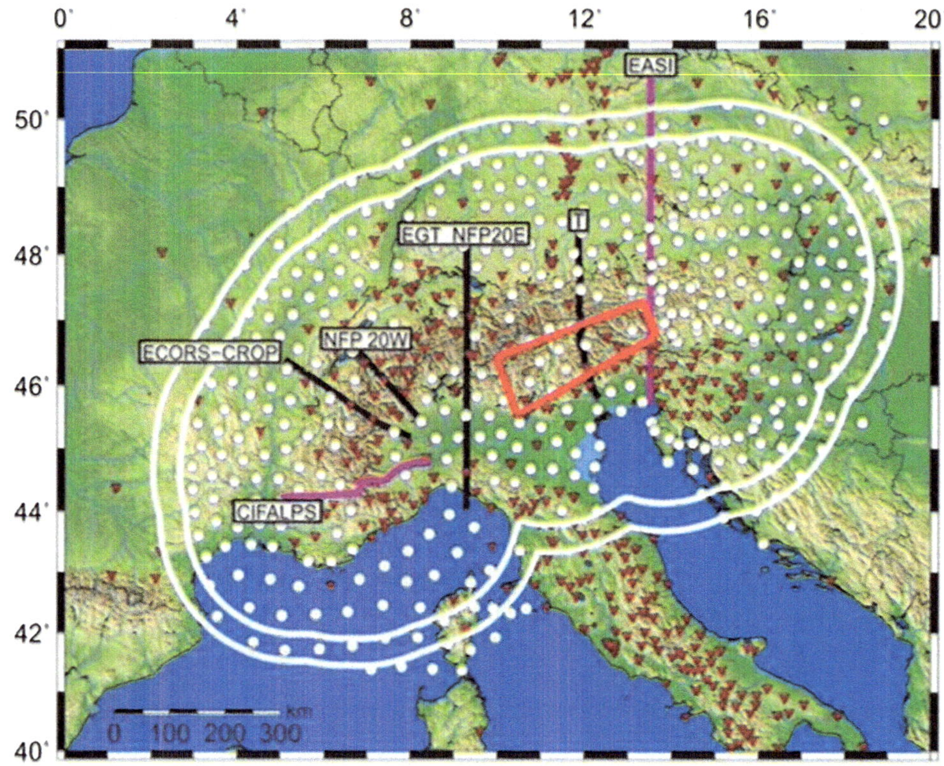

Fig. 1 The Alpine domain with seismological experiments. **a** White lines delimit the international backbone network with all 579 broadband seismometers. **b** Purple lines define the Alp Array passive seismic swaths, with densified station configurations: EASI—Czech-Austria-Swiss swath, CIFALPS—China-Italy-France swath. **c** Black lines designate the active seismic experiments of the 1980s and 90s, including T, the TRANSALP profile. **d** Orange rectangle determines the Swath D, with densified station networks (Fig. 2)

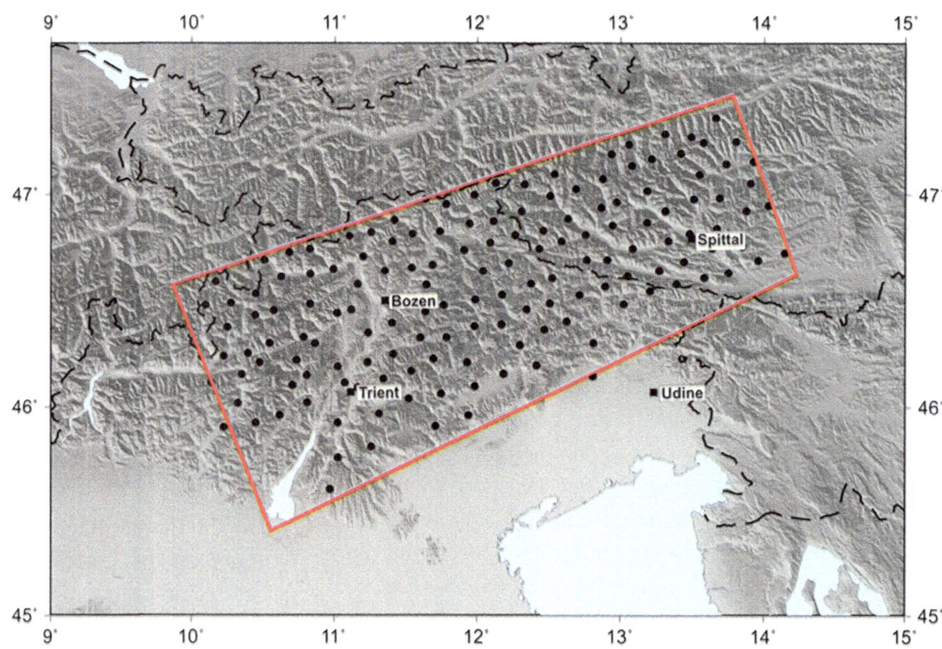

Fig. 2 153 BB-stations placed in the Central/Southern Alps

The 4D-MB takes advantage of the Alps' unique exposure of different stages of orogenesis, ranging from the ongoing continental collision and indentation in the east, to the post-collisional rebound in the west. Our focus is placed on integrated geophysical-geological studies along a swath of 153 closely spaced broadband seismometers deployed in the seismo-tectonically most active part of the orogen (Eastern and Southern Alps, Fig. 2). The first phase of the project is still underway (2017–2020, while the second phase is planned for the interval (2020–2023).

3 Results

We have now installed over 150 broadband stations in a dense network across the Central Alps.

4 Discussion

The data, to be drawn from this deployment, would serve as an input for more than twenty projects involved in the 4D-MB scheme.

5 Conclusions

In a first place, the local and regional earthquakes' drawn results, as recorded in the network, will be displayed.

Reference

1. Hetényi, G., et al.: The AlpArray seismic network: a large-scale European experiment to image the alpine orogen. Surv. Geophys. **2** (5), 99–110 (2018). https://doi.org/10.1007/s10712-018-9472-4

Tethyan and Alpine Controls on the Crustal Architecture of the Tunisian and Algerian Atlas and Tell: Needs for Coupled Deep Seismic Soundings and Tomography

François Roure, Sami Khomsi, Dominique Frizon de Lamotte, and Rémi Lepretre

Abstract

National and international deep seismic reflection profiles provide unique controls on the Moho and crustal architecture lying beneath sedimentary basins, tectonic wedges and adjacent forelands, mainly predominant in North America and Europe. Recent and still ongoing 3D mantle and crustal tomography programs, as based on 3D passive seismic recordings, have also documented the shape of the entire lithosphere persisting underneath the Pyrenees, the Alps and the Apennines. They have also helped in recording the lateral seismic anisotropies and vertical seismic reversals lying in these tectonic wedges and their adjacent forelands. The lack of such deep controls in regard of the North African tectonic wedges and adjacent forelands still preclude any prospection of deep plays. Actually, thorough seismic sounding profiles at crustal and lithospheric scales would be of huge interest for an effective monitoring of the deep architecture, geodynamic and thermal evolution of North African sedimentary basins to take place. Such an undertaking would be useful for unlocking new exploration objectives, and an efficient documentation of the active fault systems' deep architecture to be achieved. The aim of this keynote and related roundtable will be to kindle the interest of both national and international research teams, agencies and companies as to the importance of such topics, and to investigate the possibilities for ambitious North African crustal imagery projects to be launched.

F. Roure (✉)
IFPEN, Rueil-Malmaison, France
e-mail: francoisroure84@gmail.com

S. Khomsi
King Abdulaziz University, Jeddah, Saudi Arabia

D. F. de Lamotte · R. Lepretre
University of Cergy-Pontoise, Cergy-Pontoise, France

Keywords

Lithospheric and crustal architecture • Algerian and Tunisian Tell and Atlas • Deep seismic soundings Passive seismic and tomography

1 Introduction

So far, only a single, old refraction profile has been recorded in regard of the Tunisian context [1]. The most recently registered deep seismic reflection profiles have been recorded by means of Western Geco and refraction-wide angle profiles of the Spiral program, and are restricted to the Algerian offshore [2, 3]. Onshore, however, several conventional industry (shallow) reflection profiles have been recorded with respect to both of the Tunisian and Algerian Tellian and Atlas domains. Still, they could hardly provide any information going beneath the Triassic salt, the basement architecture as well as the extent and nature of the Paleozoic basins, which remain, for the most part, conjectural.

2 Crustal and Lithospheric Structures of the Pyrenees, Alps and Apennines

National and international deep seismic reflection profiles have provided unique controls on the Moho and crustal architecture of tectonic wedges and adjacent forelands in the Pyrenees, Alps and Apennines [4–6], (Figs. 1 and 2). For instance, these profiles have well documented the Late Cretaceous to Eocene fan-shaped thrust systems of the Pyrenees, Paleogene and Neogene thrusts of the Alps and, above all, the Apennines' Plio-Quaternary compressional structures. They have also imaged the negative and positive inversion of former thrusts, i.e., the negative inversion of Carboniferous thrusts during the Permian collapse of the Variscan orogen in the conjugate forelands of the Pyrenees.

F. Rossetti et al. (eds.), *The Structural Geology Contribution to the Africa-Eurasia Geology: Basement and Reservoir Structure, Ore Mineralisation and Tectonic Modelling*,
Advances in Science, Technology & Innovation, https://doi.org/10.1007/978-3-030-01455-1_3

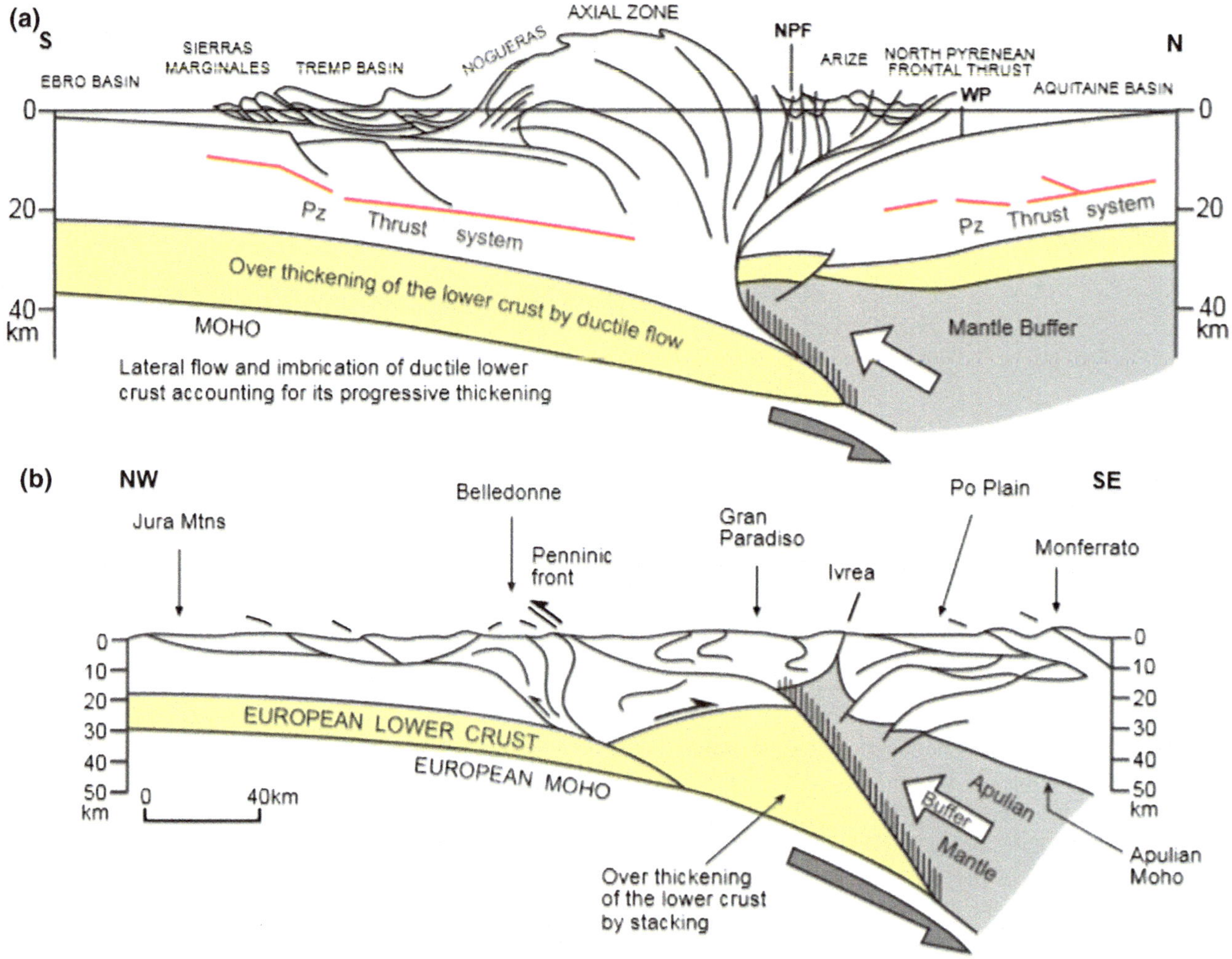

Fig. 1 Crustal structure of the Pyrenees and Alps along the Ecors and crop profiles (after [4, 5])

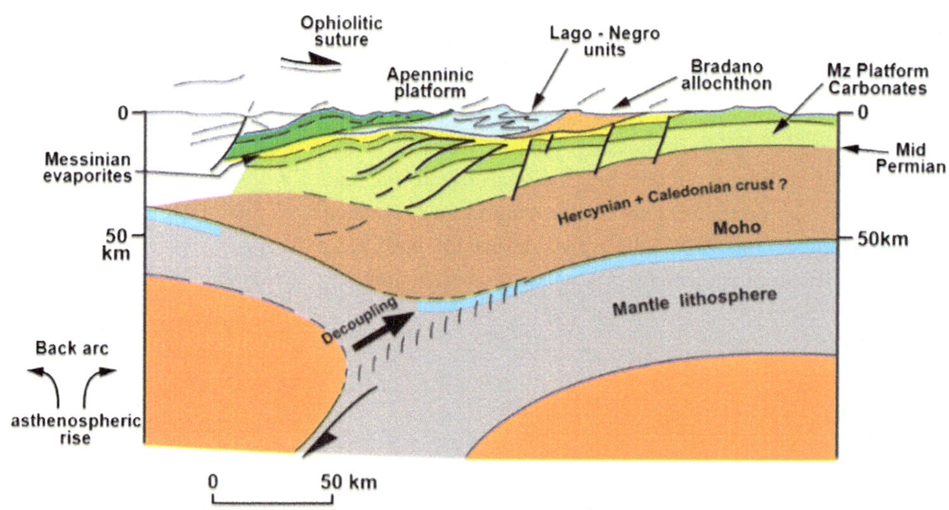

Fig. 2 Lithospheric structure of the Southern Apennines (after [6])

The same imaging has been performed on the negative inversion of Pyrenean thrusts during the Oligocene opening of the Gulf of Lions, and the strong control exerted by Mesozoic normal faults, as inherited from the Tethyan rifting during subsequent Alpine inversions in the Pyrenean and Alpine forelands.

More recent and still ongoing 3D mantle and crustal tomography programs, based on 3D passive seismic recording, have also helped document the lithospheric architecture lying beneath these orogens [7–9]. These newly drawn tomographic profiles have also outlined the persistence of intracrustal lateral seismic anisotropies and vertical seismic reversals in the Alpine and Pyrenean forelands as well as beneath the Po Plain and Northern Apennines. Thus, they helped in providing additional constraints on the main intracrustal interfaces, such as the major tectonic features (old-higher velocity rocks thrusted on top on young-slower velocity units), the base of the sediments (top basement), the carbonates and clastics transition (boundary between the former passive margin sequence and younger clastics of the subsequent flexural sequence) (Figs. 3 and 4).

A compilation of the entirety of deep seismic and industry profiles, lying beneath the Po Valley basin, has also been applied to build a 3-D structural model at crustal scale. Thus, it helped in better understanding the connection between the still active thrusts of the Northern Apennines and the location of focal mechanism of recent earthquakes [10].

3 Positive and Negative Inversions in the Algerian and Tunisian Tell and Atlas

Pioneer field geologists have for long recognized the occurrence of tectonic windows lying beneath the Tellian allochthon in Algeria, where parautochthonous platform units are locally exposed to the surface in the core of wide anticlines' nappes [11, 12]. Tentative crustal-scale cross-sections have also been proposed to persist between the Western Mediterranean basin and the Saharan Platform, suggesting various structural styles for the basement lying underneath the Algerian Tell and the Saharan Atlas, though they remain only conjectural, due to lacking deep seismic controls [13, 14].

Regional structural cross-sections, calibrated against field data and seismic profiles, have now better documented the thin-skinned architecture of the Tellian allochthon [15–18], thick-skinned foreland inversion in the adjacent Atlas domain, where both Jurassic and Cretaceous normal faults have been inverted [19, 20], along with the complex halo-kinesis involving the Triassic salt in North Algeria and Tunisia [21, 22].

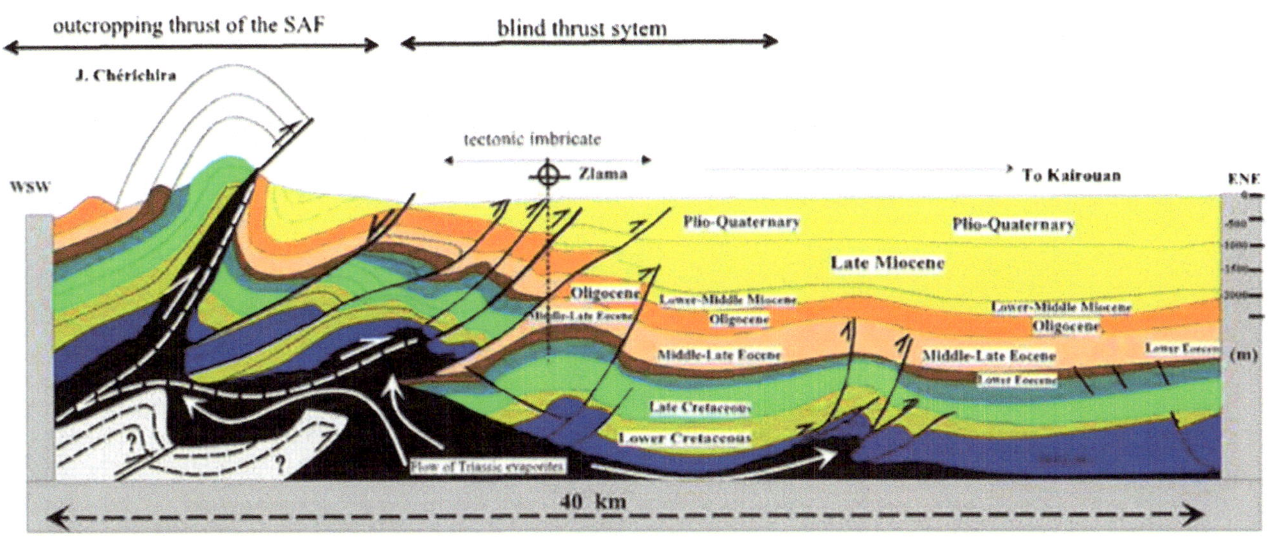

Fig. 3 Structural sketch of the frontal structures of the Tunisian Atlas (after [19])

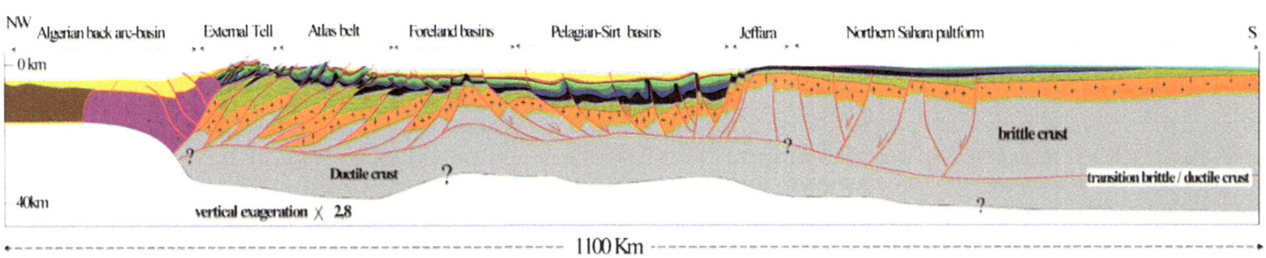

Fig. 4 Tentative crustal scale cross-section of Alpine Tunisia and its autochthonous foreland (after [19])

However, the infra-salt architecture and sedimentary, metamorphic or crystalline nature and overall structural style of the infra-Triassic basement still remain unconstrained, due to the lack of deep seismic imagery.

Controlling both of the crustal and lithospheric architecture of Alpine North Africa would be of prime importance for an effective exploration of energy resources and prediction of seismic hazards. At this stage, we do not know whether the underthrust basement is imbricated or, rather, poorly deformed. Nothing is known as to whether it is still tilted towards the north in northern Tunisia and Algeria, or rather tilted towards the southeast, due to slab detachment and coeval unflexing of the foreland lithosphere.

4 Conclusions

As it is the case with the northern side of the Mediterranean, the North African margin has been strongly impacted with successive compressional and extensional events, which have progressively shaped the foreland architecture. The lack of deep controls still preclude any prospection of deep plays, and deep seismic soundings profiles would be of huge interest for unlocking such novel exploration objectives.

Our predictive crustal cross-sections, relevant to the Tunisian and Algerian onshore segments of the Maghrebides (Tell and Atlas), should stimulate the interest of both national and international research teams, agencies and companies, and initiate discussion debates among these communities for an ambitious North African crustal imagery project to be launched.

References

1. Morelli, C., Nicolich, R.: A cross-section of the lithosphere along the European Geotraverse southern segment (from the Alps to Tunisia). Tectonophysics **176**(1–2), 229–243 (1990)
2. Bouyahiaoui, B., Sage, F., Abtout, A., Klingelhoefer, F., Yelles-Chaouche, K., Schnürle, P., Marok, A., Déverchère, J., Arab, M., Galve, A., Collot, J.Y.: Crustal structure of the Eastern Algerian continental margin and adjacent deep basin: implications for late Cenozoic geodynamic evolution of the Western Mediterranean. Geophys. J. Intern. **201**, 1912–1938 (2015)
3. Arab, M., Rabineau, M., Déverchère, J., Bracene, R., Belhai, D., Roure, F., Marok, A., Bouyahiaoui, B., Granjeon, D., Andriessen, P., Sage, F.: Tectonostratigraphic evolution of the Eastern Algerian margin and basin from seismic data and onshore-offshore correlation. Mar. Petrol. Geol. **77**, 1355–1375 (2016)
4. Roure, F., Choukroune, P., Berastegui, X., Munoz, J.A., Villien, A., Matheron, P., Bareyt, M., Séguret, M., Camara, P., Deramond, J.: Ecors deep seismic data and balanced cross-sections: geometric constraints on the evolution of the Pyrenees. Tectonics **8**(1), 41–50 (1989)
5. Roure, F.: Crustal architecture, thermal evolution and energy resources of compressional basins. Geol. Belg. **17**(2) (2014)
6. Roure, F., Casero, P., Addoum, B.: Alpine inversion of the North African margin and delamination of its continental lithosphere. Tectonics **31** (2012). https://doi.org/10.1029/2011TC002989
7. Wang, Y., Chevrot, V., Monteiller, V., Komatitsch, D., Mouthereau, F., Manatschal, G., Sylvander, M., Diaz, J., Ruiz, M., Grimaud, F., Benahmed, S., Pauchel, H., Martin, R.: The deep roots of the Western Pyrenees revealed by full waveform inversion of teleseismic P waves. Geology **44**(6), 475–478 (2016)
8. Potin, B.: Les Alpes occidentales: tomographie, localisation de séismes et topographie du Moho. Ph.D. Grenoble University, 232 pp (2016)
9. Zhao, L., Paul, A., Guillot, S., Solarino, S., Zheng, T., Aubert, C., Salimbeni, S., Dumont, T., Schwartz, S., Zhu, R., Wang, Q.: First seismic evidence for continental subduction beneath the Western Alps. Geology **43**, 815–818 (2015)
10. Turrini, C., Angeloni, P., Lacombe, O., Ponton, M., Roure, F.: 3-D seismotectonic in the Po valley basin, Italy. Tectonophysics **661**, 156–179 (2015)
11. Mattauer, M.: Etude géologique de l'Ouarsenis oriental, Algérie. Public du service de la carte géol. d'Algérie **17**, 550 pp (1958)
12. Vila, J.M.: La chaîne alpine d'Algérie orientale et des confins algéro-tunisiens. Thèse Doc. d'Etat, Paris VI Univ., 3 vol., 665 pp, unpublished (1980)
13. Roeder, D.: American and Tethyan thrust-belts. Gebrüder Borntraeger. Berlin, 168 pp (2008)
14. Benaouali-Mebarek, N., Frizon de Lamotte, D., Roca, E., Bracene, R., Sassi, W., Faure, J.L., Roure, F.: Post-cretaceous kinematics of the Atlas and Tell systems in central Algeria: early foreland folding and subduction-related deformation. C. R. Geosci. **338**, 115–125 (2006)
15. Vially, R., Letouzey, J., Bénard, F., Haddad, N., Desforges, G., Askri, H., Boudjema, A.: Basin inversion along the North African margin, the Saharan Atlas. In: Roure, F. (ed.) PeriTethyan Platforms, pp. 79–117. Editions Technip, Paris (1994)
16. Bracene, R., Frizon de Lamotte, D.: Origin of intraplate deformation in the Atlas system of western and central Algeria: from Jurassic rifting to Cenozoic-Quaternary inversions. Tectonophysics **357**, 207–226 (2002)
17. Tricart, P., Torelli, L., Argnani, A., Rekhiss, F., Zitellini, N.: Extensional collapse related to compressional uplift in the Alpine Chain off Northern Tunisia (Central Mediterranean). Tectonophysics **238**(1–4), 317–329 (1994)
18. El Euchi, H., Saidi, M., Fourati, L., El Maherssi, C.: Northern Tunisia thrust belt: deformation models and hydrocarbon systems. In: Swennen, R., Roure, F., Granath, J. (eds.) Deformation, fluid flow and reservoir appraisal in foreland fold-and-thrust belts, AAPG Memoir, Hedberg Series, vol. 1, pp. 371–390 (2004)
19. Khomsi, S., Frizon de Lamotte, D., Bedir, M., Echihi, O.: The Late Eocene and Late Miocene fronts of the Atlas belt in eastern Maghreb: integration in the geodynamic evolution of the Mediterranean domain. Arab. J. Geosci. **9**(15), 125–142 (2016)
20. Gharbi, M., Espurt, N., Masrouhi, A., Bellier, O., El Amjed, A.: Style of Atlasic deformation and geodynamic evolution of the Southern Tethyan margin, Tunisia. Mar. Petrol. Geol. (2015). http://dx.doi.org/10.1016/j.marpetgeo.2015.07.020
21. Vila, J.M., Ghanmi, M., Ben Youssef, M., Jouirou, M.: Les glaciers de sel sous-marins des marges continentals passives du NE du Maghreb (Algérie-Tunisie), et de la Gulf Coast (USA). Eclogae Geologicae Helvetia **95**, 347–380 (2002)
22. Masrouhi, A., Bellier, O.M., Koyi, H.: Geometry and structural evolution of the Lorbeus diapir, northwestern Tunisia: polyphase diapirism of the North African inverted passive margin. Int. J. Earth Sci. **103**(3), 881–900 (2014)

Impact of Geodynamics on Tectonic and Climatic Risk Instabilities in Interior Seas: Insights of Land-to-Sea Study on Caribbean Pathways (Haiti). Potential Application on Mediterranean

Nadine Ellouz-Zimmermann

Abstract

The Caribbean Sea can be described as an intercontinental sea, which is presently squeezed between the two major American continental blocks. In comparison, the Mediterranean Sea which occupied several Ma ago a comparable position, at present registers a more advanced stage as it has been nearly closed between the African and the Eurasian converging continental plates. It is proposed to examine how tectono-sedimentary processes interact under active convergent tectonic regimes and how they impact the land-to-sea evolution, on these two examples of interior seas. The objectives are to analyze, and propose a hierarchy for the forcing parameters, geodynamics, local tectonics and seismicity, sedimentology, seawater circulation and climate (wind, rain falls.), in order to constrain the dynamics of the margins.

Keywords

Tectonics • Hazards • Destabilization Caribbean and mediterranean interior seas

1 Introduction on the Geodynamics of the Caribbean and Mediterranean Seas

On a global scale, the North and South American continental plates are smoothly converging and, in between, the available space for the Caribbean composite plate is continuously decreasing, thus favoring the westward lateral transfer along two transpressive boundaries. Laterally, the Eastern and Western borders of the Caribbean Plate are bounded by subduction zones where oceanic plates of various Mesozoic ages disappear.

The Mediterranean system has a comparable environment, triggered by a dominant transpressive/convergent regime, with an oblique strike-slip fault system along N African margins and around the Apulian and Anatolian blocks. The strong variation in the distribution of the subsiding and uplifting zones, the shortening or collapsing tectonics associated with the variable thickness and dissolution of the Messinian salt, introduce others parameter in the Mediterranean domain (Fig. 1).

2 Land-to-Sea New Acquisition Around Hispaniola

Within the last years, after the 2010 big earthquake in Haiti, the northern plate boundary of the Caribbean plate has been intensively studied around Hispaniola by different groups. The N Caribbean plate motion there is distributed along two left lateral major strike-slip faults, that are expressed by transtensive Troughs which are locally inverted. During the last studies, these large transtensive/transpressive fault segments have been observed offshore in different positions, especially in the two straits: Windward (to the N, between Cuba and Haiti) and Navassa (to the S, between Jamaica and Haiti) passages. The prolongation of the southern fault zone has also been described on land along the Southern Peninsula in Haiti.

In Haitian waters, several studies have been conducted from 2012 to 2017, during the Haiti-SIS and Haiti-BGF campaigns, in the vicinity of the two large fault systems. Three offshore campaigns produced a lot of data, bathymetry, seismic profiles, magnetics cores and fluids. Meanwhile, on land studies have been also conducted by IFPEN, especially in the Haitian Southern Peninsula, along the transpressive Ridge, which represents the eastward development of the southern strike-slip zone (observed offshore). This new data allows us to image both the present-day situation with respect to the risk assessment, but also allows us (1) to restore the past geometry along regional

N. Ellouz-Zimmermann (✉)
IFP New Energy, 1-4 av. De Bois-Preau,
92852 Rueil Malmaison Cedex, France
e-mail: nadine.ellouz@ifpen.fr

F. Rossetti et al. (eds.), *The Structural Geology Contribution to the Africa-Eurasia Geology: Basement and Reservoir Structure, Ore Mineralisation and Tectonic Modelling*,
Advances in Science, Technology & Innovation, https://doi.org/10.1007/978-3-030-01455-1_4

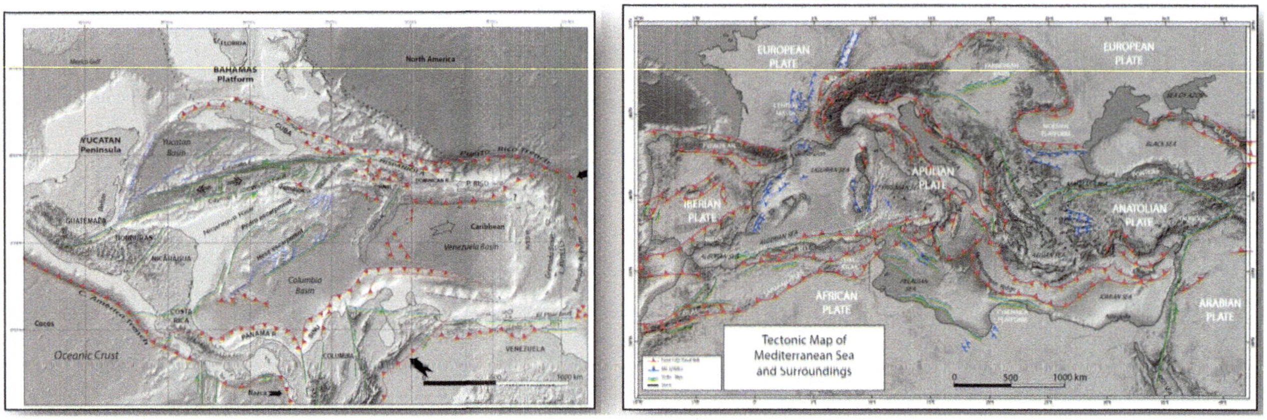

Fig. 1 Tectonic sketches of the Caribbean plate environment (left), and Mediterranean plates environment (right)

cross-sections, and (2) describe the impact on paleo-topography dynamics.

3 Results

3.1 Offshore Results

In Fig. 2, the tectonic inversion of the Neogene basins has been evidenced close by the Windward passage. It illustrates the relationships between tectonics and the subsequent change of sediment drainage/deposition areas, mass water circulation and relative isolation of the adjacent basin.

Same type of morphological evolution linked to tectonic uplift has been observed on the southern fault zone in the Navassa passage.

3.2 Onshore Results

Between these major faults, strain partitioning is expressed by the development of a F&T belt, which emerged and

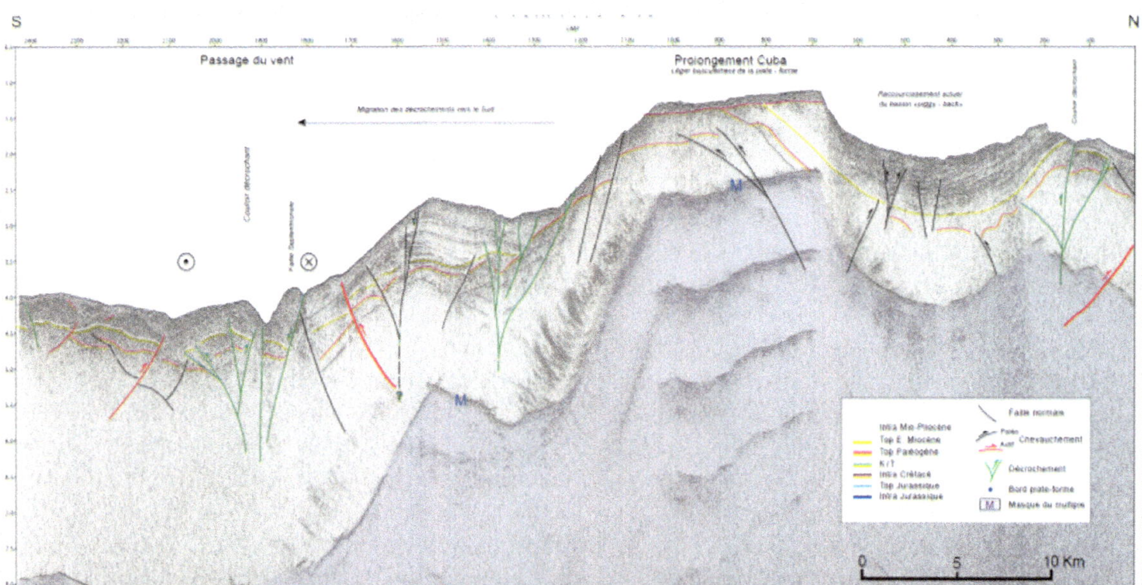

Fig. 2 Interpretation of the N-S Haiti-SIS cross section across the windward passage

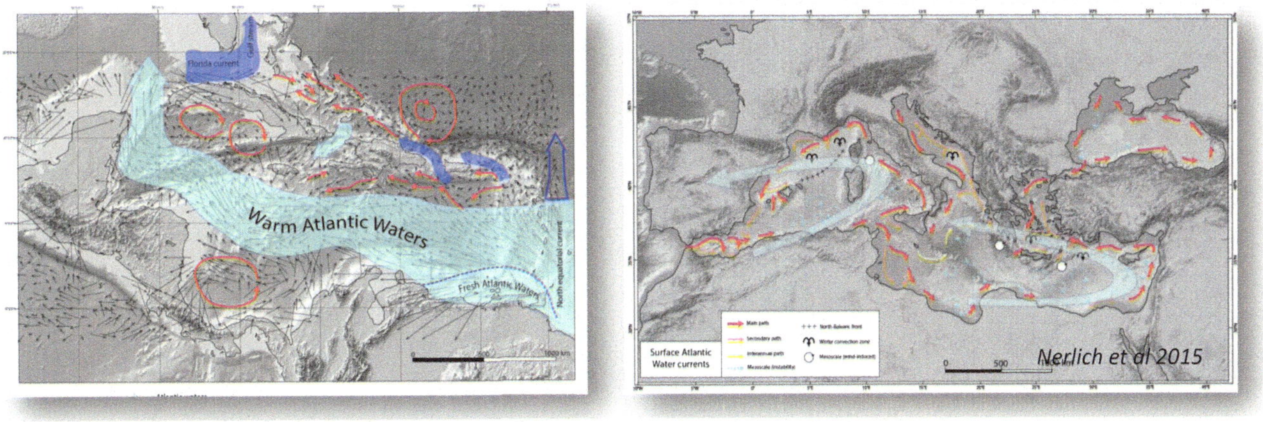

Fig. 3 Mass water surface circulation across Caribbean (left) and Mediterranean (right) seas. Note the strong difference between the Atlantic surface currents of the "still open" interior sea, and the "nearly closed" Mediterranean system

developed as a large restraining bend, initially expressed by S-W propagating thin-skin tectonics, and at present evolves toward a thick skin deformation mode. We compare the large offshore anticlinorium of la Gonâve with the large-scale deformation, deeply rooted and described onshore.

The connection between the deep tectonics and the dynamics of the onshore topography is spectacular in Haiti. Coastal parts of the shelf close to the related strike slip ridges or basins, are in permanent destabilization, due to uplift/erosion cycles and have become particularly sensitive to the seismicity and anthropic effects (deforestation).

4 Impact of the Tectonics on Coastal Instabilities

In Fig. 3, surface water masses circulation are illustrated showing the Caribbean circulation: Atlantic waters are mainly entering across the Antilles passages, are warmed up by 5–6 °C inside the interior sea along a single westward surface current, and then escape along surface currents northwards. The studied Windward and Navassa straits, which are tectonically active and uplifted, do not register strong surface currents, and are probably filtering the mass water exchanges, and thus preserve the climatic recordings.

In the "nearly closed" Mediterranean system, the surface distribution of Atlantic waters is more complex and displays two intra Mediterranean large gyres. The escape of the Atlantic/Mediterranean mixed and heated waters is not active at the surface but is transferred to the deepest part of the Gibraltar strait.

5 Conclusions

It is clear that the topography dynamics which integrate tectonic and sedimentological processes will impact the "regular" life of the coastal areas. As both Caribbean and Mediterranean domains are bordered by abrupt and local highly seismically active margins, multidisciplinary studies are needed in order to be predictive for coastal and margins destabilization and associated societal risks. At Present, and particularly on the Mediterranean southern margins, the N. Tunisian Onland-Offshore deep structuration is not correctly imaged, as well as the coastal and shelf morphology. It is important to note that the connection between western and eastern Mediterranean is of crucial importance, as the Pelagian platform starts to separate the Tunisian shelf from the Sicilian one along a transcurrent fault system, punctuated by present-day volcanism. On the current maps, only the surface Atlantic waters can migrate eastward. The escape of Mediterranean waters can only occur at depth. In this situation, the (episodic) closure of the Sicilian-Tunisian strait, a catastrophic event, will seriously impact the stability of the southern N. African coasts as well as the confinement degree of the eastern Mediterranean Sea (as has already occurred in the Late Miocene).

New bathymetric, seismic and coring acquisition are strongly needed, especially in the North of Tunisia and in the Siculo-Tunisian Channel, in order to document the connection between deep and surface Dynamical evolution, and to analyze their impact on water circulation and risk assessment. These new results have to be analyzed under different climatic scenarios, which will be especially sensitive in the "partially closed"

Caribbean and "nearly closed" Mediterranean domains.

Eduard Suess and the Essence of Geology

A. M. Celal Şengör

Abstract

Eduard Suess was probably the greatest geologist who ever lived. His epoch-making, four-volume *Das Antlitz der Erde* (The Face of the Earth), published between 1883 and 1909 and translated into English, French, Spanish and partly into Italian, is generally considered to be the herald of modern global geology. Before Suess, geology was practiced as a form of local science, deducing ideas for the behavior of the earth using local observations not tied to one another in a general synthesis, except in such general statements as the uniformitarian principle. There were also general syntheses, but those were a priori attempts assuming a general theory and then applying it to selected observations with a view to establishing the theory assumed. Suess abandoned both approaches: he attempted to collect and connect with one another world-wide observations within the framework of Constant Prévost's version of the theory of contraction with a view to testing its deductions. He had two main aims: (1) the delineation of the trend-lines of the world's mountain systems, in other words their structure, to understand whether as an ensemble they made sense in the framework of the contraction theory. He said there were older 'plans' of the trend-lines indicating that the present face of the earth was a production of the Phanerozoic and that the face of the earth wore other aspects in the past, but all generated by the processes familiar to us from their presently-active representatives. His approach was thus actualistic, very much Lyellian, although he differed from Lyell in admitting the existence of global events. At the end of his work he announced that the contraction theory did not seem sufficient to explain all the peculiarities of the mountain belts, especially the enormous shortening they displayed. He did not assume how the mountain belts would behave using the contraction theory. He first worked out how they behaved using the field evidence and then checked whether his results were compatible with the theory. (2) Suess' second concern was the explanation of global stratigraphy. He was surprised that the geological time-scale, worked out in such a small place as Europe, should be applicable world-wide. He assumed that the divisions of the geological time table were natural units and from this concluded that some global phenomenon must cause their global validity. This phenomenon he found in global sea-level changes, which he called 'eustatic movements'. He worked out a table of global sea-level changes and showed that they were also responsible for the changes in the biosphere (the term 'biosphere' was introduced by him in 1875) causing the diversification and extinction of organisms. Suess showed that geology can only be properly understood if the details of contingent events are worked out and their mutual causative relations established. He was a critical rationalist in Popper's sense and never fell in love with his own hypotheses.

Keywords

Eduard Suess • Behaviour of the earth • Geological methodology • Mountain-building • Sea-level changes

1 Introduction

Eduard Suess (1831–1914) was probably the greatest geologist who ever lived. He, for the first time, showed how geology had to be done to attain its object, namely understanding the structure and the history of the planet earth. Before him, classical works by such founding geologists as James Hutton (1726–1797), Abraham Gottlob Werner (1749–1817), Georges Cuvier (1769–1832), William Smith (1769–1839), Leopold von Buch (1774–1853), Charles Lyell (1797–1875) and Élie de Beaumont (1798–1874) were

A. M. C. Şengör (✉)
İTÜ Faculty of Mines, Department of Geology, Eurasia Institute of Earth Sciences, Ayazağa, 34469 Istanbul, Turkey
e-mail: sengor@itu.edu.tr

F. Rossetti et al. (eds.), *The Structural Geology Contribution to the Africa-Eurasia Geology: Basement and Reservoir Structure, Ore Mineralisation and Tectonic Modelling*,
Advances in Science, Technology & Innovation, https://doi.org/10.1007/978-3-030-01455-1_5

of two kinds: they either worked in limited areas (mostly in Europe), developed explanatory theories of what they thought they saw and generalised their results to the whole world or they assumed a global theory a priori and collected selected data to confirm it. All, except Élie de Beaumont fit into the first category and Élie de Beaumont alone represents the second. Since antiquity, geology has always worked in these two ways. Anaximander developed his ideas from his local observations around Miletus and Plato from his religious world-picture. Suess chose neither, although he had started with local observations around Prague and later Vienna. He was an avid field worker and an excellent observer. However, even in his earliest publications we note something else: his command of the relevant international literature and his inclination to comparison. For example, when he was writing about brachiopods, he began with a survey of their living relatives, their behaviour and their environments. He then compared his deductions from the observations on fossils brachiopods with the recent ones. When he became interested in mountain structure, he not only studied the Alps of his homeland Austria, but all the mountain ranges of continental Europe and the Central Mediterranean. When he did the stratigraphy of the Cainozoic deposits of the local Vienna Basin, he collected data on all similar deposits throughout Europe and even western Asia. This comparative activity led him to two momentous conclusions: (1) mountain belts were not symmetrical structures uplifted by a central intrusion and (2) young stratigraphic sequences showed a remarkable uniformity of composition and structure at great distances which made it impossible to assume that the transgressions and regressions recorded by them resulted from local vertical motions of continents. Instead of immediately publishing these findings in terms of 'general rules', he decided to check his inferences on a global scale. This required a global survey of the available field observations. This survey took about 40 years to complete and generated two books: *Die Entstehung der Alpen* (formation of the Alps), published in 1875 and which is really a survey of all mountain ranges in Eurasia and *Das Antlitz der Erde* (1883–1909 in four volumes), a global survey of geological data with the purpose of checking his two deductions mentioned above (Fig. 1).

Fig. 1 Eduard Suess (1831–1914)

2 Mountain Structure and the Plan of the Trend-Lines on the Globe

By 1873 Suess had concluded that the large mountain ranges of our globe, what we today call orogenic belts, have asymmetric internal structures and are formed by lateral shortening, not by primary uplift as it had previously been presumed. He noticed that the average direction of the strike of planar structures in them (beds, foliations, etc.) and the direction of elongation of intrusions and commonly volcanic alignments follow definite trends and these trends display map-view structures such as sharp bends in trend (syntaxes), horsetail shaped divergences and convergences in the trend of individual branches (virgations), cross-cutting of various trend-lines (linkages) and gentle bends around older massifs (deflections) in a way that suggests that these massifs had somehow hindered the progress of folding. Suess depicted these trends with single curvilinear lines and called them trend-lines (Fig. 2). In young mountain chains, trend-lines were mostly parallel with the topographic lie of the chain. From the disposition of trend-lines Suess concluded that horizontal motions in making mountains were considerable. Later he even asserted that entire continents 'flowed', a conclusion adopted by Frank Bursley Taylor in 1910 and by Émile Argand very shortly afterward and used extensively in his now-classic *La Tectonique de l'Asie* (1924).

But what was the motor of such horizontal motions?

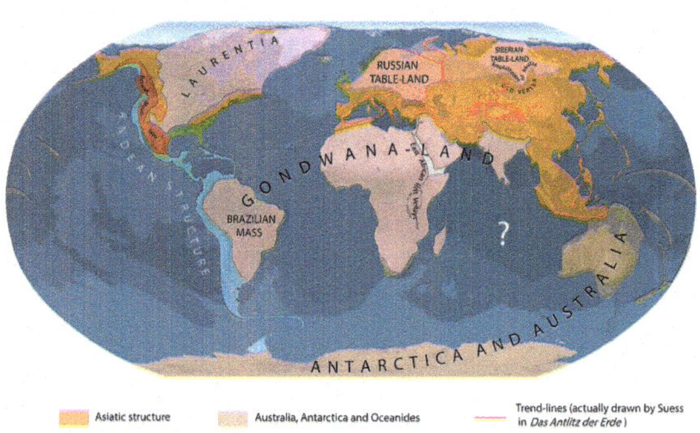

Fig. 2 A tectonic map of the earth drawn after the descriptions of Suess in *Das Antlitz der Erde*. Only those trend-lines are shown that were actually drawn by Suess. He also described verbally many others. Those are not included here

3 Sea-Level Changes, Origin of Ocean Basins and the Source of Mountain Building Movements

In the sixties of the nineteenth century, Suess worked on the later Cainozoic stratigraphy of the Vienna Basin. He corresponded with others who specialized in stratigraphy working on similar sequences from around the Mediterranean to southern Russia. He was astonished to find that similar sequences at similar elevations were widespread all the way from Vienna to the Aral Sea. He further noted that Quaternary marine terraces marking ancient strand lines looked entirely independent of the structure of the coasts onto which they were etched. Suess thought that such uniformity could not be explained by the bobbing up-and-down of the continents as it was then generally believed. He thought that sea-levels had to have changed. Part of his time for the next quarter of a century was devoted to checking this hypothesis using world-wide data. In 1888 he presented his results in the second volume of the *Antlitz*: the sea-level had indeed gone up and down independent of any continental vertical oscillations, causing transgressions and regressions, for which he presented a time-table (Fig. 3). He called these sea-level variations 'eustatic movements' and argued that they were the cause of the world-wide correlatability of the stratigraphic sequences.

But what had caused these sea-level changes?

At the time, global thermal contraction was considered a *vera causa* for tectonic movements and Suess adopted it. But contraction-based tectonic theory had two main variants: One, defended by Élie de Beaumont, held that contraction caused tangential shortening creating large folds that appeared as ocean basins (later to be interpreted as 'geosynclines') and continents and episodically some of these basins collapsed under compression to form mountain ranges. The other, championed by Constant Prévost (1787–1856) ascribed the origin of ocean basins to radial subsidence and the mountain belts to the reaction of these subsidences along their margins. Suess thought that world-wide observations were more in line with the deductions of Prévost's version of the

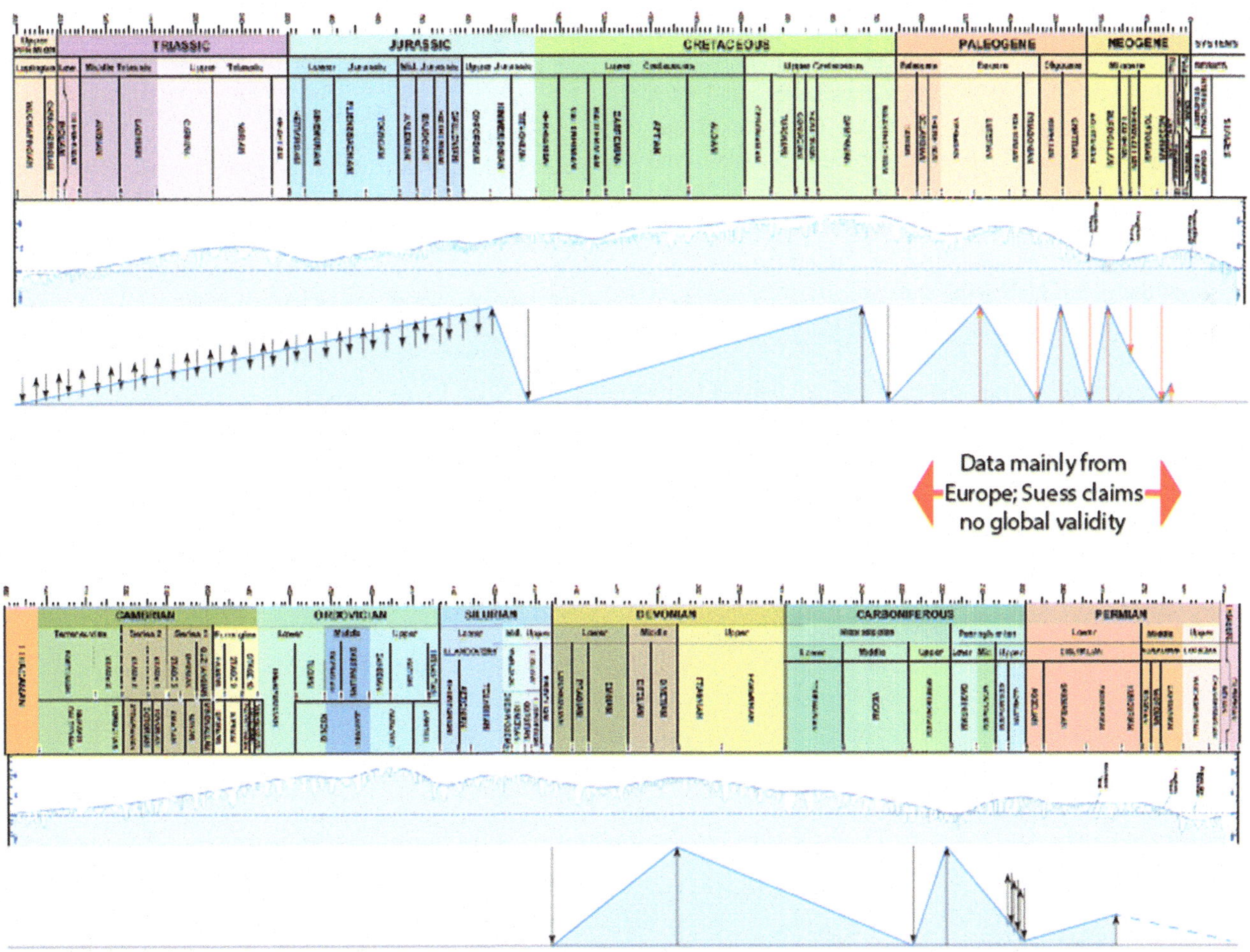

Fig. 3 A modern sea-level curve for the Phanerozoic compared with Suess' time-table for global regressions and transgressions from [5]

theory and made it the basis of his own ideas. But to account for the structure of the mountains, Suess had to assume various levels of detachment in the crust; rock masses above these detachments were only locally and irregularly glued to the masses below. This enabled Suess to explain the simultaneous shortening in the 'external' parts of the mountains, i.e., toward which the main tectonic transport occurred and extension and vulcanicity in the 'internal' parts. Every time an oceanic subsidence took place, that event enlarged the capacity of ocean basins world-wide causing a rapid regression and subsequent sedimentary infilling of the subsidence triggered a slow transgression.

4 Conclusion: How to Deal with Contingent Natural Events with a Critical Rationalist Approach

The general form of Suess' theory of terrestrial tectonics did not survive. But the deductions he made from his vast database, such as the asymmetry of mountain belts, extensional origin of rift valleys, foredeeps, Atlantic- and Pacific-type continental margins, listric faults, strike-slip faults, collapse of large mountain chains by extension, eustatic movements, relation of vulcanism to plutonism as laid bare by what he called 'denudation series' and such entities in the past of the earth as Gondwana-Land, Angara-Land, Tethys, Laurentia, Brasilia, Russian and Siberian table-lands all retain their validity and found themselves a ready-made home in plate tectonics.

Why was Suess so successful and why did his theory of global tectonics fail?

His success resulted from his painstaking analysis of all kinds of global geological data. He realised that unless the entire globe is considered geologically, there would be little hope of understanding its behaviour. Although geological events are contingent, Suess found Ariadne's thread in connecting geological phenomena by using the comparative anatomy of geological structures and their history. He repudiated a priori reasoning, but valued hypotheses as tools to spur observations realising at the same time that they had to be discarded when observations demanded it. His philosophy was what Hawking and Mlodinow called a 'model-driven realism'. He never fell in love with any of his own hypotheses and constantly changed his mind, which has not made reading his works easy.

The only reason for the failure of his global tectonics was the almost total ignorance at the time of the structure and the rock content of the ocean floors. Suess also could not accept isostasy, which later led to Wegener's theory of continental drift. A combination of isostasy with Suess' and Wegener's ideas led Argand to a global tectonic synthesis that in the continents very much resembles what we consider correct today.

Further Reading[1]

1. Şengör, A.M.C.: Eduard Suess' relations to the pre-1950 schools of thought in global tectonics. Geol. Rundsch. **71**, 381–420 (1982)
2. Şengör, A.M.C.: Die Bedeutung von Eduard Suess (1831–1914) für die Geschichte der Tektonik. *Berichte der Geologischen Bundesanstalt*, **51**, 57–72 (2000)
3. Şengör, A.M.C.: Eduard Suess and global tectonics: an illustrated 'short guide'. Aust. J. Earth Sci. (Suess special issue), **107**, 6–82 (2014)
4. Hofmann, T., Bloeschl, G., Lammerhuber, L., Piller, W., Şengör, A. M.C.: The Face of the Earth—The Legacy of Eduard Suess: Edition Lammerhuber, Vienna, 104 pp (2014)
5. Şengör, A.M.C.: The founder of modern geology died 100 years ago: the scientific work and legacy of Eduard Suess. Geosci. Can. **42**, 181–246 (2015a)
6. Şengör, A.M.C.: Eduard Suess Geologist—Catalogue of the Exhibition: Das Österreichische Kulturforum İstanbul and İstanbul Teknik Üniversitesi, İstanbul, 69 pp (2015b)

[1]What I relate above in greatly compressed form has been elaborated in the listed publications, which also have extensive bibliographies.

Part II

Basement Geology

Variation of Subsurface Resistivity in the Pan-African Belt of Central Africa

Dieudonné Bisso, Zakari Arétouyap, Georges Nshagali Biringanine, Philippe Njandjock Nouck, and Jamal Asfahani

Abstract

50 Vertical Electrical Soundings (VES) and 180 Horizontal Soundings (HS) were carried out in the Pan-African geological setting of Central Africa in order to explain the behavior of the electrical signal in this context. Results highlight the presence of very high resistivity values in some locations (greater than 300 kΩm) and large amplitude of resistivity dropping in other sites. Further geological and tectonic investigation shows that higher resistivity values match with the gneiss-granitic nature of the substratum while lower values are due to the presence of faults, shears, and alteration of ultramafic rocks as well as the presence of sedimentary rocks and deformation phases in the region. Those low-resistivity areas can be exploited in the groundwater exploration processing.

Keywords

Apparent resistivity • VES • HS • Pan-African Groundwater

1 Introduction

The basement abounds of countless natural resources (mining, energy, water, etc.) necessary to improve the conditions of human beings. These resources, of a fundamental importance, are usually arranged randomly and discreetly. For this reason, there is a relative difficulty of access for humans. To detect, locate, characterize and quantify these resources, the geophysicist may start with geo-electrical methods, namely in this case, VES and HS. Because of their reliability and efficiency, these methods have been widely used by many scientists and researchers since the 1960s [1–3].

Indeed, the behavior of an electrical signal depends on the geological settings of the investigated area. The Pan-African is a special geological context characterized by geological formations the age of which spans from the Paleoarchean to the late Neoproterozoic (Fig. 1). The Archaean (Congo Craton) and Paleoproterozoic (Central African belt) evolution of the crust are followed during the Neoproterozoic by the Pan-African orogeny from which emanates the Pan-African belt of Central Africa that extends from Congo to Brazil [4]. The main objective of this article is to clarify the vertical and horizontal variations of the electrical signal in the Pan-African particular geological context, using VES and HS.

2 Materials and Methods

2.1 Ohm's Law

Ohm's law in its general form is written as expressed in Eq. 1:

$$\vec{j} = \sigma \vec{E} \quad (1)$$

where $\vec{j}$ is the current density per unit area; σ the electric conductivity and $\vec{E}$ the electric field.

Across an ohmic conductor or resistance R carrying a current of intensity I at a voltage U, this law is written in its macroscopic form given by Eq. 2.

$$U = RI \quad (2)$$

The resistance of a medium is strongly dependent on its geometry but gives no direct information on the nature and

D. Bisso · Z. Arétouyap (✉) · P. Njandjock Nouck
Faculty of Science, University of Yaounde I, P.O. Box 812 Yaounde, Cameroon
e-mail: aretouyap@gmail.com

G. Nshagali Biringanine
Department of Geology, Official University of Bukavu, 570, Bukavu, Democratic Republic of Congo

J. Asfahani
Applied Geophysics Division, Atomic Energy Commission, P.O. Box 6091 Damascus, Syria

F. Rossetti et al. (eds.), *The Structural Geology Contribution to the Africa-Eurasia Geology: Basement and Reservoir Structure, Ore Mineralisation and Tectonic Modelling*, Advances in Science, Technology & Innovation, https://doi.org/10.1007/978-3-030-01455-1_6

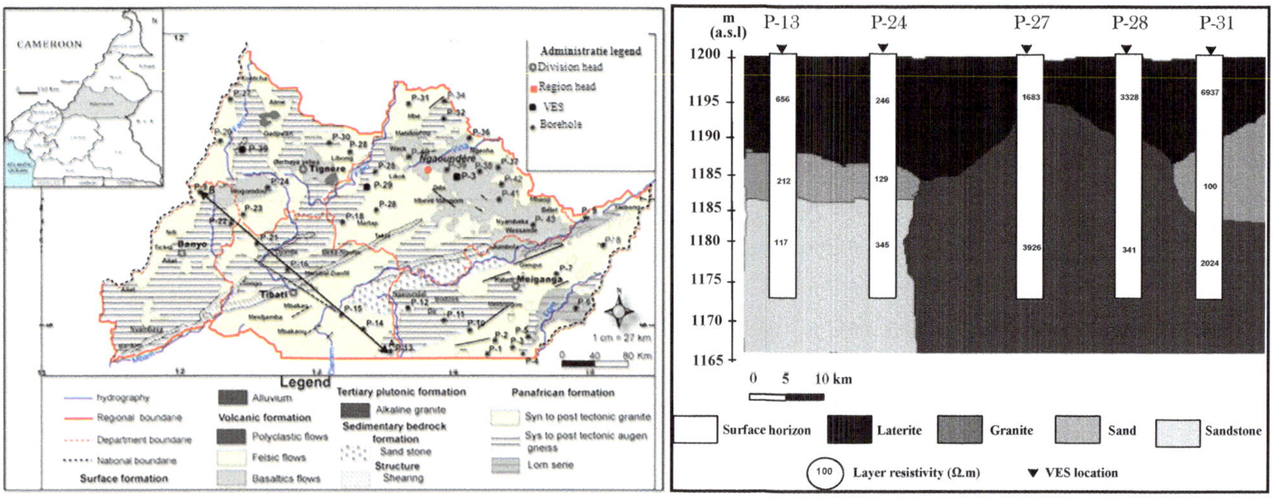

Fig. 1 Simplified geological map of the study area [5] and an interpretative geological cross section along the A-B profile

dimensions of the material crossed by the electric current. It is, therefore, necessary to introduce the concept of resistivity.

2.2 Resistivity

The resistivity ρ of a soil is strongly related to its intrinsic properties and depends mainly on the particle size and porosity, water content, clay content, and mineralogy.

In the case of a quadrupole measuring device with two injection electrodes A and B and two receiving electrodes M and N, the apparent resistivity ρ_a is a function of the potential difference measured between M and N, and a geometric factor, denoted by k defined by Eq. 3 [6].

$$k = \frac{2\pi}{\left(\frac{1}{AM} - \frac{1}{BM} - \frac{1}{AN} + \frac{1}{BN}\right)} \tag{3}$$

Thus, the apparent resistivity is expressed in Eq. 4.

$$\rho_a = k\frac{U_{MN}}{I} \tag{4}$$

3 Results and Discussion

3.1 Vertical Variation

Figure 2 illustrates the variation of the apparent resistivity in function of AB/2.

3.2 Horizontal Variations

At each point of the remarkable fall of apparent resistivity, we associate a crack or an alteration of the basement. Horizontal soundings are then conducted in order to estimate the scale of the alteration (Fig. 3). It is observed that the amplitude of variation of apparent resistivity along a given profile is huge and can sometimes reach a thousand ohm-meters.

Remarkable drops of apparent resistivity can be explained on the one hand by the presence of two main phases of deformation in the region [7] and on the other hand by the presence of faults and shear zones highlighted by the studies

Fig. 2 An example of sounding curves obtained in the study area

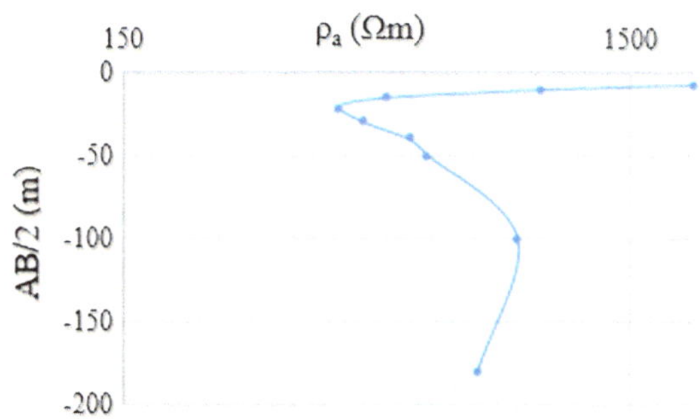

N	ρa	h	d
1	3633	7.7	7.7
2	127	0.79	8.49
3	4140	1.68	10.2
4	65	20	30.2
5	36.5		

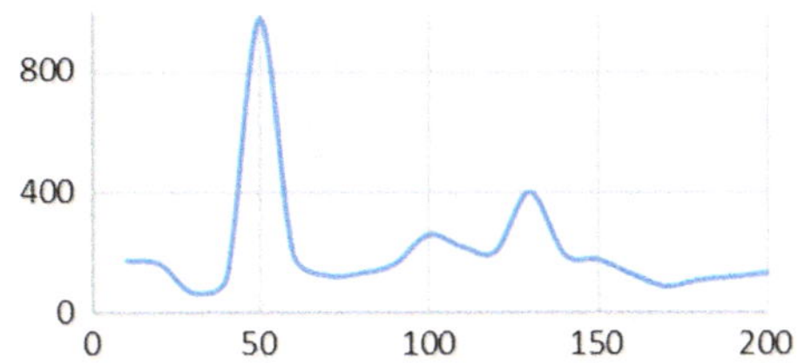

Fig. 3 An example of resistivity variation along a horizontal profiling

of Dumont [8]. Indeed, the Central Cameroon Shear zone is a major structure that would continue into the Gulf of Aden [9] and extend to the Pernambuco fault in Northeast Brazil. On the other hand, this can be explained by the presence of sedimentary rocks and hydrothermal alteration of ultramafic rocks in the region [10].

Those low-resistivity points are of utmost interest in the field of hydrogeology because they are apparently associated with high water-potential, and are important to groundwater exploration and evaluation [1, 3].

4 Conclusions

The Pan-African realm is a geological setting essentially marked by highly resistive materials because of its geological background. However, one can observe abrupt and huge resistivity drops because of some tectonic events like shears, cracks, and faults. Resistivity values range from 15 Ω to 300 kΩ. These low-resistivity points can be exploited for hydrological ends in all regions worldwide where the Pan-African is extended.

References

1. Jones, P.H., Buford, T.B.: Electrical logging applied to groundwater exploration. Geophysics **16**, 115–139 (1951)
2. Frohlich, R.K.: The influence of lateral inhomogeneities on the dipole methods. Geophys. Prospect. **12**, 314–325 (1968)
3. Asfahani, J.: Geoelectrical investigation for characterizing the hydrogeological conditions in semi-arid region in Khanasser valley. Syria. J. Arid Environ. **68**, 31–52 (2006)
4. Yonta Ngoune, C.: Le contexte géologique des indices de talc de la région de Boumnyebel (Chaîne panafricaine d'Afrique Centrale, Cameroun). Ph. D. thesis, Univ. Yaounde I & Nancy (2010)
5. Maréchal, A.: Géologie et géochimie des ressources thermominérales du Cameroun. Doc. ORSTOM **59**, 169–176 (1976)
6. Reynolds, J.M.: An Introduction to Applied and Environmental Geophysics. Wiley, London (1997)
7. Ngako, V., Affaton, P., Njonfang, E.: Pan-African tectonics in northwestern Cameroon: implication for the history of western Gondwana. Gondwana Res. **14**, 509–522 (2008)
8. Ferre, E.C., Deresis, J., Bouchet, J.L., Lar, A.U., Peucat, J.J.: The Pan-African reactivation of Eburnean and Archean provinces in Nigeria: structural and isotopic data. J. Geol. Soc. Lond. **153**, 719–728 (1996)
9. Dumont, J.F.: Identification par télédétection de l'accident de la Sanaga (Cameroun). Sa position dans le contexte des grands accidents d'Afrique centrale et de la limite nord du craton congolais. Géodynamique **1**, pp. 13–19 (1986)
10. Njonfang, E., Ngako, V., Moreau, C., Affaton, P., Diot, H.: Restraining bends in high temperature shear zones: "The Central Cameroon Shear Zone", Central Africa. J. Afr. Earth Sci. **52**, 9–20 (2008)

New Insights into the Tectonic Windows of the Eastern Desert of Egypt: An Integrated Field and Remote Sensing Approach

Zakaria Hamimi, Mahmoud Hegazy, and Hatem Aboelkhair

Abstract

This work is an integrated field and ASTER remote sensing approach for remapping and interpreting tectonic windows of the Eastern Desert of Egypt. These tectonic windows include the Hafafit, Sibai, Shalul, Meatiq and Beitan gneissic domes. The L1B data of the five areas are used, taking into consideration to avoid ASTER scenes after April, 2008 at which ASTER SWIR data show saturation of values and severe striping. False color composites (FCC) of the non-correlated bands 7, 3, 1 in RGB of the studied areas revealed a good match with field relations and observations. The relative good spatial and spectral resolution of ASTER imagery played an important role in distinguishing subtle spectral differences between the rock units outcropping in the studied areas. Field relations and observations collected from, at least three, cross-strike traverses in each dome, together with results obtained from microstructural investigations of the oriented samples and ASTER image processing led to accurate mapping of these domal structures. The results indicate that these structures of the Eastern Desert of Egypt are formed in a variety of tectonic settings, rather than in one setting, varying from (1) fold interference pattern involving multiply deformed sheath folds (Hafafit), (2) metamorphic core complex development due to regional extension (Meatiq and Sibai), (3) interference folding by single progressive deformation episode (Shalul) to (4) interference folding by separate sequential deformation events (Beitan).

Keywords

Gneissic domes • Metamorphic core complexes
Pan-African belt • Arabian-Nubian shield

1 Introduction

The Eastern Desert of Egypt is traditionally subdivided into three main provinces: Northern Eastern Desert, (NED), Central Eastern Desert (CED) and Southern Eastern Desert (SED). These provinces are juxtaposed along two megashears: the Qena-Safaga Shear Zone separates NED from CED, and Idfu-Mesa Alam Shear Zone splits CED from SED. The NED is occupied by voluminous granitoids, together with slightly deformed-unmetamorphosed Dokhan Volcanics and a post-amalgamation Hammamat volcano-sedimentary sequence. The CED includes gneisses-migmatites-sheared granitoids and remobilized equivalents outcropping as elliptical domal-like structures. In addition to volcanosedimentary succession (mainly volcanogenic metagreywackes those of the CED with the exception of high percentage of gneisses and migmatitic gneisses, as well as the ophiolitic meta ultramafics that form conspicuous tectonically transported nappes easily traced on the satellite imagery. Gneissic and migmatitic rocks of the SED occupy much larger and more complexly shaped areas associated with batholiths of foliated granodiorite [1].

Within the frame of the basement-cover model, the Neoproterozoic belt of the CED and SED is considered as consisting of two major tectonostratigraphic units: (1) older medium to high-grade gneissic association (lower crustal infrastructure or tier-1) and (2) younger Pan-African ophiolitic association, comprising low grade metamorphic ophiolitic, arc volcanic and meta-volcaniclastic assemblages (upper crustal ophiolitic association suprastructure or tier-2). Both units are overlain by molasse-type Hammamat sediments that are deposited in foreland, intermontane and strike-slip basins, and Dokhan Volcanics, and intruded by

Z. Hamimi (✉)
Department of Geology, Faculty of Science, Benha University, Benha, 13518, Egypt
e-mail: yahiahamimi@gmail.com

M. Hegazy
Nuclear Materials Authority, Cairo, Egypt

H. Aboelkhair
Geology Department, Faculty of Science, Damietta University, P.O. Box 3417 New Damietta, Egypt

F. Rossetti et al. (eds.), *The Structural Geology Contribution to the Africa-Eurasia Geology: Basement and Reservoir Structure, Ore Mineralisation and Tectonic Modelling*,
Advances in Science, Technology & Innovation, https://doi.org/10.1007/978-3-030-01455-1_7

numerous granitoids and gabbroic-dioritic complexes. They are juxtaposed along low angle Pan-African sheared contacts, marked by thick mylonite zones. The gneissic association rocks (gneisses, gneissic granites, migmatites, amphibolites and remobilized equivalents) emerge and exhume in the form of tectonic windows and culminations typified by Hafafit, Sibai, Shalul, Meatiq and Beitan structures. Their origin and significance are much debatable in terms of category of structural and tectonic models. They have been regarded as an example of metamorphic or magmatic core complexes and fault-bounded core complexes. Some workers interpreted Sibai gneissic complex in the light of a non-basement-cover model as folded nappes and syn-kinematic gneissic sheets concurrent with the nappe stacking event (≈700–650 Ma); i.e. it is not a core complex.

The present study is an integrated field and remote sensing approach aimed at the re-mapping of the studied tectonic windows, where the exiting maps are not detailed enough, and to decipher the tectonic evolution and structural setting of these domes.

2 Materials and Methods

2.1 Office Works

Office works include the following steps:

- Gathering all previous data.
- Converting maps and data into GIS format.
- ASTER-based discrimination of lithologic varieties outcropping in the studied tectonic windows (Table 1).
- Automatic lineament extraction using digital elevation data.
- Verification of the obtained results from the remote sensing analysis.

2.2 Ground Truth

A forty days-field program was carried out in the studied tectonic windows. Several traverses and selected points were completed in order to study the geology and structure and to collect measurements from all available structural elements. Oriented samples and other geologic and structural data were collected at a great number of observation points throughout the different rock units in the concerned tectonic windows. Overprinting relations between structural fabrics have given much more attention during the field work.

2.3 Lab Work

This has been done through the following steps:

1. Constructing detailed geological maps for the studied tectonic windows based on the results of ASTER data processing and the collected field observations and relations.
2. Microstructural investigations of the oriented samples.
3. Set-up models illustrating the tectonic setting of each tectonic window.

3 Results and Discussion

False color composites of the non-correlated bands 7, 3, 1 in RGB of the studied areas revealed a good match with the data obtained from our filed work. The relative good spatial and spectral resolution of ASTER imagery played an important role in distinguishing subtle spectral differences between the infracrustal gniesses, migmatites and remobilized equivalents, and the overlying ophiolitic mélange

Table 1 ASTER data used in present study

Area	Acquisition date and time	Granule ID	ASTER product	Cloud cover
Sibai	March 29, 2004 08:30:15.37 am	ASTL1B 0403290830150707010028	L1B (radiance at sensor)	0%
Shalul	October 20, 2006 08:35:35.13 am	ASTL1B 0610200835350706300078		
Meatiq	April 3, 2000 08:45:51.53 am	ASTL1B 0004030845511403280001		
Hafafit	October 30, 2003 08:24:17.25 am	ASTL1B 0310300824170807251049		
Beitan	February 21, 2005 08:23:42.49 am	ASTL1B 0502210823421403280009		

lithologies. Principal component analysis (PCA) transformed false color data composite of the first three components, PC1, PC2, PC3 in RGB highlighted the subtle differences between rock units of similar mineral composition within the study areas with more crisper colorful appearance and more correlated with field investigations and geologic maps than reflectance bands combination. PCA color composites played an important role in selecting spectral endmembers for the successive maximum likelihood supervised classification, decreasing spectral mixing of different components.

Results obtained from field relations and observations show that the contacts between the previously mentioned rock units are tectonic rather than stratigraphic. Two prominent megashear trends are prevailing; NW (to NNW) and NE (to ENE). These shearing trends are akin to the Najd-related shear corridor in the Eastern Desert of Egypt. The microstructural investigations of the oriented samples collected from the concerned areas reflect the sense of shearing of both trends, where the NW (to NNW) oriented shearing is sinistral and the NE (to ENE) shearing is dextral. Both shearing trends are not conjugated as previously mentioned.

4 Conclusions

The main points emerging from the present study are:

- Detailed geologic maps are obtained for the studied gneissic domes based on processing of remote sensing data and intensive filed work.
- The studied gneissic domes form in a variety of tectonic settings, rather than in one setting, varying from (1) fold interference pattern involving multiply deformed sheath

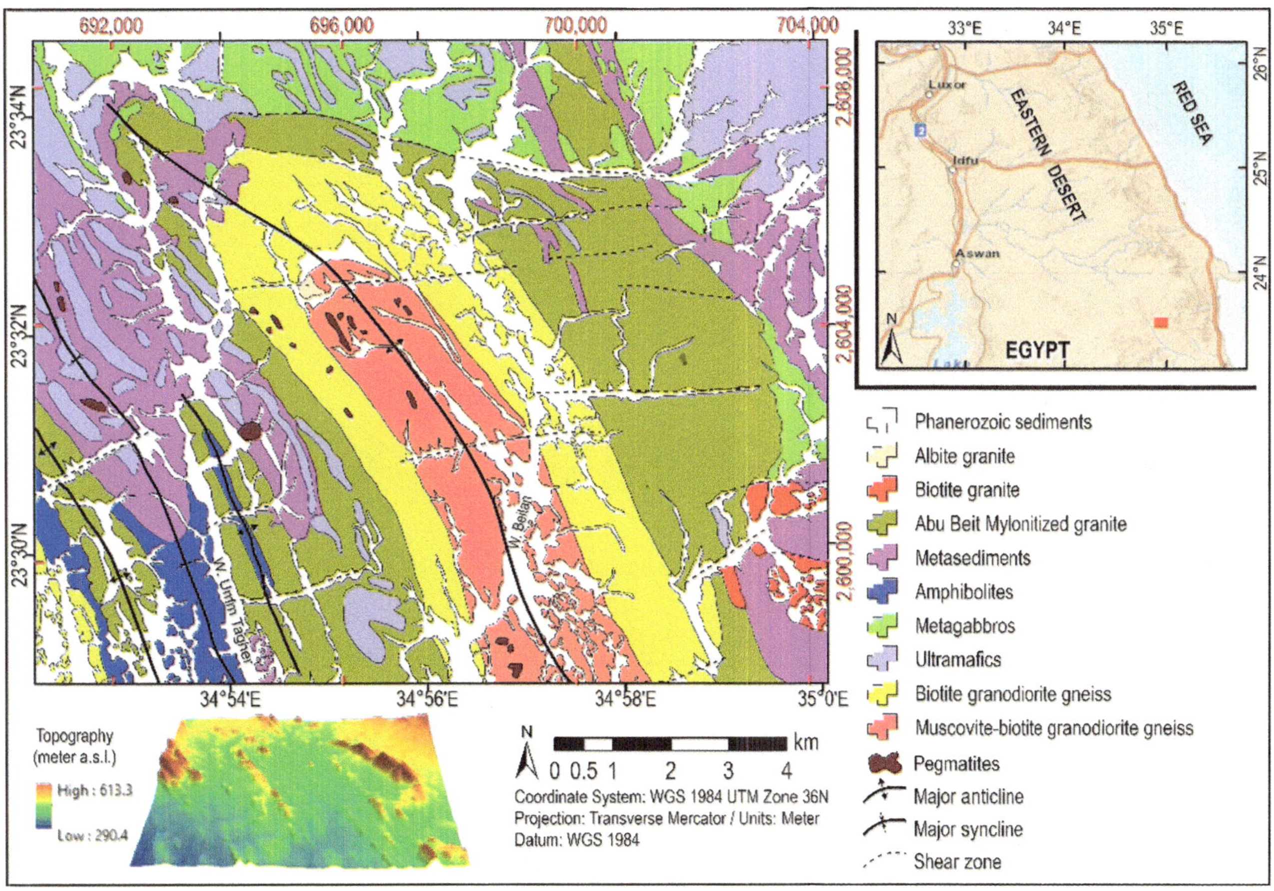

Fig. 1 Geologic map of Beitan tectonic window

folds (Hafafit), (2) metamorphic core complex development due to regional extension (Meatiq and Sibai), (3) interference folding by single progressive deformation episode (Shalul) to (4) interference folding by separate sequential deformation events (Beitan, Fig. 1).

Reference

1. Fowler, A.R., Osman, A.F.: The Sha'it-Nugrus shear zone separating Central and South Eastern Deserts, Egypt: a post-arc collision low-angle normal ductile shear zone. J. Afr. Earth Sci. **53**, 16–32 (2009)

Development of Quartz Ribbons in Felsic Granulites Under Strong Coaxial Deformation in the Highland Complex of Sri Lanka

Bhathiya Athurupana, Jun Muto, and Hiroyuki Nagahama

Abstract

Major lineation and foliation in the metamorphic rocks in the Highland Complex (HC) of Sri Lanka have been interpreted as a result of strong coaxial deformation (D_2) concurrent with the peak metamorphism (>800°C). Quartz ribbons define the structural fabrics in quartzofeldspathic granulites (QFG). We carried out microstructural analysis of a QFG sample from the HC. We found that the crosscutting of ribbons by scapolite reaction corona corresponds to the peak metamorphism. This crosscutting relationship indicates that the D_2 has taken place earlier than the peak metamorphism. The prolate ellipsoidal shape of ribbons manifests pure shear strain geometry that matches the D_2 kinematic framework. The quartz ribbons were formed by the expense of D_1 fabric by grain boundary migration during coaxial D_2. Hence, rheological implications of felsic granulites concurrent with the peak granulite conditions are misleading. Also, care must be taken when interpreting the tectono-metamorphic history of granulite terrains of Gondwana.

Keywords

Quartz deformation • Felsic granulites • Quartz LPO • Lower continental crust

1 Introduction

The Highland Complex (HC) of Sri Lanka is a lower crustal exposure of the East-African orogen. The HC has undergone granulite facies metamorphism [1] and multi-phase ductile deformation [2]. In the quartzofeldspathic rocks of HC, quartz occurs as ribbons and they form the stretching lineation (L) and tectonic foliation (S). This penetrative fabric development was acquired during strong coaxial deformation coeval with peak granulite metamorphism [2]. Such quartz ribbon formation in high-grade shear zones is also considered as the fabric development due to strong deformation by dominant grain boundary migration [3]. The Sri Lankan granulite was subjected to multiple coaxial deformation events [2] and early deformation fabrics and the strain geometry have a significant effect on ribbon formation mechanisms.

In order to understand the ribbon formation mechanism under multi-phase deformation events, we carried out microstructural analysis on the quartzofeldspathic granulite (QFG) from Sri Lanka. Here, we discuss deformation conditions, strain geometry, deformation mechanisms, and effects of early fabrics on ribbon formation, and finally the implication on the rheological interpretation of the granulitic lower crust.

2 Geological Setting, Sample, and Methods

The Precambrian terrain of Sri Lanka consists of four litho-tectonic units: Wanni Complex (WC), Kadugannawa Complex (KC), Highland Complex (HC) and Vijayan Complex (VC) (Fig. 1a). Lithologies of the HC are meta-sedimentary and meta-intrusive rocks metamorphosed at ~900 °C and 0.8–1.0 GPa [1]. Among several ductile deformation events (e.g., D_1–D_6; [2]), D_1 and D_2 events occurred during the prograde path. D_1 is identified as the formation of a foliation (S_1), which is preserved in rootless folds (F_2) and garnet porphyroblasts under the greenschist facies conditions [2]. D_2 is the strongest event coeval with peak metamorphism and responsible for the L_2–S_2 tectonites [2].

Oriented samples were collected from the hinge zone of Digana cuspate antiform between the Dumbara and Huluganga synformes (Fig. 1a). Elongated quartz in the sample demarcates the L_2. We defined a hand specimen coordinate with respect to the kinematic references S_2 and L_2 (Fig. 1b). Three thin sections were prepared (Fig. 1b; XZ, YZ, and

B. Athurupana (✉) · J. Muto · H. Nagahama
Tohoku University, Sendai, 980–8578, Japan
e-mail: bhathiyaat@dc.tohoku.ac.jp

F. Rossetti et al. (eds.), *The Structural Geology Contribution to the Africa-Eurasia Geology: Basement and Reservoir Structure, Ore Mineralisation and Tectonic Modelling*,
Advances in Science, Technology & Innovation, https://doi.org/10.1007/978-3-030-01455-1_8

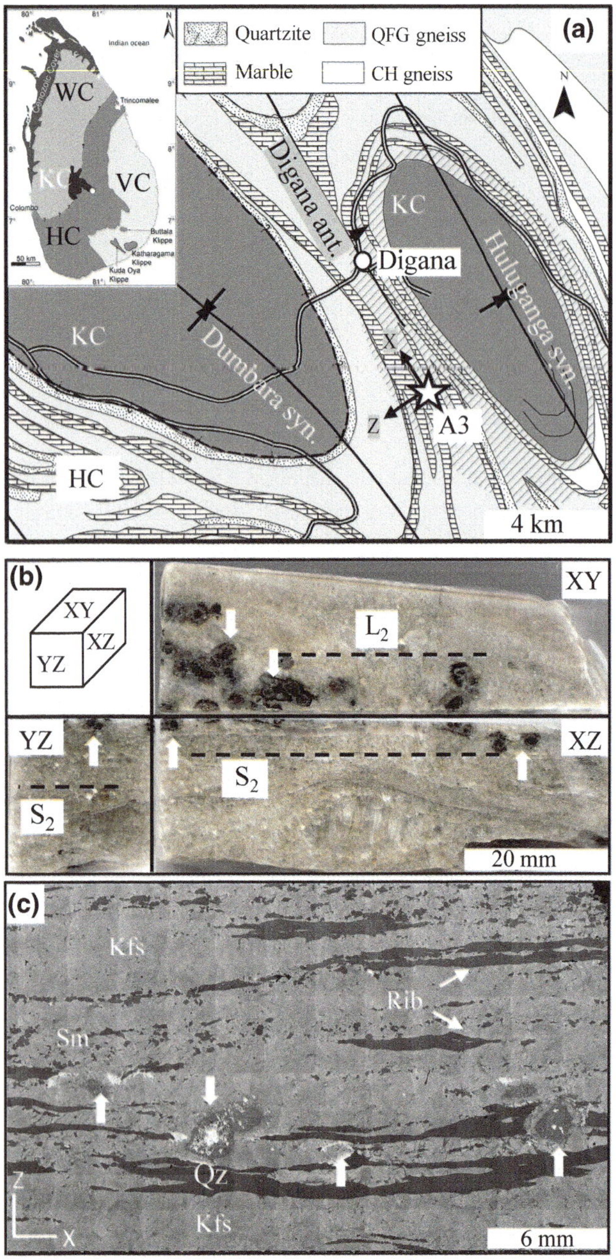

Fig. 1 a Geological map of study area and sample location (07.2615N/80.7488E). **b** Sample, kinematic reference frames, and thin sections geometry. **c** BSE image of the XZ thin section. White arrows show the garnet brake-down and scapolite forming reaction coronas that cross-cut quartz ribbons

XY) for observations. Digitized photomicrographs and backscattered electron (BSE) images were used to measure and calculate equivalent diameter (D_{eq}), aspect ratio (R), and grain boundary lobateness [4] of quartz and matrix feldspar grains. Surface orientation distribution functions (ODF) were calculated by SURFOR [4]. Lattice preferred orientation (LPO) were measured by HKL Electron backscattered diffraction system and plotted as poles to planes in equal area lower hemisphere projections.

3 Results

The sample contains K-feldspar (65%), quartz (25%), plagioclase (<5%), pyroxene ($\sim$3%), scapolite (1%) and garnet (<1%). Abbreviation for mineral names are used hereafter [5]. The matrix is made of K-feldspar (Kfs) which is in sanidine (Sa) composition (Or_{59}). Plagioclase (Pl) appears with ortho-pyroxene (Opx) as clusters of reaction products around garnet (Gt) porpyroblast (Fig. 2a). The granoblastic Pl grains do not show any deformation fabrics. Scapolite (Scp) coronas surround Pl clusters (Fig. 2a) and the both Pl and Scp coronas cut across quartz ribbons in three dimensions (Fig. 1b, c).

Quartz is present as elongated grains (ribbons; D_{eq} 0.5–4 mm and R $\sim$5) and isolated small grains (isolated quartz; D_{eq} 0.05–0.04 mm and $R \sim 1.5$) in the matrix (Fig. 1c). The ribbons show three dimensionally prolate ellipsoidal shapes, joined to form large polycrystalline ribbons (Fig. 1c).

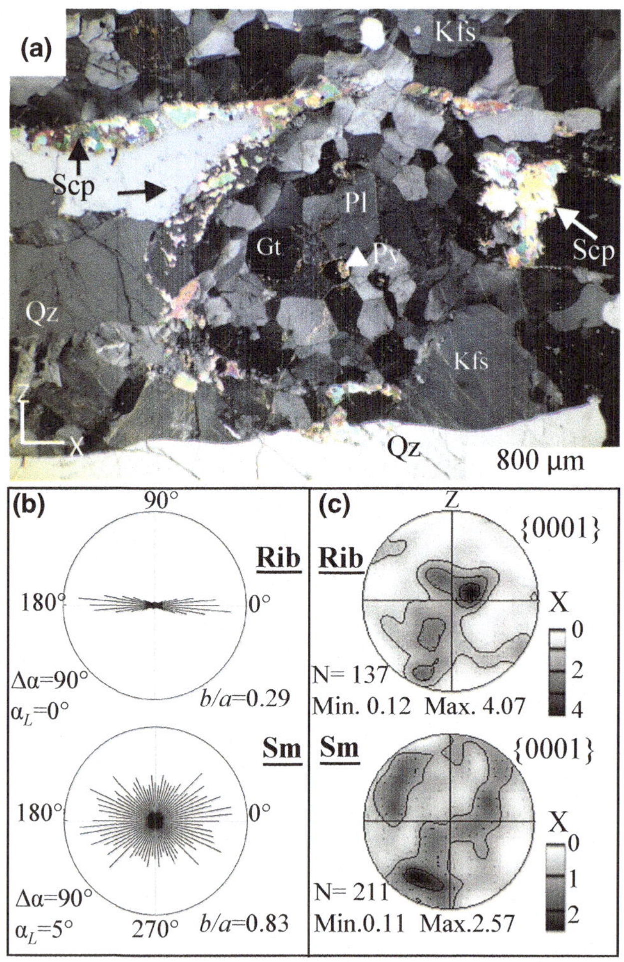

Fig. 2 a Photomicrograph of crosscutting reaction corona of garnet breakdown and scapolite. **b**, **c** Length-weighted rose diagrams (SURFOR) and *c*-axis pole figure of ribbons (Rib) and isolated quartz grains (Sm)

Both ribbons and isolated quartz do not show any evidence of internal strain. Within some ribbons, there are remnants of quartz grains that have different crystallographic orientations. The ribbons show higher grain boundary lobateness compared to a perfect elliptical shape. This could be an indication of grain boundary migration (GBM). The orthorhombic symmetry of surface fabric ODFs of both quartz types (Fig. 2b) are also consistent with the pure shear strain geometry of the D_2 deformation. Dominant Y maxima with minor Z maxima LPOs of ribbons (Fig. 2c) correspond to the activation of prism <*a*> and basal <*a*> slip systems. While the isolated quartz grains show Z maxima corresponding to the activation of basal <*a*> slip. In addition, inequigranular (D_{eq} 0.5–1.1 mm) sanidine matrix owns interlobate grains with high grain boundary lobateness also indicating GBM.

4 Discussion

Scp appears above 850 °C and 0.8 GPa [6] which corresponds to the peak metamorphic conditions of the granulite facies. This reaction consumed Pl which is a product of prograde Gt break-down reaction (Gt + Qz → Pl + Opx) approximately at ~800 °C and 0.9 GPa of the *P-T-t* path of the HC [1, 2]. Hence, this crosscutting relationship with ribbons indicates that the ribbon formation (D_2) has taken place before the peak metamorphism of the HC (<800 °C).

The quartz LPOs indicate dislocation creep as the dominant deformation mechanism in both quartz ribbons and isolated grains. Activation of prism <*a*> slip system manifest deformation temperatures of 600–750 °C [7]. Moreover, the high-temperature sanidine (>550 °C; [8]) matrix with intense grain boundary migration (GBM) is consistent with the deformation temperature manifest by prism <*a*> slip. On the other hand, the activation of basal <*a*> slip system in isolated quartz indicates temperatures below 600 °C [7], being consistent with the condition of D_1 deformation [3]. Therefore, dominating prism <*a*> slip system in quartz ribbons indicates fabric transition from basal <*a*> during the D_2 event. This transition is associated with the GBM at high strain [9]. High lobateness and remaining grains similar to isolated quartz within ribbons support the idea of GBM of quartz aggregates.

Hence, our findings show that the elongated quartz ribbon formation has occurred at temperature >600 °C where GBM was the dominant recrystallization mechanism. The final shape of ribbons represents the strain of both D_1 and D_2 events. The crosscutting of the ribbons by Scp and Gt brake-down reactions implies that D_2 event was prior to the peak metamorphism. The recovery during the D_2 or later could be the cause for no internal strain in both ribbons and isolated quartz. The idea of strong coaxial deformation at the peak metamorphism should be reevaluated in terms of the tectono-metamorphic history of the Gondwana terrains. Moreover, the rheological implications of QFG based on D_2 fabrics could be carefully addressed because the lower deformation conditions of D_2 than peak conditions and the effect of the D_1 fabrics on the D_2 fabrics.

5 Conclusions

Our observations clarified that the D_2 deformation indicative of strong coaxial deformation predated the peak metamorphism of the HC and there is no sign of penetrative deformation coeval with peak metamorphism. The difference in microstructures (grain geometry and LPO patterns) between quartz ribbons and isolated quartz grains were likely caused by the GBM under the strong deformation event during the prograde path of the HC. Both deformation events contributed to the development of elongated quartz ribbons that form the S-L tectonites in QFG. Hence, the high strain fabrics such as ribbons cannot be considered as a result of a single strong deformation event contemporaneous to peak metamorphism in granulite terrains. This could affect the interpretation of tectono-metamorphic evolution of high-grade terrains.

References

1. Raase, P., Schenk, V.: Petrology of granulite-facies metapelites of the Highland Complex, Sri Lanka: implications for the metamorphic zonation and the *P–T* path. Prec. Res. **66**, 265–294 (1994)
2. Kehelpannala, K.V.W.: Deformation of a high-grade Gondwana fragment, Sri Lanka. Gond. Res. **1**, 47–68 (1997)
3. Passchier, C.W., Trouw, R.A.J.: Microtectonics. Springer, Berlin (2005)
4. Heilbronner, R., Barrett, S.: Image analysis in Earth Sciences: microstructures and textures of Earth materials. Springer, Berlin (2014)
5. Whitney, D.L., Evans, B.W.: Abbreviations for names of rock-forming minerals. Am. Mine. **95**, 185–187 (2010)
6. Newton, R.C., Goldith, J.R.: Stability of the end-member scapolites: $3NaAlSi_3O_8 \cdot NaCl$, $3CaAl_2Si_2O_8 \cdot CaCO_3$, $3CaAl_2Si_2O_8 \cdot CaSO_4$. Zeit. für Krist. **143**, 333–353 (1976)
7. Kurz, W., Fritz, H., Tenczer, V., Unzog, W.: Tectonometamorphic evolution of the Koralm Complex (Eastern Alps): constraints from microstructures and textures of the "Plattengneis" shear zone. J. Struct. Geol. **24**, 1957–1970 (2002)
8. Bowen, N.L., Tuttle, O.F.: The system $NaAlSi_3O_8$-$KAlSi_3O_8$-H_2O. J. Geol. **58**, 489–511 (1950)
9. Muto, J., Hirth, G., Heilbronner, R., Tullis, J.: Plastic anisotropy and fabric evolution in sheared and recrystallized quartz single crystals. J. Geophy. Res.: Sol. Ear. **116**, 1–18 (2011)

Tectonic Evolution of the Metamorphic Core Complex Around the Kotah and Loe Sar Domes, Swat, North Pakistan

Asghar Ali and Hikmat Salam

Abstract

The Swat Valley rocks encompass at least four tectonic and three metamorphic events since the initial collision between the Indian Plate and Kohistan Island Arc (KIA). The early S_1 preserved in garnets and microlithon of the S_2 are obscured at the outcrop scale by D_2. S_2 foliations are preserved in garnet porphyroblasts and the main matrix foliations. Post D_2 domes exhumed S_2 tectonic fabric and the metamorphic core complex of the Indian Plate in the Kotah and Loe Sar Domes. These domes are wrapped by the Swat Tourmaline Granite (STG) that was developed by anatectic processes. S_2 parallel kyanites indicate that the D_2 tectonic event developed under upper amphibolite facies metamorphic conditions before the Kotah and Loe Sar Domes culmination. D_3 resulted in S_2 crenulations, L_3 intersection lineations and syn–S_3 growth of biotite, garnet, staurolite, kyanite and weakly developed sillimanite. The upper amphibolite facies metamorphic conditions around the domes and greenschist facies conditions along the Main Mantle Thrust (MMT) indicate that the high grade metamorphic rocks are exhumed by the Indian Plate basement rocks protrusion in the cores of the domes. Episodic reaction textures preserved in porphyroblasts and index minerals growth yielded a clockwise P–T path for the Alpurai Group Metasediments (AGM).

Keywords

Metamorphism • Exhumation • Anatexis
Domes • Main mantle thrust • P–T path

A. Ali (✉) · H. Salam
Department of Geology, University of Peshawar, Peshawar, Khyber-Pakhtunkhwa, Pakistan
e-mail: asghar.ali@uop.edu.pk

1 Introduction

The Neo–Tethys subduction beneath the KIA and Eurasian Plate resulted in the actively evolving and uplifting the Himalayan Orogenic Belt. This orogen, which extends for ~2300 km from Afghanistan to Burma presents a classic example of present day continent–continent collisional orogenic belt (Fig. 1, [1]). The intervening Neo–Tethys subduction, which started in the Mesozoic under the KIA completed in the Eocene by the Indian Plate collision with the KIA along the MMT, which is equivalent of the Indus–Tsangpo Suture Zone. Onward the final collision is followed by the Late Miocene domal structures, Plio–Pleistocene anatexis and ~5 Ma granulite facies metamorphic conditions in the Nanga Parbat region. Similarly, in the west of the Nanga Parbat Syntaxis, the Swat region is followed by domal structures and metamorphism after the initial collision between the Indian Plate and KIA. The northernmost margin of the Indian Plate comprises a discontinuous series of gneissic domes/metamorphic core complexes [2]. In active collision orogenic belts deeply buried basement rocks exhumed very rapidly due to crustal extension. The young actively evolving Central and Eastern Alps document ca. 22–12 Ma metamorphic domes [1]. Similarly, the northern portion of the actively evolving Himalayan Orogenic Belt has been experienced extension since its initial collision [3]. The Himalayan gneissic domes, metamorphic core complexes, south Tibet detachment system and associated ~36–32 Ma anatectic leucogranite are believed to be formed due to crustal thickening followed by extension. In such scenarios mesoscopic and microscopic structures provides a robust tool to elucidate relationships between regional scales with multiple deformation and metamorphic events. This study aims to recognize multiple deformation and metamorphic events, domal structure culmination, metamorphic isograds and metamorphic facies associated with the continent–continent collision belts.

F. Rossetti et al. (eds.), *The Structural Geology Contribution to the Africa-Eurasia Geology: Basement and Reservoir Structure, Ore Mineralisation and Tectonic Modelling*, Advances in Science, Technology & Innovation, https://doi.org/10.1007/978-3-030-01455-1_9

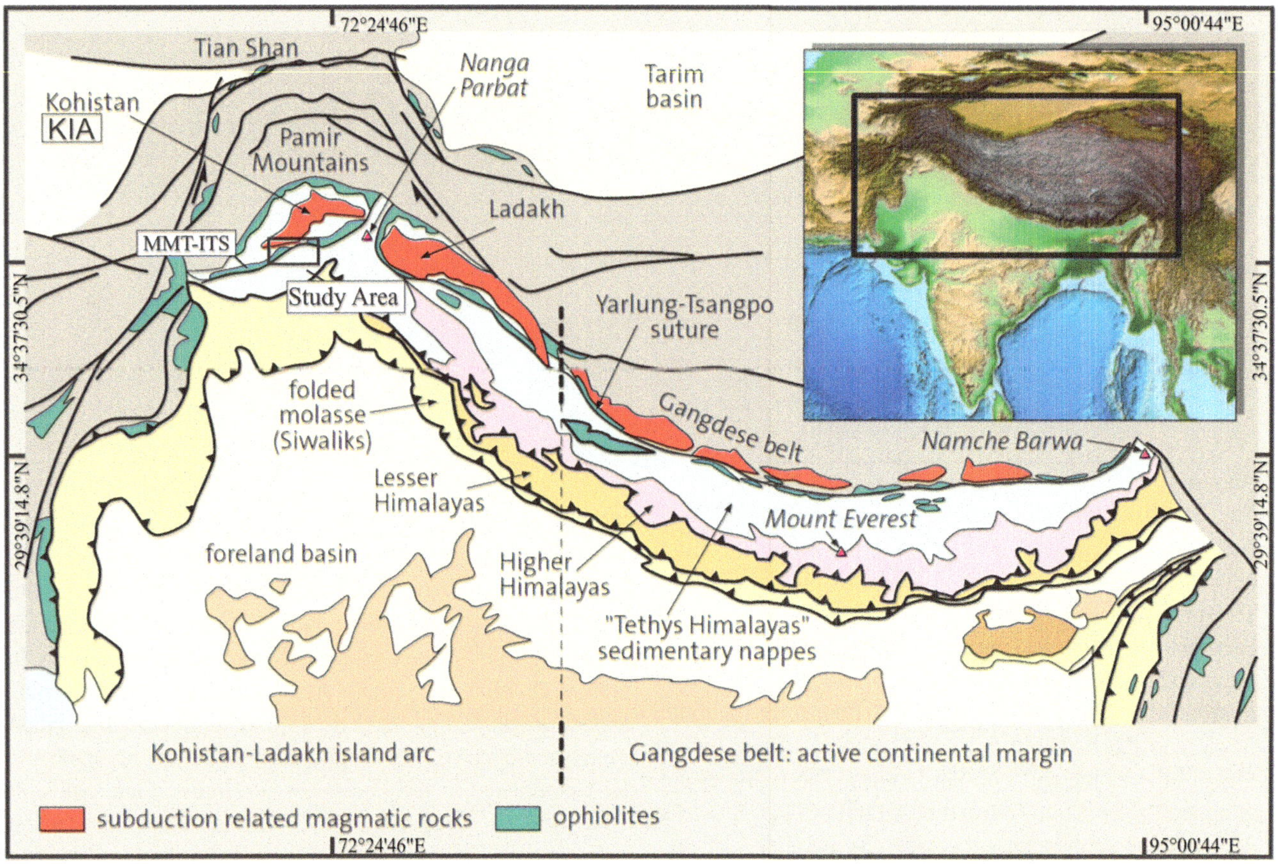

Fig. 1 Regional tectonic map of the E–W trending Himalayan Orogenic Belt showing the location of the study area (After [1])

2 Regional Geology

The Swat valley predominantly consists of multiply deformed and metamorphosed Precambrian to Early Mesozoic rocks [4]. The Cambrian to Lower Permian coarse-grained granitic gneisses intruded the Precambrian Manglaur Formation. These rocks are unconformably overlain by the Permian to Late Mesozoic poly-deformed and metamorphosed AGM. These rocks preserved early history of deformation since Eocene [5]. Major E–W structures preserved the head on collisional history, that are rotated along their eastern limits to N–S in the western hinterland zone by younger deformational events.

3 Methods

3.1 Mesoscopic Structures

D_2 folds and S_2 are visible across the Swat geological map. In the central part of the Swat Valley the main S_2 tectonic fabric is bulged upward by isostatic uplift of the Kotah and Loe Sar Domes. The contact zone between the AGM and Swat Granitic Gneisses (SGG) are wrapped up by the STG. The D_3 deformation event, which is constrained from the D_3 folds and L_3 intersection lineations microscopically and mesoscopically overprints the main S_2 tectonic fabric. The plunge of D_3 folds and L_3 are controlled by S_2 dip amount and direction. As S_2 is wrapped around the Kotah and Loe Sar Domes, L_3 plunge also changes around these regional scale domes. The wrapping of S_2, L_2 and change in plunge direction of L_3 around the Kotah and Loe Sar Domes are also evident from stereoplots. At places the main S_2 fabric, L_2 stretched mineral lineations and L_3 are overprinted by top to NE and top to SW verging asymmetric mesoscopic shear folds.

3.2 Petrography and Microstructures

Oriented samples with respect to true north were collected from the northernmost active continental margin of the Indian Plate and MMT suture zone. 40 thin sections were made from the AGM, Chakdara Granite, STG, SGG and MMT suture zone. Episodic growth of garnet porphyroblasts nicely preserved S_1, S_2 an S_3 foliations. S_1 foliations are truncated by the well-developed micro to mesoscopic S_2. S_1, which is also preserved as crenulated cleavage in the microlithon of S_2 was not observed at the outcrop scale. S_2 in the garnets are locally truncated by the S_3. S_3 that are preserved in garnets are continuous with the S_3 in the matrix. Kyanite inclusions were observed in the S_1, S_2 and S_3 garnets and in the matrix crenulations. Staurolite inclusions were observed in the rims of garnets, which is synchronously developed with S_3. Staurolites preserved D_3 folds, S_3 and crenulated S_2 foliations. Syn–S_2 fibrolitic sillimanite growth is crenulated by the later S_3. S_3 parallel tourmaline grains were observed in the STG and at contact zones of the AGM with the STG. Biotite and muscovite are concentrated in the S_2 and S_3. Retrograde chlorite replaces garnet, staurolite, kyanite, biotite and muscovite.

4 Discussion

The Loe Sar and Kotah Domes exposed the Precambrian Manglaur Formation and Cambrian SGG. These domes apart from exposing the upper amphibolite facies rocks also overprinted the well developed S_2 tectonic fabric. The S_2 parallel kyanite grains indicate that kyanite growth in the Swat region pre-dated the Kotah and Loe Sar Dome culmination. Kyanite grew during the northward subduction of the Indian Plate beneath the KIA. Subduction associated crustal thickening, high geothermal gradient and anatectic process that produced STC at the SGG and Marghazar Formation interface resulted in early stage positive buoyancy, extension and exhumation of kyanite bearing rocks.

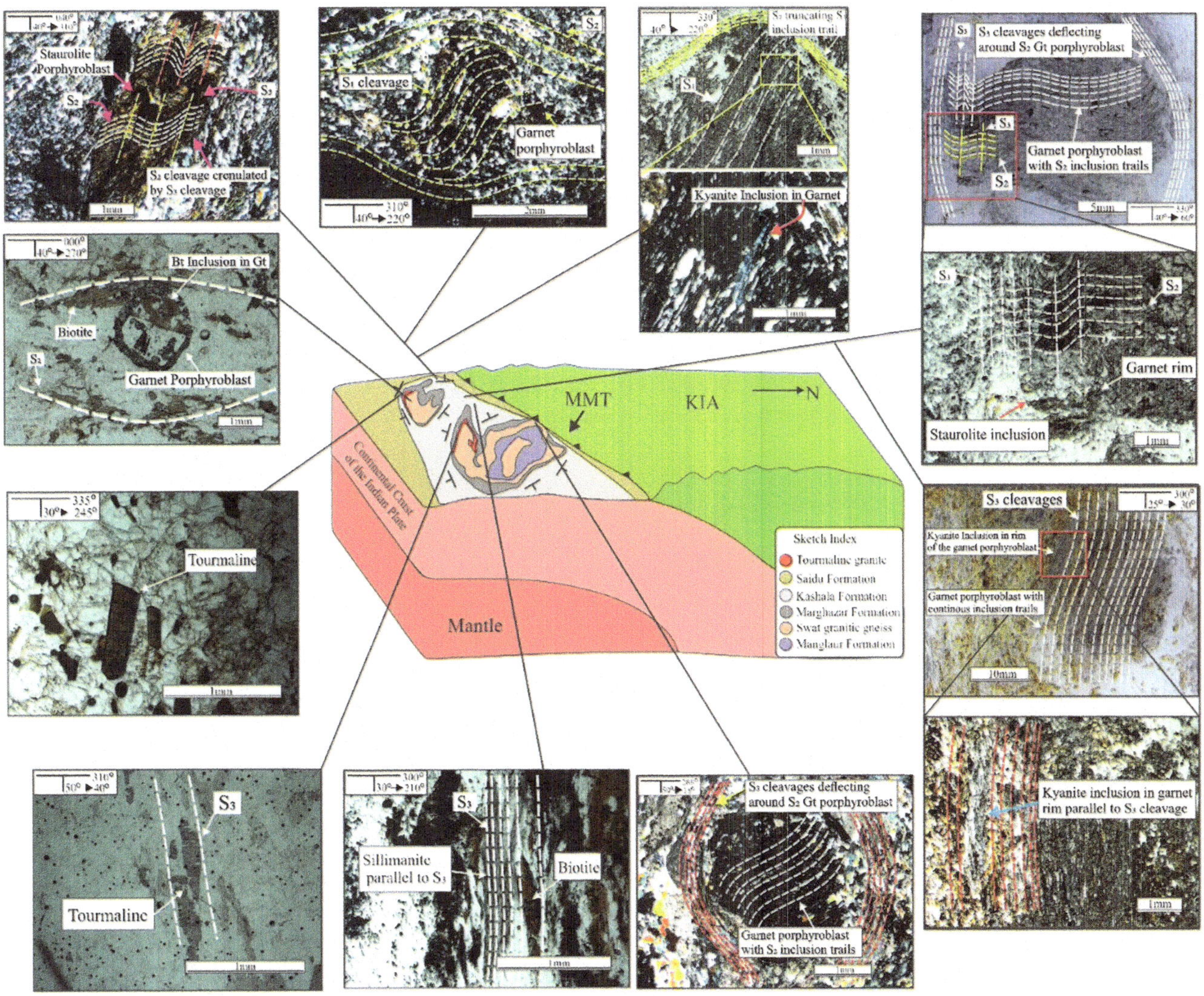

Fig. 2 Schematic model of the study area showing MMT, Swat stratigraphy and exhumed domes. The photomicrographs depicting episodic textures and mineral growth

Positive buoyancy related to partial crustal melting, higher geothermal gradient and extension led to the Indian Plate resistance to subduction and exhumation of the Kotah and Loe Sar Domes. Successive development of S_3, S_3 parallel tourmaline in the STG, L_3, syn-S_3 garnet rim, staurolite, kyanite and weak sillimanite plus retrograde chlorite growth respectively require elevated geothermal gradient and exhumation that could nicely be explained by subduction related crustal thickening and uplifting of the Indian Plate. The main tectonic fabric in the region is overprinted by asymmetric dextral and sinistral ductile shear zones, which are also characterized by localized cataclasites.

Metamorphic zones were constrained by mapping metamorphic index minerals. The grade of metamorphism increases towards the core of the Kotah and Loe Sar Domes and reduces with greenschist facies metamorphic conditions along the MMT–Melange complex (Fig. 2). The Barrovian metamorphic conditions around the domes are related to the exhumation of the domes that are facilitated by the subduction related crustal extension of the Indian Plate. Episodic reaction textures preserved in porphyroblasts and in the matrix in the form of crenulated and crenulation cleavages yielded clockwise P–T paths for the rocks exposed in the Swat Valley.

5 Conclusions

Continent–continent collision followed by extension brought the Barrovian metamorphic rocks to the surface by exhumation, away from domes the rocks are characterized by greenschist facies metamorphic conditions. Crustal thickening and positive buoyancy of the relatively hot Indian Plate resulted in Kotah and Loe Sar Domes culmination and formation of anatectic tourmaline bearing granite at the SGG and Marghazar Formation interface. Episodic subduction and crustal thickening of the Indian Plate due to ongoing collision with the KIA in the north resulted in garnet rim, staurolite, kyanite and weak sillimanite growth.

References

1. Frisch, W., Meschede, M., Blakey, R.: Plate Tectonics, Continental Drift and Mountain Building, pp. 1–212. Springer, Berlin (2011)
2. Yin, A.: Cenozoic tectonic evolution of the Himalayan orogen as constrained by along–strike variation of structural geometry, exhumation history, and foreland sedimentation. Earth-Sci. Rev. **76**, 1–131 (2006)
3. Zhang, J., Santosh, M., Wang, X., Guo, L., Yang, X., Zhang, B.: Tectonics of the northern Himalaya since India-Asia collision. Gondwana Res. **21**, 939–960 (2012)
4. Dipietro, J.A., Ahmad, I., Hussain, A.: Cenozoic kinematic history of the Kohistan fault in the Pakistan Himalaya. Geol. Soc. Am. Bull. **120**, 1428–1440 (2008)
5. Larson, P.K., Ali, A., Shresta, S., Soret, M., Cottle, M.J., Ahmad, R.: Timing of metamorphism and deformation in the Swat valley, northern Pakistan: Insight into garnet-monazite HREE partitioning. Geosci. Front. (in press)

Archean Continental Crust Beneath Mauritius, and Low Oxygen Isotopic Compositions from the Malani Rhyolites, Rajasthan, (India): Implication for the Greater Malani Supercontinent with Special Reference to South China, Seychelles and Arabian-Nubian Shield

Naresh Kochhar

Abstract

Recent discovery of Archean xenocrystic zircons (3.2 Ga) from Mauritius indicating provenance from BGC, Rajasthan (India), Kazakhstan, Mongolia, Seychelles, Arabian - Nubian shield and Tarim support the assembly of the Greater Malani Supercontinent. Further, recent oxygen isotopic data of zircons from the Malani Rhyolite terrain, Rajasthan lend support to South China connection with India during late Proterozoic.

Keywords

Malani magmatism • Rajasthan • Ring structure
Oxygen isotops • Supercontinent

1 Introduction

Malani Magmatism

The Trans Aravalli—Block (TAB) is unique in the geological evolution of the Indian shield as it marks a major period of anorogenic (A-type), 'Within Plate', high heat producing (HHP) magmatism represented by the Malani igneous suite of rocks (MIS). The Neoproterozoic Malani igneous suite (55,000 km^2; 732 Ma) comprising peralkaline (Siwana), metaluminous to mildly peralkaline (Jalor), and peraluminous (Tusham and Jhunjhunu) granites with cogenetic carapace of acid volcanic (welded tuff, trachyte, explosion breccia and perlite etc.) are characterized by volcano-plutonic ring structures and radial dykes. The suite is bimodal in nature with minor amounts of basalt, gabbro and dolerite dykes. The representatives of Malani suite also occur at Kirana Hills, and at Nagar Parkar, Sindh, Pakistan (Fig. 1). The Siwana ring structure (30 km in EW 25 km in NS direction) is the most spectacular feature of the Thar Desert (Fig. 2). The Malani magmatism is controlled by NE SW trending linements (zones of extension and high heat flow) of fundamental nature (mantle) and owes its origin to mantle plume (For a review, [5–7, 11, 12, 17]).

The volcano-plutonic ring structures, high heat flow, elevated basement rocks (gravity anomaly) 120 km across low velocity anomaly beneath the TAB are symptomatic of plume activity in the region. The location of low velocity anomaly beneath Sarnu-Dandali area marks the site of the proposed Malani plume.

2 The Greater Malani Supercontinent

The period ca. 732 Ma B.P. marks a major Pan–African tectano–magmatic event of widespread magmatism of alkali granites and comagmatic acid volcanic (anorogenic, A-type) in the Trans- Aravali block of the Indian shield, central Iran, Somalia, Nubian-Arabian shield and Madagascar and south China, Somalia. In view of the widespread ca. 750 Ma alkaline magamatism, which is so widespread, and well documented in TAB, central Iran, Arabian-Nubian shield, Madagascar and S. China, Somalia, Seychelles, Siberia, Tarim, Mongolia, Kazakhstan [10, 14–16, 18] Kochhar [12] proposed that all these micro continents were characterized by common crustal stress pattern, rifting, thermal regime, Strutian Glaciation and subsequent desiccation and similar paleolatitudinal positions which could be attributed to the existence of a super continent-the Greater Malani Super continent (Fig. 3), (Table 1). This assembly and subsequent breakup marked rift to drift tectonic environment.

N. Kochhar (✉)
Centre of Advanced Study in Geology, Punjab University, Chandigarh, 160014, India
e-mail: nareshkochhar2003@yahoo.com; nareshkochhar@pu.ac.in

F. Rossetti et al. (eds.), *The Structural Geology Contribution to the Africa-Eurasia Geology: Basement and Reservoir Structure, Ore Mineralisation and Tectonic Modelling*,
Advances in Science, Technology & Innovation, https://doi.org/10.1007/978-3-030-01455-1_10

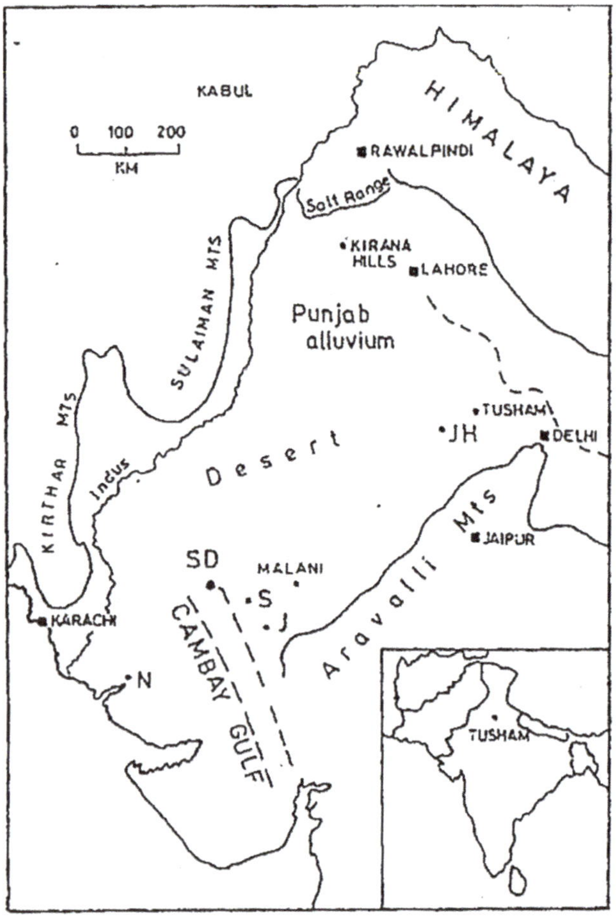

Fig. 1 Location map of Malani Igneous Suite, India. J: Jalor, JH: Jhunjhunu, S: Siwana, N: Nagarparkar, SD: Sarnu-Dandali

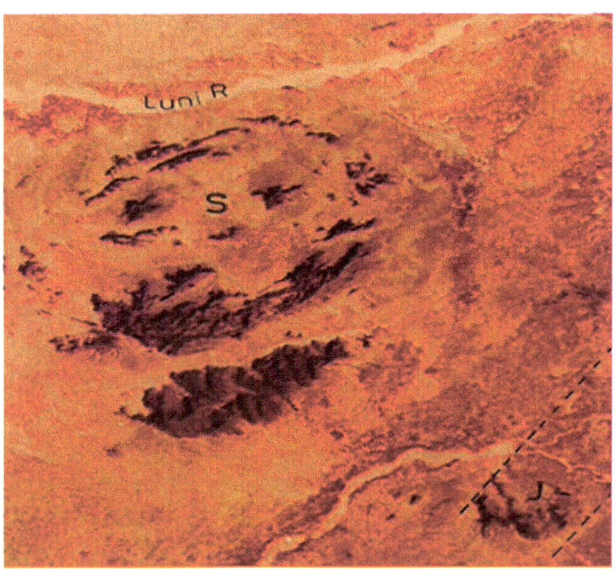

Fig. 2 Siwana ring structure (25.6 km NS, 31 km EW). J: Jalor ring structure

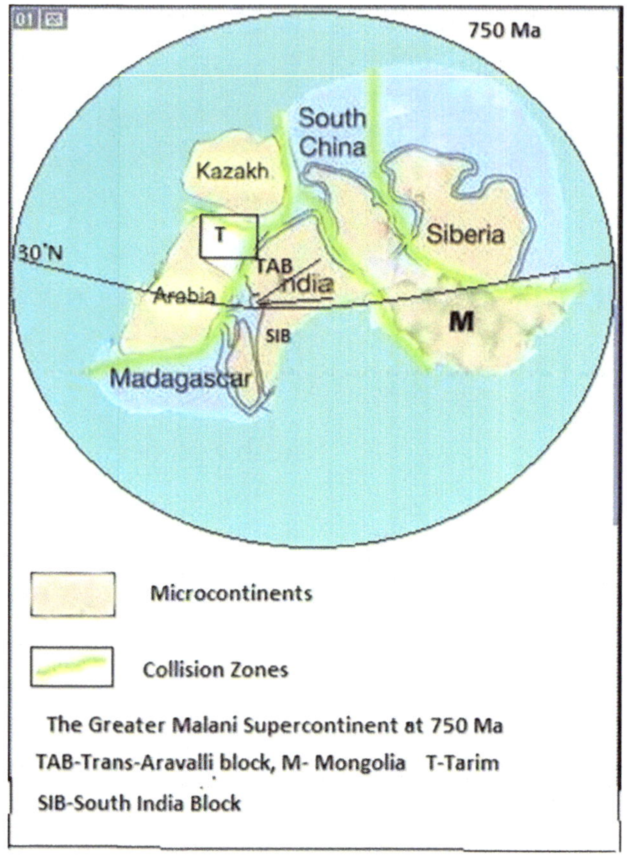

Fig. 3 Assembly of the greater Malani supercontinent at 750 Ma

3 Results

South China

In the Yangste craton of South China peralkaline to alkaline, 'within-plate' magmatic rocks (pre-rift 830–795 Ma, syn-rift 780–745 Ma) associated with coeval swarms of mafic dykes has been attributed to super plume. The syn-rift, the Chengian magmatism has been correlated with the anorogenic Malani magmatism [10, 21]. The emplacement of mafic dyke swarms initiated the breakup of Rodinia. Rounded and abraded shape of Archean zircons (3.8 Ga) from Cathysian block indicate long transport with derivation from other continental terrains linked with S. China such as India and Antartica [3]. Zheng [26] has also suggested, and based on paleomagnetic data, oxygen isotope composition and sequence stratigraphy observations, that S. China was adjacent to India and Australia in the configuration of Rodinia rather than between Australia and Laurentia. Recently Wang et al. [24] have suggested that low δ^{18}o isotope compositions of zircons from the Malani rhyolites (4.12 to −1.11‰) can be genetically linked to synchronous ^{18}o-depleted igneous rocks of South China, Seychelles and Madagascar, thus supporting

Table 1 Summary of similarities between TAB, Seychelles, Nubian-Arabian Shield, South China, Kazakhstan, Siberia, and Mongoli

	TAB NW Indian shield	Seychelles	Nubian Arabian shield	S. China Yangtze block	Kazakhistan	Siberia	Mongolia
Alkali magmatism granites	Siwana Jalor Tusham CA 745 Ma	Mahe, Ste Anna Praslin 700–755 Ma	Jabal-Al Hasasin Jabal-As SawalDahul 700–760 Ma	Chengjian 780–745 Ma	Kopa formation 775 Ma Red gneiss granite	SW MarginTataraka-Ishimila suture zone Srednetalar-aka pluton 671–774 Ma Mafic Dykes (Nersa) 740 Ma	Barguzinsky block 850–700 Ma
Protolith	Archean BGC	Archean BGC	Paleo proterozoic or older crustal precursor	–	AnrakhaiTerrane 1789 ± 0.6 Ma 2187 ± 0.5 Ma εNd (t)—6.4	–	Ust, Angara complees εNd (t) +8 to −6 Juvenile crust
Paleople	55–70°N	30°N	–	70°N	–	–	–
Strutian glaciations	Pokhran, salt range boulder bed	–	Haqf super group	Linntua and Nantua deposits	Upper part of sequence underling Karatau	NorlheinSiberna	Olokit graben
Evaporite/carbonate sequence	Hanseran, Nagaur-Bikaner Basin	–	Hormuz series	Late Sinian Late Precambrian carbonates and phosphates	South Tyan Shan	Kuonanm Black Shale	Darkhat and Khubsugul Series

the idea first given by Kochhar [10] that South China was attached to the Trans-Aravalli block of NW Indian shield during Late Proterozoic.

Paleomagnetic data

The AAPW paleomagnetic poles at ca. 800 Ma B.P. place South China at 55–70°N at par with the paleolatitudes of India during Malani Rhyolite period 'ca. 750 Ma'. Around 700 Ma B/P, India started drifting towards the tropical latitudes and around 600 Ma, India was near equator [22]. The paleopoles for the Blaini Formation indicate a medium to high latitude position for the lower diamictite (46.5°N, 109°E) but tropical latitude for the overlying cap dolostone 3.0°N–98.5°E. Detrital zircons from the base of Blaini Formation (diamictite) gave an age of 692 ± 18 Ma (207Pb/206Pb) [2].

3.1 Seychelles and Arabian-Nubian Shield

Recently Kochhar [8] has shown that the Mahe, Ste. Anne and Praslin granites of Seychelles have close resemblance to the Siwana and Jalor granites of MIS in terms of age, paleopositions, hypersolvus-subsolvus associations, Sr, Pb, Nd and oxygen isotopic compositions and the role of 3.2-Ga Archaean crust of BGC, Rajasthan. The Ga/Al versus Zr plot of Seychelles granites overlap with those of Jalor and Tusham granites (Fig. 3). These geochemical characteristics suggest that like MIS, the Seychelles granites are also 'WPG', anorogenic, A-type and do not represent Andean-type arc on the western margin of Rodinia. Based on commonalities between Arabian-Nubian shield and the Trans-Aravalli block of the NW Indian shield in terms of anorogenic magmatism with A-type geochemical signatures, ring structures, presence of Archaean crust of BGC (3.2 Ga) as protolith, and evidence of Strutian glaciation and subsequent desiccation, it has been suggested by [11, 12]; Kochhar [9] that the Arabian-Nubian shield was attached to the TAB of the NW Indian shield around 600–700 Ma in the configuration of the Malani supercontinent. Similarly there are similarities between the Arabian-Nubian shield and Central Asian Orogenic belt. According to Jahn [4], the Central Asian Orogenic Belt (CAOB) also show growth of massive juvenile crust formed my vertical accretion of non orogenic granites and their volcanic equivalents and also by later accretion of the acr assemblage in many terranes such as Mongolia, Altai- Xin Jiang, inner Mongolia, NE China, Trans- Baikalia, Kazakhstan and West Sayan and Gorny Altai of south Siberia. In this regard the evolution of the TAB crust (Jalor and Siwana ring complexes of MIS with positive Nd (T) values is very much similar to the evolution of ANS and CAOB terrains [13, 20, 25].

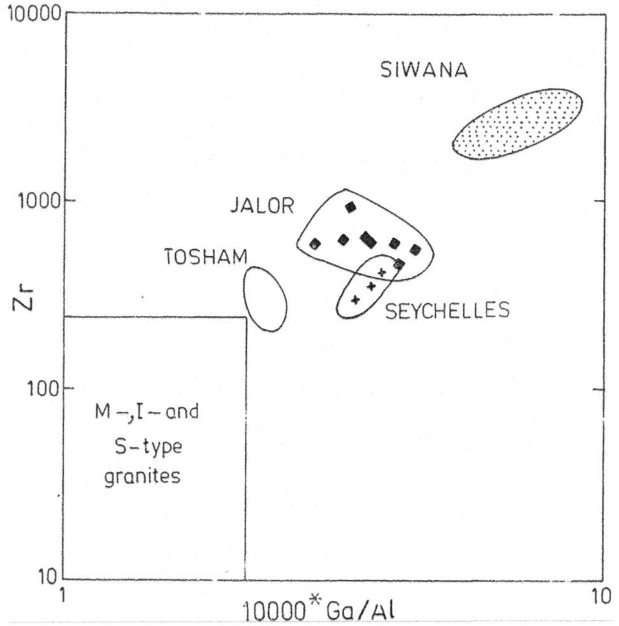

Fig. 4 A-type signatures of the Malani Granites. Also shown are fields of the Mahe and Praslin granites, Seyshelles

Recently Ashwal et al. [1] have demonstrated the existence of Archean (2.5–3.0 Ga: U/Pb Zircon) continental crust beneath Mauritus, which formed originally part of ancient nucleus of Madagascar Antongil, India (Dharwar) the Maurita continental fragment.

This recent work of Ashwal et al. [1] further support the existence of the Greater Malani Supercontinent. However they have overlooked the robust geological and isotopic data of Rajasthan.

1. The Archean crustal protolith has been well documental from BGC, Rajasthan (3.2 Ga), Kazakhstan, (Anrakhai terrain (1789±6 Ma, 2.18 ± 0.05 Ma, εNd (t)—6.4, Mongolia (Ust, Angaracomplex, εNd (t) +8 to −6), Seychelles and ANS (Archean BGC), Tarim (2.7–2.5 Ga) [19].
2. The Mahe (750 Ma) and Ste Anne (764 Ma) granite of Seychelles have been correlated with 732 Ma old hypersolvus (Siwana) and Subsolvus (Jalor) granite of Malani Igneous suite (Fig. 4). These granite have A-type geochemical signatures and owe their origin to the same protolith of Archean crust (BGC of Rajasthan) [1]. The Seychelles and Malani granite have low oxygen isotoperatios—δ^{18} O ‰ SMOW:

Siwana: −0.10 to +1.8‰
Jalore: −4.60 to +1.2‰
Seychelles: +3 to +4‰.

At 750 Ma Seychelles and Rajasthan (MIS) were only 600 km apart [19].

The evolution of the TAB crust (Jalor and Siwana ring complexes of MIS) with positive Nd (T) values is very much similar to the evolution of ANS and CAOB terrains.

4 Discussion and Conclusion

Kochhar [10] was the first to propose that South China was attached to the Trans Aravalli block of the NW Indian shield during late Proterozoic in the configuration of the Malani Supercontinent. Recent oxygen isotope data given by Wang et al. [24] have confirmed the assembly of South China in the Malani Supercontinent. However the plume related anorogenic, 'within plate', A- type geochemical signatures of the Malani magmatism [7, 23] does not support the extensional supra subduction zone model along the western margin of Rodinia proposed by Wang et al. [24]. Similarly, recent discovery by Ashwal et al. [1] of 3.2 Ga Archean crust beneath Mauritius further support the Seychelles, Madagascar and Arabian shield connection with the NW Indian shield in the configuration of the Greater Malani Supercontinent.

References

1. Ashwal, L.D., Wiedenbeck, M., Torsvik, T.H.: Archean zircons in Miocene oceanic hotspot rocks establish ancient continental crust beneath Mauritius. Nat. Commun. (2017). https://doi.org/10.1038/ncomms14086
2. Etienne, J.L., Allen, P.A., Le Guerrove, E., Heaman, L., Ghosh, S. K., Islam, R.: The Blaini formation of the lesser Himalaya, NW India. Geol. Soc. Lond. Mem. **36**, 47–355 (2011)
3. Hoffmann, M., Linnemann, U., Rai, U., Bockers, S., Gartner, A., Sagwl, A.: The India and S. China craton at the margin of Rodinia-Synchronous magmatism revealed by LA-ICP-MS Zircon analysis. Lithosphere **123**, 176–187 (2011)
4. Jahn, B.M.: The Central Asia orogenic belt and growth of the continental crust in the Phanerozoic. Geol. Soc. Lond. Spec. Publ. **226**, 73–100 (2004)
5. Kochhar, N.: Attributes and significance of the A-type Malanimagmatisim, NW peninsular India. In: Deb, M. (ed.) Crustal Evolution and Metallogeny in the Northwestern Indian Shield, Chapter 9, pp. 183–217. Narosa Publishing House, New Delhi (2000)
6. Kochhar, N.: Signatures and significance of the Pan-African thermo-tectonic event in the Indian subcontinent. Bull. Indian Geol. Assoc. **34**, 36–42 (2001)
7. Kochhar, N.: Anorogenic magmatism, mantle plume and assembly of the Late Proterozoic Malani supercontinent, NW Indian shield. Gondwana Res. Group. Misc. Publ. No. 12, pp. 23–27 (2001b)
8. Kochhar, N.: Geological evolution of the Trans- Aravali Block (TAB) of NW Indian shield: constraints from the Malani Igneous suite (MIS) and its Seychelles connection during later Proterozoic. Geol. Surv. India. Spec. Publ. No. **84**, 247–264 (2004)
9. Kochhar, N.: The Malani supercontinent: Middle East connection during Late Proterozoic. 6th International Conference on the Geology of Middle East, Al-Ain, UAE abstract, page 230 (2006)

10. Kochhar, N.: Was Yangtze craton. S. China attached to the Trans-Aravali block of the NW Indian shield during Late Proterozoic? Curr. Sci. **92**, 295–297 (2007)
11. Kochhar, N.: A- type Malani magmatism NW Peninsular India. In: Singhvi, A.K., Bhattacharya, A., Guha, S. (eds.) Glimpses of Geosciences Research in India. The Indian Report to IUGS 2004–2008, pp. 176–181. Indian National Science Academy, New Delhi (2008a)
12. Kochhar, N.: A type Malani magmatism: signatures of the Pan-African event in the North West Indian shield and assembly of the late Proterozoic Malani Supercontinent. Geol. Surv. Ind. Spl. Pub. **91**, 112–126 (2008)
13. Kochhar, N.: The Malani supercontinent: Middle East connection during late Proterozoic. In: Srivastava, K.L. (ed.) Economic Mineralization, pp. 15–25. Scientific Publishers (India), Jodhpur (2009)
14. Kochhar, N.: Were Siberia, Mongolia, Kazakhstan part of the Malani supercontinent during late Proterozoic? In: 8th International Symposium on Gondwana to Asia, IAGR Conference Series vol. 12, pp. 39–41 (2011)
15. Kochhar, N.: The greater Malani Supercontinent: Siberia, Mongolia, Kazakhstan and Tarim connection during late Proterozoic. In: Rodinia 2013: Supercontinent Cycle and Geodynamics Symposium, Moscow University, Russia, Abstract-vol. 44 (2013a)
16. Kochhar, N.: The Greater Malani supercontinent: South China, Siberia, Mongolia, Kazakhstan and Tarim connection during the Neoproterozoic. In: Precambrian Evolution and Deep Exploration of the Continental Lithosphere 7–9 Oct 2013, Beijing, China, pp. 51–57. Abstract Volume IAGR. Conference Series No. 15 (2013b)
17. Kochhar, N.: The Malani Supercontinent. In: Shrivastava, K.L., Srivastava, P.K. (eds.) The Frontiers of Earth Science, Chapter 6, pp. 122–136. Scientific Publishers (India) (2015)
18. Kochhar, N.: The Greater Malani supercontinent. American Geosciences Institute file. A dynamic earth, 3238, T30, P3 (2016)
19. Kochhar, N.: Archean continental crust beneath Mauritius: implications for the greater Malani Supercontinent. Geol. Soc. Australia, Program and Abstract, no. **12**, 43–44 (2017)
20. Kröner, A., Windley, B.F., Badarch, G., Tomurtogo, O., Hegner, E., Grushka, S., Khain, E.V., Demoux, A., Wingate, M.T.D.: Accretionary growth and crust formation in the Central Asian Orogenic belt and its comparison with Arabian-Nubian Shield. Geol. Soc. Am. Mem. **200**, 181–209 (2007)
21. Li, Z.X., Evans, D.A.D., Zhang, S.: A 90° spin on Rodinia: possible casual links between the Neoproterozoic supercontinent, super plume, true polar wander and low latitude glaciations. Earth Planet Sci. Lett. **220**, 409–421 (2004)
22. Pooranchandra Rao, G.V.S., Mallikarjuna, J., Chako, S.T., Subrahmanyan, K.: Paleo-magnetism of Baghain sandstone, Kaimur Group. J. Ind. Geophys. Union. **1**, 41–48 (1997)
23. Singh, A.K., Singh, R.K.B., Vallinayaagam, G.: Anorogenic acid volcanic rocks in the Kundal area of the Malani igneous suite, NW India: geochemical and petrogenetic studies. Jour. Asian. Earth. Sci. **27**, 544–557 (2006)
24. Wang, W., Cawood, P.A., Zhou, M.-F., Pandi, M.K., Xia, X.-P., Zhao, J.-H.: Low-$\delta^{18}O$ Rhyolites from the Malani igneous suite: a positive test for South China and NW India Linkage in Rodinia. Geophys. Res. Lett. 10298–10305 (2017). https://doi.org/10.1002/2017gl074717
25. Yarmolyuk, V.V., Kovalenko, V.I., Kovach, V.P., Rytsk, E., Kozakov, L.K., Kotov, A.B., Salnikova, E.B.: Results of geochronological, isotopic and geochemical investigation of Riphean and Vendian- Cambrian complexes in the Central Asian foldbelts. Earth Sci. **411**, 1184–1189 (2006)
26. Zheng, Y.F.: Position of South China in a configuration of Neoproterozoic supercontinent. Chin. Sci. Bull. (2003) https://doi.org/10.1360/04wdoi53

Part III

Fluid-Rock Interaction, Hydrothermalism and Ore Deposits

The Lithological, Stratigraphical, Structural Control of Quartzites Hosting Sulphide Mineralisation in the Proterozoic Shillong Group, Meghalaya, India

Sumit Kumar Mitra

Abstract

The White Quarzite belonging to the Lower Shillong Group, overlying the older Archeans, in the Proterozoic Shillong Group of rocks host the sulphide mineralization. The sulphide minerals are argentiferous galena and subordinate auriferous arsenopyrite, loellignite, gersdorffite, safflorite, pyrite and pyrohotite occurring as bed - like or lens - like ore bodies along the compositions bands. Surface geochemical prospecting revealed Pb values of (maximum) 29%, silver values (maximum) of 0.3%, and gold values ranging from 50 to 260 ppb. The lithological, stratigraphic and structural controls of the sulphide mineralization have been established. The syn-sedimentary/syn-genetic Pb (+Ag) mineralization was remobilized during the F_1 folding. The Pb/(Pb + Zn) ratio of 0.93 and co-existence of detrital monazite and zircon suggests deposition of the sediments in shallow marine condition. The S– isotopic analysis of the samples have revealed δ^{34} S values with an average of +9.3‰. The positive δ^{34}S value shows derivation of sulphur due to sulphate reduction within a limited reservoir in a sedimentary domain and bears no mantle signature. Sulphide ore minerals are crystalline and unzoned, indicating crystallaization under equilibrium condition. Mineral assemblage of the host rock and mineralized zones indicate biotite-grade of regional metamorphism. Pb–Pb age of the galena being between 1530 and 1550 Ma interpreted the age of the mineralization.

Keywords

Shillong group • Sulphide mineralisation
Lithological, stratigraphical, structural control

1 Introduction

The Shillong—Mikir massif represents NE prolongation of the Chotonagpur Gneissic belt of the Indian continent. This consists of a basement complex made up of gneisses and migmatites with enclaves of amphibolites, rarely 'BIF' (banded iron formation) and patchy distribution of high-grade granulitic supracrustal rocks. The basement is unconformably overlain by siliciclastic Proterozoic cover of the Shillong Group (Fig. 1).

Search for sulphide mineralization within the Shillong Group spanned mostly over a 40 km stretch of the shear zone (Tyrsad—Barapani) (Fig. 2). Details prospecting and drilling was carried out by GSI (Geological Survey of India) between 1974 and 1980 to study the base metal potentiality, without encountering economically significant sulphide mineralization. Studies indicated that the sulphide mineralization occurs as specks, disseminations, patches and stringers along the foliation, joints and fissures. Pyrite is the dominant mineral with subordinate amounts of arsenopypite, pyrrhotite, chalcopyrite, sphalerite and galena in the phyllites and carbonaceous slates. Recent discovery of sulphide occurring along the compositional bands of the White Quartzites of Lower Shillong Group requires the assessment of the geological conditions (structure, stratigraphy, geochemistry) leading to sulpide ore mineralisation.

2 Methods

The present work was carried out through extensive field work consisting of structural-geological mapping, geochemistry analyses, EPMA studies, sulphur isotope analysis and dating of sulphides.

S. K. Mitra (✉)
Geological Survey of India, 284/1A NSC.
Bose Road, Kolkata, 700047, India
e-mail: sumit0224@rediffmail.com

F. Rossetti et al. (eds.), *The Structural Geology Contribution to the Africa-Eurasia Geology: Basement and Reservoir Structure, Ore Mineralisation and Tectonic Modelling*,
Advances in Science, Technology & Innovation, https://doi.org/10.1007/978-3-030-01455-1_11

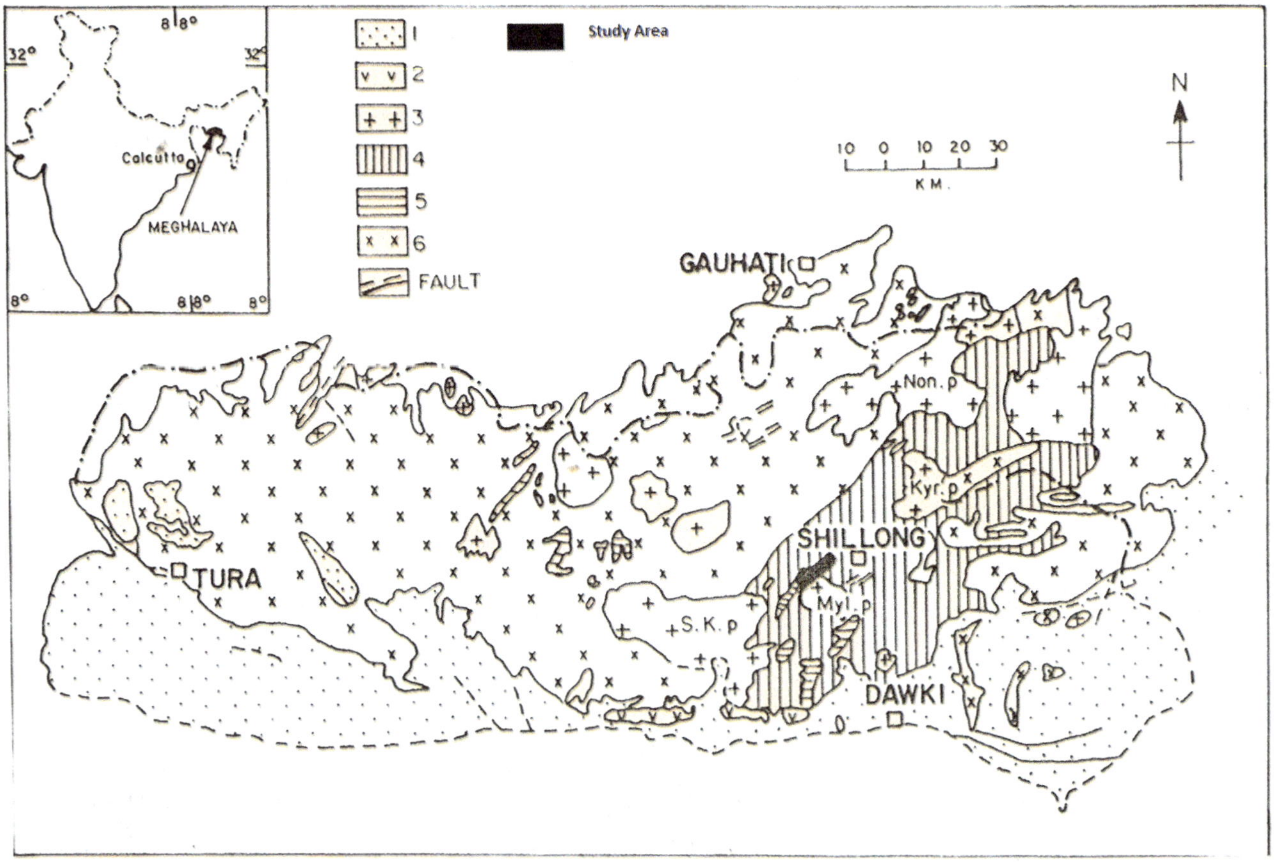

Fig. 1 Generalised geological map of Meghalaya (after Mazumder [2], 1 = Cretaceous Tertiary sediments: 2 = Syl Traps: 3 = porphyritic granite: 4 = Shillong Group: 5 = mafics: 6 = Precambrian Gneissic Complex: S.K.p. = South Khasi Pluton: Non.p. = Nongpoh pluton: Kyr.p. = Kyrdem pluton: Myl.p. = Mylliem pluton. Inset: Location of the Shillong

3 Results and Discussion

The occurrence of the ores within the quartzites is the result of the deposition in an open tidal beach environment. The restriction of the minerals to the white quartzites of the lower Shillong Group only suggests a stratigraphic as well as lithological control. The presence of detrital heavy minerals like garnet, monazite, and zircon also corroborates the hypothesis of a depositional regime.

The occurrence of the ores along the S_0–S_1 in the white quartzite as bed-like or lens-like bodies, concentrated more in the hinges of the F_1 recumbent folds reflects a syn—F_1 mobilisation of the pre—F_1 sulphides. The geometry of the ore body is controlled by the F_1 phase of deformation; the ore itself is pre—F_1. The ore bodies seem to have been reshaped by later F_2 and F_4 deformations, without any apparent mobilization during this period [3].

EPMA results (Table 1) shows concentration of Ag in galena and Au in arsenopyrite, loellingite, gersdorffite, safflorite. The correlation matrix drawn from the surface geochemical samples analysis also shows good correlation of Ag with Pb, Au with As.

Pb/(Pb + Zn) ratio around 0.9822 for the mineralized portions and 0.738 for the ferruginised layers associated with the mineralized zones suggest a syn-genetic/syn-sedimentary Sandstone-Lead deposit genetic model akin to 'Quartzite hosted model' by Bjorlykke and Sanster [1] has been suggested.

The ore minerals are dominantly galena (Fig. 3a) with subordinate arsenopyrite (Fig. 3b), loellingite, gersdorffite, safflorite and varying amounts of chalcopyrite, sphalerite (Fig. 3c), pyrite, pyrrhotite, cubanite, covellite and minor sulphosalt of Ag. Most of the ore minerals are crystalline and unaltered, and unzoned indicating that they have crystallized under an equilibrium condition. Metamorphism of the host

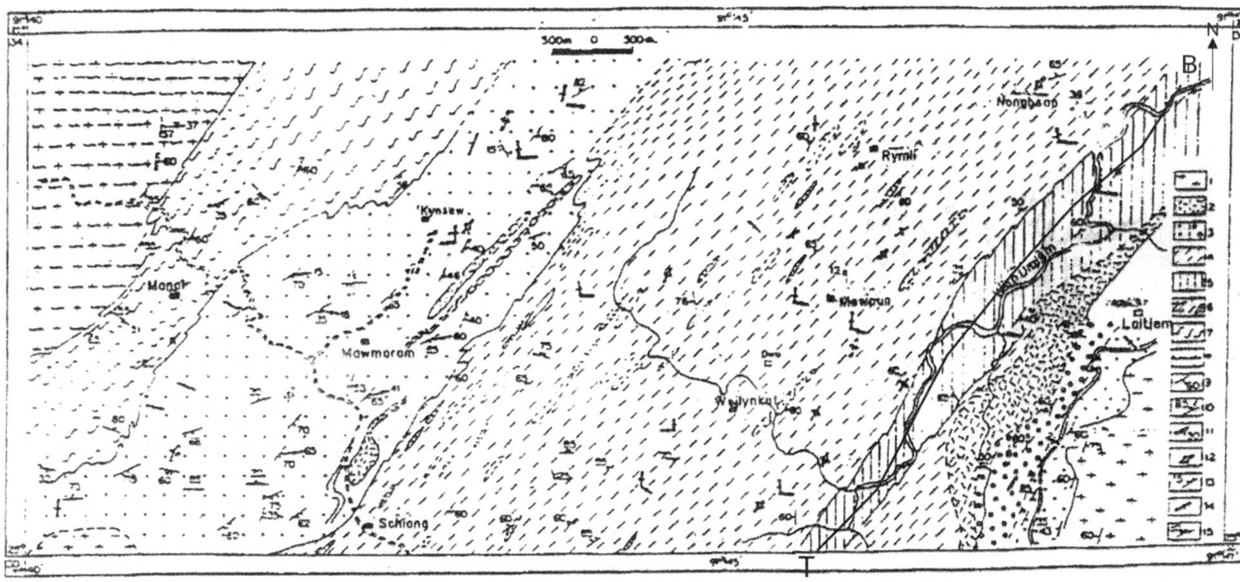

1 = Granite, 2 = Khasi Green stone, 3 = Quartzite and conglomerate, 4 = Phyllite, 5 = Slate, 6 = Phyllite with quartzite and quartz-sericite schist, 7 = Mica schist, 8 = Granite gneiss, 9 = Bedding, 10 = S_1 schistosity, F_1-F_2 linear structures, 11 = F_1 axial plane, 12 = F_2 axial plane, 13 = F_3 axial plane, 14 = F_4 axial plane, 15 = Joint plane. T—— B = Tyrsad- Barapani Shear Zone

Fig. 2 Geological map of Shillong Group around Shiong-Manai, Meghalaya

Table 1 EPMA results of ore minerals

Element	Galena	Arsenopyrite	Loellingite	Gersdorffite	Safflorite
S	12.75	18.65	1.45	16.44	0.77
Fe	3.50	32.61	24.63	8.17	12.80
Co	0.00	0.08	1.68	8.94	7.93
Ni	0.00	0.06	1.31	16.52	7.48
Cu	0.05	0.00	0.00	0.14	0.00
Zn	0.00	0.00	0.00	0.00	0.00
As	0.00	49.55	72.04	49.27	72.05
Ag	0.67	0.00	0.00	0.07	0.00
Sb	0.14	0.03	0.02	0.15	0.13
Au	0.08	0.14	0.09	0.06	0.08
Pb	81.35	0.00	0.06	0.04	0.26
Bi	0.01	0.00	0.00	0.00	0.00
K	0.00	0.00	0.00	0.02	0.01
Total	98.55	101.12	101.28	99.82	101.51

rock as well as arsenopyrite-loellingite-pyrrhotite/pyrite association indicates that biotite-in isograd of regional metamorphism has been reached.

Sulphur isotope analysis, revealed $\delta^{34}S$ value with an average of +9.3 permil. The positive $\delta^{34}S$ value shows a wide variation due to sulphate reduction within a limited reservoir. Thus, it indicates mineralisation in a sedimentary domain and bears no mantle signature [4], and also indicates that the mineralization is essentially syn-sedimentary.

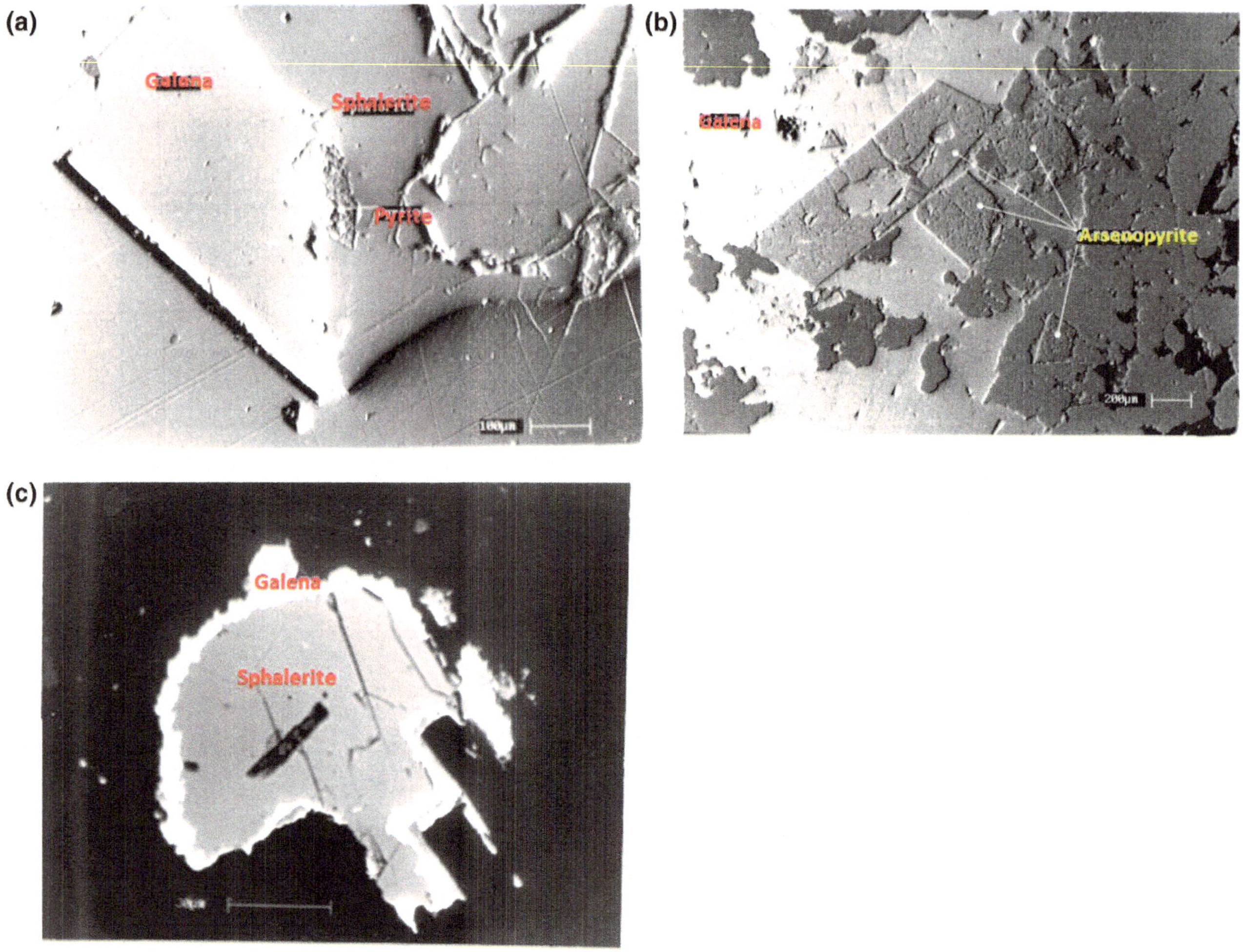

Fig. 3 Back scattered electron images of ore mineral (**a** Galena, pyrite, sphalerite association, **b** Tabular grain of arsnopyrite with galena, **c** Sphalerite rimmed by Galena.)

4 Conclusions

- The ores, preferentially occur within the White Quartzite only, the quartzite being the result of deposition in an open tidal beach. The presence of detrital heavies like garnet, monazite, zircon also corroborates with the depositional regime.
- The occurrence of mineralisation within the White Quartzite of Lower Shillong Group, therefore, suggest a stratigraphic as well as lithological control.
- The concentration of the sulphide occurring in the hinge area of the large-scale F_1 recumbent/reclined folds suggest a syn-tectonic mobilization of the pre-F_1 ores with a structural control has been established.
- EPMA studies along a correlation matrix drawn from the surface geochemical samples show partitioning of Ag with galena and Au with arsenopyrite, loellingite, gersdorffite, safflorite.
- Most of the ore minerals are crystalline and unzoned indicating that they have crystallized under an equilibrium condition. Metamorphism of the host rocks as well as arsenopyrite-loellingite-pyrrhotite/pyrite association indicates a biotite-grade of regional metamorphism.
- The positive $\delta^{34}S$ value shows a wide variation due to sulphate reduction within a limited reservoir in a sedimentary domain and bears no mantle signature, and also indicates that the mineralisation is essentially syn-sedimentary.
- A syngenetic/synsedimentary ore genetic model akin to 'Quartzite hosted model' by Bjorlykke and Sanster [1] has been suggested based on the depositional regime as well as Pb/(Pb + Zn) ratio of 0.93.
- Based on the Pb–Pb Age of 1530–1550 Ma of galena, the maximum age of the Shillong Group is suggested to be greater than 1550 Ma.

References

1. Bjorlykke A., Songster, D.F.: Eco.Geol. 75th Annuv. pp. 179–213 (1981)
2. Mazumder, S.K.: A summary of the precambrian geology of the Khasi Hills. Meghalaya. Misc. Pub. Geol. Surv. India **23**(2), 311–324 (1976)
3. Mitra, S.K.: Control of lead-silver mineralisation. Mawmaram Area, East Khasi Hills Meghalaya. India Minneral **53**(3&4), 223–234 (1999)
4. Nielsen, H.: Sulfur Isotopes. In: Jager, E., Hunziker, J.C. (eds.) Lectures in Isotope Geology, pp. 283–312. Springer-Verlag, Berlin, Heidelberg (1979)

Epithermal Hg, Sb and Au Deposits in the Continental Rift Zones of the Menderes Massif, Western Anatolia, Turkey

Nevzat Özgür

Abstract

Within the Menderes Massif the continental rift zones of the Büyük Menderes, Küçük Menderes and Gediz were formed by extensional tectonic regimes from Early to Middle Miocene which generally strike E-W and are represented by a great number of epithermal Hg, Sb and Au mineralizations, geothermal waters and volcanos from Middle Miocene to recent. The epithermal mineralizations and geothermal waters are related to faults which strike preferentially NW-SE and/or NE-SW and locate diagonal to the E-W general strike of the continental rift zones. These faults are probably generated by compressional tectonic stress which leads to the deformation of uplift between two extensional rift zones. The epithermal Hg, Sb and Au mineralization are located in the eastern part of the continental rift zone of the Küçük Menderes within the Menderes Massive. The ore mineralizations of these deposits are associated with Paleozoic mica schists. At the surface, the host rocks are intensively altered by the interaction with the circulation of geothermal fluids. Therefore, the ore fields can be recognized by a distinct color change of the host rocks. The hydrothermal alteration is clearly noticeable at the surface which is distinguished by phyllic, argillic and silicic alterations zones. The isotopic ratios of $\delta^{18}O$ and $\delta^{2}H$ in fluid inclusions of quartz and stibnite samples from Hg deposit of Halıköy, Sb deposit of Emirli and arsenopyrite-Au deposit of Küre in the continental rift zone of the Küçük Menderes show a similarity with active geothermal systems. By using geological, geochemical, isotope geochemical, ore and rock microscopical and microthermometric methods, these epithermal Hg, Sb and Au deposits in the continental rift zone of the Küçük Menderes have been modeled genetically and can be considered as fossil geothermal systems.

Keywords

Turkey • Western anatolia • Menderes massif • Continental rift zones • Epithermal Hg • Sb • Au deposits

1 Introduction

The Anatolian and Aegean micro plates control the plate tectonic position of the Eastern Mediterranean area between the Eurasian and African plates. This plate tectonic development results in the lifting of the Menderes Massif in Western Anatolia, Turkey shows a dome structure due to compressional tectonic features from Oligocene to Middle Miocene [1]. From Early Miocene to Middle Miocene, the continental rift zones of the Büyük Menderes, Küçük Menderes, and Gediz were formed by extensional tectonic features, which generally strike E-W and are represented by a great number of Hg, Sb, and Au mineralizations and thermal waters in connection with volcanic rocks from Middle Miocene to recent (Fig. 1). The Hg, Sb and Au mineralizations and thermal waters are related to faults, which strike preferentially NW-SE and NE-SW, diagonal to the general strike of the continental rift zones. These faults, representing a multitude of Hg, Sb and Au mineralizations and thermal waters, are probably generated by compressional tectonic stress, which leads to the deformation of uplift between two extensional continental rift zones. The investigated Hg, Sb and Au deposits of Halıköy, Emirli and Küre (Fig. 1) represent typical examples of epithermal mineralizations.

The aim of this paper is to present the geological context of Hg, Sb and Au mineralizations in the continental rift zone of the Küçük Menderes within the Menderes Massif and to compare them genetically. In particular, the following aspects are investigated: (i) relationship between Hg, Sb and Au mineralizations as fossil geothermal systems, active geothermal systems, tectonic features and volcanism,

N. Özgür (✉)
Department of Geological Engineering, Faculty of Engineering, Suleyman Demirel University, Isparta, Turkey
e-mail: nevzatozgur@sdu.edu.tr

F. Rossetti et al. (eds.), *The Structural Geology Contribution to the Africa-Eurasia Geology: Basement and Reservoir Structure, Ore Mineralisation and Tectonic Modelling*,
Advances in Science, Technology & Innovation, https://doi.org/10.1007/978-3-030-01455-1_12

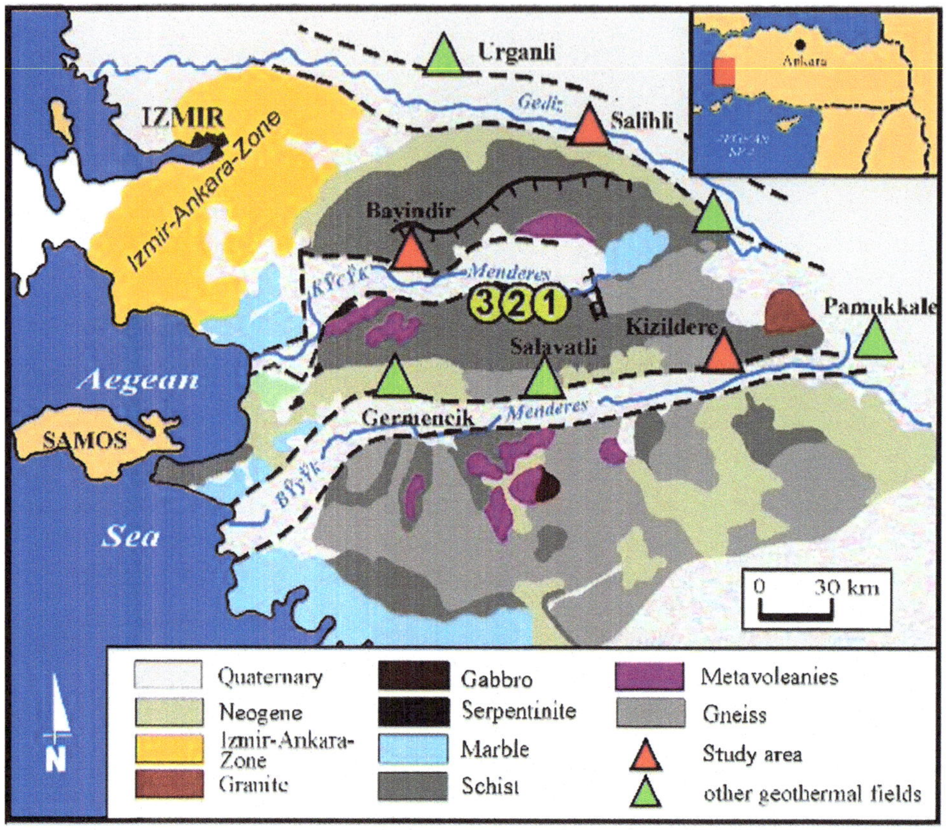

Fig. 1 Epithermal ore fields in the rift zone of the Küçük Menderes within the Menderes Massif. 1: Hg deposit of Halıköy; 2: Sb deposit of Emirli; 3: Au deposit of Küre [1–3]

(ii) fluid-rock interactions, (iii) geochemical, hydrogeochemical and isotope geochemical features of active and fossil geothermal systems, (iv) fluid inclusion studies in ore and gangue minerals and (v) source, transport and deposition of Hg, Sb and Au deposits.

2 Geologic Setting

The Hg, Sb and Au mineralizations of Halıköy, Emirli and Küre occur 25 km S and SE of the town of Ödemiş in SE part of the rift zone of the Küçük Menderes. The metamorphic rocks of the Massif are (i) Precambrian to Cambrian core series consisting of high-grade schists, gneisses, granites and metagabbros and (ii) Ordovician to Paleocene cover series composed of mica schists, phyllites, metaquartzites and metagabbros (Fig. 1). The first tectono-metamorphic event took place during the Cambrian under amphibolite to granulite facies conditions affecting the core series. The late phase of the Variscan orogenesis also recorded in the massif postdated by the Early to Middle Triassic intrusions. Alpine high-pressure metamorphism was generated under epidote-blue schist to eclogite facies conditions overprinted by Barrovian-type metamorphism. Neogene to Quaternary post-orogenic deposits overlain the metamorphic rocks.

From Middle Miocene to recent, an intense syn-tectonic volcanism was generated during the formation of the continental rift zone in the Menderes Massif [1]. The volcanic rocks in the NE part of the Hg, Sb and Au deposits of Halıköy, Emirli and Küre in the rift zone of the Küçük Menderes are distinguished by an Rb/Sr age of 15.0 ± 0.2 Ma in Karaburç and a K/Ar age of 16.7 ± 0.5 Ma in Yenişehir [1]. These volcanic rocks are crustally derived [1]. The youngest volcanism in the Menderes Massif is characterized by the Late Pliocene volcanic rocks of Denizli and the volcanics of Kula with an age ranging from 7.5 Ma to 18 ka. The magmatic system can be considered as the heat source of the active systems in the region.

3 Hg, Sb and Au Deposits

The Hg, Sb and Au deposits of Halıköy, Emirli and Küre are related to faults which strike preferentially NE-SW and NW-SE and are related diagonal to the general strike of the rift zone of the Küçük Menderes (Fig. 1).

Strongly altered, mica schists and quartzites form the host rocks of Hg, Sb and Au vein deposits of Halıköy Emirli and Küre respectively. The Hg deposit of Halıköy of veins and

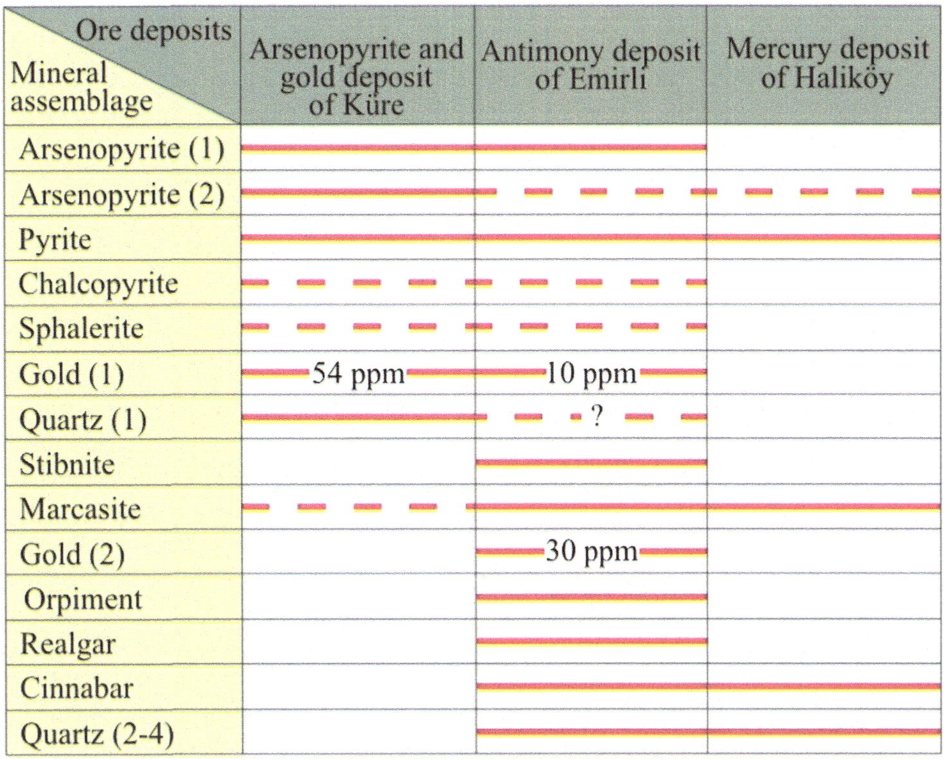

Fig. 2 Ore mineral assemblage of epithermal Hg, Sb, and Au deposits of Halıköy, Emirli and Küre in the rift zone of the Küçük Menderes [1]

veinlets contain cinnabar, metacinnabarite, pyrite, marcasite, chalcopyrite with quartz and calcite as gangue minerals (Fig. 2). The Emirli Sb deposit is made of a network of veins and veinlets and consists of pyrite, arsenopyrite, stibnite, sphalerite, chalcopyrite, tetraedrite, marcasite, orpiment, realgar, cinnabar with gangue minerals of quartz, adularia and calcite. The Au deposit of Küre represents veins and veinlets with arsenopyrite, gold, pyrite, marcasite, fahlore and gangue minerals of quartz and calcite. In the abandoned Halıköy Hg mine, there are measured ore reserves of 56.000 metric tons with a Hg mean value of 0.30%, indicated ore reserves of 210.000 metric tons with a mean value of Hg of 0.25% and possible ore reserves of 210.000 metric tons with a mean value of Hg of 0.23% [1]. The recoverable ore reserves of the Sb deposit of Emirli are estimated at 400.000 tons with an average content of 5.5% Sb. The Au deposit of Küre consists of three occurrences and shows Au contents up to 30 ppm.

The intensively altered mica schists and quartzites which form the host rocks for Hg, Sb and Au deposits, can be considered as source rocks the metals, as supported by leaching tests [1]. The hydrothermal alteration is distinguished by phyllic, argillic and silicic ± hematitization alteration zones which are comparable with those of the active geothermal systems in the continental rift zones of the Menderes Massif. This type of alteration is comparable to adularia-sericite-type-mineralization due to presence of adularia and bladed calcite crystals (Fig. 3; [1]).

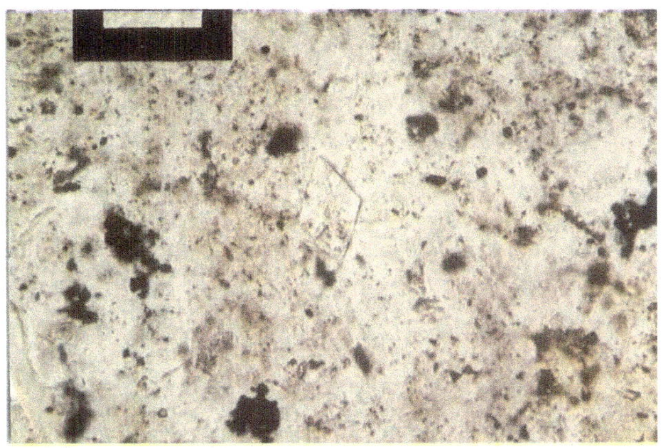

Fig. 3 Microscopical observation of rhomboidal adularia crystals in the strong silicified host rocks of the Kurşunlu Sb mineralization in the Gediz rift zone [1]. Thin section, plane polarized light [1]

4 Fluid Inclusion Study and Geochemistry

For the fluid inclusion studies, we have collected about 50 quartz samples (stage 2), which were compared with quartz samples of the crystalline massif (stage 1) and stibnite crystals of Sb mineralization. The fluid inclusion measurements were made in the Freie Universität Berlin, Germany using a Linkam THMSG 600 programmable freezing-heating stage attached to a Leitz Ortholux transmitted-light microscope. This used stage has a dynamic range from

−180 to 600 °C. The measurements in stibnite crystals were used in GeoForschungsZentrum Potsdam, Germany by a FLUID.INC SYSTEM gas-flow freezing-heating stage attached to an Olympus IR transmitted-light microscope.

Three types of primary fluid inclusions were recognized in quartz and stibnite crystals. The type-1 is an inclusion (H_2O-NaCl), which can be observed in quartz as well as in stibnite crystals and shows two phases (liquid and vapor) at room temperature [1]. The type-2 is a system of H_2O-NaCl-CO_2 $\pm$ CH_4 of which inclusions show three phases of CO_2 (CH_4)-liquid, CO_2 (CH_4)-gas and aqueous solutions at room temperature.

The type-3 is a system of aqueous vapor, which is associated with Sb deposit of Emirli and indicates boiling conditions in the system. The quartz samples are associated with ore deposits show homogenization temperatures from 180 to 300 °C and salinity ranging from 3.75 to 9.0 NaCl (eq wt%), which correspond with each other (Fig. 4). The quartz samples of the Halıköy indicate average homogenization temperature of 128 °C and salinity of 3.4 NaCl (eq wt%). The quartz samples of the Au deposit of Küre is distinguished by the homogenization temperatures from 210 to 300 °C and a salinity between 1.60 and 10.0 NaCl (eq wt %). In order to understand the composition of hydrothermal ore-forming fluids, we have analyzed in the fluid inclusions $\delta^{18}O$ and $\delta^{2}H$ by mass spectrometry, anions and cations (K^+, Na^+, Ca^{2+}, Mg^{2+}, Fe^{3+}, Al^{3+}) by ICP emission spectrometry, F^- by ion-selective electrode, Cl^- by mercury thiocyanade colorimetric analysis, SO_4^{2-} by barium sulfate turbimetric analysis, and HCO_3—by acid-base capacity titration in the State Key Laboratory for Research of Mineral Deposits, Nanjing University, PR China. The stable isotopes of $\delta^{18}O$ and $\delta^{2}H$ in fluid inclusions of quartz samples from the investigated Hg, Sb and Au deposits show similarity with that derived from the active geothermal systems (Fig. 5; [1]).

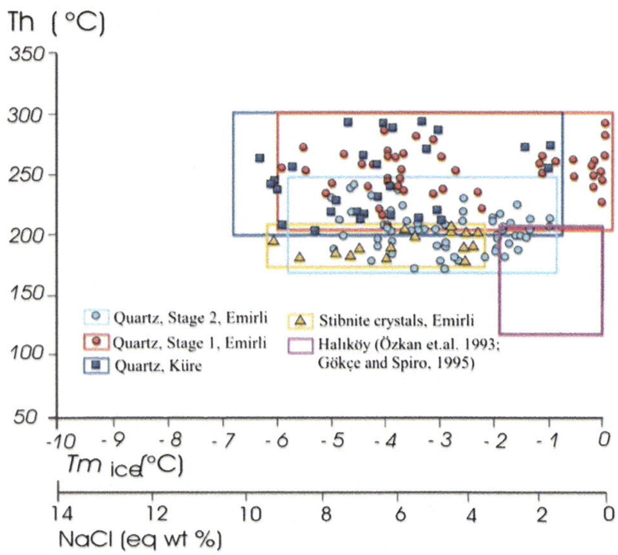

Fig. 4 Homogenization temperatures, salinity and T_{mice} in quartz and stibnite samples of the Hg, Sb, and Au deposits of Halıköy, Emirli and Küre [1, 4, 5]

Fig. 5 Plot of $\delta^{18}O$ versus δD in quartz samples from the Hg, Sb, and Au deposits of Halıköy, Emirli, and Küre [1]

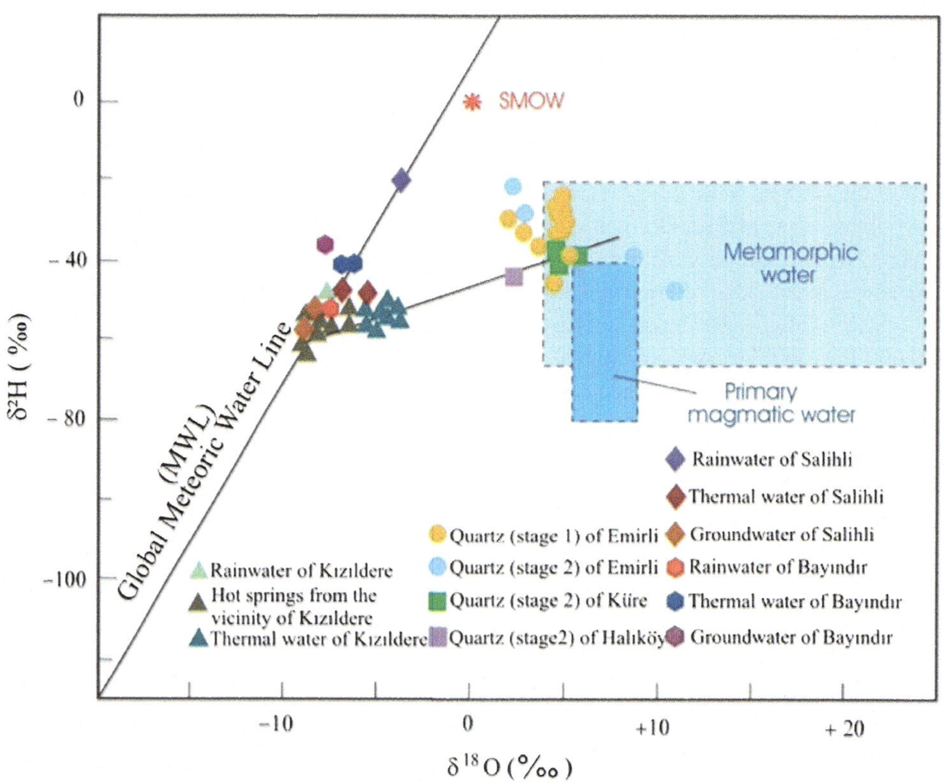

The strong deviation of the $\delta^{18}O$ from the global meteoric water line proves the intense fluid-rock interaction in the hydrothermally environment during the differences in the degree of isotope shift from the active geothermal systems to investigated Hg, Sb and Au deposits indicate a relationship between both systems in meteoric origin [1]. The anions and cations compositions in fluid inclusions of quartz crystals show Ca-(Cl)-(SO_4)-HCO_3 type exchange fluids which form a sharp contrast with them of active geothermal systems and fluid inclusion studies. In one respect, this may relate to the process of albitization by Na metasomatism in the investigated area and its environs in which Ca contents dominate during the Na contents decrease due to the same process.

5 Discussion

The homogenization temperatures of quartz and stibnite crystals range from 150 to 300 °C, which are compatible with geochemical temperatures of geothermal water reservoirs from 220 to 260 °C. The ore-forming fluids of Halıköy, Emirli and Küre show a mean value of 6.0 NaCl eq wt %; it is comparable to the salinity of active geothermal fluids. The isotope ratios of $\delta^{18}O$ and $\delta^{2}H$ in fluid inclusions of quartz crystals of Halıköy, Emirli and Küre also show a similarity with those available from the active geothermal fluids (Fig. 5). The strong deviation of the $\delta^{18}O$ values from the meteoric water line shows the intensive fluid-rock interaction in the hydrothermal environment. The trend of deviation increases linearly from the active geothermal field to the epithermal ore fields indicating a relationship between the two systems.

Ultimately, with the cooling of magma chamber, the Au mineralization of Küre was formed at temperatures below 300 °C (Fig. 6). Arsenopyrite is present in the ore mineral assemblages of investigated Hg, Sb and Au mineralizations. In connection with further cooling of magma chamber, the Sb mineralization of Emirli took place at temperatures from 180 to 250 °C. The Hg deposit of Halıköy is generated as the last mineralization at temperatures from 128 to 200 °C. The meteoric fluids percolate above permeable clastic sediments in the reaction zone of the roof area of magma chamber in depth of 2–3 km where the fluids are heated and ascend to the surface because of low density. The volatile components of CO_2, SO_2, H_2S and HCl from magma reached the geothermal water reservoir as ascending gas phases where equilibrium reactions between altered rocks, gas components and fluids took place [1]. The ascending fluids contain CO_2, H_2S and HCl particularly. Hydrothermal convection cells press the heated fluids toward the surface because of their lower density. Thus, geothermal waters ascend in tectonic zones of weakness. As geochemical

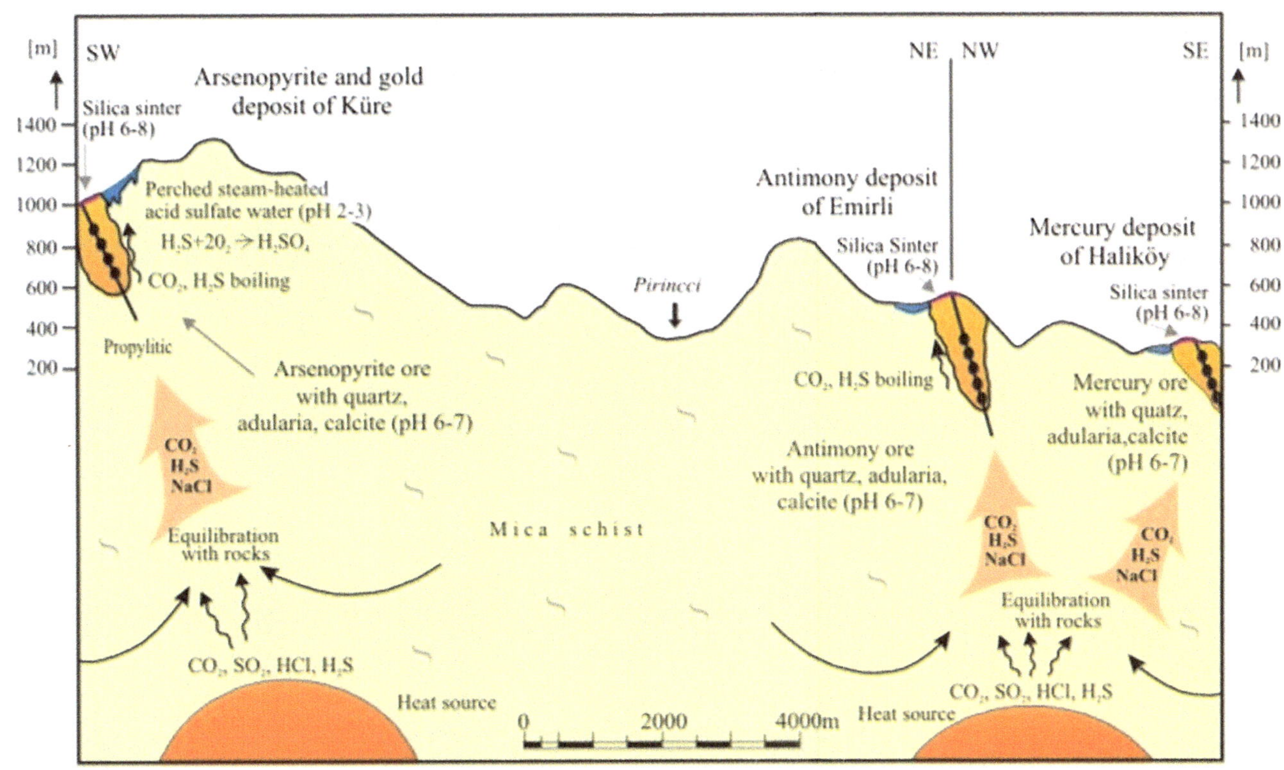

Fig. 6 Simplified genetic model of the Hg, Sb, and Au deposits of Halıköy, Emirli, and Küre [1]

pH-neutral fluids, the waters find an outlet at the surface as hot springs, gas and steam. The fluids indicate a reduced pH-neutral sphere in the reaction zone after equilibrium with host rocks.

Finally, it is concluded that (i) the fluids of Hg, Sb and Au mineralizations of Halıköy, Emirli and Küre At the subsurface spheres, the ore deposits are generated in terms of stockwork mineralizations (veins, veinlets) and gangue minerals represented by quartz, calcite and adularia [1]. The Hg, Sb and Au deposits of Halıköy, Emirli and Küre can be assigned to an epithermal type in connection with a calkalkaline volcanism in Middle Miocene age, comparable to other Hg, Sb and mineralizations in the rift zones of the Menderes Massif, similar to the epithermal Sb and Au deposits in the metallotect of Jiangnan, PR China and active and extinct geothermal systems of New Zealand, and considered as fossil equivalents of active geothermal systems.

References

1. Özgür, N.: Aktive und fossile Geothermalsysteme in den kontinentalen Riftzonendes Menderes-Massives, W-Anatolien, Türkei. Habilitationsschrift, Freie Universität Berlin, 171 p. (1998)
2. Özgür, N.: Origin of high boron contents of the thermal waters of Kizildere and vicinity, western Anatolia Turkey. Int. Geol. Rev. **43**, 910–920 (2001)
3. Özgür, N.: Geochemical signatureof the Kizildere geothermal field, western Anatolia Turkey. Int. Geol. Rev. **44**, 153–163 (2002)
4. Özkan, H. M., Spiro, B., Moon, C. J., Akçay, M., Spiro, B.C.: Genesis of stratabound and structure controlled antimony mineralization at Emirli, Menderes Massif, West Turkey. Mineral paragenesis, fluid inclusion and stable isotope studies. Geologicaln Congress, 38–39 (1993)
5. Gökçe, A., Spiro, B.: Sulfur isotope study of the antimony and mercury deposits in Beydagı (İzmir, Western Turkey) area and the origin of the sulfur in stibnite and cinnabar: Turk. J. Earth Sci. **4**, 23–28 (1995)

Rare Earth Element Signatures of the Bled M'Dena Porphyry Molybdenum-Copper System, Eglab Massif (SW, Algeria)

Karima Lagraa, Stefano Salvi, and Didier Béziat

Abstract

This work depicts the geochemical characteristics and distribution of the rare earth elements (REE) of the Bled M'Dena porphyry Mo-Cu complex which is located in the Eglab-Yetti Junction Zone (Reguibat Rise), SW Algeria. The whole-rock REE data exhibit LREE enrichment relative to HREE (Lagraa et al in J Afr Earth Sci 127:159–174, 2017 [2]), whereas the potential deposits reveal a downward pattern from LREE to HREE, suggesting a higher content of the former REE (Zarasvandi et al in Ore Geol Rev 70:407–423, 2015 [7]). The Eu anomalies are low (Eu/Eu* = 0.87–1.10) and also show relatively high La/Sm and Sm/Yb values, representing probably thickened crust and indicate the fractionation of amphibole with minor garnet. The geochemistry of the host rocks and the REE ore minerals indicate that the Mo-Cu Bled M'Dena porphyry presents similarities with other economic porphyries in the world.

Keywords

REE geochemistry • Mo-Cu porphyry • Eglab Bled M'Dena • Economic porphyry

1 Introduction

The demand for rare earth element (REE) has been increased in the last years due to their applications in a wide range of high technology applications. Several petrogenetic processes, including late-stage magmatic or hydrothermal fluids are implicated in the enrichment and distribution of the REE and may involve high salinity brines through processes such as hydrothermal fluid mobilization and/or precipitation [1].

The principal objective of this work is to document the whole-rock REE geochemistry of the intrusive rocks hosting the Mo-Cu Bled M'Dena porphyry complex. These data are compared with the world-famous porphyry rich in REE, in order to evaluate its exploitation potential.

2 Geological Setting

The Bled M'Dena porphyry Mo-Cu mineralization of Early-Proterozoic age is located in SW Algeria. It crops out within the Eglab "Junction Zone" which is located in the eastern margin of the Reguibat Rise (Fig. 1). The Bled M'Dena volcano-plutonic complex is bounded by two faults, oriented N-S and NE-SW, respectively. In this region, plutonic units predominate and they consist of quartz-monzodiorite and granodiorite of Paleoproterozoic age (cf. also [2]). The granitoids are intruded by dykes of different compositions (diorite, granodiorite), as well as by andesitic to dacitic volcanic domes. The intrusive rocks are metaluminous I-type granitoids with high-K calc-alkaline affinity compatible with a volcanic arc environment.

The Mo-Cu mineralization is mainly hosted by quartz-monzodiorite and granodiorite (locally microgranodiorite). The most important ore consists of molybdenite and chalcopyrite, which occur, with pyrite, disseminated and in stockwork veins. Less abundant sulphides and metals include galena, bornite, chalcosite, scheelite, powellite and gold. The mineralization is associated with propylitic alteration (chlorite, calcite, epidote, sericite, anhydrite) developed around mineralized veinlets and the disseminated sulphides aggregates.

K. Lagraa (✉)
University of Oran 2 Mohamed Ben Ahmed, Oran, 31000, Algeria
e-mail: lagraa.karima@univ-oran2.dz

S. Salvi · D. Béziat
University of Paul Sabatier, Toulouse, 31000, France

F. Rossetti et al. (eds.), *The Structural Geology Contribution to the Africa-Eurasia Geology: Basement and Reservoir Structure, Ore Mineralisation and Tectonic Modelling*,
Advances in Science, Technology & Innovation, https://doi.org/10.1007/978-3-030-01455-1_13

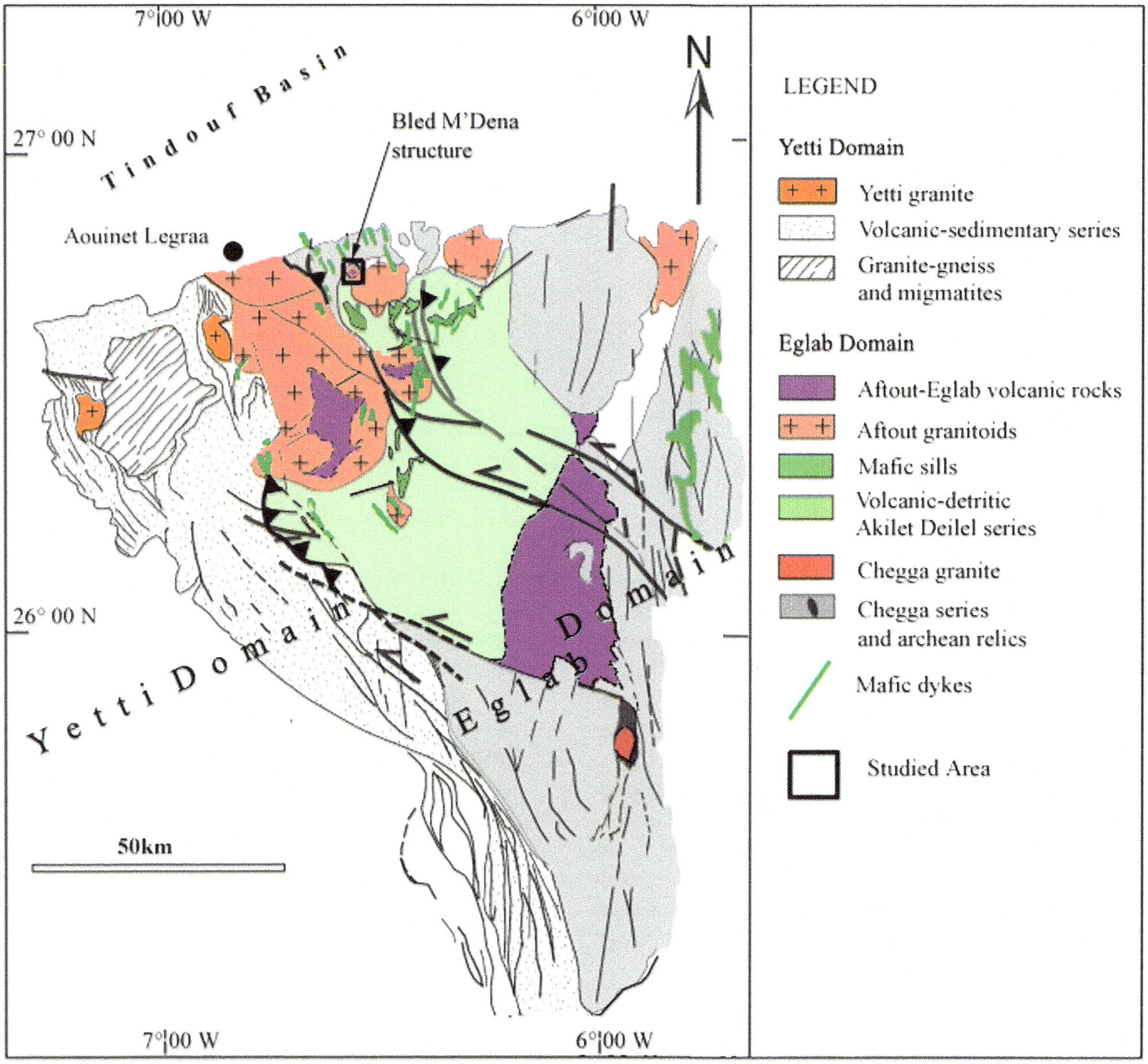

Fig. 1 Location of studied area (the Bled M'Dena circular structure) within the Yetti-Eglab junction zone (Modified after [4])

3 Methods

3.1 Whole-Rock Geochemistry

This study focuses on the REE fingerprints of the intrusive rocks of the Bled M'Dena hosting the Mo-Cu mineralization. The concentrations of REE were obtained by lithium metaborate fusion with nitric digestion performed by inductively coupled plasma-mass spectrometry (ICP–MS) at the Géosciences Environnement Toulouse (GET) laboratory, University of Toulouse (France). The detection limit is between 0.1 and 0.01 ppm [2].

3.2 Scanning Electron Microscopy (SEM)

We carried out SEM studies at the GET laboratory, a JEOL 6360LV SEM coupled to an energy dispersive X-ray spectrometer (SDD Bruker 129 eV) to acquire images in backscattered electron mode (BSE), at an acceleration voltage of 20 kV.

4 Discussion and Conclusions

Alteration assemblages of intrusive-related mineralizations may exhibit a wide range of REE distributions (Fig. 3) and REE fractionation trends, which ultimately depend on the concentrations of these elements in the rocks and fluids (e.g., [3]). Several processes are typically involved in the enrichment and distribution of the REE in altered igneous systems. These may involve remobilization by late-stage magmatic or hydrothermal fluids, which are commonly high salinity brines [1]. The Bled M'Dena pluton exhibits light rare-earth element (LREE) enriched profiles, with $(La/Yb)_N$ between 22 and 31. There are medium $(La/Sm)_N$ ratio which vary between 4.8 and 6.6 and low fractionation

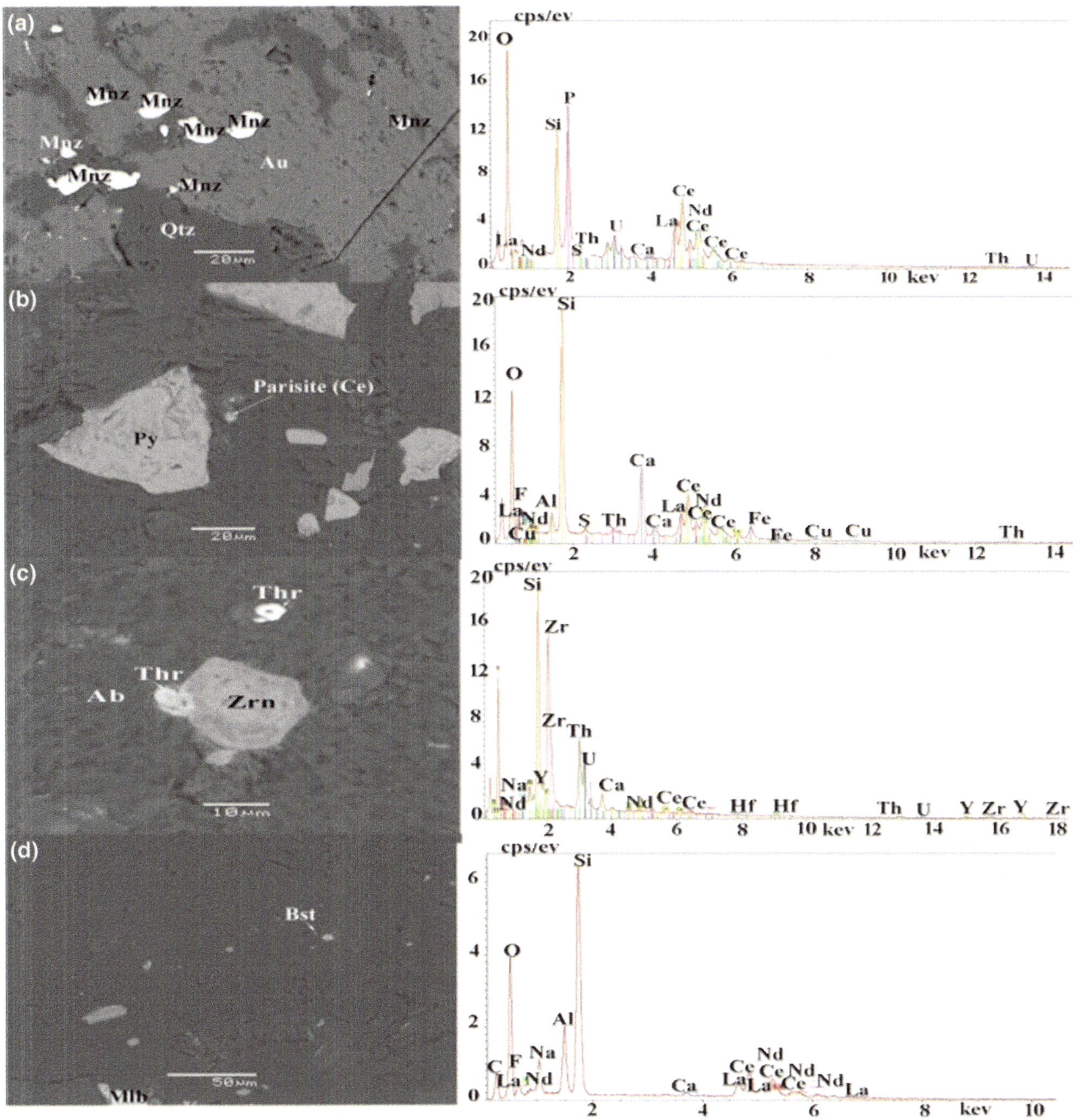

Fig. 2 Secondary electron SEM images with accompanying EDS spectra of REE-bearing minerals. *Ab: albite; Au: gold; Mlb: molybdenite; Mnz: monazite; Py: pyrite; Qtz: quartz; Thr: thorite; Zrn: zircon*

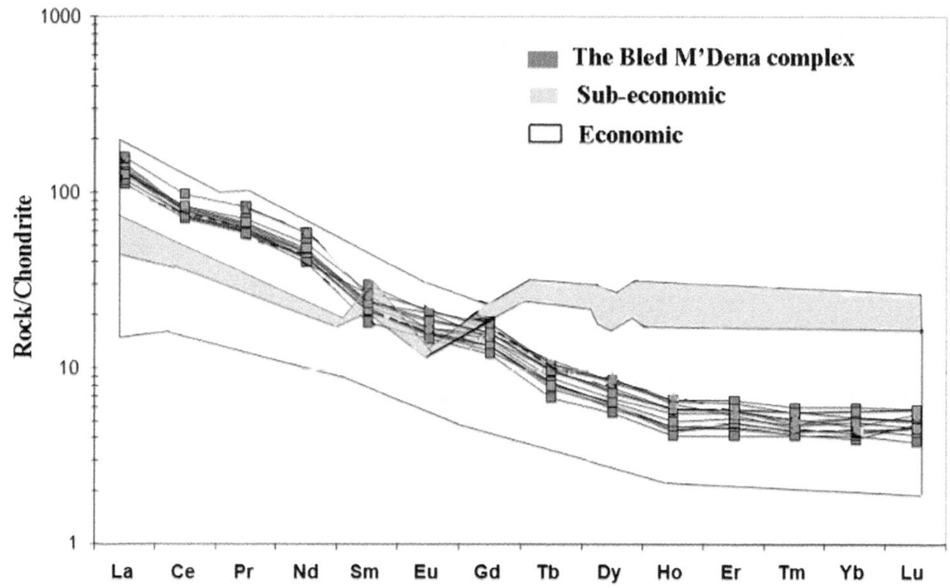

Fig. 3 Chondrite-normalized REE patterns for the Bled M'Dena intrusives compared sub-economic and economic porphyry Cu deposits. Normalized-values are from Sun and McDonough

of heavy rare earth element (HREE) ($(Dy/Lu)_N$ = 1.2–1.5). Europium anomalies are low (Eu/Eu* = 0.9–1.1; cf. [2]). These values may be evidence for a less enriched melt source and garnet fractionation [5]. Alteration assemblages of the intrusive-related mineralization at Bled M'Dena are associated mainly with sulphides as well as accessory minerals such as monazite, xenotime, bastnasite, parisite and thorite all of which may concentrate the REEs. The minerals that appear to crystallize from late-stage fluids are the fluorocarbonates (bastnasite, parisite) and phosphates (monazite) (see also [6]). Images of these minerals are illustrated in Fig. 2.

LREE enrichment and HREE depletion is a feature that is encountered in other orogenic calc-alkaline magmas, which are believed to be issued from metasomatized mantle sources. This is consistent with high Sr/Y ratios and low Yb_N concentrations, which is a common feature of adakitic-like magmas in TTG suites. The REE patterns from the Bled M'Dena porphyry Mo-Cu share similar features with other economic to sub-economic porphyry Cu systems (Fig. 3) such as the Ali-Abad and Dalli areas in Iran [7].

References

1. Goodenough, K.M., Schilling, J., Jonsson, E., Kalvig, P., Charles, N., Tuduri, J., Deady, E.A., Sadeghi, M., Schiellerup, H., Müller, A., Bertrand, G., Arvanitidis, N., Eliopoulos, D.G., Shaw, R.A., Thrane, K., Keulen, N.: Europe's rare earth element resource potential: an overview of REE metallogenetic provinces and their geodynamic setting. Ore Geol. Rev. **72**, 838–856 (2016)
2. Lagraa, K., Salvi, S., Béziat, D., Debat, P., Kolli, O.: First insights on the molybdenum- copper Bled M'Dena complex (Eglab massif, Algeria). J. Afr. Earth Sc. **127**, 159–174 (2017)
3. Lottermoser, B.G.: Rare earth elements and hydrothermal ore formation. Ore Geol. Rev. **7**, 25–41(1992)
4. Mahdjoub, Y., Kahoui, M.: The Paleoproterozoic Eglab domain (Algeria, Réguibat Rise, West African Craton); geodynamic model and ore controls; 1ère Conférence International en Algérie sur les Ressources Minérales (CIRMA) (Communication oral); Décembre 2007; Algérie (2007)
5. Richards, J.P., Kerrich, R.: Adakite-like rocks: their diverse origins and questionable role in metallogenesis. Econ. Geol. **102**, 537–576 (2007)
6. Verplancka, P.L.: The role of fluids in the formation of rare earth element deposits. Procedia Earth Planet. Sci. **17**, 758–761 (2017)
7. Zarasvandi, A., Rezaei, M., Sadeghi, M., Lentz, D., Adelpour, M., Pourkaseb, H.: Rare earth element signatures of economic and sub-economic porphyry copper systems in Urumieh-Dokhtar Magmatic Arc (UDMA), Iran. Ore Geol. Rev. **70**, 407–423 (2015)

Heavy Metals in Mine Waters of Diverse Mineralization of Polygenetic Ore-Magmatic System: Merisi Ore Cluster, Achara Region (Georgia)

Archil Magalashvili, Levan Doliashvili, Kakhaber Karchkhadze, and Akaki Nadaraia

Abstract

The distribution of heavy metals in mine waters of a single polygenetic ore-magmatic system in Merisi, Georgia, has been studied using the standard methods of sampling, Spectrophotometry and AA Spectroscopy analyses. Mining and background waters, soils, and ores have been sampled and analyzed on pH, As, Cd, Cu, Ni, Pb, Zn, Co, and Sr. The research illustrates the dependence of heavy metals distribution on pH, the morphogenetical type of mineralization, ore composition, and other geological controls. The study has shown that the distribution of heavy metals in mine waters of the single polygenic ore-magmatic system generally having a homogeneous petrochemical and hydrogeochemical background, despite the presence of various morphogenetic types of mineralization, exhibits the same general regularities as the individual types of deposits. This conclusion is preliminary and needs to be confirmed at other similar sites as well.

Keywords

Mine waters • Heavy metals • Polygenetic ore system
Hydrogeochemistry of ore deposits • Geo-ecology
Merisi ore cluster

1 Introduction

Distribution regularities of metals in draining waters of diverse ore deposit types have been substantially studied [1–3]. The given studies mainly refer to specific mineral deposits and the varieties of morphological mineralization (porphyry, vein, VMS etc.). In addition, there are relatively small ore clusters that consist of a genetically and morphologically different type of ores connected with a single ore-magmatic system affecting draining waters and subsequently the environmental status. An example of such ore deposits developed under relatively simple geological conditions is the Merisi (Fig. 1) Au-base metals ore cluster in Georgia [4–6]. The main aim of the research was to study the hydro-geoecology significance of the mining waters in the abandoned prospecting and exploration area of the Merisi ore deposits and simultaneously to test how the presence of different morpho-genetical type of mineralization influences individual deposits and overall hydrogeochemistry pattern of the site.

2 Geological Settings

Merisi polymetallic ore cluster is considered as a polygenetic ore unity developed in a relatively short period by a single ore-magmatic process in the Middle-Late Eocene [5, 6]. The sulfide deposits are plutonogenic Au-polymetallic vein (Varaza and others), porphyry Cu-Mo (Namonastrevi and Uchamba), Ag-polymetallic veinlets-disseminated (Vaio, Surnali), and high-sulfidation epithermal quartz alunite (Tsablana, not covered by this study). Each of these deposit types are connected with various stages of a single ore-magmatic process.

The main ore generating structure is polygenetic Merisi gabbro-syenite-monzonite pluton, which is intruded into massive co-magmatic volcanites. South of the aforementioned intrusive bodies there are Au-Cu-polymetallic deposits. Ba-Ag-polymetallic deposits with veinlets mineralization in tectonically crashed zones are found to the north, while porphyry Cu-Mo (?) with disseminated mineralization occurs in the central part of the area (Fig. 1).

Merisi ore cluster is situated in a humid, sub-tropical climatic zone and is characterized by an abundance of both surface and ground waters causing intensive drainage from ore-bearing masses and pollution of waters with heavy

A. Magalashvili (✉) · L. Doliashvili · K. Karchkhadze
A. Nadaraia
Ilia State University, 3/5 Kakutsa Cholokashvili Ave, Tbilisi, 0162, Georgia
e-mail: archil.magalashvili@iliauni.edu.ge

F. Rossetti et al. (eds.), *The Structural Geology Contribution to the Africa-Eurasia Geology: Basement and Reservoir Structure, Ore Mineralisation and Tectonic Modelling*,
Advances in Science, Technology & Innovation, https://doi.org/10.1007/978-3-030-01455-1_14

metals. Since there are no carbonate formations at the studied depths, and host rocks represented by monotonous volcanites are characterized with almost identical petrology, it can be presumed with high likelihood that the main impact on shallow underground and mine waters hydrogeochemistry is generated by ore bearing zones and the character of their mineralization.

3 Research Methodology

The soils, characteristic ores, and effluents of adits and heaps/montons as well as various surface waters were sampled according to guidelines provided by [7]. The Water pH has been measured in situ. The spectrophotometry analyses (HACH 3900) were made in Ilya State University and AA Spectroscopy analyses (Perkin Elmer A Analyst 200) in accredited chemical laboratory of Gamma Consulting Ltd (www.gamma.ge). All analyses were subjected to standard QC and statistical treatment procedures. For the background (sample—BCKGR in Fig. 2) the data from water streams situated in a distance, upstream from the ores were used (Fig. 1).

Varaza (Fig. 1) mining waters are sampled at two horizons distanced by 90 meters from each other vertically. In the case of Vaio and Namonastrevi (Fig. 1) the samplings were made from adits positioned at one horizon (Fig. 2). Apart from adits, underground water flow was also sampled at the same horizon in Namonastrevi. Results of analyses are statistically averaged and finally represented in various diagrams (Figs. 2 and 3).

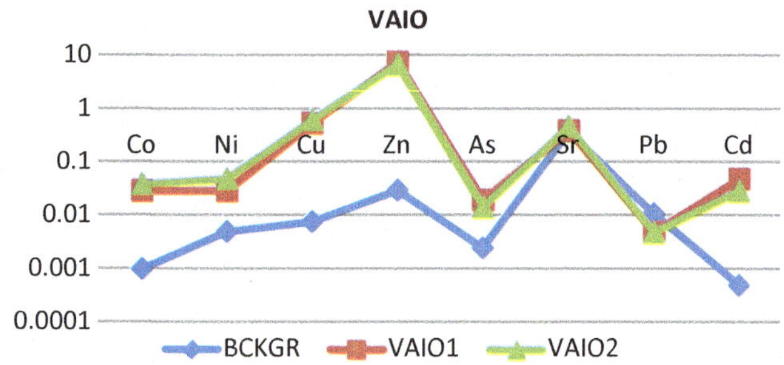

Fig. 2 Diagram showing heavy metals distribution in mining waters (VAIO1 and VAIO2) of Vaio deposit against the background water (BCKGR). Vertical axis represents concentrations (mg/l) of metals on a logarithmic scale

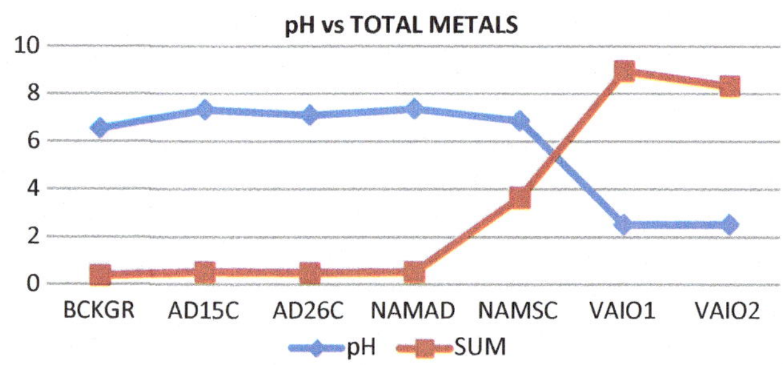

Fig. 3 Diagram showing correlation between pH and concentrations (mg/l) of total metals in different mining waters of Merisi ore cluster (BCKGR is the background water). Vertical axis shows metal concentrations and pH on the same scale

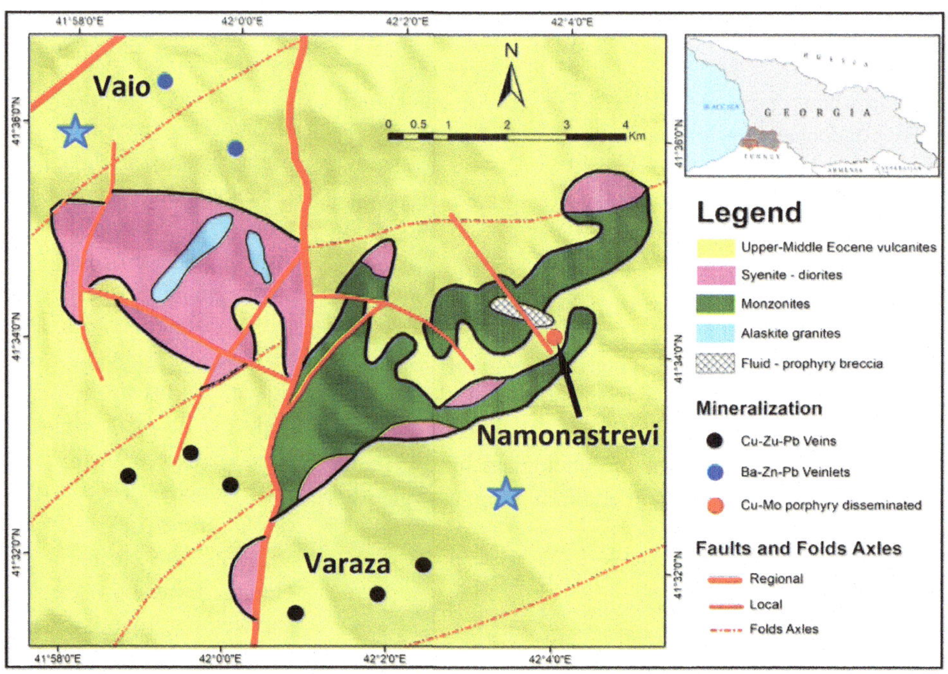

Fig. 1 Generalized geological-structural map of Merisi ore cluster based on Magalashvili et al. (2004). Blue stars represent sampling locations of background waters (sample—BCKGR)

4 Research Outcomes

The research outcomes are represented in Tables and illustrated in different diagrams showing the correlation between various metals, waters' pH and other geological controls (typical examples—Figs. 2 and 3). The average concentrations (mg/l) of analyzed metals in Vaio deposit are presented in Fig. 2. Figure 3 represents correlation between pH and total metals concentrations (mg/l) in different mining waters of Merisi ore cluster.

5 Discussion

The results of the present study show that the distribution of heavy metals in mining waters of genetically dissimilar mineralization reveals a number of regularities. For example, in Fig. 3 it can be clearly seen that the total amount of metals in Merisi waters is inversely proportional to pH value, which fully correlates with the earlier established general regularities for different sulphide ores [1–3]. However, metal distribution in draining waters of various types of mineralization show dissimilar regularities. For instance, in Namonastrevi waters, Cu and Zn are represented with smaller concentration than in "background" water. Similar regularity has been observed before in porphyry Mo mines [1]. Based on this observation it seems that Namonastrevi could be regarded as porphyry Cu-Mo or porphyry Mo type of mineralization instead of porphyry Cu.

Varaza mining waters from both horizons do not reveal increased content of metals as compared to background water which are characterized with a little more acidity (pH—6.6 vs. 7.35 and 7.13). This can be explained by the massive character and low permeability of ore veins and host rocks as well as by relatively high water abundance of adits. It should be noted that despite a significant (more than 90 m) vertical diversion the Varaza waters do not reveal zonality in Cu, Pb, and Zn distribution, which is abnormal for similar sulphide mineralization. This can be caused by previously revealed local rhythmic zonality in this deposit [5, 6].

6 Conclusions

Preliminary studies have shown that, in general, distribution of heavy metals in mine waters connected with Merisi polygenetic ore-magmatic system with relatively homogeneous petrochemical and hydrogeochemical background reveal the same regularities as each separate type of ore. The research illustrates the dependence of heavy metals distribution on pH, the morphogenetical type of mineralization, ore composition, and other geological controls.

The above conclusion is considered preliminary and needs to be confirmed at other similar sites as well.

Acknowledgements We appreciate the support for this research provided by the Democracy Commission Small Grant Program funded by the U.S. Embassy in Georgia.

References

1. Plumlee, G., Smith, K., Montour, M., et al.: Geologic controls on the composition of natural waters and mine waters draining diverse mineral-deposit types. SEG Publisher, pp. 373–432 (1999)
2. Plumlee, G.: The Environmental geology of mineral deposits. U.S. Geological Survey, Box 25046, MS 964, Federal Center, Denver, CO 80225–0046 (1999)
3. Zautashvili, B.: Gidrogeokhimia rudnikh mestorojdenii Gruzii [Hydrogeochemistry of ore deposits of Georgia]. Tbilisi, Metsniereba, p. 190 (1984)
4. Adamia, S., Zakariadze, G., Chkhotua, T., et al.: Geology of the Caucasus: a review. Turkish J. Earth Sci. **20**, 489–544 (2011)
5. Magalashvili, A., Kikava, A., Tuskia, T., Kvatashidze, R.: Meriskii gornorudniy raion Adjarii (zapadnaia gruzia) [Merissi mining region in Adzharia (Western Georgia)]. Mining Journal, N4, pp. 71–74 (2004). http://www.rudmet.ru/journal/1091/article/18011/. Accessed 10 June 2018
6. Magalashvili A.: Zonalnost i genesis Merisskogo mednopolimetalicheskogo rudnogo polia [Zoning and genesis of Merisi Cu-polymetallic ore cluster]. Metsniereba, Tbilisi, 76 p. (1991)
7. Crock, J., Arbogast, B., and Lamothe, P.J.: Laboratory methods for the analysis of environmental samples. In: Plumlee, G.S., Logsdon, M.J. (eds.) The Environmental Geochemistry of Mineral Deposits, Part A. Society of Economic Geologists, Reviews in Economic Geology, vol. 6A, pp. 265–287 (1999)

Geochemical Characters of the Fluids Involved in the Genesis of the Tebessa Region Mineralizations (North-East Algeria)

Azzedine Bouzenoune

Abstract

Data from fluid inclusions and stable isotopes (O, C, S) have been used to characterize the fluids involved in the peridiapiric mineralization genesis in the Tebessa area. Iron ore mineralization (siderites and ankerites) occurred at temperatures of about 120 °C by substitution of Cretaceous carbonate rocks while epithermal polymetallic mineralization (Pb-Zn, Cu, Ba and F) occurred at relatively higher temperatures ranging from 100 to 250 °C from basinal brines expelled from sedimentary series by different episodes of diapiric activities and by neogen compressions. The carbon of iron carbonates are of mineral origin, they come from the reuse of the carbon in the limestone host rocks. The sulphide sulfur is produced by thermochemical reduction of the sulphates of the Triassic evaporates.

Keywords

Fluid inclusions • Stable isotopes • Iron carbonates • Sulphides • Tebessa

1 Introduction

One of the main geological features that characterizes the Saharan Atlas mountain chain at the Algerian-Tunisian borders is the presence of Triassic bodies in the form of classic "diapirs". An examination of the distribution map mineral deposits in this region reveals a close spatial relationship between Triassic outcrops and mineral concentrations (Fig. 1). The salt domes and associated "cap-rocks" as well as the surrounding sedimentary formations constitute a geological setting whose economic interest is well established [1, 2]. The importance of these halokinetic and/or diapiric structures lies in their production of salts (halite, NaCl or sylvite KCl), sulfur (S) and their ability to favor the creation of lithological receptacles and structural traps for hydrocarbons and metalliferous concentrations (Pb, Zn, Cu, Fe, Ba, Sr).

2 Geological Setting

The geological formations outcropping the eastern Saharan Atlas range from the Triassic to the Quaternary, with no outcrop of Jurassic formations. All of these formations with a thickness exceeding 6 km, are folded into the so-called atlasic synclines and anticlines (NE-SW) and are intersected by NW-SE oriented grabens (Fig. 1). The geology of this region is also characterized by Triassic evaporite domes whose structures are considered diapiric [3, 4]. Some of these structures, however, have recently been reinterpreted as "salt glaciers" [5].

The interaction between salt tectonics and sedimentation is therefore fundamental in any exploration of these resources. In the area of Tebessa, the first signs (remobilization of Triassic debris in sedimentation, variation of thicknesses and facies…) of halokinetic mobilization of Triassic evaporites in a distensive context are dated as early as Aptian [4, 6]. These dome structures control the geometric (extensions and thickness) and lithological (facies) characters of the Albo-Aptian sedimentary series that constitute the largest host-rock of mineralization in this area. The indices of this halokinetic activity are also known from the Vraconian and Senonian [3, 7]. The Neogene compression has remobilized these Triassic evaporates complicating the existing structures. The indices of this remobilization are known to date back to the Eocene, Miocene and Quaternary periods.

Thus, the diapiric structures of the Triassic evaporates generated in this part of the Saharan Atlas have strongly contributed to the creation of suitable conditions for ore deposits genesis. Among these conditions are favorable

A. Bouzenoune (✉)
Laboratoire de Génie Géologique, Mohamed Seddik Benyahia University, 18000 Jijel, Algeria
e-mail: bouzenoune@univ-jijel.dz

F. Rossetti et al. (eds.), *The Structural Geology Contribution to the Africa-Eurasia Geology: Basement and Reservoir Structure, Ore Mineralisation and Tectonic Modelling*,
Advances in Science, Technology & Innovation, https://doi.org/10.1007/978-3-030-01455-1_15

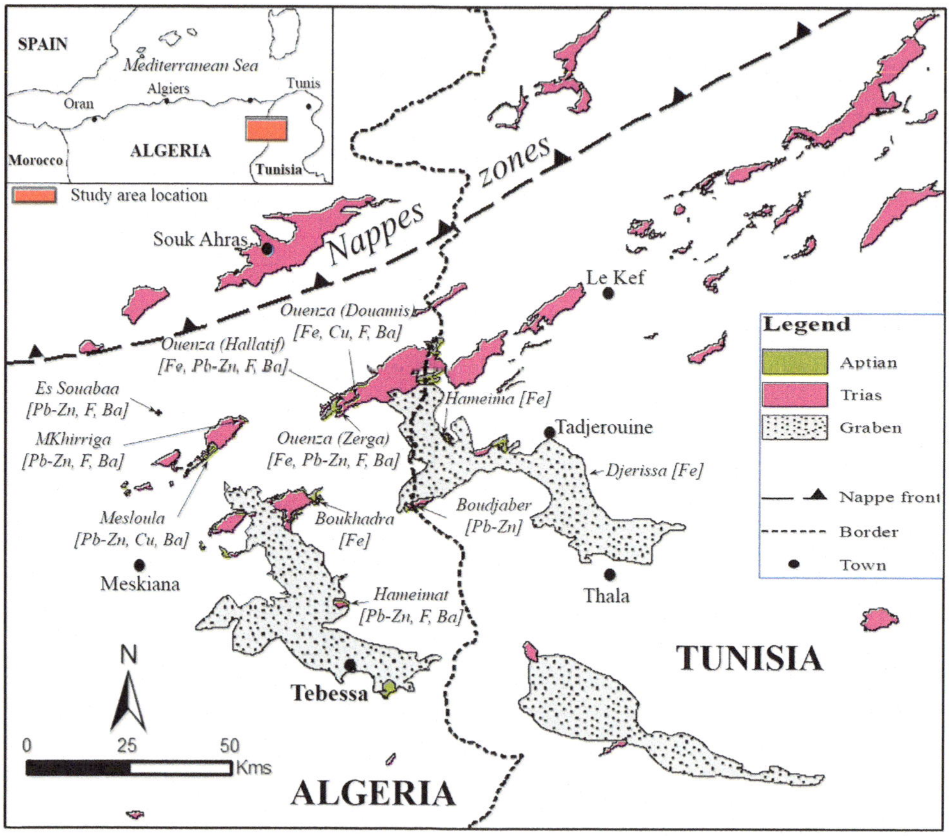

Fig. 1 Schematic map showing the location of the main mineralizations of the Tebessa area and their spatial relationship with the outcrops of Triassic evaporites

lithologies (limestones which are more or less dolomitized), paleogeographic discontinuities (such as the vraconian discontinuity), intense fracturing and brecciation of the rocks which hosts the majority of the mineralization. This salt tectonics has also generated very favourable structures for the circulation of the mineralizing fluids by creating drains (contact between the diapirs and their host rocks) and geothermal gradients through the deformation of isotherms around the saliferous domes. Finally, they have become a potential source for sulphide sulfur and probably metals.

The Tebessa area contains numerous mineralized sites in Fe, Pb-Zn, Cu, Ba, F, P some of which are in exploitation (namely the iron ore deposits of Ouenza, Boukhadra and Khanguet el Mohad, phosphate deposits of Jebel Onk district). The polymetallic mineralizations (Pb-Zn, Cu, Ba, F) are all hydrothermal and are mainly located in paleogeographic and/or structural traps (Mesloula, Boudjaber, M'Khirriga, Essouabaa, Hameimat, M'Zouzia, Ouasta…) while iron ores (siderites, ankerites and iron oxides) are of metasomatic origin, replacing the Aptian carbonate rocks. Phosphatic mineralization is of sedimentary origin.

3 Results

Mineralization in Tebessa area has been, in recent years, the subject of numerous studies that have provided valuable geochemical data such as fluid inclusions and stable isotopes. In fact, this has contributed significantly to broadening the geochemical database for this region [3, 8–14].

3.1 Fluid Inclusion Data

The results of the microthermometric measurements carried out on the fluid inclusions trapped in the fluorites gave homogenization temperatures (Th) between 60 and 260 °C. The melting temperatures of ice (Tmi) vary between −5.4 and −26 °C corresponding, according to Bodnar [15] equations, to salinities between 8.41 and 29% equivalent NaCl. The fluid inclusions of the barite have a Th between 130 and 160 °C while the Tmi varies between −0.4 and −22 °C corresponding to salinities of the order of 26% equ.

NaCl [12]. Gangue minerals such as calcite and quartz contain fluid inclusions which gave a Th between 90 and 160 °C for calcite and 120 and 260 for quartz. The Tmi is between −5 and −25 °C for calcite corresponding to salinities of 8 to 26% equ. NaCl. The Tmi of the fluid inclusions trapped in quartz varies from −5 to −26 °C.

3.2 Stable Isotope Data (O, C, S)

Oxygen and carbon isotopic compositions of the Cretaceous carbonate rocks (micrite) range from −0.77 to −7.91‰/PDB for oxygen and between –2.80 and +4.60‰/PDB for carbon (Fig. 2). The calcite, a gangue mineral associated with mineralization, shows $\delta^{18}O$ values between −7.50 and −12.3‰/PDB and $\delta^{13}C$ values between −4.80 and +8.0‰/PDB. Carbonates iron mineralizations (siderites and ankerites) show $\delta^{18}O$ values between −7.81 and −4.37‰/PDB and $\delta^{13}C$ values between −0.75 and +1.40‰/PDB for siderites whereas the ankerites are characterized by $\delta^{18}O$ values between −8.65 and −6.83‰/PDB and $\delta^{13}C$ values between −0.23 and +1.72‰/PDB. The isotopic composition of sulphide sulfur (essentially galena, sphalerite) and sulphates (gypsum of triassic evaporates and barite) vary between −1.20 and +16.20‰ for sulphides and between +17.90 and +24.0 for barite. The sulphates of Triassic evaporates show a range of $\delta^{34}S$ values between +13.4 and +16‰.

4 Discussions

Fluid inclusions and stable isotopes (O, C, S) for the studied different mineral phases and their main carbonate host-rocks have allowed, in a first stage, to specify the physical and chemical conditions under which various mineralizations have taken place. They have also been used to clarify the narrow spatial and genetic relationship between the triassic evaporates and their multiphase halokenetic and diapiric dynamics on one side and the mineralizing processes on the other.

The obtained fluid inclusions data argue for a polyphase mineralization brought about by relatively "hot" brines (salinities 10–30% equ. NaCl, temperatures 100–250 °C). These high salinities are explained by the contribution of the triassic evaporate halite (Ouenza) intercepted at 100 m in depth over a thickness of 200 m. In the absence of a magmatic event in the geological context of the region, the recorded temperatures can be linked to an abnormally high geothermal gradient induced by the high thermal conductivity of these salt extrusions. The halokinetic motion of these Triassic

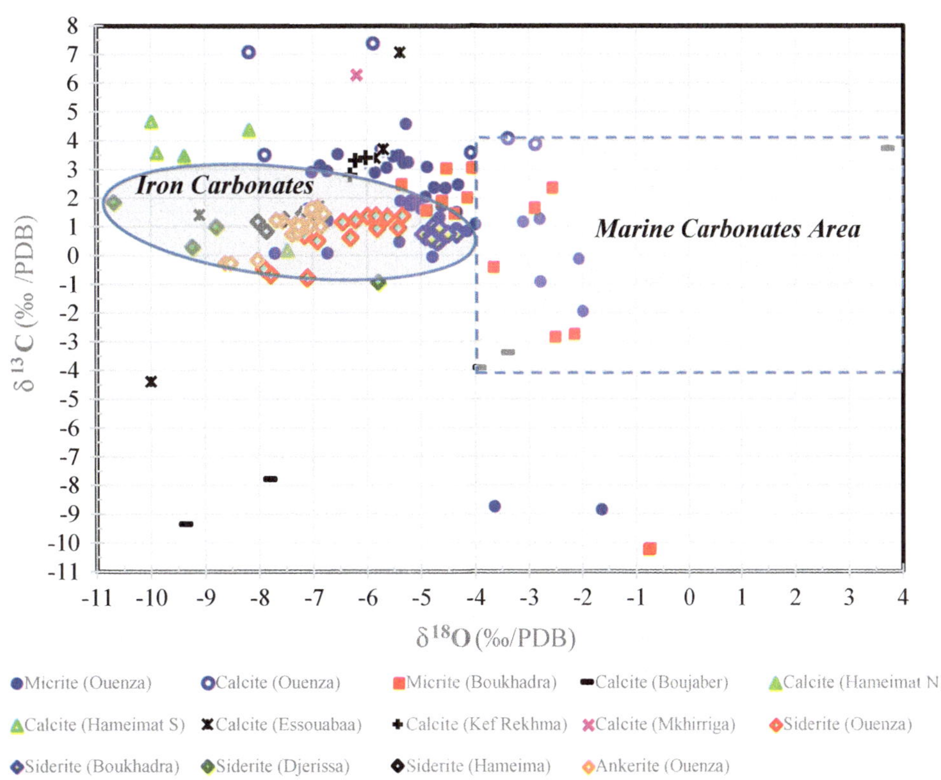

Fig. 2 Diagram $\delta^{18}O$ versus $\delta^{13}C$ of carbonates associated with the Tebessa area mineralization (data are from [3, 10, 12, 16, 14])

evaporates during the Mesozoic distensive period and their diapiric extrusion during the Neogen compressive phases are two salient tectonic features of this region. These tectonics have had an impact on mineralization processes by controlling the petrographic nature of some host rocks (aptian facies), by creating paleogeographic traps (i.e.: Vraconian unconformity) and structural traps (various fractures) and by generating pressure gradients that have participated in the triggering of the fluid circulation and the expulsion of the so-called "basinal brine" type of mineralization fluid such as the Mississippi Valley Type (MVT) deposits.

Stable isotope data has provided information on the carbon origin of carbonates (siderite and ankerite) and the origin of sulfur sulphide (sphalerite and galena) of polymetallic mineralization. They have also made it possible to determine the temperature of the carbonate mineralization.

Indeed, for iron carbonates (siderites and ankerites) and calcites associated with mineralization of deposits in Tebessa area, the majority of $\delta^{13}C$ values are in the range of $\delta^{13}C$ values of marine carbonates [−4 to +4‰/PDB]. We can therefore say that the carbon of these calcites is of mineral origin. It would come from the re-use of carbon from the carbonate rocks that host the mineralization. The few values above +4‰/PDB could be explained by the effect of temperature on the fractionation of carbon isotopes between calcites and the mineralizing fluids which would exceed 100 °C [17].

In view of the $\delta^{18}O$ values of iron carbonates and calcites associated with the mineralization, it is clear that these carbonates would have precipitated either at high temperatures, or from fluids depleted in ^{18}O such as meteoric waters. Both possibilities can obviously be combined. Given the temperatures obtained by the study of fluid inclusions which are higher than 100 °C, it is logical to attribute the depletion in ^{18}O of these carbonates to the effect of the temperature of the fluids from which they were formed.

The results of isotopic analyses of the Triassic gypsum sulfur of Tebessa area show $\delta^{34}S$ values which reflect typical values of isotopic compositions of sulfur in the waters of the Triassic Sea. They show that these sulphates are isotopically homogeneous and they have not undergone any interactions since their deposition. They reflect the isotopic composition of the sea at that time in the region. The $\delta^{34}S$ values for all sulphides indicate a common source of sulfur. The latter probably derives from Triassic sulphates deduced through the reduction of SO_4 in the sulphates of Triassic evaporates. The studies carried out in this diapirs region have clearly shown that there is a direct relationship between the emplacement of Triassic masses and the deposition of polymetallic mineralization [2, 3, 4, 10–13, 17–19]. These mainly sulphide polymetallic mineralizations consist of an early sulphide paragenesis followed later by many others, all with relatively homogeneous $\delta^{34}S$ values. This suggests that the different sulfur episodes are genetically linked and that the sulfur derives from the same source, i.e. Triassic evaporate sulphates. In view of the emplacement temperature of these mineralizations (>100 °C), the reduction of the sulfur of the Triassic sulphates by biological processes is excluded. On the other hand, the thermochemical reduction could have taken place where the reduced sulfur in H_2S, poor in heavy sulfur isotope (S^{34}), is integrated in the structure of the sulphides, while the $(SO_4)^{2-}$ remaining in the fluid gets enriched in heavy sulfur isotope and is deposited in the form of sulphate (barite). This mechanism, which is involved in the genesis of the diapir mineralization, has indeed been demonstrated in several deposits on the basis of stable isotope and fluid inclusions data [10, 12].

5 Conclusions

The mineralization in the area of Tebessa is the result of a polyphase geological history marked by the deposit of a thick Mesozoic marno-carbonate sedimentary series. This Neritic sedimentation has been repeatedly disturbed by the halokinetic activity of Triassic evaporates that induced paleogeographic discontinuities as well as the control of facies constituting the host of mineralization. These sedimentary series were initially folded during the Cenozoic eras into a set of NE-SW oriented synclines and anticlines, most often intersected by NW-SE graben.

The main geological phenomena that have marked the geological history of this region have also generated fluids that are at the origin of the mineralization processes. The geochemical data (fluid inclusions and stable isotopes) acquired in recent years have made it possible to characterize these fluids and to highlight the genesis conditions of these mineralizations.

The data of the fluid inclusions showed that these mineralizations take place under epithermal conditions at temperatures between 100 and 250 °C by brines with salinities of up to 26% equ. NaCl.

Carbon isotopic compositions of iron carbonates and associated hydrothermal calcites indicate a mineral origin of carbon. It would have come from the reuse of the sedimentary carbon of the Cretaceous host-rocks. Oxygen compositions indicates that these same carbonates were formed at relatively high temperatures (>100 °C). Sulfur isotopes have confirmed the genetic link between sulphates of Triassic evaporates, sulphate (barite) and sulphide (galena, sphalerite, pyrite) mineralizations. The thermo-chemical reduction of Triassic sulphates is the mechanism used to produce the sulfur involved in the genesis of these baritic and polymetallic sulphide mineralizations.

References

1. Pohl, W., Amouri, M., Kolli, O., Scheffer, R., Zachmann, D.: A new genetic model for the North African metasomatic siderite deposits. Mineral Deposita **21**, 228–233 (1986)
2. Rouvier, H., Perthuisot, V., Mansouri, A.: Pb-Zn deposits and saltbearing diapirsin southern Europe and North Africa. Econ. Geol. **80**, 666–687 (1985)
3. Bouzenoune, A.: Minéralisations péridiapirique de l'Aptien calcaire: les carbonates de fer du gisement hématitique de l'Ouenza (Algérie orientale). Thèse Doct, d'Univ. Paris VI. P 206, France (1993)
4. Dubourdieu, G.: Etude géologique de la région de l'Ouenza (confins algéro-tunisiens). Thèse. Sci. Paris. Publ. Serv. Carte géol. Algérie, n.s, Bull., n°10, 659 p. (1956)
5. Vila, J.M.: Halocinèse distensive albienne à «glacier de sel» sous-marin et plissements tertiaires du secteur Ouenza-La djebel-Méridef. Bull. Serv. Géol. Algérie **7**(1), 3–34 (1996)
6. Thibieroz, J., Madre, M.: Le gisement de sidérite du Djebel El Ouenza (Algérie) est contrôlé par un golfe de la mer aptienne. Bull. Soc. Hist. Nat. Afrique du nord, Alger **67**(3–4), 126–150 (1976)
7. Aoudjehane, M., Bouzenoune, A., Rouvier, H., Thibiéroz, J.: Halocinèse et dispositifs d'extrusions du Trias dans l'Atlas saharien oriental (NE algérien), pp. 273–287. Géol. Médit, Marseille, XIX (1991)
8. Diane, E.-B., Perthuisot, V.: Première approche des conditions de genèse des minéralisations à Pb-Zn du Djebel Mesloula (Est algérien): étude micro thermométrique des inclusions fluides primaires de la calcite. Bull. du Serv. Géol. de l'Algérie. **7**(1), 59–69 (1996)
9. Paraire-Akrour, H.: Fluid inclusion of the F-Ba-Pb late paragenesis of borders of the Ouenza, Mesloula, Hameimet ed Dahra diapirs (N-E Algeria). In: Pagel, M., Leroy, J.L. (eds.) Source, Transport and Deposition of Metals, pp. 219–222. Balkema, Rotterdam (1991)
10. Salmi-Laouar, S.: Contribution à l'étude géologique et géochimie des isotopes stables (S, O, C) des minéralisations polymétalliques (Zn, Pb, F, Ba, Fe, Hg) de la «zone des diapirs» du nord de Tébessa (NE algérien). Nouvelle Thèse de Doctorat, Univ. Badji-Mokhtar. Annaba, Algérie, 190 p. (2004)
11. Laouar, R., Salmi-Laouar, S., Sami, L., Boyce, A.J., Kolli, O., Boutaleb, A., Fallick, A.E.: Fluid inclusion and stable isotope studies of the Mesloula Pb-Zn-Ba ore deposit, NE Algeria: Characteristics and origin of the mineralizing fluids. J. Afr. Earth Sc. **121**, 119–135 (2016)
12. Sami, L.: Caractérisation géochimique des minéralisations à Pb-Zn, F, Ba, Cu, Fe et Hg des confins Algéro-tunisiens. FSTGA., USTHB Bab Ezzouar Alger, Thèse doctorat (2011)
13. Bouhlel, S., Leach, D.L., Craig, A.J., Marsh, E., Salmi-Laouar, S., Banks, D.A.: A salt diapir-related Mississippi Valley-type deposit: the Bou Jaber Pb-Zn-Ba-F deposit, Tunisia: fluid inclusion and isotope study. Miner. Deposita **51**, 749–780 (2016)
14. Aït Abdelouahab, D.J., Bouzenoune, A., Préat, A.: Les isotopes stables du carbone et de l'oxygène des carbonates (calcaires et sidérites) du gisement de fer de Boukhadra (Algérie nord-orientale). Bulletin du Service géologique National, **22**(3), 381–395 (2011)
15. Bodnar, R.J.: Revised equation and table from determining the freezing point depression of H_2O-NaCl solution. Geochim. Cosmochim. Acta **7**, 683–684 (1993)
16. Adjali-Aïssaoui, S.: Structuration et genèse des gisements de fer carbonaté du Jebel Jerissa et du Jebel Hameima (NV/ de Tunisie). Pétrographie des carbonates, Minéralogie et étude de la matière organique, appliquées aux gîtes de couverture. Thèse Doct. Univ. Tunis, II, 185 p. (1990)
17. Bouzenoune, A., Lecolle, P.: Pterographic and geochemical arguments for hydrothermal formation of the Ouenza siderite deposit (NE Algeria). Miner. Deposita **32**, 189–196 (1997)
18. Bouzenoune, A., Rouvier, H., Thibieroz, J.: Trias de l'Ouenza: contexte diapirique, zonation minéralogique et conséquences métallogéniques. Bull. Serv. Géol. Algérie **6**(1), 3–24 (1995)
19. Otmanine, A.: Les minéralisations en fluorine, barytine, Pb, Zn et fer sidéritique autour du fossé de Tébessa-Morsott. Relation entre la paléogéographie albo-aptienne, diapirisme, structure et métallogénie. Thèse Doct. 3ème cycle, Univ. P. et M. Curie, Paris VI, 220p. (1987)
20. Price, P.E., Kyle, J.R.: Genesis of salt dome hosted metallic sulfide deposits : the role of hydrocarbons and related fluids. In : W.E. Dean (eds.) Proceeding of the Symposium on the Organic and ore Deposits. Denver region Explor. Geologists Soc. pp.171–184 (1986)
21. Madre, M.: Contribution à l'étude géologique et métallogéniques du Djebel Ouenza (Est-algérien). Thèse 3ème cycle, Paris, France, 98p. (1969)

Quartz-Rich Fault Rocks as Potential High Purity Quartz Source (I) in a Sequence of the Central Cameroon Shear Zone (Etam Shear Zone): Geology and Structure

Cyrille Sigue, Amidou Moundi, Jean Lavenir Ndema Mbongue, Akumbom Vishiti, Arnold Chi Kedia, Cheo Emmanuel Suh, and Jean Paul Nzenti

Abstract

The Etam shear zone is the NNE-SSE to NE-SW trending fault zone with coexistence of a sinistral shearing mylonitic corridor and a brecciated quartz vein. The mylonitic zone is evidence of progressive deformation. Primary minerals include $Qtz_1 + Kfs_1 + Pl_1 + Bt_1$. These minerals increasingly and gradually reduce from the wall rock towards the shear zone center. Quartz recrystallization also gradually increases in the same direction. Microstructures observed include: recrystallization, sub-grain boundary, myrmekites and undulose extinction. Mica-chlorite-epidote assemblages define the greenschist assemblage, but the observation of amphibole in the mylonite and the presence of flame perthite in feldspar suggest a transitional greenschist-amphibolite metamorphic condition in the Etam Shear Zone. The brecciated quartz vein has dimensions of about 2.5 × 0.5 km and shows several quartzitic blocks, mainly composed of milky to transparent quartz clasts and a siliceous matrix. Quartz clasts are pure and the observation of thin sections of these quartz under polarized microscope shows grains free of microstructures or displays small micro quartz veins and rare feldspar inclusions. These are some criteria, which can lead to a preliminary assessment of the purity of quartz in Etam area.

Keywords

Etam • Fault zone • Quartz vein • Pure quartz
Inclusions

1 Introduction

An abundant literature in fluid-flow mechanisms in structural features within a shear zone is available [1, 2]. Alteration in brittle–ductile shear zones is mainly driven by syn-tectonic fluids which are associated with a variety of weakening mechanisms [3]. The process of grain size reduction through fracturing during cataclastic deformation often releases Si through breakdown of albite or the formation of chlorite ± epidote and phengite [4]. Fluid saturated in Si can then circulates by diffusional transport [5] and form small quartz veins or by mobile hydrofracturing with very ascent fluids flow and form large quartz veins [6].

In southwestern Cameroon numerous quartz-rich rocks of metamorphic and fault-related origin have been described [7]. These rocks lie essentially along the Central Cameroon Shear Zone (CCSZ). Although this structure has witnessed numerous structural and economic studies [8], no previous studies have focused on these quartz-rich fault rocks whose contribution to the overall evolution of the shear zone and potential role as High Purity Quartz (HPQ) source is of more than local significance.

This study which is typically descriptive with preliminary geochemistry data examines the mineralogy and textural context of fault rocks and associated wall rock within the ESZ. The preliminary results obtained can constitute the commencement of the assessment of the potential HPQ in these rocks.

C. Sigue (✉) · J. L. N. Mbongue · A. Vishiti · A. C. Kedia
C. E. Suh
Economic Geology Unit, Department of Geology, University of Buea, P.O. Box 63 Buea, Cameroon
e-mail: siguecyrille@gmail.com

C. Sigue · J. L. N. Mbongue · J. P. Nzenti
Laboratory of Geosciences for Internal Formations and Applications, Department of Earth Sciences, Faculty of Science, University of Yaoundé I, Messa-Yaoundé, P.O. Box 3412 Cameroon

A. Moundi
Department of Earth Sciences, University of Yaoundé I, P.O. Box 3412 Messa-Yaoundé, Cameroon

A. Vishiti
Higher Institute of Science, Engineering and Technology, Cameroon Christian University Institute, P.O. Box 5 Bali, North West Region, Cameroon

F. Rossetti et al. (eds.), *The Structural Geology Contribution to the Africa-Eurasia Geology: Basement and Reservoir Structure, Ore Mineralisation and Tectonic Modelling*,
Advances in Science, Technology & Innovation, https://doi.org/10.1007/978-3-030-01455-1_16

2 Methods of Study

The Etam shear zone was mapped in details. Representative mylonitic and quartz samples were cut and polished down to ~ 30 µm in thicknesses in preparation of thin sections in the IRGM laboratory, Nkolbisson-Yaounde. The polished samples were examined with the aid of a polarized light microscopy at the Geology laboratory, University of Buea. Petrographic studies involved the description of mineralogical features, microtectonics and alteration pattern. Structural data were displayed as stereoplots and the regional geological maps compiled to show the principal shear zone. All the data were built into a GIS platform in the Remote sensing Unit University of Buea (both field and laboratory) for easy retrieval and comparative studies.

All chemical analysis measurements were conducted by the ICP-MS technique using an X series II (Thermo) instrument, equipped with CCT (collision cell technology) in Analytical Geochemistry Laboratory, Campinas State University-Brazil.

3 Results: Texture, Mineralogy, Microstructures and Geochemistry

3.1 Wall Rock, Ductile Zone and Brittle Zone

The wall rock is a weakly deformed elongated (4 km long and 1.8 km wide) plutonic body, oriented N040E. The rock is massive, coarse grained, and made up of magmatic constituents (Kfs_1, Qtz_1, Pl_1 and Bt_1). Recrystallization produced 2% of quartz (Qtz_2), alteration of Kfs_1 produced albite and chlorite. The ductile zone consists of medium to coarse-grained protomylonite, medium grain mylonite and ultramylonites wrapped by mylonitic bands. Qtz_2 and Qtz_3, Pl_2, Bt_2 increased in volume in expense of Qtz_1, Pl_1 and Bt_1 through recrystallization. Microstructures include: Macro-fracturing, undulose extinction, Bulging (BLG) recrystallisation, Flame perthite, Myrmekitic and minerals fish. The brittle zone shows a large quartz vein (0.150×2.5 km), brecciated (Fig. 1a). Blocks of this vein are composed mainly of milky to transparent quartz clasts, <0.5 mm to 8 cm in size (Fig. 1 b and c). Microstructures are restricted to microfractures and rare Mineral small inclusion of feldspar (Fig. 1d–f).

3.2 Whole Rock Geochemistry by ICP-MS

ICP-MS analysis shows quartz composition range with 98.46–99.75 wt% SiO_2. Very low concentrations of the overall other major elements (Al_2O_3 = 0.01 to 0.15 wt%, Fe_2O_3tot = 0.1 to 0.59 wt%, MgO = 0.01 to 0.02 wt%, TiO_2 = 0.002 to 0.006 wt% Na_2O = 0.01 to 0.02, K_2O = 0.01 wt%). Trace elements and REE are under detection limit (LD), Except for: Cr (55–220 ppm), Zr(3–4 ppm), Ba (3–4 ppm), La(2–4 ppm), Ce (3–4 ppm), Pr (0.03–0.05 ppm) and Nd (0.1–0.2 ppm).

3.3 Structural Context of the Shear Zone Deformation

Quartz shows evidence of crystal-plastic deformation, while K-feldspar and plagioclase are deformed in a plastic and brittle manner. ESZ define an oblique-slip with a mylonitic schistosity, trending N10°E in the south of the shear zone although turn elsewhere to N40°E towards the north. Asymmetric boudin, C-type, C'-type and S/C structures define a sinistral strike-slip sense of shear. During brittle deformation, ESZ experienced a N060E to N065E fracture event, formed earlier in the zone follow by N120E to N140E and N100E to N110E (Fig. 1a) [9].

4 Discussion

Crystal-plastic deformation and recrystallization are associated with micro-fracturing in ESZ suggesting the greater role played by stress. Metasomatic and alteration reactions particularly released quartz and K^+, Na^+, Ca^+, Fe^+, Al^+ [3, 4]. The presence of several quartz veins in ESZ suggest

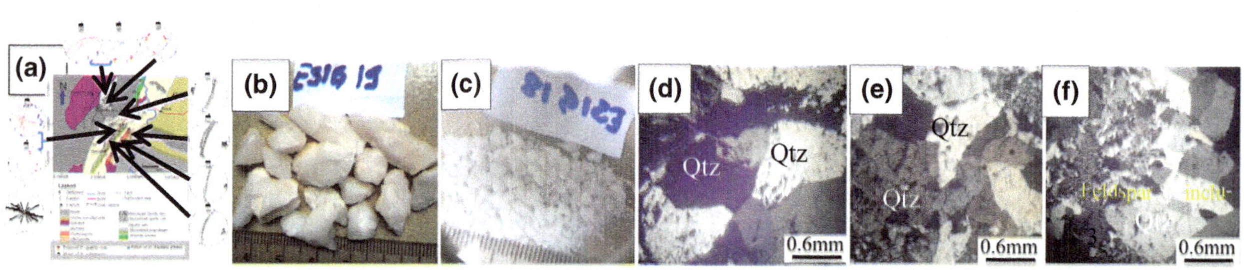

Fig. 1 **a** Geologic map of Etam, **b** and **c** photograph of Milky quartz, **d** and **e** photomicrograph of milky quartz free of mineral inclusion, **f** photomicrograph with mineral inclusion

therefore, a strong mobility of hydrothermal fluid saturated in silica. Si is dissolved during cataclasis and precipitate in strain shadow and fractures. The deformation mechanism is suggested to be similar to that of the Cameroon Central Shear Zone (CCSZ) [7, 10] and Wyangala shear zone, Australia [3]. The metamorphic conditions are similar as indicated in Tombel Graben [7]: green-schist facies with epidote, sericite, neocrystalized quartz and chlorite assemblage, low temperature grain boundaries ($\sim$300 °C), but the albitization of K-feldspar by the formation of perthite suggest a low-medium (400–500 °C) conditions metamorphism. The coexistence of such microstructures at different grades suggests that the deformation started in the presence of a residual melt and ended after the full crystallization of the rock at high temperature. The sinistral strike-slip sense of shear has recorded in the CCSZ follow by dextral shear sense [7, 11]. The N010E to N040E trend of mylonitic corridor in ESZ confirms how anastomosing is the CCSZ. Similar age (neoproterozoic) has been recorded in Panafrican fold belt in Cameroon and in the environment of the HPQ deposit is Brazil [12]. In the optical microscope, the amount of inclusions in Etam quartz is very small. Preliminary ICP-MS data shows very low amount of all elements, and high SiO_2 contain. These suggest an optimistic result of HPQ source. The only problem is probably the brittle nature of this quartz which can enhance the formation of fluid inclusions that causes the formation of bubbles during fusion processes and negatively affect the purity of the quartz [12].

5 Conclusions

We show in this study that quartz in veins are milky to transparent, with no significant minerals or fluid inclusions observed in optical microscope. This minor amount of impurities in this quartz is an indicator of their purity, which allows assessing the quartz vein as potential HPQ source in the further studies.

References

1. Ferry, J.M., Gerdes, M.L.: Chemically reactive fluid flow during metamorphism. Ann. Rev. Earth Planet. Sci. **26**, 255–287 (1998)
2. Streit, J.E., Cox, S.F.: Fluid infiltration and volume change during mid-crustal mylonitization of Proterozoic granite, King Island, Tasmania. J. Metamorph. Geol. **16**, 197–212 (1998)
3. Spruzeniece, L., Piazolo, S.: Strain localization in brittle–ductile shear zones: fluid-abundant versus fluid-limited conditions (an example from Wyangala area, Australia). Solid Earth **6**, 881–901 (2015)
4. Suh, C.E., Dada, S.S.: Fault rocks and differential reactivity of minerals in the kanawaViolaine uraniferous vein, N.E. Nigeria. J. Struct. Geol. (1997)
5. Widmer, T., Thompson, A.B.: Local origin of high pressure vein material in eclogitefacies rocks of the Zermatt-Saas zone, Switzerland. Am. J. Sci. **301**, 627–656 (2001)
6. Bons, P.D.: Development of crystal morphology during unitaxial growth in a progressively widening vein: I. The numerical model. J. Struct. Geol. **23**, 865–872 (2001)
7. Njome, M.S., Suh, C.E., Ghogomu, R.T.: A microstructural approach to interpreting the structural setting of the Tombel graben, south western Cameroon. Geo. Acta **2**, 181–200 (2003)
8. Mosoh Bambi, C.K., Frimmel, H.E., Zeh, A., Suh, C.E.: Age and origin of Pan-African granites and associated U-Mo mineralization at Ekomedion, southwestern Cameroon. J. Afr. Earth Sci. **88**, 15–37 (2013)
9. Njome, M.S., Suh, C.E.: Tectonic evolution of the Tombelgraben basement. Southwestern Cameroon Episodes, **28**, 37–41 (2005)
10. Dawaï, D., Tchameni, R., Bascou, J., Wangmene, S.A., Tchunte, P.M.F., Bouchez, J.L.: Microstructures and magnetic fabrics of the Ngaoundere granite pluton (Cameroon): Implications to the late-Pan-African evolution of Central Cameroon Shear Zone. J. Afr. Earth Sci. **129**, 887–897 (2017)
11. Ngako, V., Affaton, P., Nnange, J.M., Njanko, T.H.: Pan-African tectonic evolution in central and Southern Cameroon: transpression and transtension during sinistral shear movements. J. Afr. Earth Sc. **36**, 207–214 (2003)
12. Santos, M.F.M., Fujiwara, E., Schenkel, E.A., Enzweiler, J., Suzuki, C.K.: Quartz sand resources in the Santa Maria Eterna formation, Bahia, Brazil: a geochemical and morphological study. J. South Am. Earth Sci. **53**(62), 176–185 (2015)

Travertine-Tufa Deposition in Relation with Gafsa-Jeffara Fault System: Implication on Fluid-Flow (Southern Tunisia)

Saber Idriss, Samir Bouaziz, and Soumyajit Mukherjee

Abstract

Travertine and Tufa are continental carbonate deposits that originate from deep geothermal systems supersaturated by bicarbonates. These non-marine carbonates are characterized by complex interactions between biological, chemical and physical processes that make them interesting recorders of climatic and environmental changes. Previous works have shown that exhumed travertine affected by active tectonics could be an ideal tool in understanding the relation in a fault-fluid flow. The principal purpose of this work is to carry out for a the first time a detailed study on representative rock samples of Jebel El Mzar (Southern Tunisia, province of Kirchaou): "a thermogene travertine and tufa" related study to the Gafsa-Jeffara fault system activity which is characterized by diverse facies. Geochemical, microstructural, petrological and petrophysical studies of Travertine/Tufa samples were carried out.

Keywords

Travertine-Tufa • Active faults • Fluid flow
Petrophysical • Geochemical

1 Introduction

Travertine refers to all non-marine carbonate precipitation formed at or near terrestrial springs, rivers, lakes, and caves and divided into two categories: Meteogene and Thermogene [7]. The meteogene travertine is usually referred to as "tufa", especially when it contains remains of micro and macrophytes, invertebrates and bacteria. Thermogene deposits of CO_2 have a deeper origin as their source is either magmatic degassing or a decarbonation processes. Such deposits are typical of tectonically active areas related to active faults [4, 6]. In this work, travertine-tufa deposits are related to the Gafsa-Jeffara fault system. The Gafsa fault is an active NW-SE strike slip fault. It passes through different lithostratigraphic formations. It extends to the SE called Tunisian south accident [5] or Jeffara fault [3] (Fig. 1).

The main objective of this work is to highlight the relationships between faults and their associated features (e.g. banded calcite veins, fractures), and fluid flow (travertine deposits) and to characterize the geochemical feature of parental fluids deposited during tectonic activity.

2 Geological Setting and Methods

2.1 Geological Setting

Located in Southern Tunisia (Government of Tataouine, Province of Kirchaou), Jebel El Mzar is an isolated massif in the plain of Touazine. Consisting essentially of tufa and travertine with white banded calcite and traces of iron and manganese. Its thickness is about 50 m. The presence of vegetal print "Sabal Major" ages it to the Aquitanian era [2]). The position of this massif at the intersection of E-W and NW-SE, highlights the existence of hydrothermal activities [2] (Figs. 1, 2).

S. Idriss (✉) · S. Bouaziz
Laboratory of Water-Energy-Environment, Department of Geology, National Engineering School of Sfax, University of Sfax, Sfax, Tunisia
e-mail: saber.idriss@enis.tn

S. Bouaziz
e-mail: samir.bouaziz@enis.rnu.tn

S. Mukherjee
Geodynamic Laboratory, Department of Earth Sciences, Indian Institute of Technology, Bombay, India
e-mail: smukherjee@iitb.ac.in

F. Rossetti et al. (eds.), *The Structural Geology Contribution to the Africa-Eurasia Geology: Basement and Reservoir Structure, Ore Mineralisation and Tectonic Modelling*,
Advances in Science, Technology & Innovation, https://doi.org/10.1007/978-3-030-01455-1_17

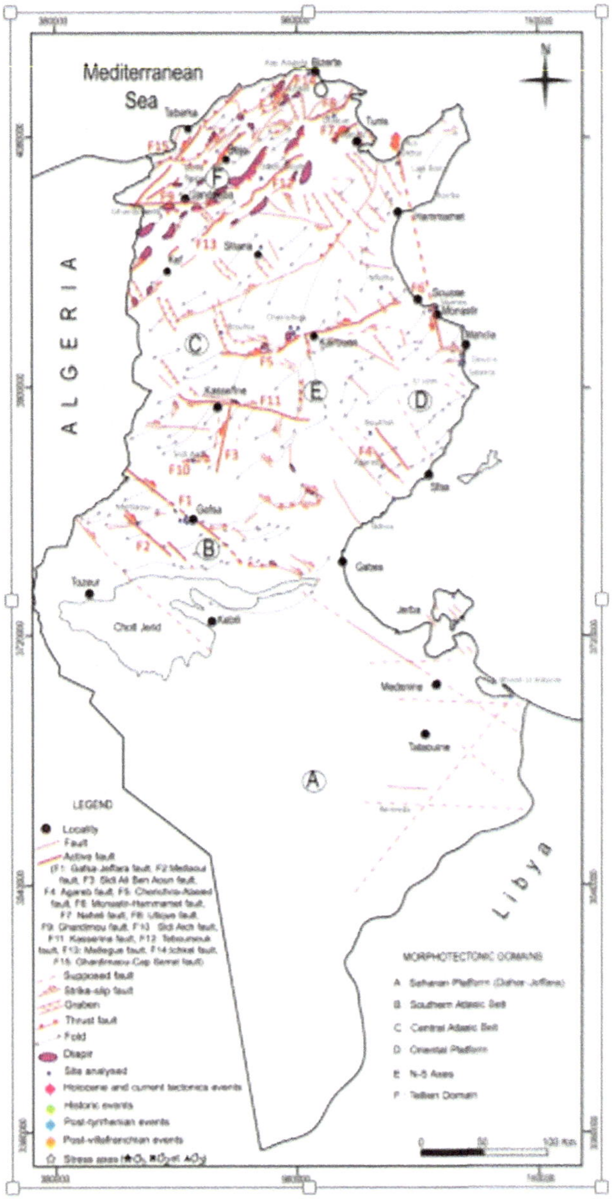

Fig. 1 Structural map of Tunisia [1]

2.2 Methods

In order to characterize these Travertine and Tufa deposits, XRD analyses were done to determine their mineralogical composition. Textural features, microfacies, diagenetic phases, and deformation microstructures were investigated by optical and scanning electron microscopes (ESEM, FEG-SEM). Trace element concentration tests were carried out based on semi quantitative analysis using X-ray fluorescence. Some samples were selected for fluid inclusion microthermometry based on the presence of sufficient number of fluid inclusions to determine the temperature and salinity of fluid precipitation. Stable C-O isotopic analysis of veins allowed the identification of the CO_2 source. Raman Spectroscopy was used for the assessment of mineral composition, fluid inclusion composition. The facies reservoir properties (porosity and permeability) were studied by 4D X-ray computer tomography (XRCT) on large slabs.

3 Results

3.1 XRF

See Table 1.

3.2 Petrography, Diagenesis, Paleontology, Deformations Microstructures, Fluid Inclusion and Raman Spectroscopy

See Fig. 3.

3.3 X-Ray Computed Tomography (XRCT)

See Fig. 4.

4 Conclusion

The mineralogical compositions of the total travertine samples mainly consists of calcite with varying amounts of quartz (i.e., detrital quartz). Fluid inclusion and Raman spectroscopy studies show that three-phase inclusion present H_2O (l), CO_2 (l) and CO_2 (g), varying in size, shape and chronology.

Fieldwork combined with a microstructural deformation study highlighted a brittle deformation due to the presence of micro-faults, cracks, joints, veining, stylolites, deformation twins etc. This confirms that the travertine-tufa deposition in Jebel El Mazar is highly controled by the fault activity and that the calcite veins act as feeding conduits.

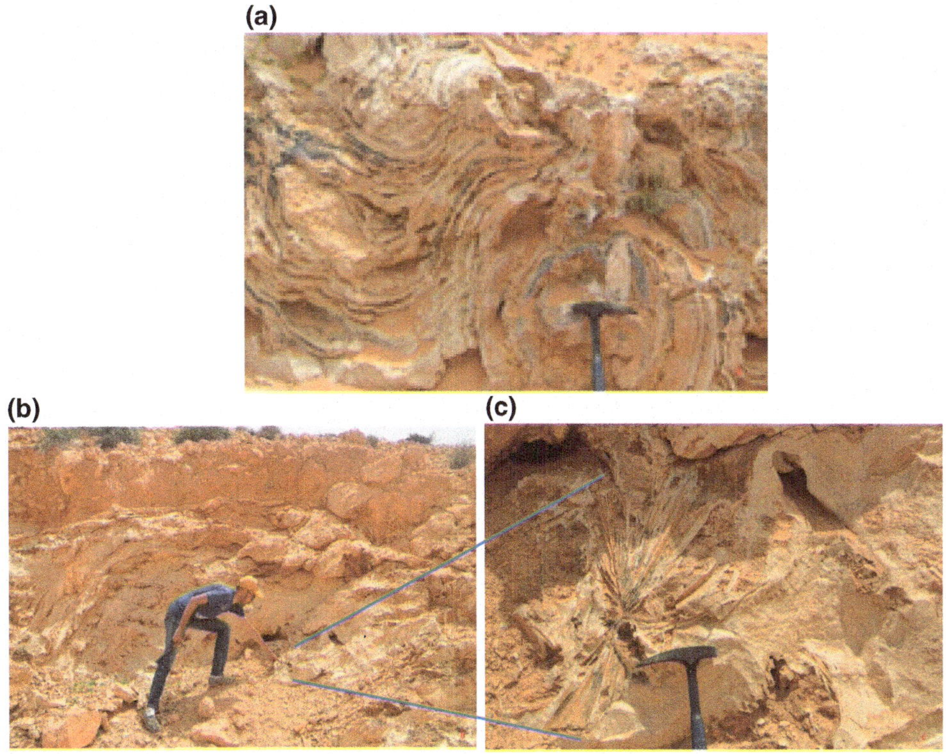

Fig. 2 **a** Circular central veining surrounded by undulated banded veins developed in Consequence of repeated episodes of Carbonate deposition within the same fracture playing as conduits, **b** and **c**: 'Sabal major': Aquitanian (Miocene) in Age

Table 1 X-ray fluorescence results of major and minor elements of sample 6 (wt%)

Na_2O	MgO	SiO_2	P_2O_5	Cl	K_2O	SO_3	Al_2O_3
0.0928	0.437	6.28	0.0596	0.0938	0.268	0.161	2.72
CaO	TiO_2	MnO	Fe_2O_3	NiO	CuO	ZnO	SrO
87	0.191	0.160	2.50	0.0107	0.0104	0.0072	0.0287

Petrography combined with a petrophysical study show different facies associations with dominant components of micrite, sparite and microsparite, and a variety of porosity types: inter and intra particle, channel, fenestral, vugs, moldic, shelter etc. The presence of fossil fauna and flora, and diagenetic processes (cementation, dissolution, recrystallization, fracturing etc.) was also revealed by the aforementioned study.

XRCT provided informations about the reservoir characteristics of travertine-tufa (porosity, permeability, interconnectivity etc.). The comparison between the XRCT results of different samples showed the impact of diagenesis and microstructure deformation on porosity and permeability (e.g., Stylolitisation causes decreases in the porosity, permeability).

The presence of fauna, flora, and the repetitive alternation of colors confirm that the travertine-tufa deposition is controlled by the biota and climate.

The travertine-tufa of Jebel El Mazar is an exhumed hydrothermal system and could be a reservoir analog for active hydrothermal systems and a key for future hydrothermal exploration.

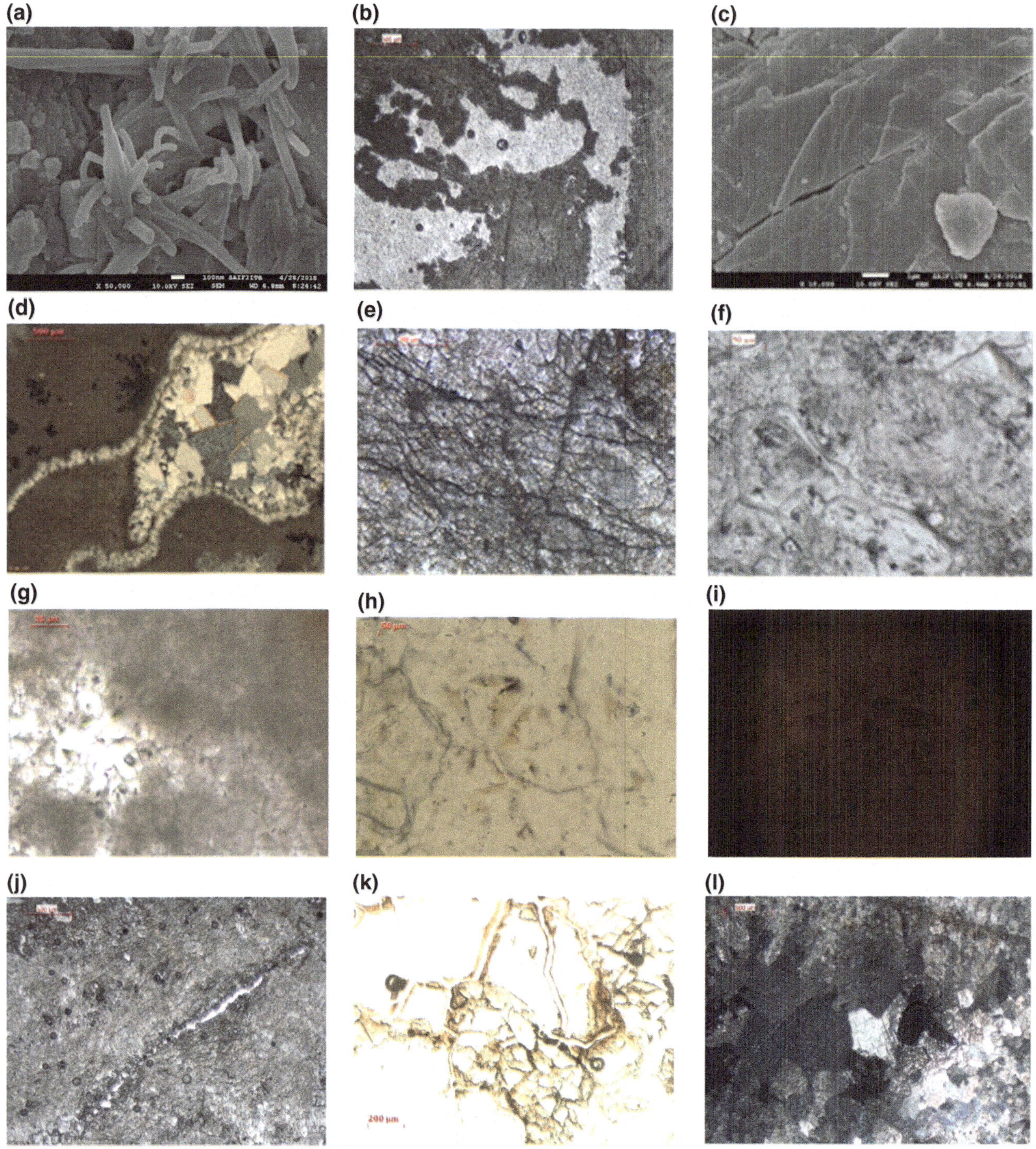

Fig. 3 **a** A SEM image of Shrub Travertine displaying bacterial rods encased within calcite crystals. **b** Micrograph showing fenestral fabric with elongated spar-filled pores in micritic sediment (PPL). **c** SEM image showing the geometry and shear sense of Riedel Y-S shears. **d** Micrograph showing calcite cement (the size of the crystals increases away from the initial substrate of the sparry mosaic, The intercristalline boundaries in the mosaic are made up of plane interfaces and a high percentage of enfacial junctions among the triple junctions (XPL). **e** Thin-section photomicrograph showing solution seam stylolite with residue accumulation. **f**, **g**, **h** Examples of Fluid inclusion in Calcite (Vapour-Liquid inclusion), **i** Photomicrographs of Fluid inclusion in Calcite, during Raman Spectrometry analysis. **j** A thin section of photomicrograph showing stylolite-associated porosity (PPL). **k** A thin section of photomicrograph showing inter-particle, intra-partice and channel porosities (PPL). **l** A thin section of photomicrograph showing deformation twins in calcite (XPL)

Fig. 4 **a** Picture shows the sample 8, cut into 2 cm cube for XRCT scanning. **b** Sample 8 in three-dimensionality illustrate fabrics, pores, and overall structure. **c** XRCT image gives information about the pore shapes, volumes and connections. *Note* darker gray (or Black) shades represent lower optical densities (porous phase) and lighter gray shades represent the solid phase

References

1. Bahrouni, N., Bouaziz, S., Soumaya, A., Ben Ayed, N., Attafi, K., Houla, Y., El Ghali, A., Rebai, N.: Neotectonic and seismotectonic investigation of seismically active regions in Tunisia: a multidisciplinary approach (2013)
2. Bouaziz, S., Ben Salem, H.: Carte Geol. Tunisie 1:100000, Feuille Kirchaou. Office Nation. des Mines (1990)
3. Bouaziz, S.: Etude de la tectonique cassante dans la plateforme et l'Atlas saharien (Tunisie méridionale); évolution des paléochamps de contraintes et implications géodynamiques. Thèse Doc. Etat, Univ. Tunis II, Tunisia, 485 p. (1995)
4. Brogi, A., Alçiçek, M.C., Yalçıner, C.Ç., Capezzuoli, E., Liotta, D., Meccheri, M., Rimondi, V., Ruggieri, G., Gandin, A., Boschi, C., Büyüksaraç, A., Alçiçek, H., Bülbül, A., Baykara, M.O., Shen, C.-C.: Hydrothermal fluids circulation and travertine deposition in an active tectonic setting: insights from the Kamara geothermal area. Western Anatolia, Turkey (2016)
5. Castany, G.: L'accident Sud tunisien, son âge et ses relations avec l'accident Sud atlasique d'Algérie. C. R. Acad. Sci. Paris **238**, 916–918 (1954)
6. Henchiri, M., Ben Ahmed, W., Brogi, A., Cihat Alçiçek, M., Benassi, R.: Evolution of Pleistocene travertine depositional system from terraced slope to fissure-ridge in a mixed travertine-alluvial succession. Jebel El Mida, Gafsa, southern Tunisia (2017)
7. Pentecost, A.: Travertine, 1st ed. Springer, London, UK,; ISBN 978-1-4020-3606-4 (2005)

The Tsavorite (Gem Quality Vanadium Grossularite) Minerogenetic System in East Africa

Walter L. Pohl

Abstract

When Supercontinent Rodinia began to break up at ~800 Ma, wide rifts formed evolving into a passive margin of the Mozambique Ocean. At one stage, organic-rich pelites with gypsum (the later host rock of tsavorites) were interbedded with siliciclastics, carbonates, and bimodal volcanic rocks. Ocean closure, subduction, collision and welding of various microcontinents and volcanic arcs formed the Mozambique Belt and at ~550 Ma, ended with the assembly of Supercontinent Gondwana. In East Africa, deformation and metamorphism climaxed at 615–600 Ma Fritz et al. (J. Afr. Earth Sci. 86:65–106, 2013 [1]). At peak metamorphic conditions, anatexis of siliciclastic metasediments produced abundant metamorphic mobilizates such as pegmatites and quartz veins. In the black schist hosting tsavorites, however, partial melting only produced small fractions of a supercritical hydrous liquid. Most of this liquid remained within the host but migrated over short distances into structural traps of different origin; turbid green grossular crystallized from the melt and transparent tsavorite from exsolved fluids.

Keywords

East Africa • Mozambique belt • Metasediments • Anhydrite • Graphite • Supercritical hydrous melt • Lateral secretion

1 Introduction

Modern metallogenetic and minerogenetic research aims at understanding the whole source—transport—trap system. Source regions are rarely accessible and identification mostly rests on geochemical models. Here, we describe a case where partial melting secreted from high-grade metamorphic rocks. Remaining within the source lithology it migrated short distances into low-pressure structural traps where gemstones formed. The aim of this paper is to establish the minerogenetic system for tsavorite (gem quality vanadium grossularite) deposits to be applied to exploration and mining.

2 Methods

We modify the template of the metallogenetic or mineral systems analysis (MSA), which is a holistic genetic investigation identifying all factors and processes that contribute to the origin of a mineral deposit, similar to petroleum systems [2]. In practice, mineral systems analysis can be conceived as the modern high-technology application of metallogeny by employing GIS and probability science. For exploration, target models are based on the underlying mineralization processes and their mappable features. The critical processes acting together to form a hydrothermal ore deposit, for example, may include (a) the establishment of an energy gradient to drive the system (e.g. intrusions); (b) the generation of hydrothermal fluids (e.g. dehydration); (c) the extraction of metals and chemical ligands for metal complexation from suitable sources (anomalous geochemistry); (d) the transport of fluids and metals from source regions to traps (e.g. effective flow channels such as transcrustal structures); (e) deposition of metals triggered by chemical and physical processes that affect fluids migrating through traps; and (f) the preservation of mineral deposits through time.

3 Results

3.1 Results: Setting of Tsavorite Gems

In the Mozambique Belt in SE Kenya near Mwatate [3], and in neighboring Tanzania [4], a Neoproterozoic high-grade

W. L. Pohl (✉)
Austrian Academy of Sciences, Dr. Ignaz Seipel-Platz 2, 1010 Vienna, Austria
e-mail: ecogeo13@gmail.com

F. Rossetti et al. (eds.), *The Structural Geology Contribution to the Africa-Eurasia Geology: Basement and Reservoir Structure, Ore Mineralisation and Tectonic Modelling*,
Advances in Science, Technology & Innovation, https://doi.org/10.1007/978-3-030-01455-1_18

metasedimentary suite hosts gemstone grossular, in the trade called tsavorite. The gems are emerald green grossulars (GGr) colored by reduced V^{3+} and Cr^{3+} and characterized by the formula $[Ca_3\ (Al, V, Cr)_2(SiO_4)_3]$ [5]. Country rocks of GGr are thinly banded calcareous, pelitic, evaporitic and graphitic metasediments, that are interbedded with calcite marble, paragneiss, quartzites, felsic and mafic metavolcanics [3, 6]. The immediate host rocks of GGr in SE Kenya are calcsilicate-graphite-sillimanite-muscovite schists.

3.2 Results: A Minerogenetic System for Tsavorite (Gem Quality Vanadium Grossularite) Deposits

(1) Deposition of the tsavorite source rocks took place in an evaporitic mixed-carbonate-siliciclastic stratigraphy; the setting may have been similar to modern continental sabkhas/playas [7]. Generally, continental sabkhas are part of a system ranging from alluvial fans of the basin margin to ephemeral or perennial salt lakes. Boron isotopes validate the terrestrial setting in East Africa [8]. Nodular or enterolithic anhydrite, dolomitization, structureless algal peat, and cyanobacterial mats characterize evaporitic beds in playas. In many of the East African tsavorite mines, spherical or ellipsoidal bodies of graphitic diopside fels of different diameters (10–100 cm) mark the tsavorite horizons [3]. Most likely, the spheres were early diagenetic dolomite concretions within calcareous organic-rich mud beds.

(2) After major deformation and at peak metamorphic P/T ($\sim 800 \pm 50$ °C and 10 ± 1 kbar) [9], the GGr host rocks recrystallized. Dehydration caused partial anatexis that produced a supercritical hydrous silicate melt (liquid I), the water content of which reached 50%, as a function of the fraction of hydrous minerals. Minerals crystallizing from this melt include macroscopic turbid green grossular, quartz, calcite, and graphite (often in large bladed books). In turbid green grossular, microscopic melt inclusions (MIs) consist of diopside, anorthite, scapolite-Ca, trilithionite and tiny grains of titanite, lanthanite, fersmite, and zircon, as well as graphite, sulfides and sulfosalts such as patronite, violarite, hessite, tennantite, and arzakite [10].Other MIs contain little silicate and water, but much native S and H_2S; they trapped a liquid (II) that unmixed from the first. At high temperatures (> 650°C) such systems are highly reactive and corrosive [11]. The reduced sulfur is believed to be derived from anhydrite in the rock [8]. Clearly, the supercritical, reduced and acidic liquids dissolved elements from the dehydrating, melting black shale, including all constituents needed to form grossular and tsavorite.

(3) Turbid green grossular that contains MIs of liquid I and liquid II side by side suggests that unmixing took place during the migration of the melt through the rock matrix following a down-pressure path toward traps, similar to the lateral secretion model of vein mineralization. Often, traps are fissures in pressure shadows near spherical calc-silicate bodies; elsewhere, faults locate the gemstones. At this stage, the tectonic background is one of generally mild compressive strain. Veins and pockets display a paragenesis of quartz, calcite, pyrite, massive green vanadium grossular and finally, gem quality transparent tsavorite. The latter crystallized from co-existing sulfurous liquid II [8] and an aqueous fluid that is characterized by moderate Na_2CO_3 and low NaCl concentrations; fluid inclusions contain natron and REE carbonate daughters [10].

4 Discussion

The current genetic model of tsavorite gemstones needs to be amended and revised. This model proposes that tsavorites crystallized from metamorphic devolatilization fluids [4, 5, 8, 12]. Based on recently published data [10], it is clear that the main stages of green grossularite and tsavorite gem formation include: (i) deposition of gypsiferous organic-rich mudrock; (ii) high T/P metamorphic segregation of anatectic supercritical corrosive melt; the distance of migration probably being measured in metres only; and (iii) the exsolution of an aqueous and an immiscible sulfurous fluid while fissures opened providing space for growth of idiomorphic gems.

Exploration and mining guides resulting from the new comprehension of the minerogenetic system may support a transition from artisanal to modern operations: In SE Kenya, the metapelitic-evaporitic Lualenyi Member extends over a distance of ca. 100 km, punctuated by about two dozen gemstone mines [12]. In between, tsavorite prospectivity is high. The prospective graphite horizon(s) could be located by airborne electromagnetic surveys at high resolution. Gypsum after anhydrite, probably an indicator mineral, can be detected by airborne hyperspectral mapping, or on the ground, by a portable infrared field spectrometer. Concretions and their pressure shadows being preferred gem sites, the distribution of the tough spheres within softer graphite schist could be mapped by 3D seismic imaging. In exploration and in mining, complex folding of the graphite horizon [3] and a possible lateral change of sedimentary facies [7] must be accounted for. Mining might be more efficient if open pits are established in order to expose the host bed and to reduce extraction costs.

5 Conclusions

Similar to novel hydrocarbon resources hosted in tight rock, tsavorite gems are found in metapelitic source beds of a peculiar evaporitic organic-rich composition. Moderate synchronous strain near peak metamorphic P and T created low-pressure traps that attracted segregating supercritical hydrous, reduced and acidic anatectic melt, from which turbid green grossular was precipitated. Grossular gems (tsavorites) formed from hydrous and sulfurous fluids that exsolved from the melt. The processing system was metamorphogenic lateral secretion [13].

References

1. Fritz, H., Abdelsalam, M., Ali, K.A., et al.: Orogen styles in the East African Orogen: a review of the Neoproterozoic to Cambrian tectonic evolution. J. Afr. Earth Sci. **86**, 65–106 (2013)
2. Hagemann, S.G., Lisitsin, V.A., Huston, D.L.: Mineral system analysis: quo Vadis. Ore Geol. Rev. **76**, 504–522 (2016)
3. Pohl, W.L., Nauta, W.J., Niedermayr, G.: Geology of the Mwatate Quadrangle and the Vanadium Grossularite Deposits of the Area (with a Geological Map 1:50,000). Kenya Geol Survey Report No. 101, 1–55, Nairobi (1979)
4. Feneyrol, J., Giuliani, G., Demaiffe, D., et al.: Age and origin of the tsavorite and tanzanite mineralizing fluids in the Neoproterozoic Mozambique Metamorphic Belt. Can. Mineral. **55**, 763–786 (2017)
5. Feneyrol, J., Giuliani, G., Ohnenstetter, D., et al.: Worldwide tsavorite deposits: new aspects and perspectives. Ore Geol. Rev. **53**, 1–25 (2013)
6. Pohl, W.L., Horkel, A.: Notes on the geology and mineral resources of the Mtito Andei-Taita area (Southern Kenya). Mitt. Oesterr. Geol. Ges. **73**, 135–152 (1980)
7. Warren, J.K.: Sabkhas, saline mudflats and pans. In: Warren, J.K. (ed) Evaporites: sediments, resources and hydrocarbons, chapter 3, pp. 139–220. Springer, Berlin–Heidelberg (2006)
8. Giuliani, G., Dubessy, J., Ohnenstetter, D., et al.: The role of evaporites in the formation of gems during metamorphism of carbonate platforms: a review. Miner. Deposita. **53**, 1–20 (2018)
9. Jacob, J.-B., Martelat, J.-E., Goncalves, P., Giuliani, G., et al.: New P-T-X conditions for the formation of gem tsavorite garnet in the Voi area (southwestern Kenya). Lithos. **320–321**, 250–264 (2018)
10. Thomas, R., Rericha A., Pohl, W.L., Davidson, P.: Genetic significance of the 867 cm^{-1} out-of-plane Raman mode in graphite associated with V-bearing green grossular. Mineralogy and Petrology (online) https://doi.org/10.1007/s00710-018-0563-1 (2018)
11. Gorbaty, Y.E., Bondarenko, G.V.: The physical state of supercritical fluids. J. Supercrit Fluid **14**, 1–8 (1998)
12. Martelat, J.-E., Paquette, J.-L., Bosse, V., et al.: Chronological constraints on tsavorite mineralizations and related metamorphic episodes in Southeast Kenya. Can. Mineral. **55**, 845–865 (2017)
13. Pohl, W.L.: Economic Geology, Principles and Practice: Metals, Minerals, Coal and Hydrocarbons—An Introduction to Formation and Sustainable Exploitation of Mineral Deposits. Wiley, Oxford (2011)

Sedimentary Structure Interpretation, Grain Size Analysis, and Stochastic Modeling of P_2O_5 Grades of Kef Essennoun Phosphorite Deposit (Northeast Algeria): Implication for Paleogeographic Setting and Paleoprocesses

Mohamed Dassamiour, Hamid Mezghache, Otmane Raji, and Jean-Louis Bodinier

Abstract

The Kef Essennoun phosphorite (northeast Algeria) is a layer with an average thickness of 35 m, dipping 10–15° southwards and bounded to the west by the Ain Fouris shoal. On the outcrop, the deposit shows internal layering forming large-scale, high-angle cross bedding. In order to investigate the phosphorite depositional and paleogeographic settings, grain size analysis and Sequential Gaussian Simulation of P_2O_5 content from 29 drill holes were carried out. Grain size analysis reveals the transport and reworking of phosphates by paleocurrents. The simulation indicates the elongation of the phosphate formation in the direction N63°, parallel to the Ain Fouris shoal. These results are consistent with sedimentological and oceanographic processes involving transport and migration of phosphates as a submarine dune, as well as interaction of the paleocurrents carrying phosphates with the Ain Fouris shoal.

Keywords

Grain size analysis • Sequential gaussian simulation
Phosphorite deposition • Submarine dune
Ain Fouris shoal

M. Dassamiour (✉)
Department of Earth Sciences, Institute of Architecture and Geosciences, Ferhat Abbas Setif 1 University, 19000 Setif, Algeria
e-mail: dassamiour.m@univ-setif.dz

H. Mezghache
Department of Geology, Faculty of Earth Sciences, Badji Mokhtar - Annaba University, 23000 Annaba, Algeria

O. Raji · J.-L. Bodinier
Geology and Sustainable Mining, Mohammed VI Polytechnic University, Benguerir, Morocco

J.-L. Bodinier
Géosciences Montpellier, Université de Montpellier, 34095 Montpellier, France

1 Introduction

Sedimentary structures formed during or after the deposition, represent manifestations of the physical and biological processes that operate in depositional environments, where physical properties and fluid mechanics control sediment transport and deposition. Stochastic modeling has become increasingly applied in geoscience to simulate a variety of structures, allowing visualization and better understanding of geological processes.

The Kef Essennoun phosphorite deposit, belonging to Djebel Onk Basin, is located in the northeast of Algeria (Fig. 1). The deposit occurs as a layer forming a monocline with an average thickness of 35 m and dipping 10°–15° southwards. The paleogeographic map for the Ypresian period shows that the Kef Essennoun deposit was situated southwest of the Kasserine Island and east of the Ain Fouris shoal, elongated northeast—southwest [1, 2].

On the outcrop, the Kef Essennoun phosphorite shows a peculiar sedimentary structure related to poorly understood depositional conditions. This paper aims to investigate the depositional and paleogeographic settings using grain size analysis and Sequential Gaussian Simulation (SGS) of the P_2O_5 grade distribution.

2 Materials and Methods

A total of 23 samples, denoted S-2 to S-24, were collected on the Kef Essennoun quarry front, within the principal phosphate layer (from bottom to top) and perpendicularly with a step of 1 m using a channel sampling method (1 m length, 20 cm width and 10 cm thick for each). Each sample of about 500 g was oven dried at 80 °C for three hours, and then grains were liberated by attrition without fragmentation. The grain size was measured by standard sieves (ranged

F. Rossetti et al. (eds.), *The Structural Geology Contribution to the Africa-Eurasia Geology: Basement and Reservoir Structure, Ore Mineralisation and Tectonic Modelling*,
Advances in Science, Technology & Innovation, https://doi.org/10.1007/978-3-030-01455-1_19

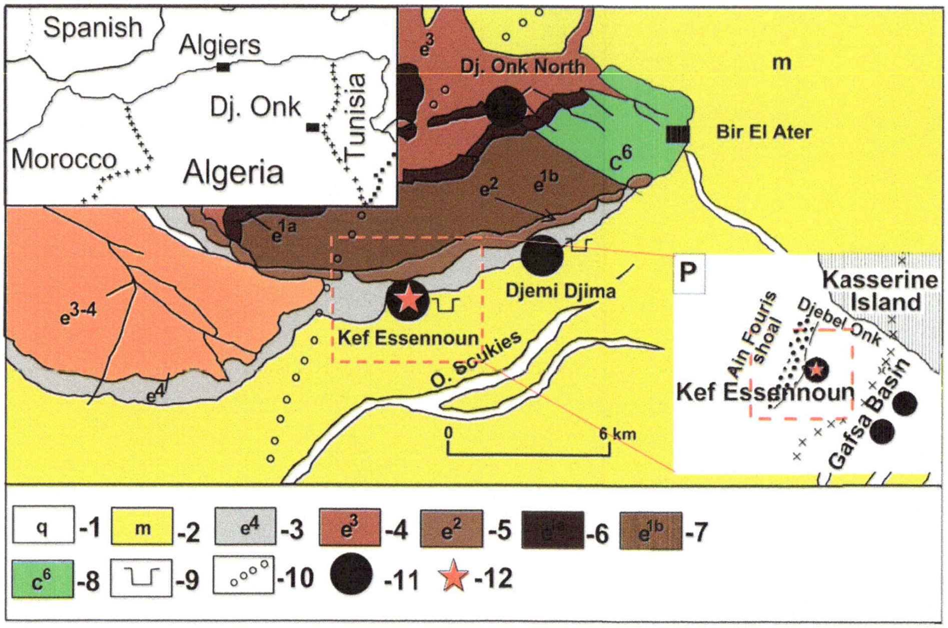

Fig. 1 Schematic geological map and location of Kef Essennoun deposit [3]. 1—Silty alluvial deposits (Quaternary); 2—Sand and clay (Miocene); 3—Marl and limestone (Lutetian); 4—Marl and limestone with Phosphate level (Ypresian); 5—Marl, limestone, dolomite and main phosphate layer (Thanetian); 6—Coquina limestone and marl (Montian); 7—Marl and limestone (Danian); 8—Limestone (Maastrichtian); 9—Open pit; 10—Ain Fouris shoal; 11—Phosphate deposits; 12—Sampling positions; P—Part of the paleogeographic map (Ypresian) [1]

from 4 mm to 0.063 mm) with automatic sieve shakers for 15 min. Grain-size parameters were calculated according to Folk and Ward (1957) formulae using GRADISTAT v8 and data are presented as weight % using φ grain size units. The cumulative probability curves of Visher [4] and the Passega C-M diagram [5] were used to determine transport modes.

Sequential Gaussian Simulation (SGS) of P_2O_5 grades was used to obtain a 3D model of the phosphorite deposit. The P_2O_5 content data from 29 vertical drillholes were determined on 961 drill-core of 1 m each using spectrophotometric method (SEAL Analytical Technicon AutoAnalyzer II). The simulation grid was chosen to be 100 m by 50 m laterally and 10 m vertically. One hundred realizations of P_2O_5 spatial distribution were generated using SGS.

3 Results

3.1 Grain Size Analysis

Based on the results of grain size analysis, the Kef Essennoun phosphorites reveal slight variability. They range from poorly to moderately fine and medium sands. The hydrodynamic partition includes traction (average 4.70%), and suspension (average 31.42%), but saltation was the dominated transport mode (average 80.88%). Most samples in the Passega C-M diagram plot in the Q-P field, with only two samples in the P-O field. This indicates predominant transport under a suspension regime with some rolling and subordinate transport by rolling with a contribution of suspension.

3.2 Stochastic Modeling of P_2O_5 Grades

The preliminary step in SGS is the variographic analysis. Experimental variograms of P_2O_5 grades calculated in various directions with a tolerance angle of 20° were fitted with an exponential (for N45° and N90°) and a spherical (for N135°) model. SGS model shows the arrangement of the phosphorite formation into elongating structure with an average direction of N63°, with a length of 2800 m and a width of 1380 m.

4 Discussion

Based on grain size analysis the studied phosphorites are sands characterized by predominant saltation transport mode with some graded suspension and saltation. This pattern is indicative of paleocurrent intensity fluctuations. The Passega C-M diagram confirms these results and indicates an environment characterized by high-energy conditions. SGS of P_2O_5 grades suggests the arrangement of the phosphorite formation into a northeast trending elongated structure.

Evidence such as (1) the sandy texture of the phosphorites, (2) the shallow marine environment (3) the large-scale cross stratification representing dune forest bed, (4) the elongated structure determined by SGS of P_2O_5 grades, (5) the predominance of saltation during phosphate material transport, and (6) the cross stratification dip indicating a NE flow direction, parallel to the Ain Fouris shoal elongation, suggest a submarine dune pattern for the transport and migration of the phosphorites. As many studies reflect that the dynamics of subaqueous dunes inferred environment of high-energy conditions [6, 7].

5 Conclusions

In this study, sedimentary structure interpretation, grain size analysis, and SGS are used to investigate the depositional and paleoenvironment settings of Kef Essennoun phosphorites, in northeast Algeria. The deposit of about 35 m thick was bounded to the west by the Ain Fouris shoal. The phosphate material is classified as fine to medium sand. The hydrodynamic partition shows that the saltation was the dominant transport mode. The interaction of the paleocurrents with the Ain Fouris shoal topography delineated the transport direction. In addition, oceanographic conditions such as high deposition energy, seafloor morphology, and availability of phosphate material of a specific grain size range have led to the transport of these phosphorites as a submarine dune. The shoal formed a trap for the deposition of the phosphorites in a depression on its slope. As a result, the deposit developed as an elongated layer, parallel to the shoal elongation.

References

1. Chaabani, F.: Dynamique de la partie orientale du bassin de Gafsa au Crétacé et au Paléogène: Étude minéralogique et géochimique de la série phosphatée éocène (Tunisie méridionale). Ph.D. thesis, University of Tunis II,Tunisia (1995)
2. Dass Amiour, M., Mezghache, H., Elouadi, B.: The use of three physico-chemical methods in the study of the organic matter associated with the sedimentary phosphorites in Djebel Onk Basin, Algeria. Arabian Journal of Geosciences **6**(2), 309–319 (2013)
3. Kassatkine, Y., Yahyaoui, A., Chatilov, S.: The works of prospecting and assessment on phosphate executed in 1976–1978 in the mining district of Djebel Onk. Internal report, SONAREM (Société Nationale de Recherche et d'Exploration Minière), Algeria (1980)
4. Visher, G.S.: Grain size distributions and depositional processes. J. Sediment. Res. **39**, 1074–1106 (1969)
5. Passega, R.: Grain size representation by CM patterns as a geological tool. J. Sediment. Petrol. **34**, 830–847 (1964)
6. Besio, G., Blondeaux, P., Brocchini, M., Vittori, G.: On the modeling of sand wave migration. J. Geophys. Res. **109**(C04018), 1–13 (2004)
7. Ashley, G.M.: Classification of large scale subaqueous bedforms: a new look at an old problem. J. Sediment. Petrol. **60**, 160–172 (1990)

Part IV

Reservoir Geology, Structure and Stratigraphy

Clay Minerals in Fluvial Shaly Sandstone of Upper Triassic Reservoir (TAGS)—Toual Field SE Algeria: Identification from Wireline Logs and Core Data

El Hadi Mazouz, Messaoud Hamimed, Abdelouahab Yahiaoui, Mohamed Said Benzagouta, Mohamed Khodja, Nada Achi, and Joëlle Duplay

Abstract

The present investigation attempts to highlight clay minerals existing in fluvial shaly sandstones belonging to the Upper Triassic reservoir (Rhaetian) using the French abbreviation "TAGS" for Trias argilo-gréseux supérieur. This reservoir is localized in the Toual Field SE in Algeria. In absence of core samples, three types of clay minerals are detected by wireline log analysis and core data (porosity and permeability measurements). The Thomas and Stieber petrophysical model (shale volume vs. total porosity) was applied to differentiate allogenic clay (laminated and structural clays) from authigenic clay (dispersed clay). This model clearly shows that dispersed clay type has a leading presence in the reservoir's sandstone. The Neasham petrophysical model (porosity versus permeability) allowed us to find that authigenic illite (a dispersed type of clay) is dominant in TAGS sandstones. This model also shows that permeability is highly influenced by authigenic illite whereas porosity is less affected.

Keywords

Clay minerals • TAGS • Toual field • Upper triassic

1 Introduction

The French abbreviation TAGS for "Trias argilo-gréseux supérieur" is the upper Triassic (Rhaetian) reservoir well known in the Saharan platform. It is located in the Toual field (covering about 5×10 Km^2) in the south-east of Algeria. 25 wells have been identified across the reservoir from 1962 till present time. TAGS structure is represented by a SSW-NNE oriented anticline resulting from the Alpine orogeny [1]. With an average of a hundred of meters in thickness, TAGS is divided into two main parts: lower TAGS composed of fluvial channels deposits, and upper TAGS which consist of sand lenses incorporated in thick muddy intervals attributed to shallow marine sabkha. The TAGS reservoir lower seal (*Trias Carbonaté*) and upper seal (*Lias argileux and salifère*) are both considered as shallow marine deposits—sabkha [1–3]. Rare papers have investigated the diagenesis features of TAGS, only information about the depositional environment and general reservoir information are found. The presence of illite and halite in TAGS sandstone as authigenic minerals is indicated by Mazouz et al. [4]. In order to help bridge this gap, this paper contributes, using other methods, to clarify the existence of authigenic clay in TAGS sandstone.

2 Methods

Usually rock minerals are identified by microscopic techniques. In the absence of physical samples, wireline logs and core data are made use of for this purpose. This technique has given good results in many surveys [5–10].

Three main clay types are distinguished in the sandstone: laminated and structural clays are commonly allogenic where dispersed clay is authigenic [11, 12].

To recognize clay types, present in TAGS reservoir sandstone, we tried to apply the Thomas and Stieber model

E. H. Mazouz (✉) · M. S. Benzagouta · N. Achi
University of Larbi Ben M'Hidi, 04000 Oum-El-Bouaghi, Algeria
e-mail: emazouz@yahoo.fr

M. Hamimed
Larbi Tebessi University, 12000 Tebessa, Algeria

A. Yahiaoui
Mostapha Benboulaid (Batna-2) University, 05000 Batna, Algeria

M. Khodja
Sonatrach Research Department, 16000 Hydra, Algiers, Algeria

J. Duplay
LHyGes Strasbourg University, Strasbourg, France

F. Rossetti et al. (eds.), *The Structural Geology Contribution to the Africa-Eurasia Geology: Basement and Reservoir Structure, Ore Mineralisation and Tectonic Modelling*, Advances in Science, Technology & Innovation, https://doi.org/10.1007/978-3-030-01455-1_20

[13] (Fig. 1, middle). This model consists of a cross-plot of Shale volume (Vsh) versus Total porosity (Φ_T).

Based on available data from only two wells (well14 and well15, Fig. 2, left), application of the above cited model is given in Fig. (1, left and right). The following parameters are calculated or taken from logs (Figs. 3 and 4) before proceeding to the model application (Table 1):

Φ_s: Clean sandstone porosity is the maximal porosity in the studied interval, it is taken directly from porosity log Φ_T (Track 3, Figs. 3 and 4, red curve). Φ_{sh}: Shale porosity, is read from Φ_T log (in front of the highest value of V_{sh}: Track 3, Figs. 3 and 4, green curve). The cutoff used for V_{sh} is 50%.

Neasham model [14] (Fig. 2, middle) is used to distinguish between three sub-types of dispersed clays and their relation with the quality of the reservoir. The used data were porosity and permeability core laboratory measurements (Φ & K).

3 Results

The Thomas & Stieber model [13], applied for TAGS sandstones, is shown in Fig. 1(left and right), it allows us to conclude that dispersed clays are dominant in the TAGS reservoir.

Based on core porosity and core permeability of two wells (well8 and well9, Fig. 2 left) a cross plot is built as shown in Fig. 2 (right); Neasham [14] divides dispersed (authigenic) clays into three mineral types (Fig. 2, middle):

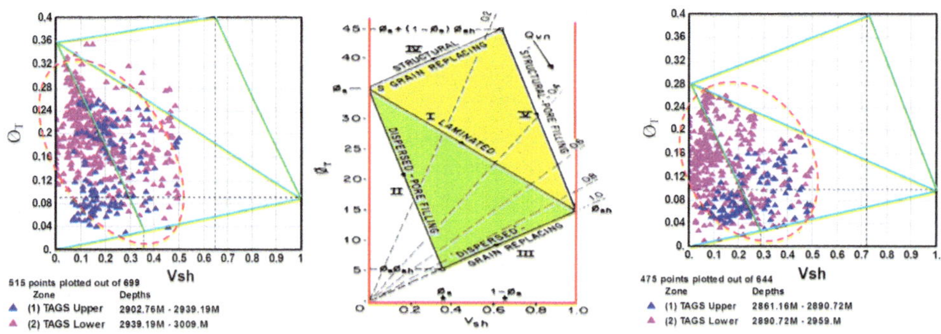

Fig. 1 Thomas and Stieber original model [13] modified from [15] (middle cross-plot) and application model for TAGS sandstones in well14 and well15 (left and right cross-plots)

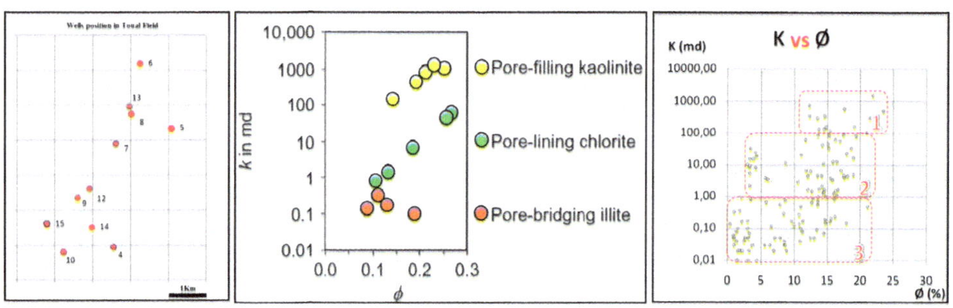

Fig. 2 Toual field wells position (left map), Dispersed clays types modified from Neasham model [8] (middle cross-plot) and application model for TAGS sandstones (right cross-plot)

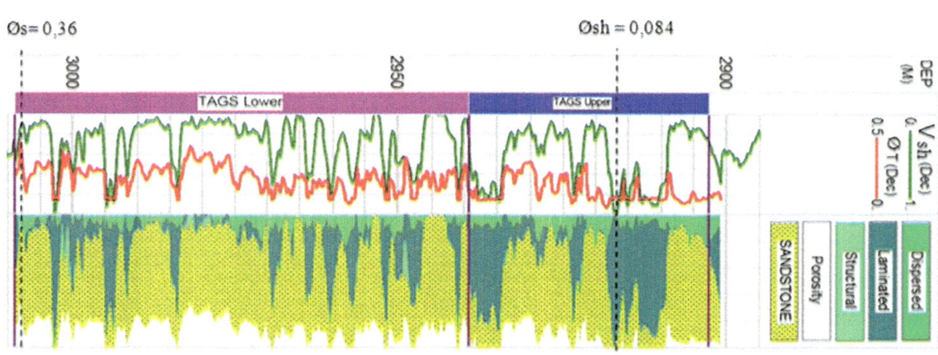

Fig. 3 Composite log of clay types of TAGS reservoir (Well14)

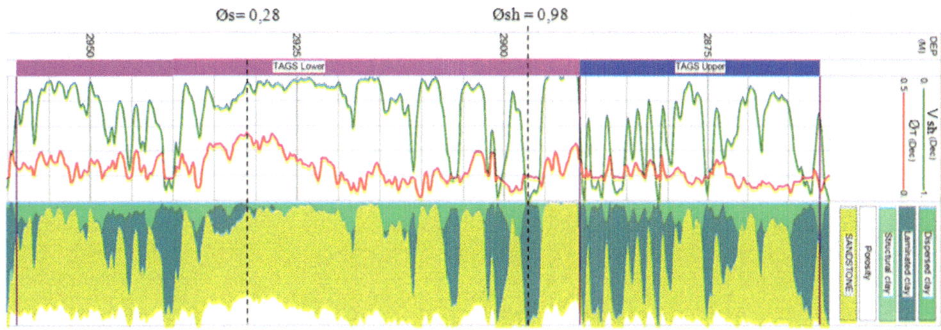

Fig. 4 Composite log of clay types of TAGS reservoir (Well15)

Table 1 Parameters used to build Thomas and Stieber model

Parameters	Well 14	Well 15
Φ_s	0.360	0.280
Φ_{sh}	0.084	0.098
$1-\Phi_s$	0.650	0.720
$\Phi_s + (1-\Phi_s) \times \Phi_{sh}$	0.405	0.350
$\Phi_s \times \Phi_{sh}$	0.030	0.027

(1) Pore-filling clay (mainly kaolinite),
(2) Pore-lining clay (mainly chlorite),
(3) Pore-bridging clay (mainly illite).

Figure 2 (right) shows three zones representing three types of dispersed clays, the most populated zone is the third one (Illite zone).

The composite log of well14 and well15 shown respectively in Figs. 3 (Track 4) and 4 (Track 4) confirms the presence of dispersed clay (authigenic) in TAGS sandstone. The Laminated clay type is abundant and structural clay is rare. Clean sandstone porosity (Φ_s) and Shale porosity (Φ_{sh}) used in the Thomas and Stieber model are mentioned on same Figs. (3 and 4).

4 Discussion

The Thomas and Stieber model [13] applied for TAGS sandstone indicates that dispersed clays are the main type of clay present in this reservoir. Structural clay type is occasional whereas laminated clay is presented using detrital clays sheets which act as secondary seals in TAGS reservoir.

Dispersed clays are represented mainly by illite and secondarily chlorite. Illite, indicated in the Neasham model [14] (Fig. 2, left, zone 3), greatly influences the reservoir's permeability ($0.01 < K < 1$ md). Porosity ($0\% < \Phi < 21\%$) is relatively less influenced by this clay type when it comes to permeability.

5 Conclusions

The investigation of authigenic clay minerals without physical samples (direct data) is possible but it is not an easy issue. The present work attempts to extract maximum information from indirect data (well logs and core measurements). Authigenic clay minerals are mainly identified as dispersed clays (illite). The TAGS reservoir quality deterioration by illite was confirmed. Permeability is extremely reduced by illite. Porosity reduction is less observed. Laminated detrital clay sheets acting as secondary seals inside TAGS reservoir can reduce vertical permeability. The presence of laminated clay acting as barriers or baffles to vertical flows cannot be confirmed by the present study.

References

1. Askri, H., Belmeheri, A., Benrabah, B., Boudjema, A., Boumendjel, K., Daoudi, M., Drid, M., Ghalem, T., Docca, A.M., Ghandriche, H., Ghomari, A., Guellati, N., Khennous, M., Lounici, R., Naili, H., Takherist, D., Terkmani, M.: Geology of Algeria, pp. 1–93. WECSH&SLB, Algiers (1995)
2. Zeroug, S., Bounoua, N. Lounissi, R., Zeghouani, R., Djellas, N., Kartobi, K., Tchambaz, M., Abadir, S, Simon, P., Fuller, J.: Petroleum geology of Algeria, pp. 1–489. WECSH&SLB, Algiers (2007)
3. Galeazzi, S., Point, O., Haddadi, N., Mather, J., Druesne, D.: Regional geology and petroleum systems of the Illizi-Berkine area of the Algerian saharan platform: an overview. Mar. Pet. Geol. **27**, 143–178 (2010)

4. Mazouz, E., Hamimed, M., Yahiaoui, A., El-Ghali, M.A.K.: Prediction of diagenesis and reservoir quality using wireline logs: Evidence from the Upper Triassic (Raethian) fluvial reservoir TAGS—Toual field, GassiTouil area, SE Algeria. J. Fundam. Appl. Sci. **9**(2), 808–828 (2017)
5. Allen, J.R.: Prediction of permeability from logs by multiple regression. Trans, SPWLA (1979)
6. Wendt, W.A., Sakurai, S., and Nelson, P.H.: Permeability prediction from well logs using multiple regression. In: Lake, L. W., Carroll, H.B. Jr. (eds.) Reservoir Characterization, pp. 181–222. New York City (1986)
7. Bloch, S.: Empirical prediction of porosity and permeability in sandstones. AAPGBulletin. **75**(7), 1145–1160 (1991)
8. Yan, J.: Reservoir parameters estimation from well log and core data: a case study from the North Sea. Petrol. Geosci. **8**, 63–69 (2002)
9. Chow, J.J., Li, M.-C., Fuh, S.: Geophysical well log study on the paleoenvironment of the hydrocarbon producing zones in the Erchungchi Formation. Hsinyin, SW Taiwan. TAO, **16**(3), 531–543, (2005)
10. Lai, J., Wang, G., Chai, Y., Ran, Y.: Prediction of Diagenetic facies using well logs: evidences from upper triassic yanchang formation: chang 8 sandstones in jiyuan region, ordos basin, china oil & gas science and technology—Rev. IFP Energies Nouvelles **71**, 34 (2016)
11. Heslop, A.: Porosity in shaly-sands. In: SPWLA Sixteenth annual logging symposium, 1975, Imperial Oil Limited Calgary, Alberta, 1975
12. Serra, O.: Diagraphies Différés base de l'interprétation, Mémoire1, Tome1. Services techniques Schlumberger, Paris (1979)
13. Thomas, E.C., Stieber, S.J.: The distribution of shale in sandstones and its effect upon porosity. In: SPWLA 16th Annual Logging Symposium 1975, 15p 1975
14. Neasham, J.W.: The morphology of dispersed clay in sandstone reservoirs and its effect on sandstone shaliness, pore space, and fluid flow properties. SPE-6858 (1977)
15. Juhasz, I.: Normalised QV—The key to shaly sand evaluation using the Waxman-Smits equation in the absence of core data. In: SPWLA 22nd Annual Logging Symposium, Shell International Petroleum Maatschappij, Netherlands, 1981

The Paleozoic Geology of Saudi Arabia: History, Tectono-Stratigraphy, Glaciations, and Natural Resources

Abdulaziz A. Laboun

Abstract

Siliciclastics, carbonates and minor evaporites of fluvial, eolian, glacial, and shallow and deep marine environments representing all of the Paleozoic systems are well exposed and present in Saudi Arabia. The Paleozoic successions are exposed in a great curved belt along the northern, eastern, and southern margins of the Arabian Shield in three main basins: the Tabuk, Widyan, and Wajid basins, respectively. The basins are separated by the Hail Arch and the Central Arabian Arch. Two Paleozoic glacial episodes: The Late Ordovician glaciation (Saqiyah, Sarah, and Hawban formations in the Widyan and Tabuk basins and the Sanamah Formation in the Wajid Basin) and Permo-Carboniferous glaciation (the Juwayl Formation in the Wajid Basin), are well defined and documented. Detailed regional lithologic, paleontologic, and structural field studies have resulted in dividing the succession into seven well marked mega tectono-depositional cycles: the Cambro-Ordovician (Tayma Group), Late Ordovician (Tabuk Group), Early Silurian (Qalibah Group), Late Silurian Sharawra (Group), Siluro-Devonian Huj (Group), Devono-Carboniferous Sakaka (Group), and Permo-Carboniferous Buraydah (Group). These cycles or groups are separated by regional well- defined unconformities, probably caused by global time-equivalent movements: the Assyntic, Taconic, early and maximum Acadian, and early and maximum Hercynian movements. The Paleozoic rocks are becoming a primary exploration target and their tectono-stratigraphy has therefore attracted considerable attention. Porous and permeable siliciclastic and carbonate rocks host great quantities of oil, condensate, and natural gas in several oil and gas fields in Saudi Arabia.

Keywords

Tectono-stratigraphy • Mineral resources • Arabia

A. A. Laboun (✉)
King Saud University, Riyadh, Saudi Arabia
e-mail: ibnlaboun@yahoo.com

1 Paleozoic Rocks

Cambrian to Permian deposits are exposed in a great curved belt along the northern, eastern, and southern margins of the Arabian Shield in three main basins: the Tabuk, Widyan, and Wajid basins, respectively (Fig. 1). The Tabuk and Widyan basins are separated by the Hail arch and the Wajid Basin is located south of the Central Arabian Arch. These rocks are becoming a primary exploration target and their tectono-stratigraphy has therefore attracted considerable attention. Porous and permeable siliciclastic and carbonate rocks host great quantities of oil, condensate, and natural gas in several oil and gas fields in Saudi Arabia.

2 Stratigraphy

The Paleozoic succession of the Greater Arabian Basin in the Arabian Plate, in general and in Saudi Arabia in particular is composed mainly of siliciclastics, carbonates, and minor evaporites divided into seven meg-tectonostratigraphic cycles (Fig. 2).

All Paleozoic systems, except the Carboniferous, are represented in well exposed sections in Saudi Arabia. However, the Carboniferous sediments are encountered in the subsurface. The Paleozoic beds thin onto paleohighs and effect of tectonic movements. As such, depositional facies' changes are more pronounced towards the Arabian Shield and basement arches such as the Hail Arch and central the Arabian Arch.

3 Tectono-Stratigraphic Cycles

The Paleozoic succession is severely affected by the series of major tectonic activities. Detailed regional lithologic, paleontologic, and structural field studies have resulted in dividing the succession into seven well marked mega tectono-depositional cycles. These cycles are well preserved

F. Rossetti et al. (eds.), *The Structural Geology Contribution to the Africa-Eurasia Geology: Basement and Reservoir Structure, Ore Mineralisation and Tectonic Modelling*,
Advances in Science, Technology & Innovation, https://doi.org/10.1007/978-3-030-01455-1_21

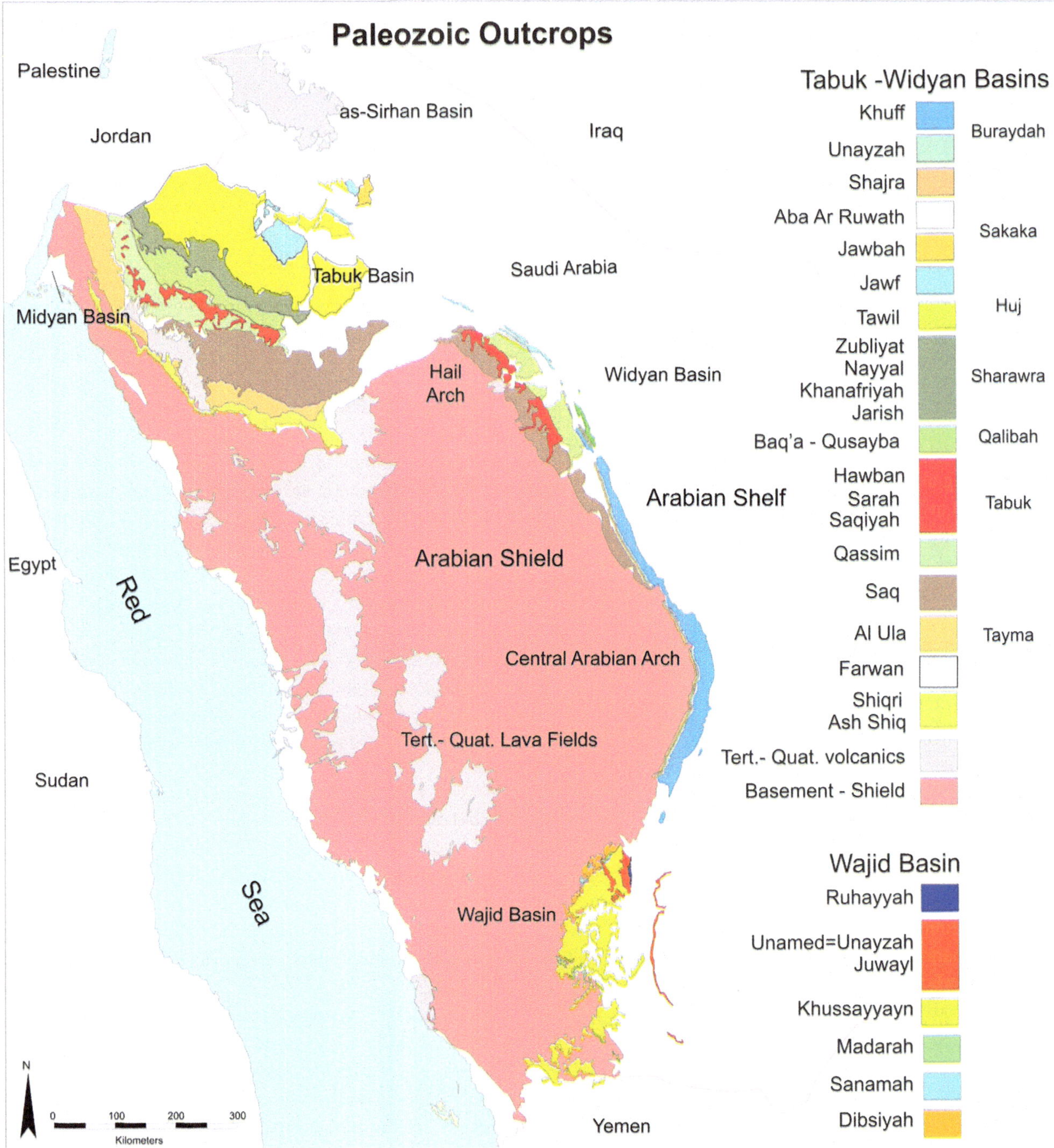

Fig. 1 Paleozoic rocks are exposed in a great curved belt along the margin of the Arabian Shield in three main basins; Tabuk, Widyan, and Wajid basins

in less tectonically affected areas where boundaries between respective cycles are marked by unconformities. In more tectonically active areas, successions are less preserved and boundaries between cycles are more complicated due to longer periods of erosion and the influence of later movements and/or non-deposition. These pronounced and regional unconformities are probably time-equivalent of major global tectonic movements such as the Assyntic, the Taconic, the Early Acadian, the Acadian, the Early Hercynian, and the Hercynian events. The lithology of the groups is extending without significant facies change into the subsurface. The tectono-stratigraphic divisions are (Fig. 2).

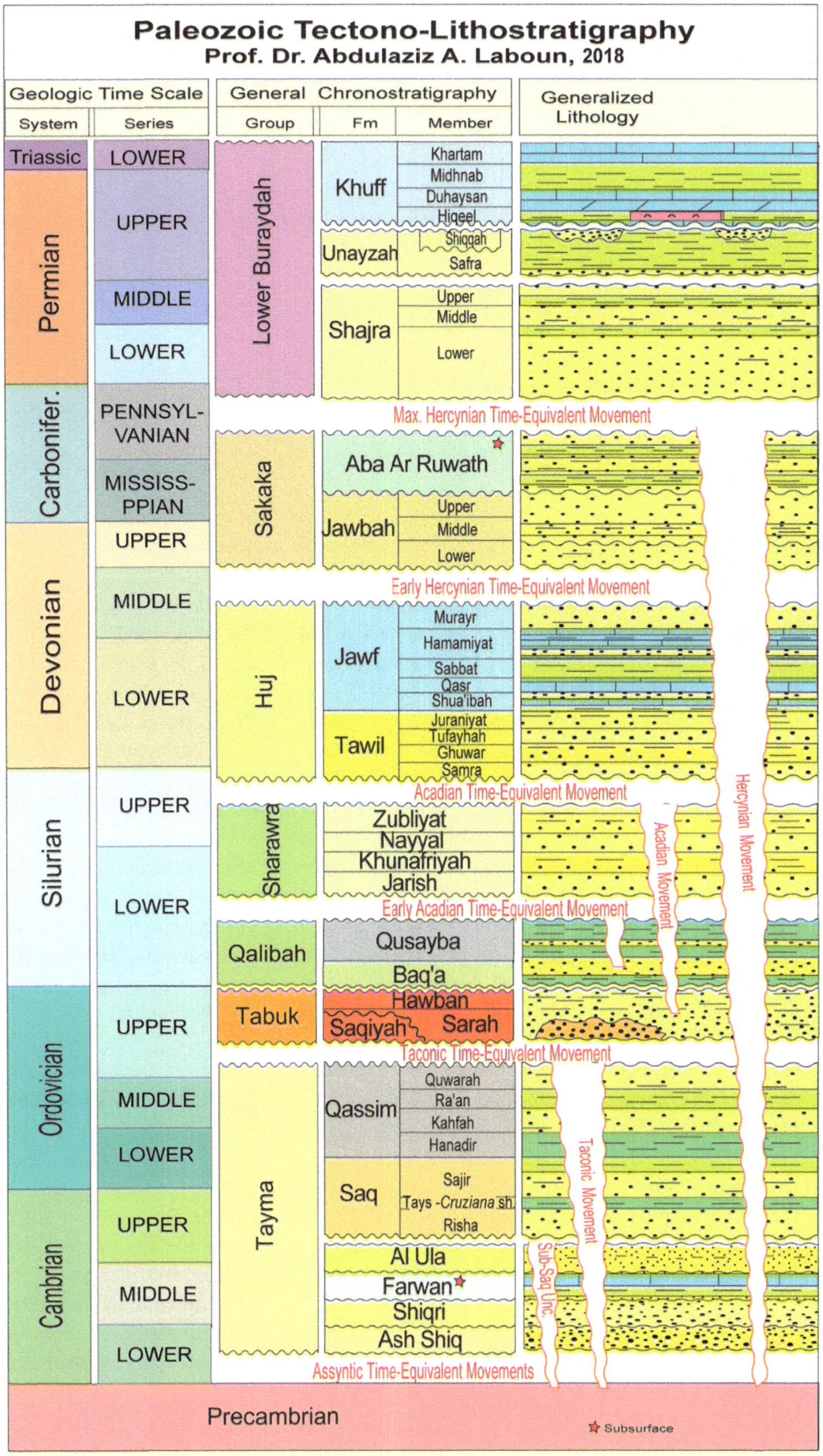

Fig. 2 Chrono-lithostratigraphic column of the Paleozoic of the Tabuk Basin (northwestern) and Widyan Basin (central) Saudi Arabia

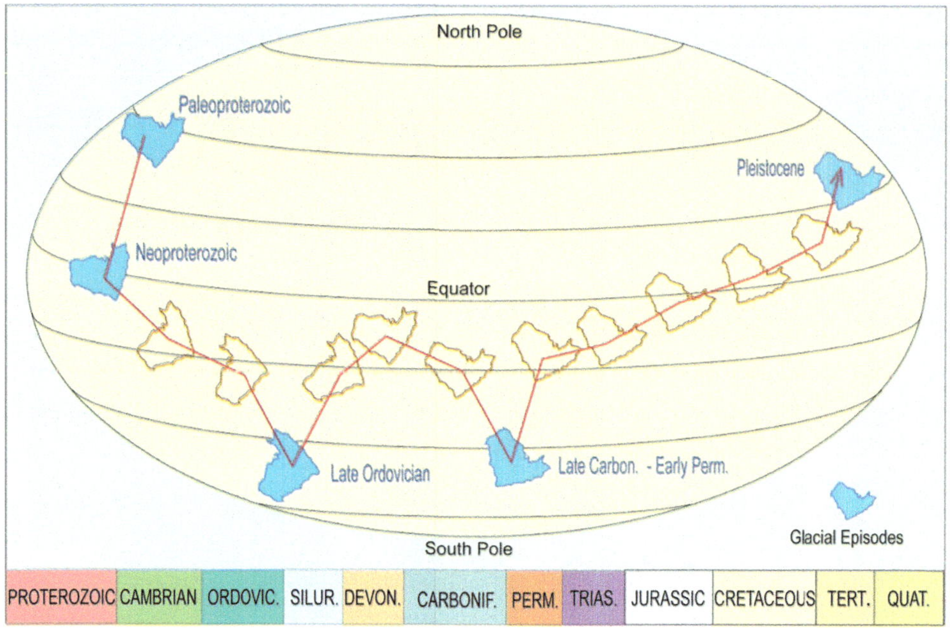

Fig. 3 Glacial episodes of Saudi Arabia

1. Cambro-Ordovician: continental and shallow marine deposits, represented by the Tayma Group (Ash Shiq, Shiqri, Farwan (subsurface [1], Al Ula, Saq, and Qassim formations) in the Tabuk Basin and only by the Saq and Qassim formations in the Widyan Basin.
2. Late Ordovician: glacial siliciclastics, the Tabuk Group (Saqiyah, Sarah, and Hawban formations) in the Tabuk and Widyan basins.
3. Early Silurian: marine and deltaic siliciclastics, the Qalibah Group (herein redefined) (Baq'a and Qusayba formations) in the Tabuk and Widyan basins.
4. Late Silurian: deltaic and shallow marine siliciclastic, the Sharawra Group (herein raised to group status) in the Tabuk Basin (Jarish, Khunafriyah, Nayyal, and Zubliyat formations and herein raised to formations).
5. Siluro-Devonian: continental and marine siliciclastics and carbonates, the Huj Group (Tawil and Jawf formations) in the Tabuk and Widyan basins.
6. Devono-Carboniferous: continental siliciclastics, Sakaka Group [2] (Jawbah and Aba Ar Ruwath formations) in the Tabuk Basin.
7. Carboniferous-Permian: continental and marine siliciclastics and carbonates, the Lower Buraydah Group (Shajra [3], Unayzah, and Khuff formations) in the Widyan Basin.

4 Glacial Episodes

The Arabian Peninsula is one of the few regions in the world where sediments of at least five main global glacial episodes are well preserved and exposed in the Arabian Shield and Arabian Shelf (Fig. 3). Two main episodes occurred during the Proterozoic, two in the Paleozoic, and one in the Cenozoic. Permo-Carboniferous glaciation is represented by the Juwayl Formation of the Fard al-Ban Group [4], in the Wajid Basin. Late Ordovician, Hirnantian glaciation is represented by the Tabuk Group (Saqiyah [4], Sarah, and Hawban formations) in the Tabuk and Widyan basins and the Sanamah Formation in the Wajid Basin, and Pleistocene glaciation is represented by the Midyan Formation [5], in the Midyan Basin (Fig. 1). Permeable glacio-aeolian, glacio-fan, glacio-fluvial, subglacial, as well as proglacial and glacio-lacustrine deposits are important hydrocarbon exploration targets. The sandstones of the Sarah Formation are found to be oil and gas reservoirs in many fields in Saudi Arabia (Fig. 4).

5 Natural Resources

The Paleozoic rocks host enormous amounts of natural resources. They include all elements of the hydrocarbon system, developed throughout the succession. Organic rich shales (mainly the "Hot shale" of the Qusayba Formation), porous and permeable sandstones (Unayzah Reservoir: Unayzah and Shajra formations), carbonates (Khuff Formation), impermeable beds such as shales and evaporites, and different types of structural and stratigraphic traps (Fig. 4) are among the resources found. Commercial oil and gas have been discovered in Paleozoic reservoirs in more than 30 fields in Saudi Arabia. Ground water has been discovered in sandstones and limestones aquifers in different parts of the Arabian Shelf.

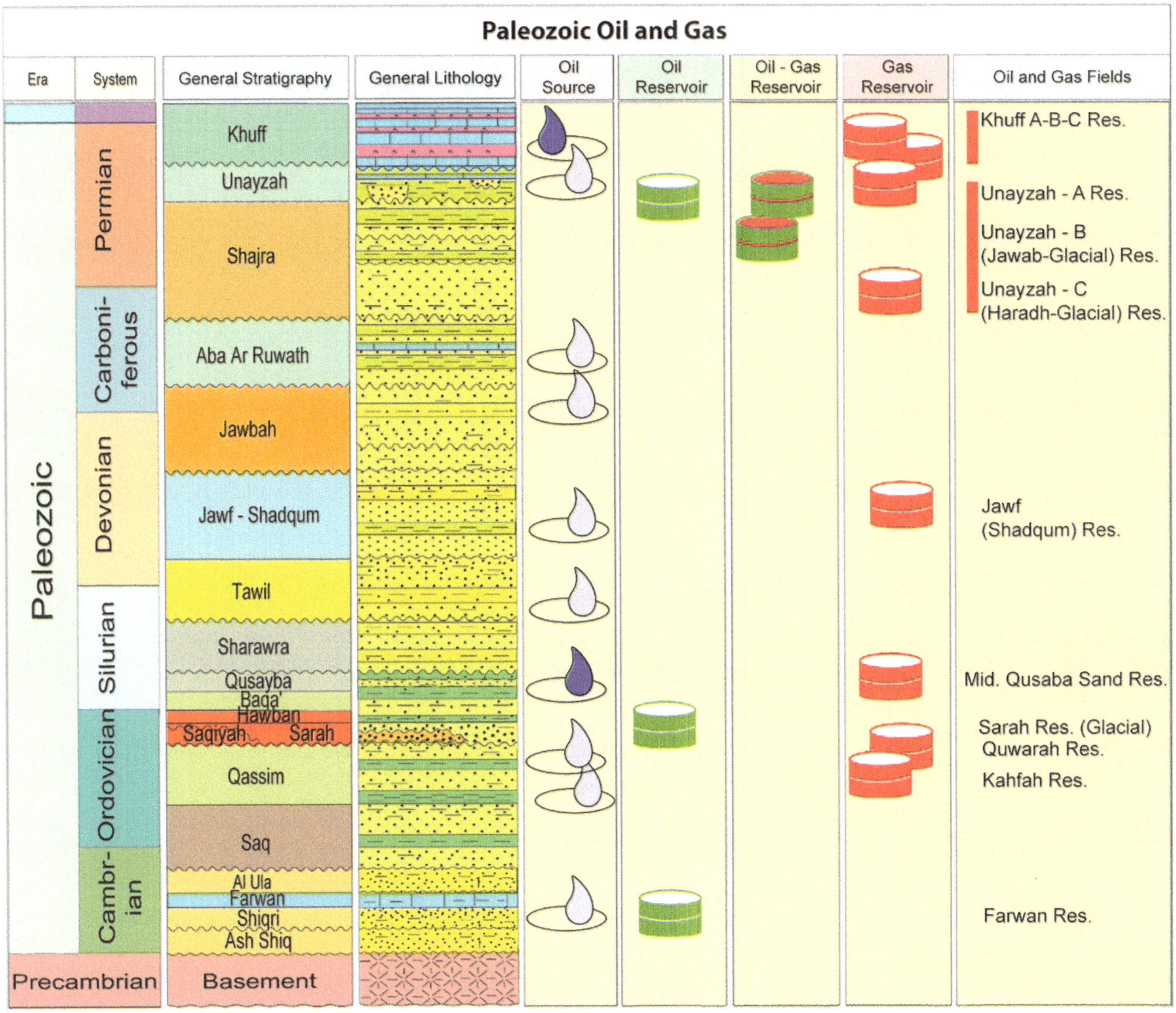

Fig. 4 Oil and gas of the Paleozoic of Saudi Arabia

References

1. Laboun, A. A.: Lexicon of the Paleozoic and Lower Mesozoic of Saudi Arabia (Part-1: Lithostratigraphic Units, Nomenclature Review), Al-Hudhud Publishers, Box 63280 Riyadh 11516 (Introduction and definition of the Farwan Formation). 1993
2. Laboun, A.A.: The Paleozoic History of Saudi Arabia (Abstract), Saudi Society for Geosciences, 12th International Geosciences Meeting. Jeddah, Saudi Arabia (2018)
3. Laboun, A.A.: The Shajra Formation: a new permo-carboniferous unit in Saudi Arabia. J. Saudi Arabia Oil and Gas Issue **17** (Formal introduction and definition of the Shajra Formation) (2010)
4. Laboun, A.A.: Saqiyah glacial facies: a new late ordovician glacial related units (Abstract), 3rd meeting of the Saudi Society for Geosciences, October 15–17, Riyadh (Introduction and definition of Saqiyah Formation) (1996)
5. Laboun, A.A.: Did glaciers exist during pleistocene in the midyan region, northwest corner of the Arabian Peninsula? Arab., Arab. J. Geosci. **5**(6), 1333–1339 (First discovery of Pleistocene glaciation in KSA, and introduction the Midyan Formation)

Paleozoic Reservoir Distribution in South-Eastern Tunisia

Maroua Ghalgaoui, Noomen Dkhaili, Kawthar Sbei, and Mohamed Hedi Inoubli

Abstract

2D-seismic reflection profiles and borehole data, in the studied area of approximately 18,850 Km^2 and reaching depths between 1200 and 4000 m, have been investigated to characterize the Southeastern Tunisia during the Paleozoic period. This study aims to enhance the geologic understing of the area by analysing the Paleozoic distribution. Three main unconformities were indentified in this area: unconformity A, between the Ordovician and Silurian series (Probably the Taconic unconformity); unconformity B, between the Silurian and Triassic series resulting from the Hercyninian Orogeney; and unconformity C, separating the Permian from Triassic deposits. The seismic interpretation has confirmed the existence of several features dominated by NW–SE trends associated with horst and graben structures. These tectonic elements have an important role in the structural and stratigraphic traps.

Keywords

Paleozoic • Unconformity • 2D-seismic • Tunisia

1 Introcuction

The study area is located to the Southeast of Tunisia and covers the regions of Tataouine and Medenine (Fig. 1). It is limited to the North by the sea shoreline, to the West by the region of Kebelli, to the East by the Libyan territory and to the South by the desert.

M. Ghalgaoui (✉) · K. Sbei · M. H. Inoubli
Research Unit of Applied Geophysics (URGAMM), Faculty of Sciences of Tunis, University Tunis-El-Manar, 2092 Tunis, Tunisia
e-mail: marwa.ghalgaoui@etudiant-fst.utm.tn

N. Dkhaili
ETAP Entreprise Tunisienne D'Activités Pétrolières, 1002 Tunis, Tunsia

The Jeffara—Dahar domain is characterized by a complex tectonic history, since it consists of two provinces with different geologic features [1, 4] with a series of compressional and extensional tectonic events [1, 2]. The most important of these has been two phases of the Herycynian orogeny during late Paleozoic [3].

This work is mainly based on the litho-stratigraphic description, series correlation and depth maps elaboration of the encountered Formations of available drilled wells. The Ordovician deposits consisting of sandstones interceded by shaly layers, constitute the primary target as referred to by several companies. In contrast, the sandstone and carbonate layers of Cambrian, Late Silurian, Late Carboniferous and Permian Formations, where they exist, constitute a potential secondary target. All cited Paleozoic reservoirs are sourced by the radioactive hot shale layer at the base of the Early Silurian Tannezuft Formation.

2 Data and Methods

Data used in this work consists of more than twenty petroleum wells and 2D seismic-reflection profiles distributed in the Jeffara—Dahar region. Well data is principally the final geological report with some associated data (log data) provided by the Tunisian oil company (ETAP) covering many blocks in Southern Tunisia. The work is focused on describing the major crossed geological units and synthetic lithostratigraphic chart and inter-well correlations establishment. Depth maps at the top of geological formations have also been elaborated.

3 Results and Discussions

The established regional chart (Fig. 2), is mainly based on analyzing and interpreting the lithostratigraphic column of drilled wells. It reveals the existence of many unconformities and sedimentary gaps which characterize most of the

F. Rossetti et al. (eds.), *The Structural Geology Contribution to the Africa-Eurasia Geology: Basement and Reservoir Structure, Ore Mineralisation and Tectonic Modelling*,
Advances in Science, Technology & Innovation, https://doi.org/10.1007/978-3-030-01455-1_22

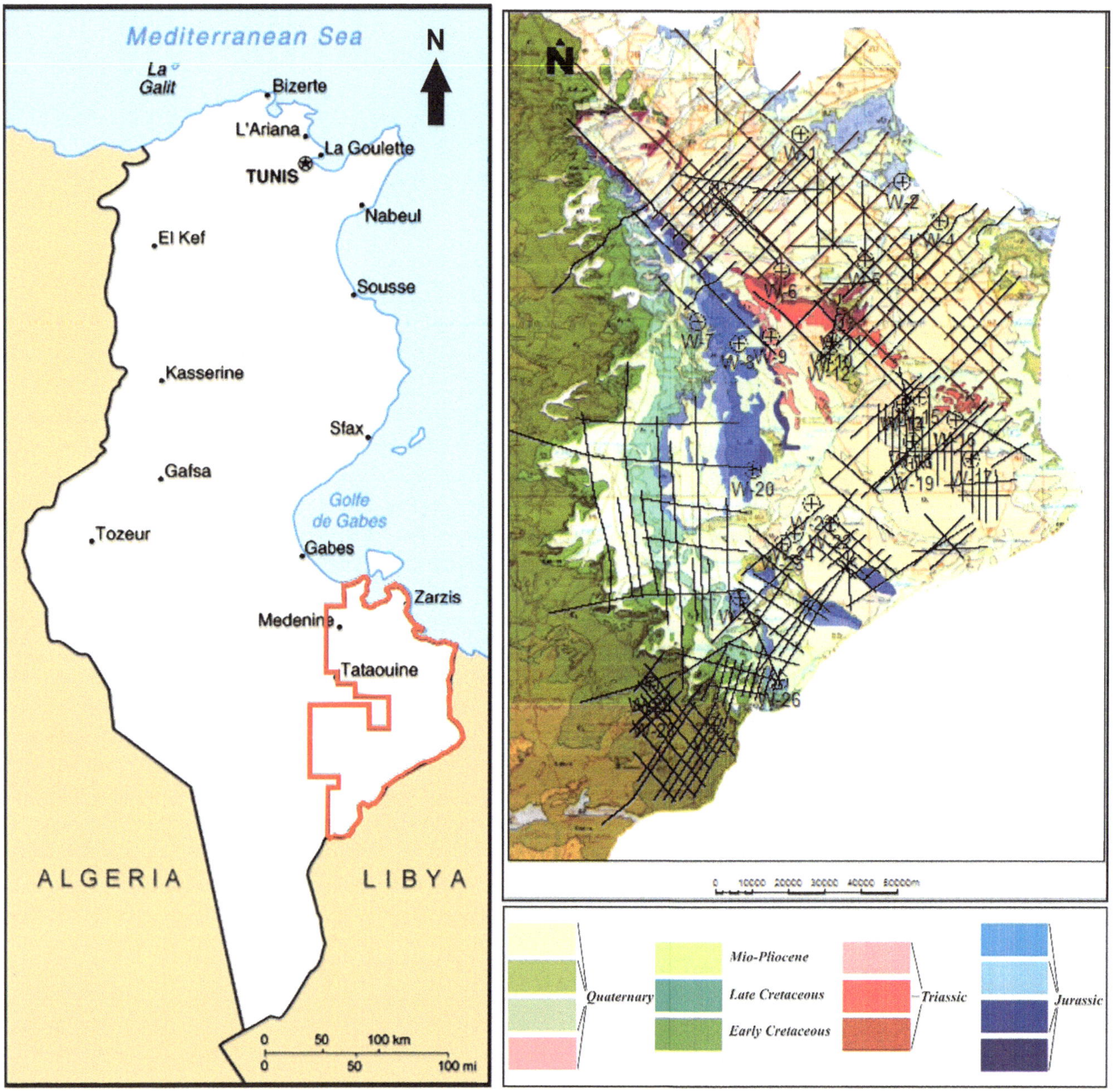

Fig. 1 Location of the study area with wells and seismic lines distribution used in this work

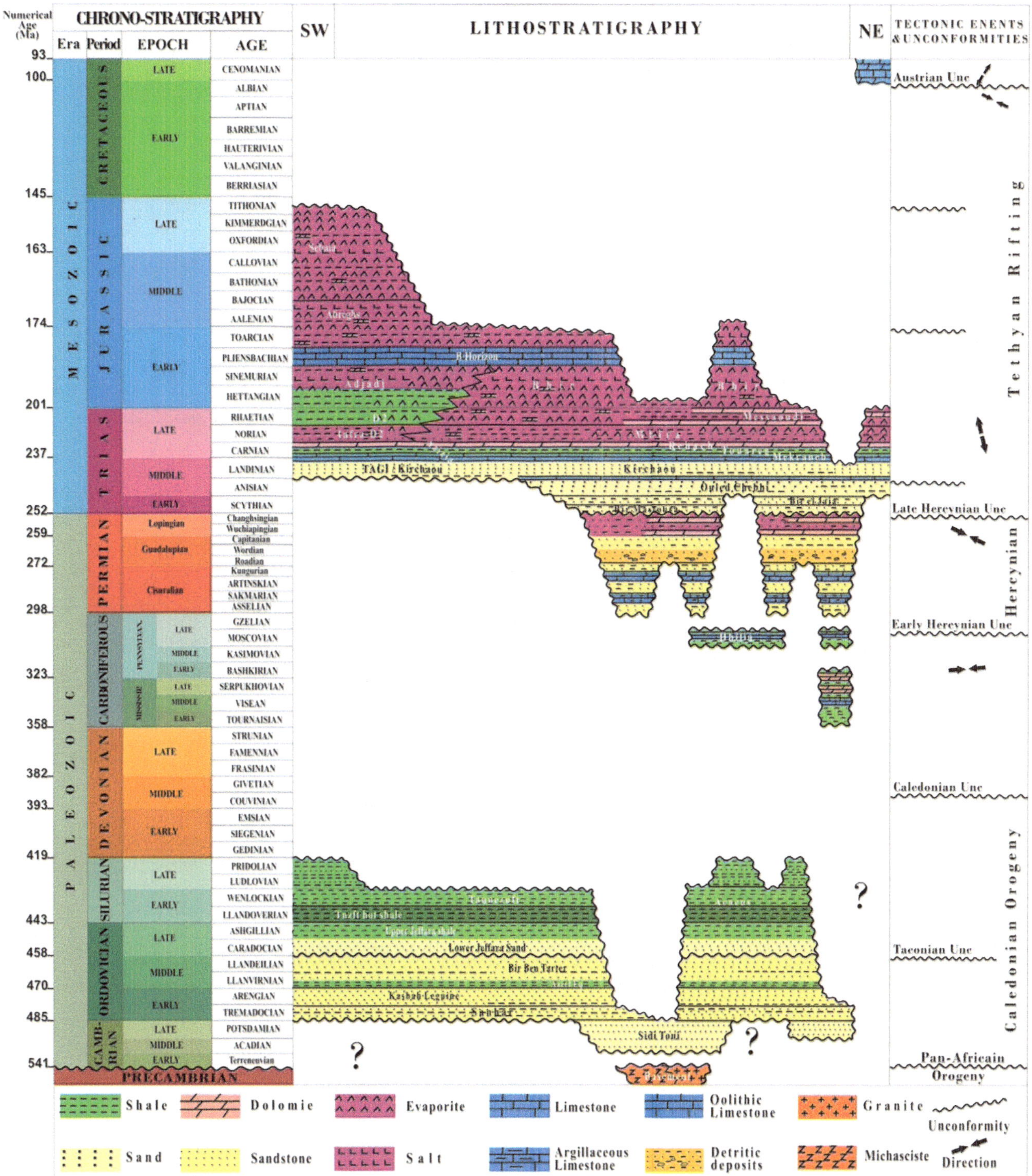

Fig. 2 Regional Lithostratigraphic chart of the Jeffara—Dahar region

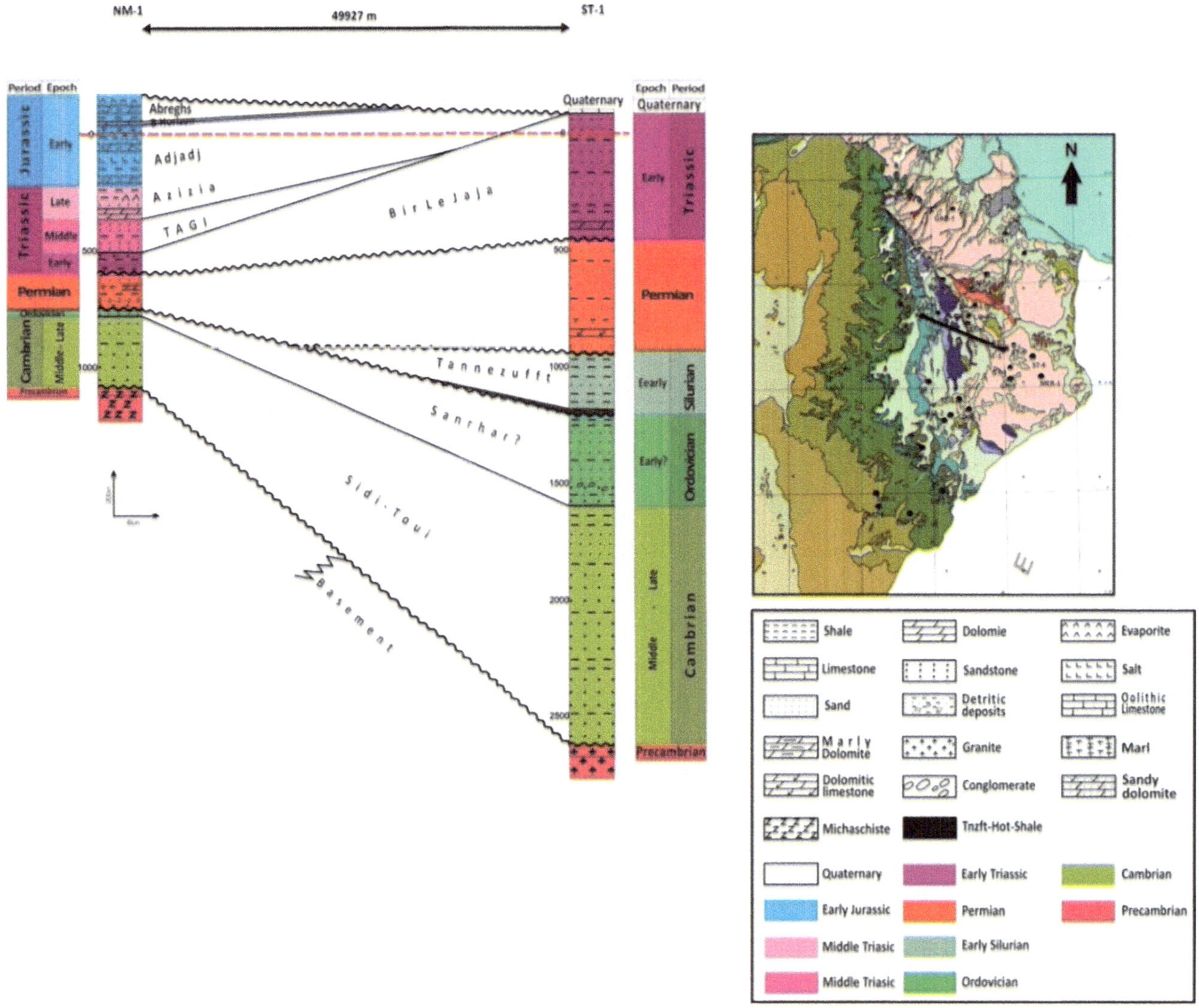

Fig. 3 NW—SE lithostratigraphical correlation

Paleozoic sediments. Starting with the oldest Pan-African unconformity, separating the Palaeozoic series from the basement, the Taconic unconformity which took place eroded the Upper Ordovician series. After this erosional surface, we meet the Caledonian unconformity onlapped by Carboniferous, the Hercynian unconformity separating Permian from the Mesozoic Triassic deposits. Finally the Austrian Unconformity characterizes Jurassic and Late Cretaceous period.

This arrangement is clearely illustrated from lithostratigraphic correlations. Therefore, the most important feature shown in the NW-SE transect (Fig. 2) is the existance of the Precambrian basement at 1253 m and the significant lateral thinning of the Cambro-Ordovician series toward the west, indicating a high structure near the NM—1 well (Fig. 3).

On the other side, the analysis of a regional NE—SW wells correlation (Fig. 4) shows an important lateral variation of the Paleo-Mesozoic deposits, especially, the Permian-Triassic series. This notable thickness variation is suggested to have been created during Late Triassic in subsiding areas.

As seen in those correlations, the impact of the tectonic manifestation took place in the arrangement of many structural and stratigraphical traps which are successively formed by compressional movements and angular unconformities.

4 Conclusion

The Jeffara Dahar—Region was affected by many tectonic events and manifestations during the Paleozoic summarized by an important number of unconformities. The occurrence of those orogenic phases made the area more specific in

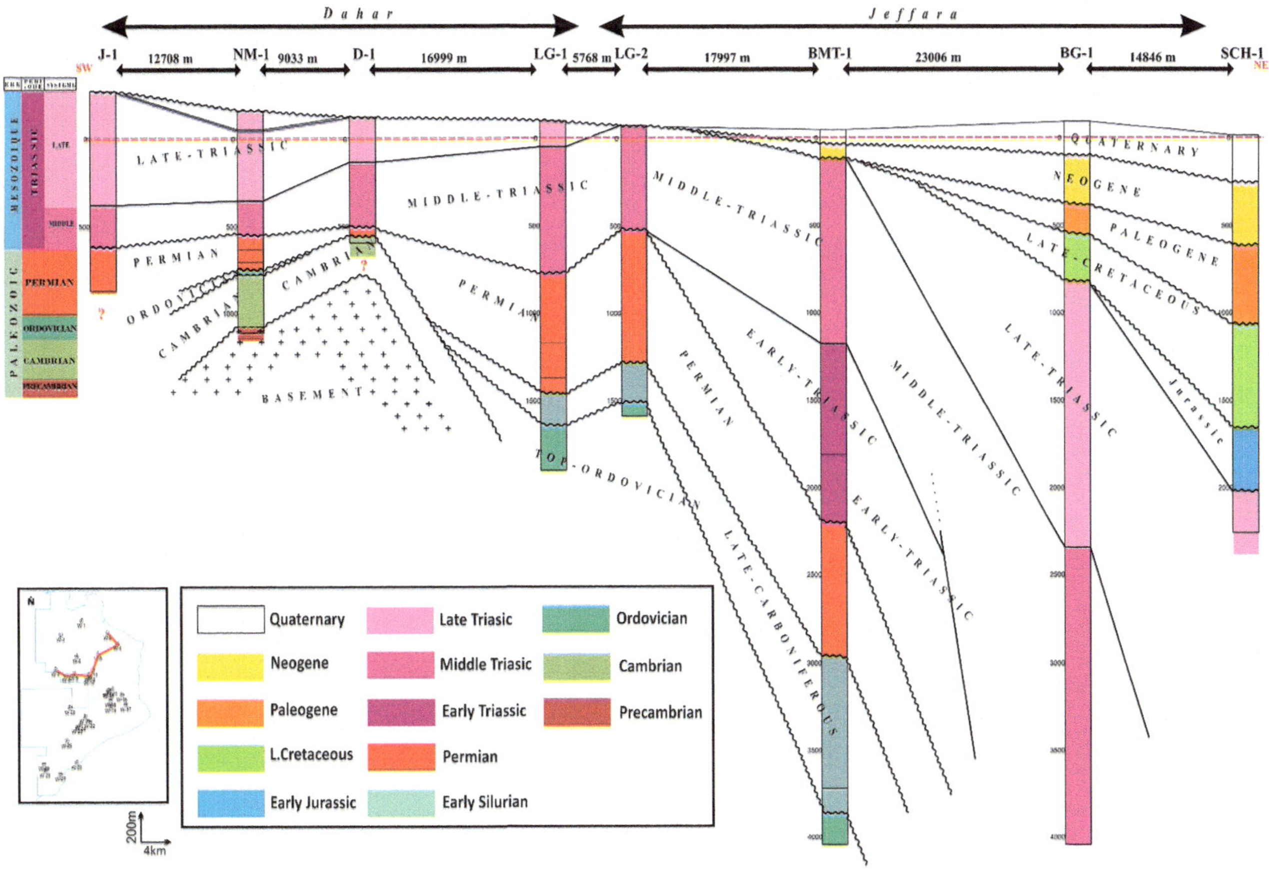

Fig. 4 NE—SW lithostragraphic correlation

terms of controlling the sediments environment of Paleozoic deposits and as a consequence resulted in the complex configuration and distribution of the Paleozoic reservoirs.

References

1. Bouaziz, S.: Palaeostress reconstruction and tectonic evolution of the Tataouine Basin (southern Tunisia).In: M.A. Sola, Worsley D. (eds.) Geological Exploration in Murzuq Basin 2000, chapter 22, pp. 1–13. Elsevier Science B.V. (2000)
2. Bouaziz, S., Barrier, E., Soussi, M., Turki, M.M., Zouari, H.: Tectonic evolution of the northern African margin in Tunisia from paleostress data and sedimentary record. Tectonophysics **357**, 227–253 (2002)
3. Echikh, K.: Geology and hydrocarbon occurrences in the Ghadames Basin, Algeria, Tunisia, Libya. In: MacGregor, D., Moody, R., Clark-Lowes, D. (eds.) Petroleum Geology of North Africa, vol. 132, pp. 109–129. Geological Society of London Special Publication (1998)
4. Gabtni, H., Jallouli, C., Mickus, K.L., Zouari, H., Turki, M.M.: Deep structure and crustal configuration of the Jeffara basin (Southern Tunisia) based on regional gravity, seismic reflection and borehole data: how to explain a gravity maximum within a large sedimentary? J. Geodyn. **47**, 142–152 (2009)

Characteristics of Bleaching Around Major Structures in a Reservoir-Cap Rock System, SE Utah, USA

Young-Seog Kim, Kyoungtae Ko, and Jin-Hyuck Choi

Abstract

Our research is focused on the bleaching patterns indicating natural CO_2-saturated water leakage to the surface, which provides the understanding of controlling factors on fluid flow processes. To examine the relationships between bleaching patterns and geologic characteristics, we investigated the characteristics of bleaching patterns around major faults (Moab and Salt Wash Faults) based on field observations in the exhumed reservoir-cap rock systems at SE Utah, USA. According to the results of our scanline survey, the deformation bands frequently observed in fault damage zones and in bleached layers compared with unbleached layers. Although almost all fault zones act as conduits for fluid flow, some fault zones filled with clay-rich gouges can impede fluid flow.

Keywords

CO_2 leakage • Fluid flow • Fault • Deformation band • CO_2 storage

1 Introduction

Understanding the mechanisms of natural CO_2 leakage through geological media is an important issue for successful geological CO_2 storage. The south-eastern part of Utah, USA is one of the best areas to study natural leakages associated with both past and active CO_2-saturated water. In particular, bleached rocks are commonly observed through color change by diagenetic fluid in exhumed reservoir-cap rock systems at a variety of scales (e.g. [2]). In this study, we investigated the characteristics of bleaching patterns associated with major structures based on field observations in SE Utah, USA.

2 Geological Overview

The study area is located in the north-western Paradox Basin of the Colorado Plateau, USA (Fig. 1a). This basin is filled with thick sedimentary sequences (Fig. 1b) with organic-rich Upper Paleozoic rocks as sources of CO_2 and hydrocarbon production (e.g. [1]). The reservoir-cap rock system in this area is well defined at a regional scale ([3]; Fig. 1c).

3 Bleaching Observation Around Major Fault Zones

3.1 Moab Fault Site

We conducted a scanline survey for bleached layers along the three segments of the Moab fault system in order to investigate the relationship between fault geometry and structural elements. Interestingly, deformation bands are only observed in the fault damage zone of the Courthouse fault. In contrast, fractures begin to become apparent in the wall rock at the Courthouse fault. In addition, deformation bands are observed within bleached layers compared with unbleached layers at the same fault damage zone (Fig. 2).

3.2 Salt Wash Graben Site

The lateral variations of contact boundary between the bleached sandstone and the unbleached sandstone were mainly observed around the secondary faults and fractures associated with the main Salt Wash fault (Fig. 3a). Most of the faults and fracture zones are bleached (Fig. 3b), indicating that these zones act as conduits for fluid flow. In

Y.-S. Kim (✉)
Department of Earth & Environmental Sciences, Pukyong National University, Busan, 48513, Republic of Korea
e-mail: ysk7909@pknu.ac.kr

K. Ko · J.-H. Choi
Geological Research Division, Korea Institute of Geoscience and Mineral Resources, Daejeon, 34132, Republic of Korea

F. Rossetti et al. (eds.), *The Structural Geology Contribution to the Africa-Eurasia Geology: Basement and Reservoir Structure, Ore Mineralisation and Tectonic Modelling*,
Advances in Science, Technology & Innovation, https://doi.org/10.1007/978-3-030-01455-1_23

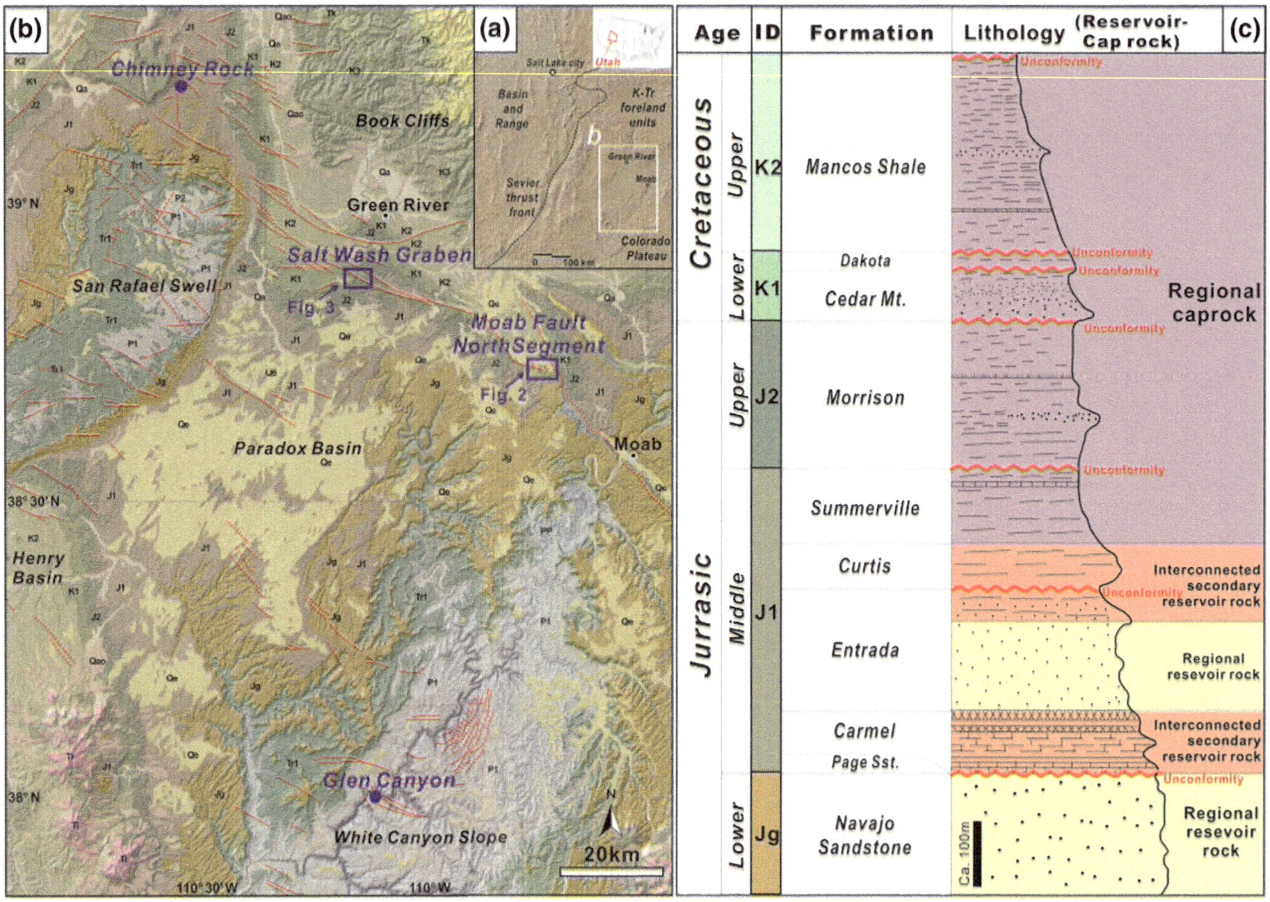

Fig. 1 **a** Location of the state of Utah, **b** simplified geological map and **c** stratigraphic column of the reservoir-cap rock system, SE Utah, USA (Modified from Ogata et al. [3])

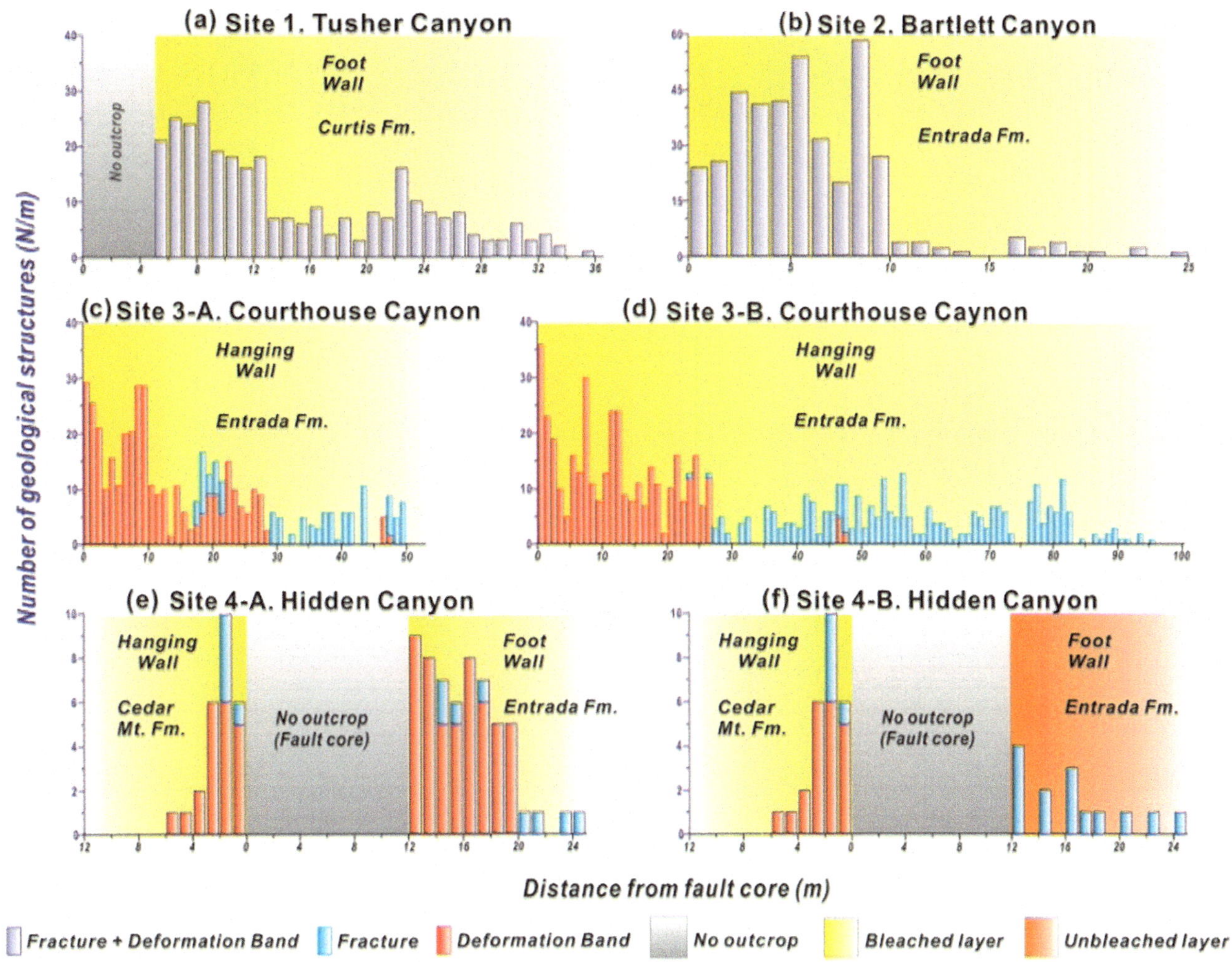

Fig. 2 Results of scanline survey along the north segment of the Moab fault system

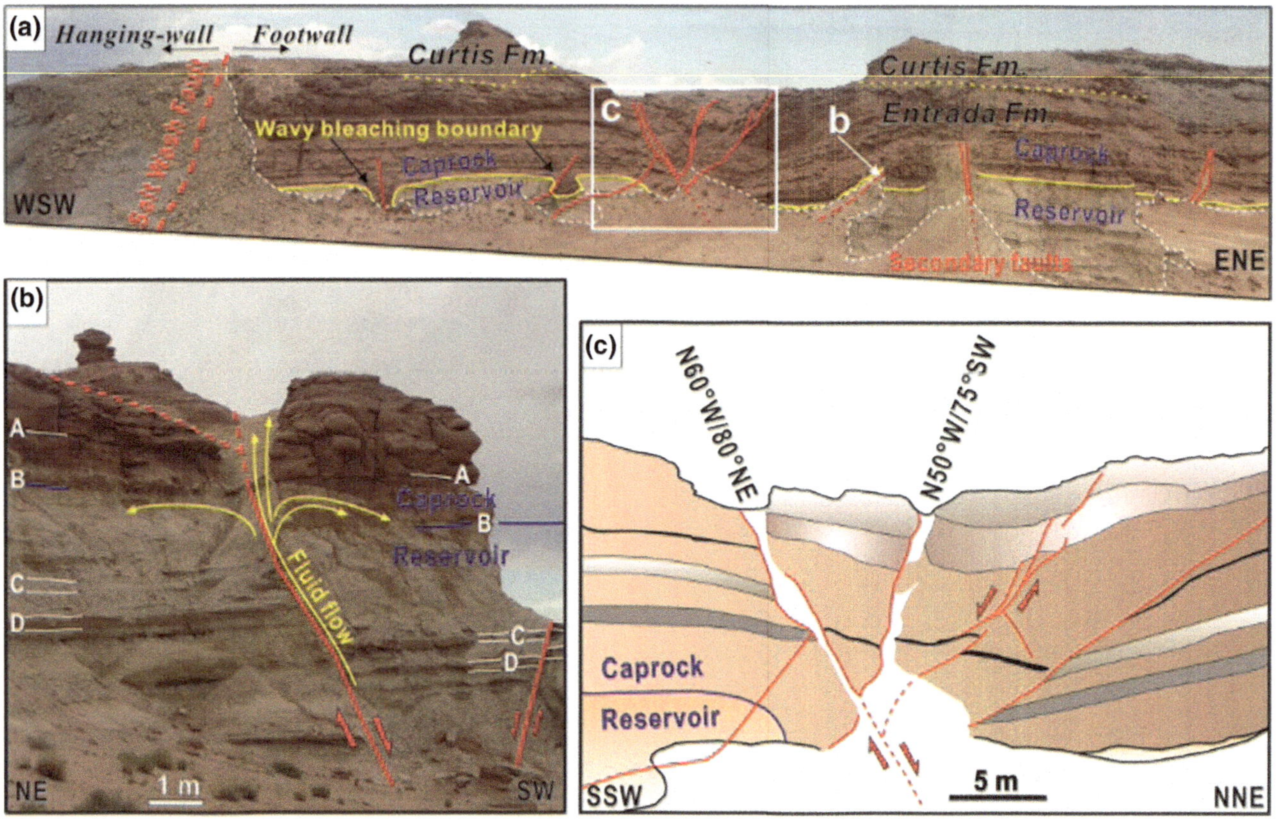

Fig. 3 Fault-related bleaching patterns in the northern side of the Salt Wash Graben

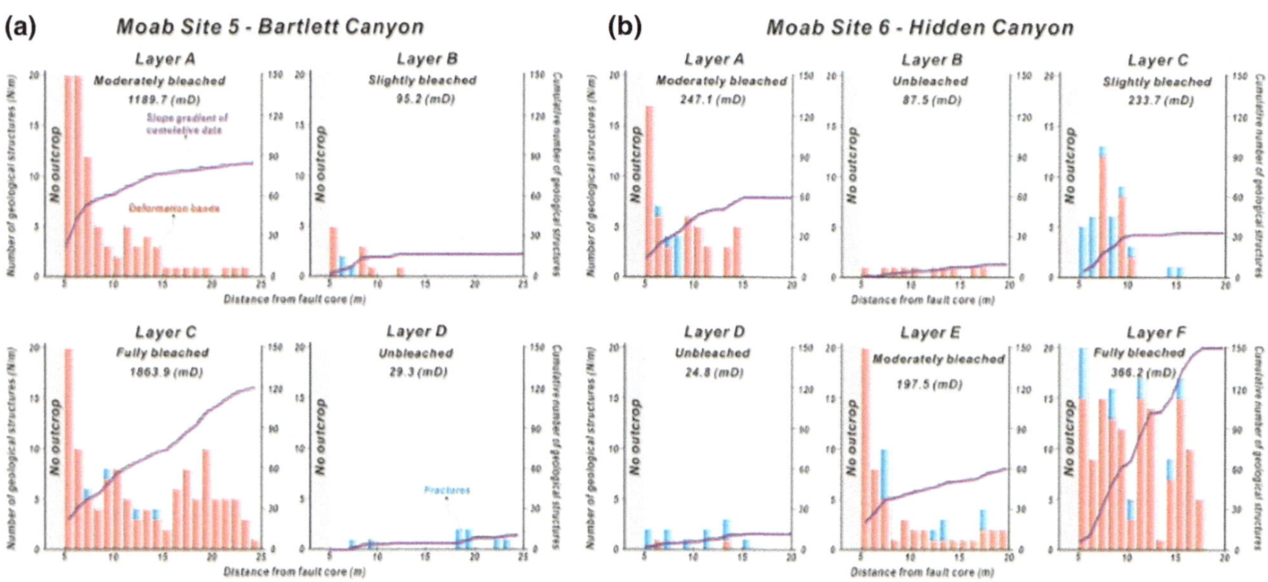

Fig. 4 Results of scanline survey and permeability measurement for each layer in the Bartlett Wash (**a**) and the Hidden Canyon (**b**) sites

contrast, some boundaries between reservoir rocks and cap rocks were depressed around the faults filled with clay gouges (Fig. 3c).

4 Discussion

Multiple layers near the Bartlett Fault Segment show variations in the degree of bleaching, providing a good opportunity to examine the main controlling factors of fluid flow. We performed multi-scanline surveys and measured the permeability of each layer at two main localities to infer the main controlling factors for different bleaching intensities: Bartlett Wash and Hidden Canyon.

As shown in Fig. 4, the distribution of geological structures is highly variable depending on the degree of bleaching in each layer as follows: (1) Fractures are rarely observed regardless of the intensity of bleaching, (2) The higher the intensity of bleaching, the denser the deformation bands are, (3) Highly bleached layers show higher permeability values than unbleached layers.

5 Conclusions

Our field observations confirm that bleaching is controlled by structural discontinuities (e.g., faults, fractures and deformation bands) as well as the lithological properties (e.g., permeability or porosity) of the reservoir-cap rock system. Although deformation bands have typically been thought to reduce the permeability of wall rock, we identified a strong positive correlation between the degree of bleaching and the intensity of the deformation band. Thus, deformation bands can be used to identify highly permeable layers that can be easily bleached. Most faults and fracture zones act as conduits for fluid flow in the study area. However, some faults filled with clay gouges impede fluid flow along such fault zones. This study could be very useful in evaluating the feasibility of geological sites for CO_2 sequestration, particularly for identifying potential CO_2 leakage in a reservoir-cap rock system. This also could be applicable to understanding fluid flow characteristics of reservoir-cap rock system in oilfields.

References

1. Chan, M.A., Parry, W.T., Bowman, J.R.: Diagenetic hematite and manganese oxides and fault-related fluid flow in Jurassic sandstones, southeastern Utah. AAPG Bull. **84**, 1281–1310 (2000)
2. Dockrill, B., Shipton, Z.K.: Structural controls on leakage from a natural CO_2 geologic storage site: Central Utah, U.S.A. J. Struct. Geol. **32**, 1768–1782 (2010)
3. Ogata, K., Senger, K., Braathen, A., Tveranger, J.: Fracture corridors as seal-bypass systems in siliciclastic reservoir-caprock successions: field-based insights from the Jurassic Entrada Formation (SE Utah, USA). J. Struct. Geol. **66**, 162–187 (2014)

Identification and Distribution of Jurassic Paleo-Reservoirs in the Central Junggar Basin, NW China

Gang Liu

Abstract

Hydrocarbon generation and the expulsion of source rocks, tectonic evolution history, reservoir-seal assemblages, hydrocarbon accumulation stages and the distribution pattern of Jurassic reservoirs are discussed in this study. The Jurassic reservoirs were formed in the Early Cretaceous at the earliest. However, most of them were destroyed owing to the Himalayan tectonic movement which created formations tilted to the south in the Paleogene. Distribution pattern of paleo-reservoirs was critical for petroleum exploration and development in the Central Junggar Basin. In the Early Cretaceous, the paleo-structure of the Jurassic Sangonghe Formation was re-established. As a result, three paleo-uplifts (Muosuowan, Lunan and Luliang) were developed during this period, which had favorable reservoir forming conditions and controlled hydrocarbon distribution. Grains with oil inclusions (GOI) method were also used to study the distribution of Jurassic paleo-reservoirs. In some Jurassic layers, the GOI index was greater than 5% while these layers were identified as non-oil layers according to the logging and testing results. Thus, these layers may be a paleo-oil-reservoirs and were structurally destroyed during the Late Palaeogene. According to the comprehensive study of reservoir forming elements as well as GOI data, Jurassic paleo-reservoirs were mainly distributed in the Mosuowan, Lunan, and Luliang paleo-uplifts in the Central Junggar Basin.

Keywords

Paleo-reservoir • Paleo-uplift • Key accumulation period GOI • Jurassic • Central junggar basin

G. Liu (✉)
Research Institute of Petroleum Exploration and Development, PetroChina, Beijing, 100083, China
e-mail: liug_2011@petrochina.com.cn

1 Introduction

The Junggar Basin is a multi-cycle superimposed basin in NW China. Altogether 6 oil and gas fields have been discovered in the Central Junggar Basin, covering an area of 2.8×10^4 km^2 (Fig. 1a). These oil and gas reservoirs are characterized by shallow burial depths (lower than 2500 m in the northern part, lower than 4000 m in the southern part), huge reserves, high efficiency of development (accounting for approximately 50% of total production in the Junggar Basin in 2014) [1]. Thus, the Central Junggar Basin is one of the most significant prospecting targets in the basin. However, no important discoveries were made in this area in the last fifteen years because of the complex oil accumulation process.

The Jurassic Paleo-structure in a critical accumulation period was re-established based on 3D seismic data. Through a comprehensive study of reservoir forming elements as well as GOI data analysis of typical wells, the distribution pattern of Jurassic paleo-reservoirs was thoroughly discussed.

2 Settings and Methods

The Permian Fengcheng(P_1f) and Xiawuerhe(P_2w) Formations in Penyijingxi Sag (Fig. 1) are main source rocks in the Central Junggar Basin [2]. Based on previous studies on hydrocarbon generation, the expulsion of source rocks, tectonic evolution history, reservoir-seal assemblage and other forming elements, the Jurassic reservoirs were formed in the Early Cretaceous at the earliest. However, most of them were destroyed owing to the Himalayan tectonic movement, which had a huge impact on the Jurassic Formations structure in the Central Junggar Basin. The formations in the northern part of this area uplifted holistically, which resulted in the Jurassic Formations becoming a monoclonal structure tilted to the south [2].

F. Rossetti et al. (eds.), *The Structural Geology Contribution to the Africa-Eurasia Geology: Basement and Reservoir Structure, Ore Mineralisation and Tectonic Modelling*, Advances in Science, Technology & Innovation, https://doi.org/10.1007/978-3-030-01455-1_24

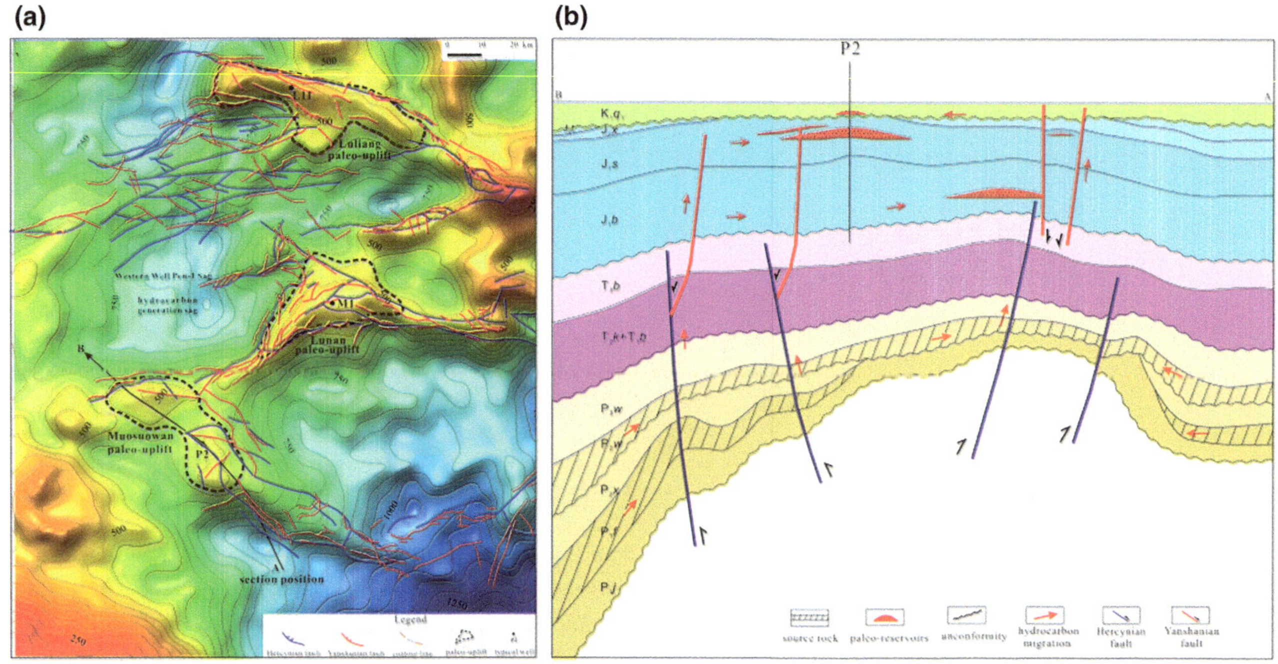

Fig. 1 Structure map (**a**) and section profile (**b**) of the Jurassic Sangonghe formation in the early cretaceous

The practical exploration showed that the main pay zones are Lower Jurassic Badaowan (J_1b) and Sangonghe (J_1s) Formations, Middle Jurassic Xishanyao Formation(J_2x). The reservoir lithology is dominated by medium-grained sandstone and glutenite.

The method of relict stratum thickness was used to re-establish the Jurassic paleo-structure. Based on 3D seismic data, the Jurassic paleo-structure in the Early Cretaceous was re-established. Several reservoir forming elements, such as hydrocarbon generating and expulsion center, conduit system, reservoir-cap assemblage, were considered to confirm the existence of paleo-oil-reservoirs in the paleo-uplifts.

Grains with oil inclusions (GOI) method was used to study reservoirs distribution. GOI is an effective tool in differentiating the oil layer from the water layer. Usually, the oil layer has a GOI ratio higher than 5% while the water layer has a GOI ratio lower than 1%. The hydrocarbon migration pathway has a GOI ratio of 1–5% [3, 4].

2.1 Results

Three Jurassic paleo-uplifts (Mosuowan, Lunan, Luliang) were identified based on the structure map of the Sangonghe Formation in the primary reservoir forming period (Early Cretaceous) (Fig. 1a). These paleo-uplifts show favorable conditions for hydrocarbon accumulation: (1) Because they are adjacent to the hydrocarbon generation depression (Pengyijingxi Sag) and the oil resource is abundant (Fig. 1 a). (2) Hercynian faults extended downwards into Permian source rocks, and also compose Y type conduit system with Yanshanian faults (Fig. 1b), which prove to be efficient conduits for oil migration. (3) Effective anticline traps and reservoir-caprock assemblages associated with fault systems have provided favorable oil-migration pathways (Fig. 1b). Therefore, the Jurassic reservoirs are obviously controlled by paleo-uplifts.

Three typical wells (Fig. 1a) with different types of oil and gas shown in the three paleo-uplifts were selected for GOI (Grains with oil inclusions) analysis. The GOI values are shown in Table 1.

The GOI values of Jurassic Formations in these 3 wells are greater than 10% (Table 1), much greater than 5%. In Well M1 in the Lunan paleo-uplift, the GOI value reaches 16.8%, which is the highest in this study. Thus, the GOI ratios indicate that oil reservoirs were once developed in the Jurassic paleo-uplifts [5]. However, the testing results in these 3 wells showed these reservoirs are water layers or oily water layers, which is inconsistent with the results reflected by GOI ratios. One reasonable explanation is that the paleo-oil-reservoirs were once developed in the Jurassic paleo-uplifts. In the Late Paleogene, the Jurassic

Table 1 GOI results of typical wells in the Central Junggar Basin

Well	Paleo-uplift	Depth/(m)	Formation	GOI/(%)	Testing result
P2	Mosuowan	4410	J_1s_2	10	Oily water layer
M1	Lunan	3327.5	J_1s_2	16.8	Water layer
L11	Luliang	2166	J_2x	11.9	Oily water layer

paleo-reservoirs were destroyed because of the Himalayan tectonic movement, which made the Jurassic Formation tilt to the south.

3 Discussion

Based on the comprehensive analysis of previous studies, the distribution, formation and evolution of paleo-uplift plays a decisive role for hydrocarbon accumulation. This role also works in the Central Junggar Basin. The Mosuowan, Lunan, and Luliang paleo-uplifts, which are near the hydrocarbon generation center, have the orientation region of oil migration and control the distribution of primary reservoirs. GOI ratios method generally is used to judge reservoirs, but for identifying paleo-reservoirs, it needs to be combined with other data and analysis, such as logging curves, testing results and regional accumulation background.

4 Conclusions

In the Central Junggar Basin, the primary Jurassic reservoirs were mainly formed in the Early Cretaceous. Most of them were destroyed in Paleogene because of the Himalayan tectonic movement, which made Jurassic Formations tilt to the south.

In the Early Cretaceous, three Jurassic paleo-uplifts (Mosuowan, Lunan, and Luliang paleo-uplifts) were formed, which had favorable reservoir forming conditions and controlled primary reservoirs distribution.

According to the oil accumulation process in this area and GOI data, the Jurassic paleo-reservoirs were mainly distributed in Mosuowan, Lunan, and Luliang paleo-uplifts in the Central Junggar Basin.

References

1. Wu, J., Meng, X., Wang, L., Cao, B., Zhang, J., Zhu, Y.: Characteristics of subtle oil/gas reservoirs and exploration strategy in Central Junggar Basin. Oil Gas Geol. **25**(6), 682–685 (2004)
2. Hu, S., Yu, Y., Dong, D.: Control of fault activity on hydrocarbon accumulation in Central Junggar Basin. Acta Petrolei Sinica **27**(1), 1–7 (2006)
3. George, S.C., Lisk, M., Eadington, P.J.: Fluid inclusion evidence for an early, marine-sourced oil charge prior to gas-condensate migration, Bayu-1, Timor Sea, Australia. Mar. Pet. Geol. **21**(9), 1107–1128 (2004)
4. Cao, J., Wang, X., Sun, P., Zhang, Y., Tang, Y., Xiang, B., et al.: Geochemistry and origins of natural gases in the Central Junggar Basin, northwest china. Org. Geochem. **53**(53), 166–176 (2012)
5. Xiao, Q., He, S., Yang, Z., He, Z., Wang, F., Li, S., et al.: Petroleum secondary migration and accumulation in the Central Junggar Basin, northwest China: insights from basin modeling. AAPG Bulletin **94** (7), 937–955 (2010)

Reservoir Properties and Facies Distribution of Mishrif Formation in Ratawi Oilfield, Southern Iraq

Furat A. Al-Musawi, Rami M. Idan, and Amani L. M. Salih

Abstract

The M. Cenomanian-E. Turonian Mishrif Formation is considered as the most extended reservoir of south Iraq. The present paper reveals the facies of this formation in Ratawi oilfield by the analysis of seven wireline logs. According to reservoir properties, i.e. effective porosity, permeability, and water saturation, the Mishrif Formation is divided into two main units, the upper (mA) and the lower (mB), separated by a shale member. These reservoir units are almost capped by two shale intervals located in the lower and upper contacts with Rumaila and Khasib formations, respectively. The reservoir properties, in terms of porosity and permeability, decrease (porosity less than 5%, permeability less than 0.001 md) to the south of Ratawi structure due to the dominance of deep marine facies, whereas the formation predicted to have good reservoir properties (porosity more than 10%, permeability more than 5 md) to other directions especially to the east, contrary to what was previously expected.

Keywords

Facies distribution • Mishrif formation • Reservoir properties • Ratawi oilfield

1 Introduction

One of the most extended formations in Iraq is the M. Cenomanian-E. Turonian Mishrif Formation, which is the second economic reservoir after Zubair Formation [1]. Mishrif Formation consists of multi-story thick carbonates deposited on a wide shelf, extended from the lagoon to the deep open marine platform [2] affected by a major Eustatic sea-level rise, started in the Middle Cenomanian [3]. The Mishrif Formation deposited over ramp-like or rimmed-shelf platforms bordering intra-shelf basins on the passive margin of the Neo-Tethys Ocean in a sub-equatorial position with warm equable climate [4]. The formation overlies the oligosteginal carbonates of the Rumaila Formation with a gradational boundary of the same period of sedimentation. Nevertheless, the upper contact is uncomfortable to the transgressive, basinal Upper Turonian–Coniacian Khasib Formation in the study area and most parts of Iraq. Mishrif Formation is composed mainly of carbonate rocks predominant by Rudists, Algal, and Foraminiferal facies which indicate neritic and shallow marine limestone, whereas, the occurrence of green Algae indicates a relatively deeper environment. There are many arguments around Mishrif Formation and its reservoir characteristics in the area of study, wherefore the objectives of this study were to explain the facies distribution, effective porosity extent, and water saturation along the formation column.

2 Geological Setting

This study focused on Mishrif Formation in Ratawi oilfield. The field was discovered in 1950 while drilling a clear seismic anomaly on top of Ahmadi Formation at about 70 km W of Basrah governorate (Fig. 1). Tectonically, the Ratawi anticline structure is placed above the Jurassic Salt Basin and over the Infra-Cambrian Salt Basin in the unstable Shelf Tectonic System [5]. In the Iraqi segments, the field is located in the Zubair subzone [6]. Ratawi oilfield is a structural dome with NS dominant direction and approximately 2 km length (Fig. 2). The N-S trend of the Ratawi field is probably due to interplay of Pre-Cambrian N-S basement faults and Infra-Cambrian salt tectonic regimes. The gentle dip may hide a steeper dip of the structural flanks

F. A. Al-Musawi · R. M. Idan (✉) · A. L. M. Salih
Department of Geophysics, College of Remote Sensing and Geophysics, Al-Karkh University of Science, Baghdad, Iraq
e-mail: Ramisc3@gmail.com

F. Rossetti et al. (eds.), *The Structural Geology Contribution to the Africa-Eurasia Geology: Basement and Reservoir Structure, Ore Mineralisation and Tectonic Modelling*,
Advances in Science, Technology & Innovation, https://doi.org/10.1007/978-3-030-01455-1_25

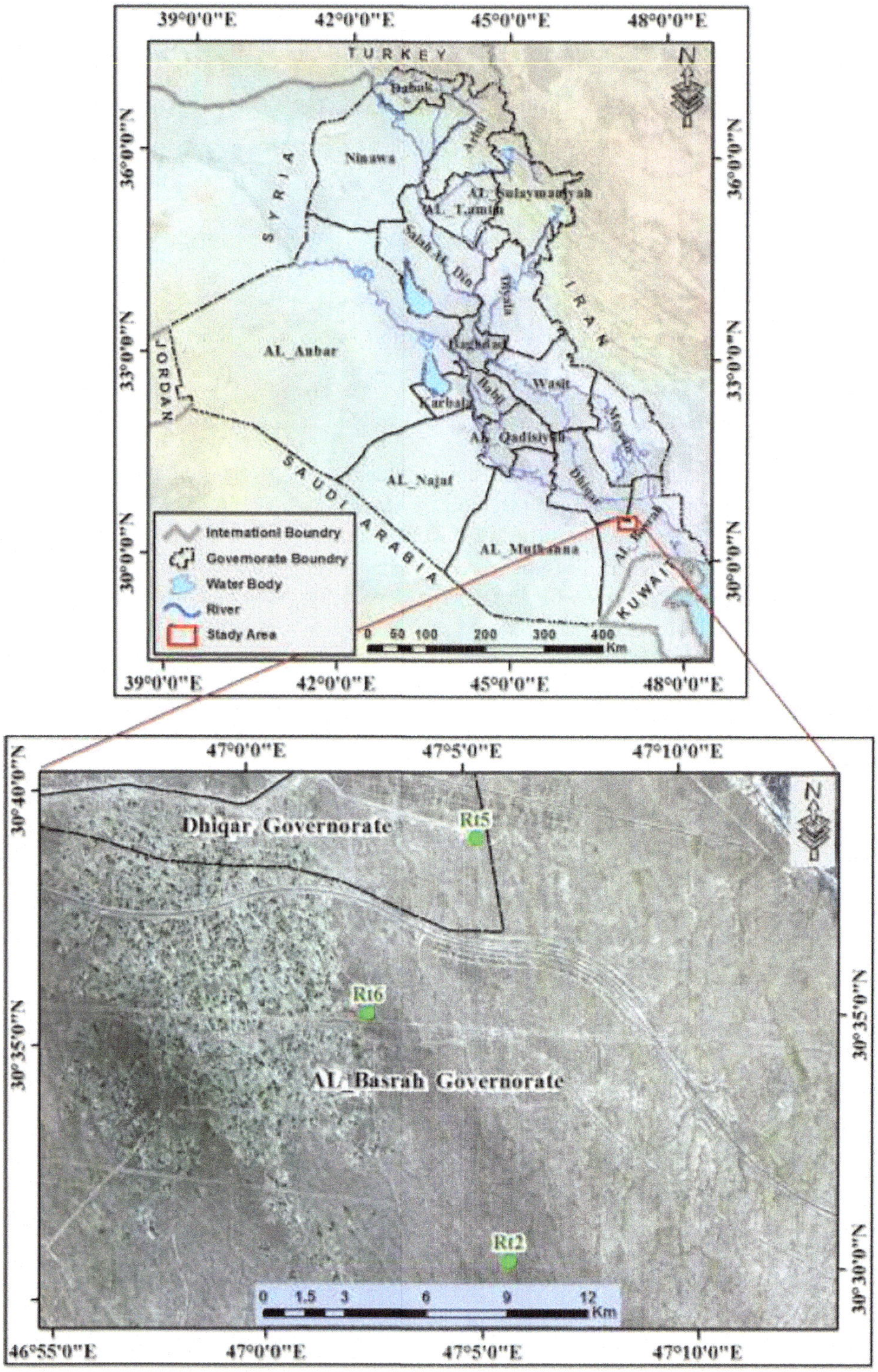

Fig. 1 Map of the study area in southern Iraq showing (Above) the location of Ratawi oilfield and (Below) the wells that investigated to this work

of the deformable Gotnia anhydrite [7]. The detailed stratigraphic correlation, seismic structural mapping and log analyses showed that this Shelf-type of anticline proves Neocomian, Albian, Turonian and Palaeocene-Oligocene pulses of structural growths. These growths characterize the multiple reservoirs giant oil fields of the Arabian Shelf and precisely southern parts of Iraq [8].

3 Materials and Methods

The data was taken from seven (7) wells in Ratawi oilfield, in terms of wells coordinate, tops, and total depth. Moreover, the velocity map was transformed into depth map to correct the well tops in respect to Mishrif Formation. The electric log set

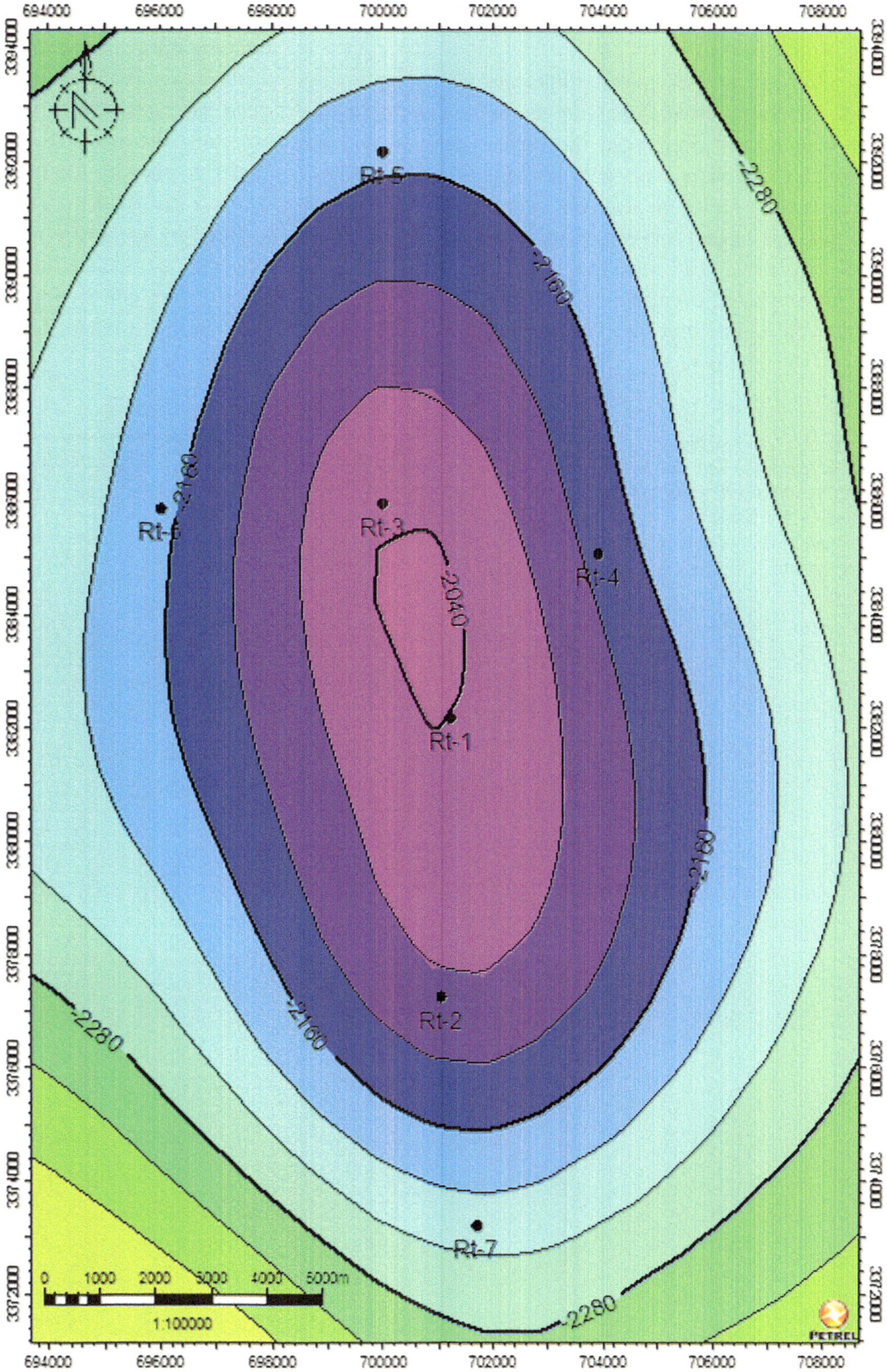

Fig. 2 Structural map on the top of Mishrif Formation in the Ratawi Oilfield, showing the thickness and wells locations. The colors refer to the depths variety (in meters)

of the Rt-5 and Rt-7 wells were treated in "didger" software to transform logs from sheets into digital data. Then the resulted LAS file were transported to "petrel software" to make the next analyses in terms of porosity, permeability, Volume of shale, formation factor, and water saturation to predict the reservoir characterization, as well as the environments and facies. Furthermore, the effective porosity and facies distribution of the formation was calculated along Ratawi structure.

4 Results and Discussions

The lithology crosses' plots results are shown in Fig. 3. Shale intervals are represented by the upper unconformable part with Khasib Formation, the barrier member, and the lower part that has contact with Rumaila Formation. These intervals characterize the seal rock, while the reservoirs (mA and mB) are capped between them. As shown in Fig. 4, the shale and/or shaly carbonate facies are restricted to the upper and lower Mishrif Formation. The barrier bed is found in all sections in Ratawi structure. The bed disappears to the south and is merged with shaly intervals of the upper contact. This wedging out pattern continues to the south of Iraq.

As shown in Fig. 5, the effective porosity decreases in the shale intervals, while it increases in the reservoir units especially the mB or lower Mishrif member. However, the effective porosity decreases in the ends of the Ratawi structure due to the relatively deep environment facies represented by carbonate mudstone and wackestone facies of the open marine environments [9]. In the mA reservoir member, the effective porosity is ranged from 10 to 15%, while in mB it ranges between 15 and 20%.

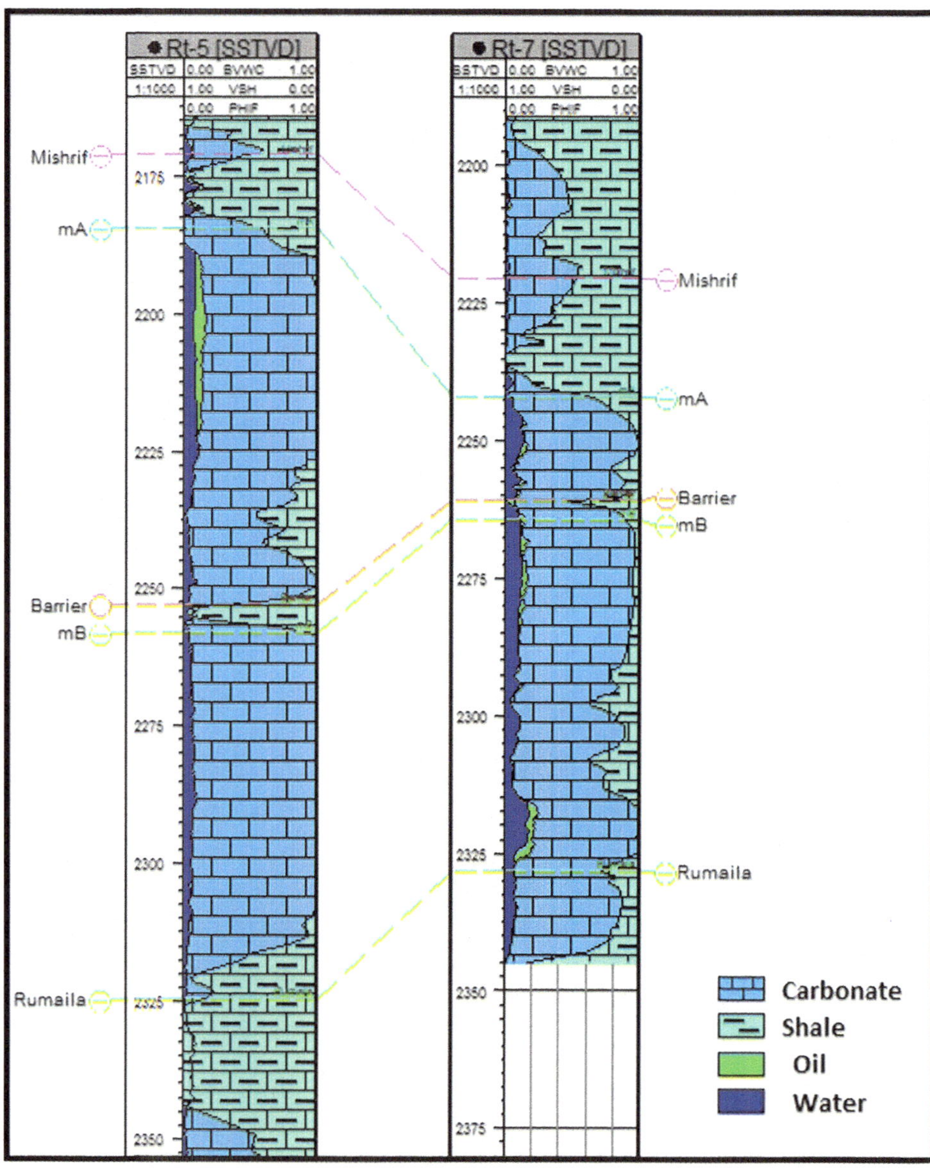

Fig. 3 N-S Cross section between the wells Rt-5 and Rt-7 shows the divisions of Mishrif Formation, the Shale to Carbonate ratio, and the oil and water content

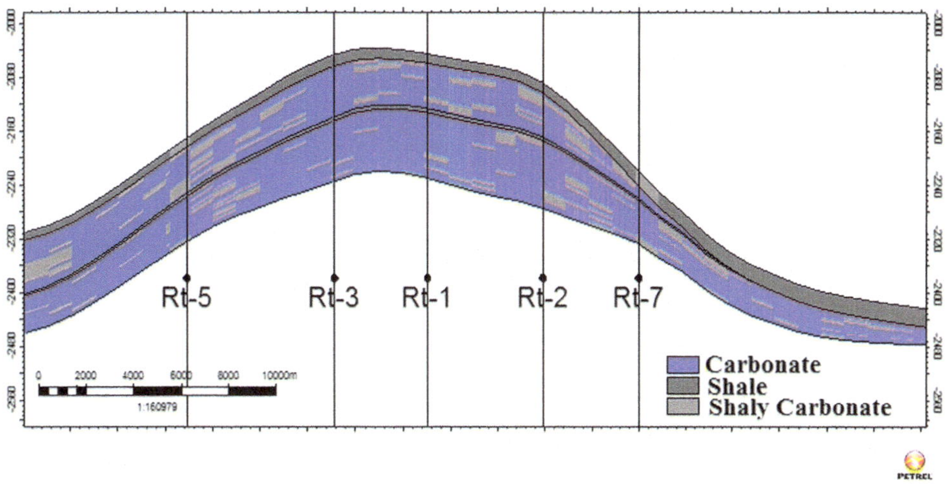

Fig. 4 N-S facies distribution along the Mishrif Formation, show the formation divisions along the Ratawi structure

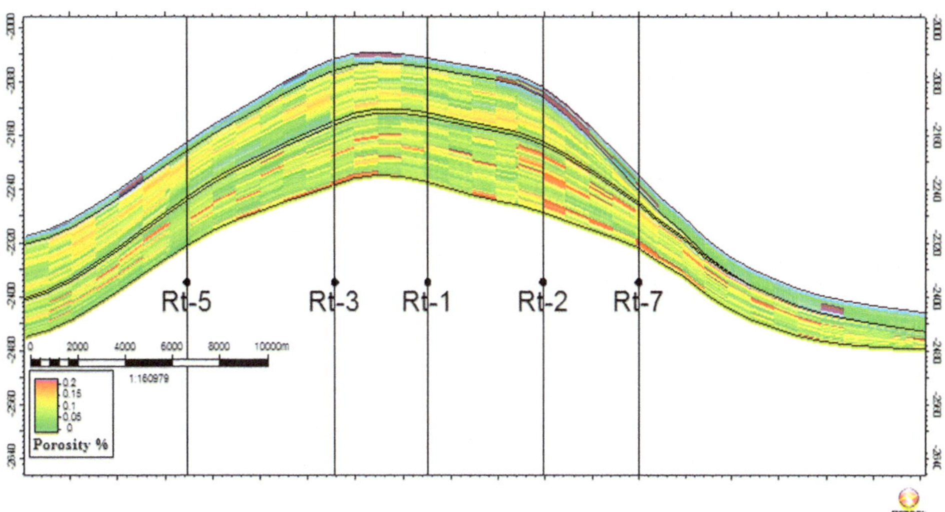

Fig. 5 N-S effective porosity distribution along the Mishrif Formation, showed the wells position

From the above results, this paper indicates that characteristics of the reservoir units of Mishrif Formation decrease toward the south of Iraq. While there is an acceptable prospect of the increase of reservoir characteristics toward the north, west, and east down to the border of Zubair subzone [10].

5 Conclusions

1. The Mishrif Formation deposited over ramp-like or rimmed-shelf platforms bordering intra-shelf basins on the passive margin of the Neo-Tethys Ocean. The Ratawi oilfield represented the deep marine environments.
2. Depending on the reservoir properties (porosity and permeability), the formation divided into two main reservoir units. The upper mA and the lower mB have porosity mean 12.5 and 17.5% respectively. While the permeability was more than 5 md to both of them.
3. These two reservoir units are capped by two shale intervals (with low porosity and permeability values) located in the lower and upper contacts with [11] Rumaila and Khasib formations, respectively, as shown in Fig. 6, and separated with shale member that have the same destructive properties.
4. The dominant facies were mudstone and wackestone of the open marine environments in Ratawi structure.
5. Reservoir units' properties of Mishrif Formation decrease toward the south of Iraq due to wedging out of the basin. While there is an acceptable prospect to increase of

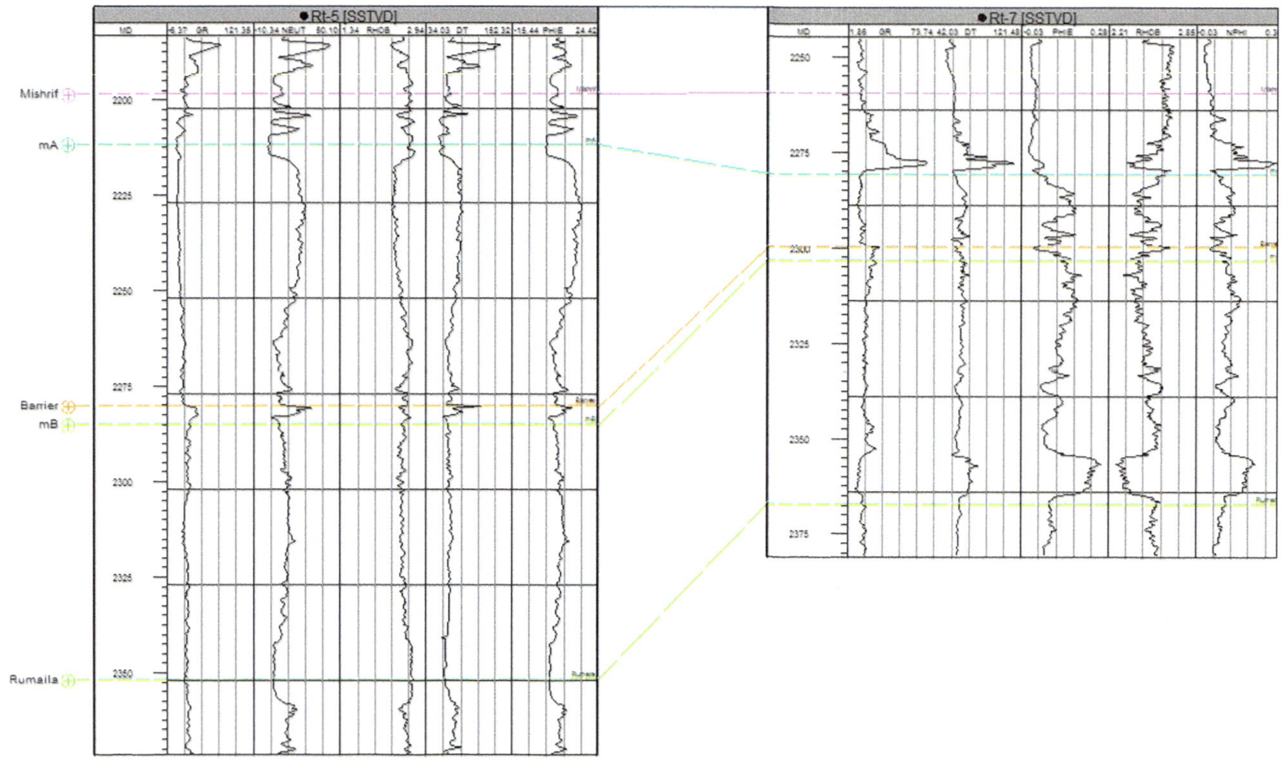

Fig. 6 Composite log of gamma ray (GR), sonic (DT), neutron (NPHI), density (RHOB), and effective porosity (PHIE)

reservoir characteristics toward the north, west, and east down to the border of Zubair subzone with Iraqi borders with Iran.

References

1. Awadeesian, A.M.R., Al-Jawed, S.N.A., Saleh, A.H.: Reservoir-scale sequence stratigraphy of Mishrif carbonates and implication to water injection strategy North Rumaila field case. Arab. J. Geosci. **8**(9), 7025–7040 (2015)
2. Van Bellen, R.C., Dunnington, H.V., Wetzel, R., Norton, D.M.: Lexique Stratigraphique International, Asie, vol 3, p. 333, Fasc.10a, Iraq, Paris (1959)
3. Gale, A.S., Voigt, S., Sageman, B.B., Kennedy, W.J.: Eustatic sea-level record for the Cenomanian (Late Cretaceous)—extension to the Western Interior Basin. USA. Geol. **36**, 859–862 (2008)
4. Aqrawi, A.A.M., Thehni, G.A., Sherwani, G.H., Kareem, B.M.A.: Mid-cretaceous rudist-bearing carbonates of the Mishrif formation: an important reservoir sequence in the Mesopotamian basin. Iraq. J. Pet. Geol. **21**(1), 57–82 (1998)
5. Sharland, P.R., Archer, R., Casey, D.M., Davies, R.B., Hall, S.H., Heward, A.P., Horbury, A.D., Simmons, M.D.: Arabian Plate Sequence Stratigraphy. GeoArabia Special Publication 2, Gulf Petrolink, Bahrain, 371 p., with 3 charts (2001)
6. Buday, T., Jassim, S.Z.: Final report and the regional geological survey of Iraq, Unpub. Report SOM. Library Tectonic Framework Baghdad 2 (1984)
7. ALSharhan, A.S., Nairn, A.E.M.: Sedimentary Basins and Petroleum Geology of the Middle East: Amsterdam, Netherlands, Elsevier Science B. V., p. 843 (2003)
8. Buday, T., Jassim, S.Z.: The Regional Geology of Iraq: Tectonism Magmatism, and Metamorphism. II. Kassab, Abbas, M.J. (eds.) Baghdad, p. 325 (1987)
9. Mahdi, T.A., Aqraw, A.A.M.: Sequence stratigraphic analysis of the Mid-Cretaceous Mishrif formation, Southern Mesopotamian Basin, Iraq. J. Pet. Geol. **37**(3), 287–312 (2014)
10. Idan, R.M., Faisal, R.F., Nasser, M.E., Al-Ameri, T.K., Al-Rawi, D.: Hydrocarbon potential of Zubair formation in the south of Iraq. Arab J. Geosci. **8**(7), 4805–4817 (2015 b). https://doi.org/10.1007/s12517-014-1569-6
11. Idan, R.M.: Total Organic Carbon (TOC) Prediction from Resistivity and Porosity Logs: A Case Study from Iraq. Bull. Iraq nat. Hist. Mus. **14**(3), 185–195 (2017)

Evaluation of Natural Fractures and Their Impact on Gas Production Using Image Logs (Sirte Basin, NE Libya)

Abedlgani Elrafadi, Mohamed Alwerfali, Walid S. Elbarghathee, Seraj Y. Bosnina, and Ali G. Najem

Abstract

The Cambrian-Ordovician Gargaf Formation, in the central part of Sirte Basin, is considered as one of the main reservoirs for gas production as in Alestiqlal field. The petrophysical properties of the reservoir revealed very low matrix porosity and permeability and can be classified as a tight reservoir. Auspiciously, the presence of natural fractures is considered as a vital element for the reservoir production. Therefore, evaluating such features becomes mandatory for Sirte Oil Company (SOC) for optimum reservoir characterization and management. Taking that into consideration, image logs were acquired by SOC in most of the drilled wells in the field. Integrated borehole case study was conducted for 10 wells to evaluate both natural and drilling induced fractures as well as the sub-seismic faults from image logs. The results represented by fractures orientation and intensity are used to examine the relationship and impact of natural and drilling induced fractures on gas production. The identified present day maximum horizontal stress indicated that a certain trend of the natural fractures relative to the maximum horizontal stress is controlling the reservoir productivity performance of Gargaf reservoir in Alestiqlal Field.

Keywords

Sirte basin • Gargaf • Image logs • Fractures
Horizontal stress

A. Elrafadi (✉) · M. Alwerfali
Sirte Oil Company, Brega, Libya
e-mail: aalrfadi@sirteoil.com.ly

W. S. Elbarghathee · S. Y. Bosnina · A. G. Najem
Schlumberger Oilfield Services, Tripoli, Libya

1 Introduction

The Gargaf Formation is mainly represented by a thick sequence of quartzitic sandstone with an estimated thickness of about 3100 ft (~950 m). In Sirte Basin, this formation is mainly tight and characterized by presence of intensive fracture network. These fractures network have huge gas capacity where this field is produced since 1990s to date. An average from 7 to 14 MMSCFD is the daily gas rate production of each producing well in Alestiqlal Field. The field is located over onshore Libya in the north central part of Sirt Basin and generally confined between latitudes 29°N and 30°N and longitudes 19°E and 20°E. in the north of Concession-6 between Jahama and Zelten Platforms (Fig. 1). The structure is a complexly faulted Gargaf Structure High and appears as a northwest-southeast trending faulted wedge.

2 Methodology and Workflow

The work is performed based on integrated data that is available within Sirte Oil Company and can be listed as:

- conventional well logs data (Gamma Ray, Density, Neutron and Resistivity).
- borehole images represented by Fullbore Formation Microimager (FMI) and Ultrasonic Borhole Imager (UBI) tools;
- seismic sections for the area of interest;
- production data per well.

The aim of this work was to interpret manually the image data from the 10 wells available to extract the geological features to be correlated with the other data, especially the production. The workflow is represented by:

F. Rossetti et al. (eds.), *The Structural Geology Contribution to the Africa-Eurasia Geology: Basement and Reservoir Structure, Ore Mineralisation and Tectonic Modelling*,
Advances in Science, Technology & Innovation, https://doi.org/10.1007/978-3-030-01455-1_26

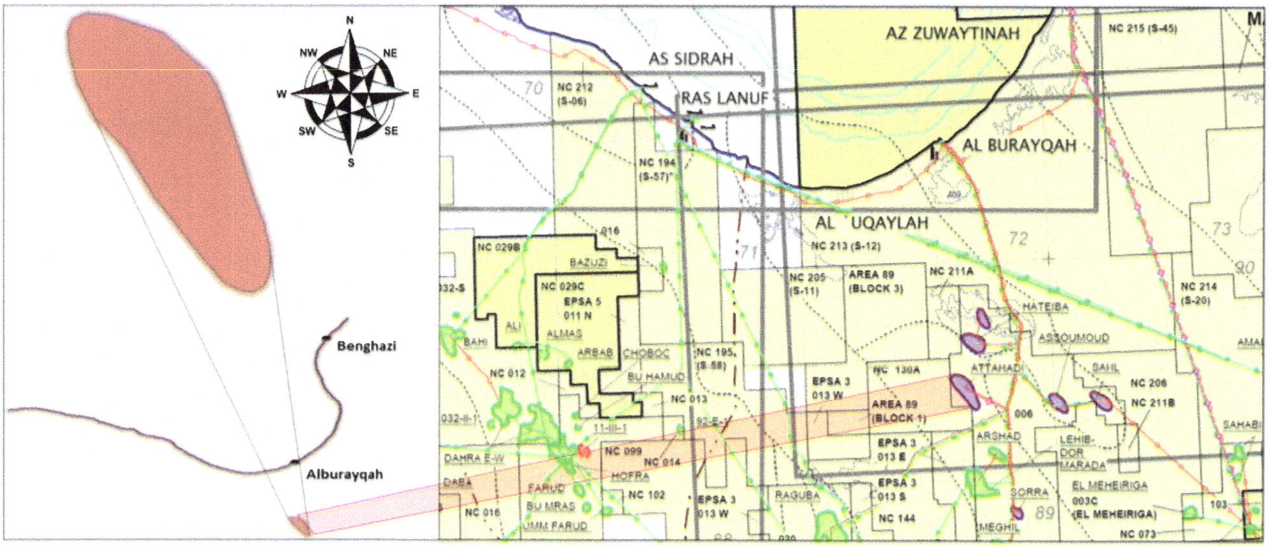

Fig. 1 Location map shows the locality of Alestiqlal field in Sirt Basin

- data collection;
- quality control;
- manual dip picking of the geological features and classification;
- fractures, faults and bedding characterization and analysis;
- single to multi-borehole analysis of the results with production data.

3 Results

In this study, more than 35,100 geological features were interpreted from the image logs including: structural bedding, natural fractures, drilling induced fractures, borehole breakouts, sub-seismic faults, and unconformity surface (Fig. 2). The natural fractures are classified based on their

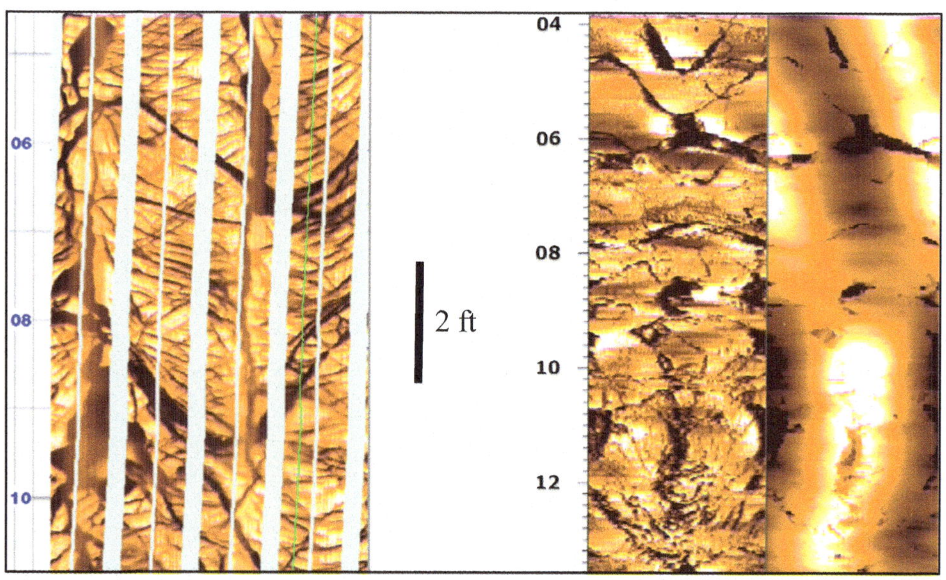

Fig. 2 Examples of natural fractures from left FMI Image Log and right UBI Image Log

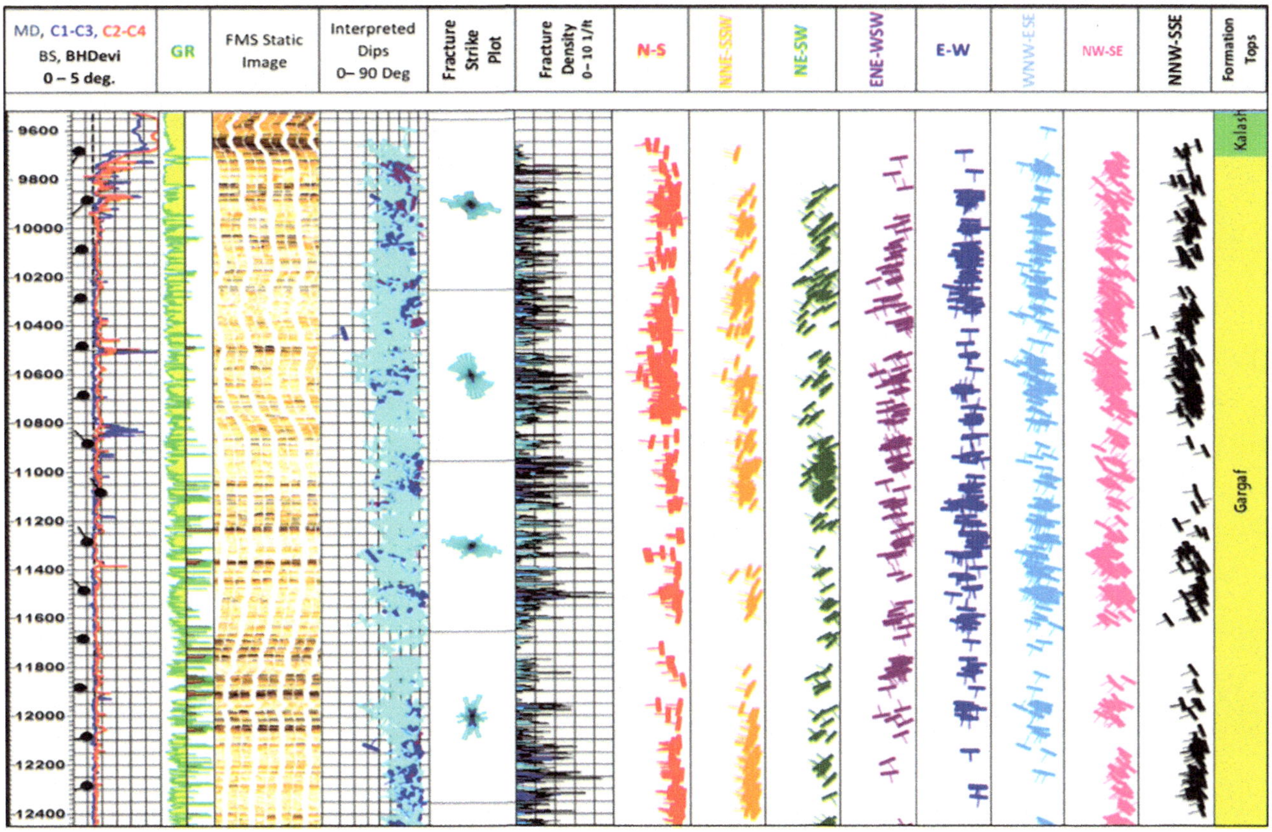

Fig. 3 Summary display of the well FF23-6 within the logged section of FMI image. The track 1 includes measured depth (MD) and borehole drift. Track 2 includes calipers (C1 & C2). Track 3 includes Gamma Ray (GR) cut-off that shows the sand in yellow and shale in brown. Track 4 includes the FMI static image. Track 5 includes all interpreted natural open fractures. Track 6 includes strike rosette for the open fractures along the logged section. Track 7 refers to the fracture density distributed throughout the well. Tracks 8–15 represent the cluster distribution of the open fractures per specific strike direction

physical appearance, apparent thickness, geometry relative to the borehole axis, and the continuity of their trace on image logs. The natural fractures manually picked are more than 10,000 fractures with different intensities from one to another (Fig. 3). Generally, the fracture distribution shows two main trends of fractures striking NE–SW and NW–SE.

4 Discussion

Even though this field has been producing since the 1990s, proper reservoir characterization did not take place. Recent efforts were made to interpret the image logs in an attempt to be correlated with the production data of the wells. More than 35,100 geological features identified within only 10 wells from the field revealed the behavior of the fractures in terms of intensity as well as direction and their contribution to production (Fig. 4). Future planning of increasing production and treating the reservoir should take into consideration the stresses identified in this study in case of directional fracturing stimulation operations.

5 Conclusion

The field study shows that the wells have higher fractures intensities with preferred direction with the maximum stress. More production rates have been recorded in FF27 and FF26 wells. In well FF27 more than 2150 natural fractures are

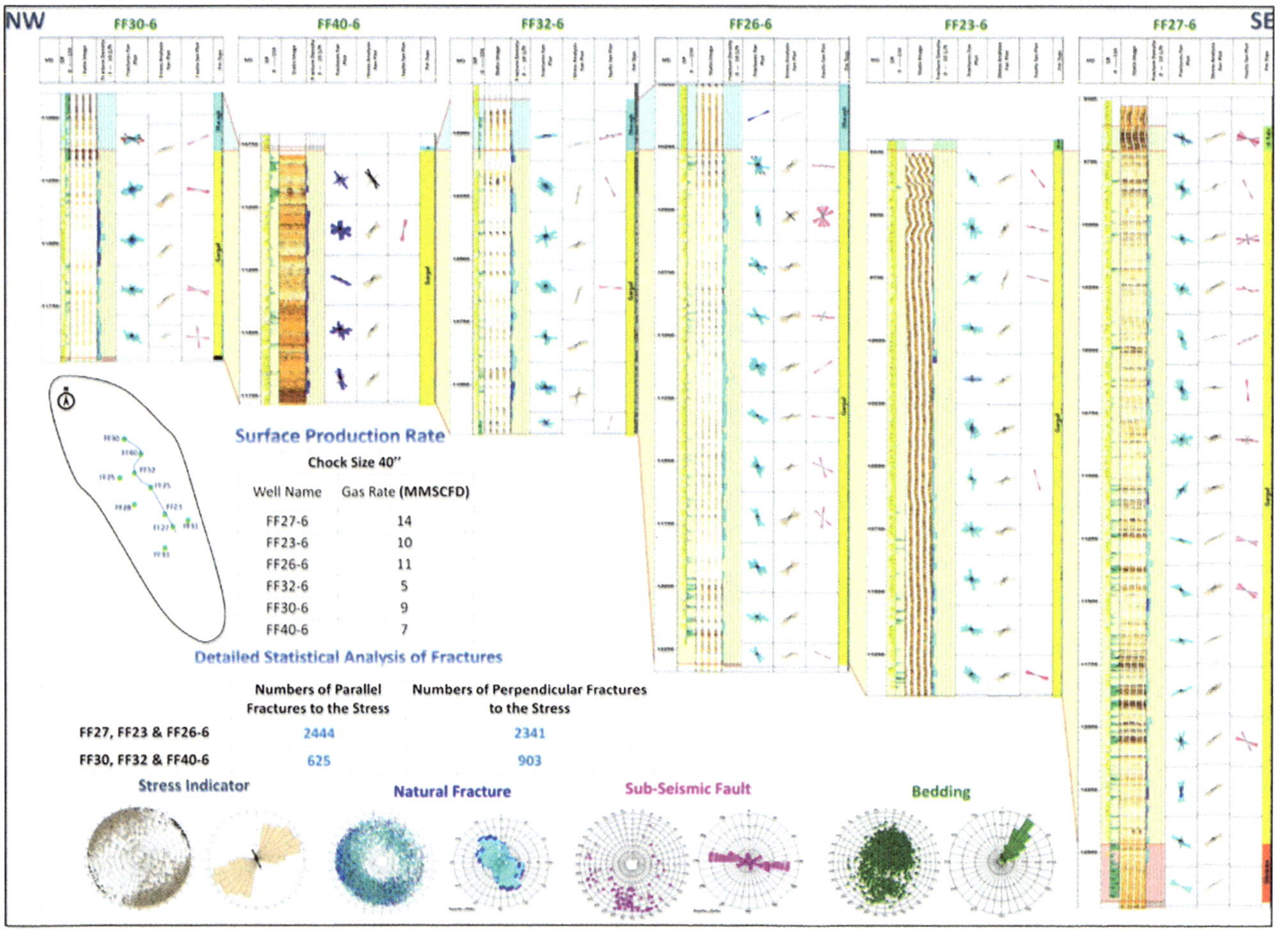

Fig. 4 Cross-correlation of the well logs with indication of the mechanical stratigraphy. It is noticed that the more fractures, especially if they are parallel to the maximum horizontal stress, the more production as in wells FF27 and FF23

picked while in well FF23 more than 1773 with an average intensity reaching more than 6 fractures per foot. In contrast, other wells such as FF40 and FF30, show around 700 fractures with an average of 4 fractures per foot.

Characterization of Stylolite Networks in Platform Carbonates

Elliot Humphrey, Enrique Gomez-Rivas, Juan Diego Martín-Martín, Joyce Neilson, Yao Shuqing, and Paul D. Bons

Abstract

Stylolites are rough dissolution surfaces that form by intergranular pressure solution resulting from burial compaction or tectonic stress. Despite being ubiquitous in carbonate rocks and potentially influencing fluid flow, it is not yet clear how the type and distribution of stylolite networks relate to lithofacies. This study investigates Lower Cretaceous platform carbonates in the Benicàssim area (Maestrat Basin, Spain) to statistically characterize stylolite network morphology in a variety of typical shallow-marine lithofacies. Stylolites in each lithofacies were sampled in the field and different measuring techniques were applied. Grain size, sorting and composition were found to be the key lithological variables responsible for the development of rough anastomosing networks. A statistical workflow, requiring only limited subsurface information, relating stylolite morphology with lithofacies can subsequently be used to predict fluid flow behavior in stylolitised reservoirs.

Keywords

Stylolite • Carbonate • Diagenesis • Lithofacies
Maestrat basin

E. Humphrey (✉) · E. Gomez-Rivas · J. Neilson · Y. Shuqing
School of Geosciences, King's College, University of Aberdeen, Aberdeen, AB24 3UE, UK
e-mail: e.humphrey@abdn.ac.uk

E. Gomez-Rivas · J. D. Martín-Martín
Departament de Mineralogia, Petrologia i Geologia Aplicada, Facultat de Ciències de la Terra, Universitat de Barcelona (UB), Martí i Franquès s/n, 08028 Barcelona, Spain

P. D. Bons
Department of Geosciences, Tübingen University, Wilhelmstr. 56, 72074 Tübingen, Germany

1 Introduction

Carbonate reservoirs are notoriously complex to develop due to spatial and temporal variations in petrophysical properties which, when combined with the presence of sub seismic-scale structures, inevitably impacts reservoir quality [1]. Structures in carbonates such as fractures and stylolites can have profound impacts on bulk rock permeability and fluid flow pathways, therefore requiring the use of predictive guides to help characterize structural influences. The morphology of stylolites varies greatly in terms of their amplitude, wavelength, spacing and connectivity, which can produce a variety of stylolite network geometries. Stylolites have been predominantly researched for their influence on fluid flow. However, little attention has been given to understanding the lithological controls on the formation of stylolite networks, their morphology and their behavior in the presence of fluids.

This study aims to statistically characterize stylolite morphology and distribution in relation to different lithofacies, providing a basic workflow to help predict stylolite network properties and their potential impact on fluid flow based on outcrop scale observations. The results of this study can help to understand the influence of lithological variations on stylolite populations at the outcrop scale, which can be utilized for inferring stylolite characteristics and subsequent impacts on fluid flow in carbonate reservoirs during production.

2 Methods

Stylolites were sampled from Aptian–Albian carbonate platform deposits in the Benassal Fm, located on the southern margin of the Maestrat Basin (Spain). Platform deposits are shallow marine and predominantly characterized by the presence of orbitolinid, rudist and coral-rich facies [2]. The Benessal Fm has been subdivided into thirteen lithofacies types, containing stylolites ranging from

F. Rossetti et al. (eds.), *The Structural Geology Contribution to the Africa-Eurasia Geology: Basement and Reservoir Structure, Ore Mineralisation and Tectonic Modelling*,
Advances in Science, Technology & Innovation, https://doi.org/10.1007/978-3-030-01455-1_27

long-parallel to anastomosing networks. Seven of these lithofacies were selected to be sampled; dolostone, framework coralline limestone, bioclastic grainstone, oolitic/peloidal grainstone, rudist floatstone, bioclastic wackestone/packstone, and spicule wackestone.

A variety of stylolite measurements were collected from different lithofacies to characterize different network morphologies. One m^2 sampling windows were used to measure the abundance, type and size of stylolites in different lithofacies. Vertical scanlines in each sampling window were used to measure stylolite spacing, amplitude and wavelength, whilst measurements relating to stylolite network connectivity and host rock 'islands' were collected using methods proposed by [3].

Stylolite intersection types were measured by abundance and density as a function of stylolite length. Intersections were classified as either *X*, *Y* or *I* [4] and tallied per window to provide a total abundance per lithofacies. The density of intersection types as a function of stylolite length was calculated for each lithofacies.

Stylolites were classified based on morphology using definitions [5] where stylolites are classified as (i) rectangular-, (ii) seismogram-, (iii) suture & sharp- (SS) and (iv) wavy-type. Stylolite types were counted per window and totaled to provide a percentage abundance of stylolite types per lithofacies.

3 Results

3.1 Statistical Analysis of Stylolite Morphology

Spacing, wavelength and amplitude measurements can typically be profiled according to whether the lithofacies is grain-or mud-supported; grainstones feature larger spacings with low amplitudes and wavelengths in comparison to sampled wackestones and packstones. Stylolites in all lithofacies were predominantly discontinuous. However, mud-supported lithofacies featured a higher proportion of *X* & *Y*-type intersection and therefore had a higher intersection density. Mud-supported lithofacies had a density of 2.6–3.2 per meter compared to 0.7–1.2 per meter in grain supported facies.

Over 50% of all stylolites in the sampled lithofacies were suture & sharp- type whilst wavy-type stylolites were found in varying abundances. Rectangular-type stylolites were found exclusively in grain-supported lithofacies. In lithofacies with a sufficient level of anastomosis, the dimensions of 'islands' shows that spicule wackestone facilitates the largest areas, whilst coralline limestone produced 'islands' which were the most lenticular-shaped.

Petrographic and morphological observations were used to create a correlation matrix, identifying relationships between stylolites and lithology. Pearson correlation coefficients between −1 and +1 were calculated for each pair of input parameters.

4 Discussion

Key relationships influencing stylolite morphology are derived from the correlation matrix of host rock lithological parameters, which can effectively evaluate different heterogeneities and their influences on a variety of stylolite morphological parameters.

A combination of grain size, composition and sorting were predominantly responsible for increasing stylolite roughness. This study shows that lithologies with 10% of grains larger than 2 mm are positively correlated with higher amplitudes, which is seen in varying extents in lithofacies with different fossil content. Lithofacies with a poorly sorted matrix are positively correlated to higher amplitude stylolites with a higher degree of anastomosis. Matrix composition can promote solubility variations, caused by mud and fossil content, producing stylolite networks that are anastomosing and feature high amplitudes with low vertical spacing measurements (Fig. 1).

Rectangular-class stylolites are abundant in grain-supported lithofacies with centimeter-scale bedding, such as oolitic grainstones, supporting previous work [5] by indicating rectangular-class stylolites propagate at layer interfaces. Our study shows that SS-type stylolites preferentially develop in a poorly sorted matrix with bimodal grain sizes to ultimately create non-uniform noise with larger pinning grains.

Isolated stylolites are most prevalent in a grain supported matrix with 10% of grains <2 mm, where grains are well sorted, and bedding is on the centimeter to decimeter scale. In contrast to this, stylolite networks require a poorly sorted mud-supported matrix with a bimodal grain size distribution where 10% of the matrix is >2 mm, which is facilitated by the presence of large fossils such as corals, bivalves and rudists. These networks have significant impacts on fluid flow, either if stylolites are sealing or if they are re-opened later on during diagenesis.

Stylolite networks are statistically characterized according to morphological variations in carbonate platform lithofacies. Multiple sources of lithological heterogeneity are identified and related to stylolite morphology. The interplay between grain size, distribution, composition and bedding scale can result in closely spaced, high amplitude stylolites that feature a high level of anastomosis. The extent of stylolite network connectivity, and subsequent interactions with fluids, can therefore be primarily controlled by lithofacies variations. This study represents the initial phase of a workflow which can quantify stylolite networks according to lithological

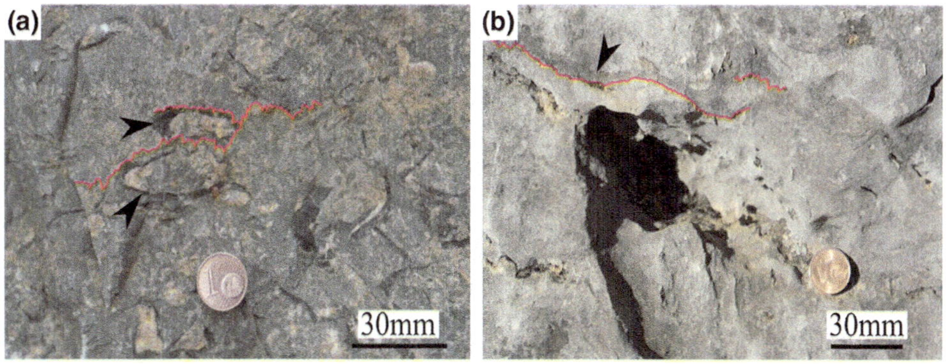

Fig. 1 **a** Y-type intersection along a rudist fossil (arrow) in a rudist floatstone. **b** Stylolite propagating along Chondrodonta boundary (arrow) in a bioclastic wackestone/packstone

parameters which can be used to seed fluid flow simulations, enabling permeability measurements to be estimated and subsequently upscaled during reservoir model building.

References

1. Agar, S.M., Geiger, S.: Fundamental controls on fluid flow in carbonates: current workflows to emerging technologies. Geol. Soc. **406**(1), 1–59 (2015). London, Special Publications
2. Martín-Martín, J.D., Gomez-Rivas, E., Bover-Arnal, T., Travé, A., Salas, R., Moreno-Bedmar, J.A., Tomás, S., Corbella, M., Teixell, A., Vergés, J., Stafford, S.L.: The Upper Aptian to Lower Albian syn-rift carbonate succession of the southern Maestrat Basin (Spain): Facies architecture and fault-controlled stratabound dolostones. Cretac. Res. **41**, 217–236 (2013)
3. Ben-Itzhak, L.L., Aharonov, E., Karcz, Z., Kaduri, M., Toussaint, R.: Sedimentary stylolite networks and connectivity in limestone: large-scale field observations and implications for structure evolution. J. Struct. Geol. **63**, 106–123 (2014)
4. Healy, D., Rizzo, R.E., Cornwell, D.G., Farrell, N.J.C., Watkins, H., Timms, N.E., Gomez-Rivas, E., Smith, M.: FracPaQ: a MATLAB toolbox for the quantification of fracture patterns. J. Struct. Geol. **95**, 1–16 (2017)
5. Koehn, D., Rood, M.P., Beaudoin, N., Chung, P., Bons, P.D., Gomez-Rivas, E.: A new stylolite classification scheme to estimate compaction and local permeability variations. Sediment. Geol. **346**, 60–71 (2016)

Mumbai High Metamorphic Basement Reservoirs—An Integrated Workflow Based Approach for Fracture Modelling and Real Time Planning and Monitoring

S. K. Mukherjee and S. N. Chitnis

Abstract

The Mumbai High field in the Western Offshore in India, is a priority area for extending the concept of fracture characterization in metamorphic basement reservoirs. The basement in the Mumbai High is hydrocarbon bearing in few areas proximal to major fault damage zones and intersections of major regional tectonic cross trends. The challenge remains in characterizing such basement reservoirs with significant heterogeneities in mineralo-facies, in situ stress fields, seismic amplitudes, fracture properties and connectivity, and flow potential. This necessitated the development of an integrated static fracture model workflow assimilating structural modeling, seismic and petro-physical interpretations for fracture drivers and geo-cellular fracture modeling, fine-tuned using geological concepts and point data extracted from well data analyses. The deterministic geo-cellular fracture model thus prepared has been calibrated with real time well observations and has been found to satisfactorily explain anomalous hydrocarbon accumulation and flow pattern in basement wells penetrated in excess of 100 m and tested in the area. The adopted workflow has helped planning for well-evaluating and exploiting basement reservoirs as well for the real time monitoring of wells.

Keywords

Mumbai high • Geo-mechanical restoration
Geo-cellular model • Fracture intensity

1 Introduction

The tectonic elements defining basement architecture of Mumbai High are defined by three major Precambrian orogenic trends in the western part of the Indian shield, the NE-SW Aravalli, ENE-WSW Satpura, and NNW-SSE Dharwar trends [1]. Chronicling of episodic reactivation of NNW-SSE trending Mumbai High East fault provides clues to periodic opening and closing of genetically- related fractures in basements in the geologic past. The gross rock association reflected from analyses of available basement cores and sidewall cores indicates dominance of Tonalite Trondhjemite gneiss (TTG), chlorite schists and granites, which are typical assemblages of the Archaean Granite greenstone terrane characterizing the Dharwar craton. Hydrocarbon accumulation in the Mumbai High basement is sporadic in nature and has demonstrated inconsistent flow behaviour. This inconsistency can be attributed to the variation in fracture intensity in the basement lithology, the orientation and aperture characteristics of fractures and the change in stress magnitudes over the field. There is a tendency for fracture height and dimension to be limited in the metamorphic formations by the layering [3].

2 Approach

In order to address heterogeneity of metamorphic reservoirs, penetrated in excess of 300 m in few wells in Mumbai High, the best plausible solution lays in the collation of separate fracture characterization workflows (Fig. 1) in order to create of a dependable static fracture model combining seismic attributes with petro-physical, geo-mechanical and geological inputs.

Image logs, dipole sonic/sonic scanner logs, ECS™ and other conventional petro-physical data of basement in a number of wells were analysed to decipher the basement

S. K. Mukherjee (✉) · S. N. Chitnis
Oil and Natural Gas Corporation Limited, Centre of Delivery—Basement Exploration, WOB Priyadarshini Sion (E), Mumbai, 400022, India
e-mail: mukherjee_sk4@ongc.in

S. N. Chitnis
e-mail: imsanju69@gmail.com

F. Rossetti et al. (eds.), *The Structural Geology Contribution to the Africa-Eurasia Geology: Basement and Reservoir Structure, Ore Mineralisation and Tectonic Modelling*,
Advances in Science, Technology & Innovation, https://doi.org/10.1007/978-3-030-01455-1_28

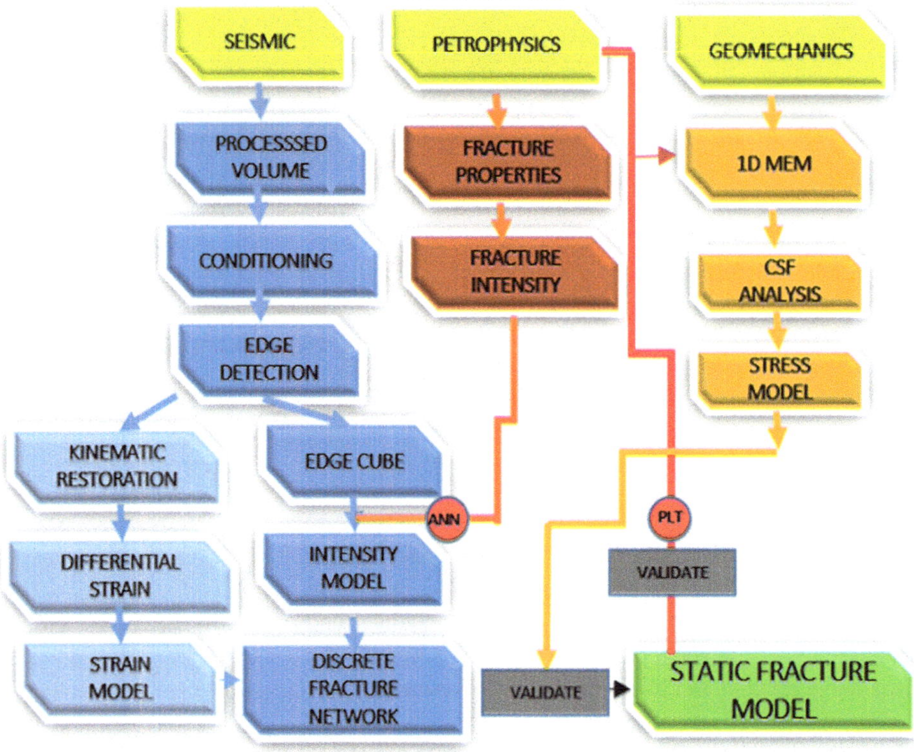

Fig. 1 Workflow for 3D static fracture model

fracture drivers, and to determine their density variation spatially with respect to geological controls.

3D seismic data, conditioned by Gaussian smoothening was used to compute various surface and volumetric attributes. Variance cube volume created from smoothened post stack seismic data was the input for computation of AntTrack^{TM} volumes which highlighted the basement discontinuity trends. Fracture drivers generated from seismic attributes and petro-physical data were correlated using artificial neural network to identify the best drivers and create a fracture intensity log. This was followed by property modelling with upscaled intensity log. The final step was integration of well data and seismic attribute analysis to build geo-cellular fracture intensity model which captured the spatial distribution of fractures and associated heterogeneities.

Geo-mechanical restoration based kinematic modelling utilising the genetic interrelationship of fractures with the structural evolution of the area was carried out to generate a predictive fracture model for the fractured basement reservoir.

Present day in situ stresses control the dilatancy factor of fractures by causing critical shear on optimal fracture orientations. SH_{max} orientation was inferred from image logs and cores, and magnitude was deduced from Leak Off Test and Extended Leak off test data of few drilled wells. 1D mechanical earth models were created for few wells for delineating critically stressed fracture zones in order to obtain an enhanced understanding to the inconsistent flow patterns from basement. To lend more accuracy to the static fracture model, the conceptual model was calibrated with collateral well data in the form of master-log data, mud loss zones, conventional and oriented cores, and validated by PLT and reservoir tests.

3 Case Study and Inferences

Analyses of drilling induced tensile fractures (DITF) and borehole breakout from image logs of basement interval in wells drilled in the area have helped decipher that the principal stress direction (SH_{max}) is oriented NNW-SSE with an estimated average value deduced as N146/326° ± 17 (Fig. 2).

This is validated by image data of wells and oriented core analyses in one well, Well-I. SH_{max} deduced in wells drilled are in broader agreement with regional SH_{max} projected in World stress map [2] for Indian sub-continent with some temporal- spatial variations. Paleostress tensors (directions and relative magnitudes) deduced from western on land field study and complemented by remote sensing lineament analyses show similar trend as observed in offshore areas [5]. Analyses of stress magnitudes suggest an overall strike slip stress regime in Mumbai High field. Out of four sets of

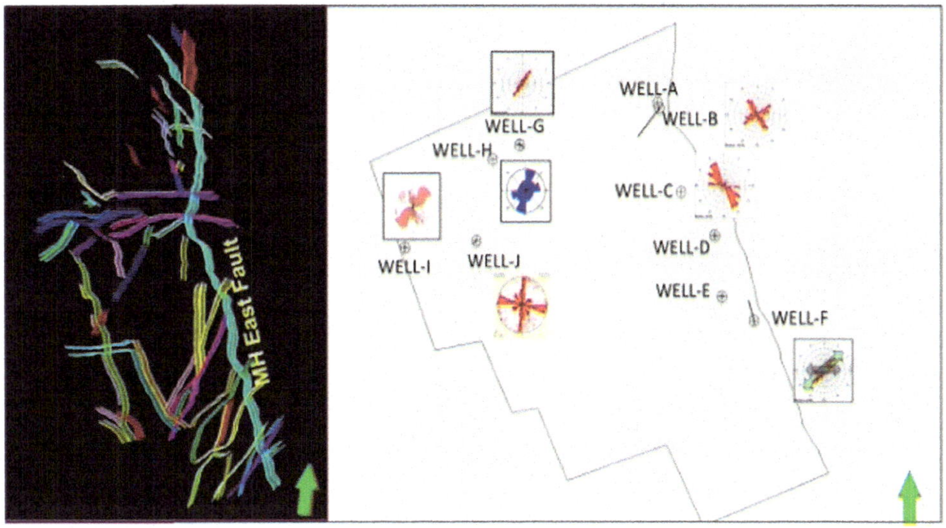

Fig. 2 Fracture strike rosettes plotted from image logs

fractures modelled, the fracture sets oriented in NW–SE direction (also the regional SH_{max}) is more prone to being critically stressed and is likely to remain open [4].

In the absence of LOT in basement, borehole breakout observations in the well-H were made use of to assume SH_{min} value ranging between 13 and 14.5 ppg as the lower limit and 14.5–16 ppg as the higher bound. SH_{max} values were estimated based on empirical relationship with a lower bound of 22–26 ppg and upper bound of 26–30 ppg. Simulation run for analyses of critically stressed fractures brought out two sets of critically stressed fractures dominantly oriented NE-SW in the well. 100 m apart in depth (Fig. 3).

Simultaneous examinations of fracture intensity model, petro-physical data and well observations and SH_{min} value of 13.95 ppg deduced from Extended Leak Off Test in WELL-I supported the CSF analysis. The bottommost zone in isolation flowed gas.

Good fracture intensity and optimal orientation of discrete fractures parallel to the maximum horizontal stress oriented in the NW-SE direction are observed in wells A and B, which have shown good flow potential from the basement. Fracture model and well data suggest absence of notable discrete fractures in WELL-C which went dry. Image and shear logs recorded in basement interval of WELL-F matched the best fracture pods predicted in the model (Fig. 4). Flowing intervals interpreted in Production log (PLT) recorded in well-G closely matched fractures predicted in model. WELL-J, drilled with an azimuth almost orthogonal to the inferred maximum horizontal stress

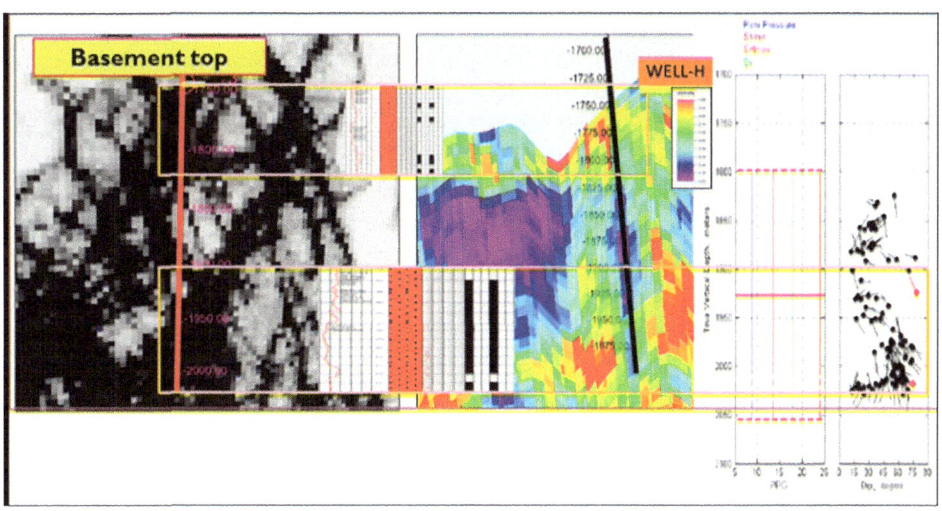

Fig. 3 Fracture characterization in WELL-H

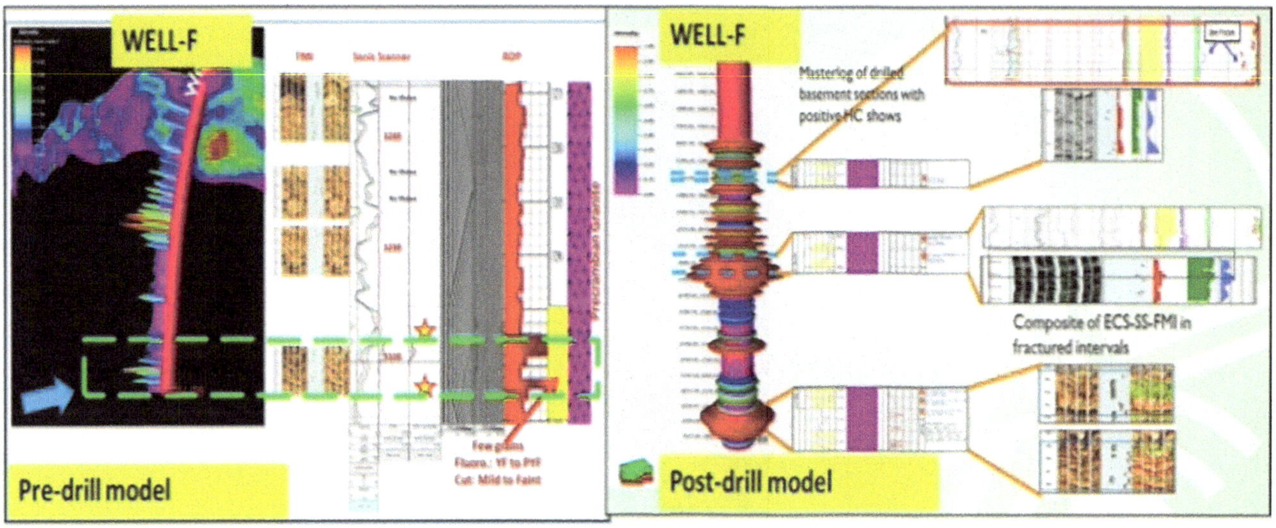

Fig. 4 Fracture characterization in WELL-F showing good calibration of well data with pre-drill and post drill fracture model

direction cut through densely fractured patches as predicted in the model and produced oil from basement.

4 Conclusions

Reservoir characteristics of the Mumbai High South fractured basement structure compare to a large extent with those at other "buried-hill" type traps reported worldwide viz., Xinlungtai field in North China basin where two main processes contribute to effective porosity: (a) weathering and mechanical reworking and (b) open fracture and joint networks. The Mumbai High basement reservoir behaves more like Type II reservoir with significant matrix support [6]. Evidences of Type I reservoir characteristics in Mumbai High much below basement top, is also evident from flow test results in a number of wells. Wells drilled based on the predictive fracture model and well path designed keeping in mind the SH_{max} orientation yielded positive results in terms of fracture prediction, interception and flow of hydrocarbons. The model facilitates real time monitoring of wells being drilled through fractured basement reservoir by constant updating of well path, facilitating course correction as required to intercept better fracture sweet pods.

Acknowledgements Authors express their sincere gratitude to ONGC for kind permission to publish this paper. Views expressed in the paper are those of authors only and not of ONGC.

References

1. Biswas, S.K.: Rift Basins in Western margin of India with special reference to hydrocarbon prospects with special reference to Kutch Basin. Amer. Assoc. Pet. Geol. Bull. **66**(10), 1497–1513 (1982)
2. Heidbach, O., Rajabi, M., Reiter, K., Ziegler, M.: World Stress Map 2016. Hosted by the Helmholtz Centre Potsdam—GFZ German Research Centre for Geosciences (2016)
3. Gutamanis, J.C.: Basement Reservoirs-A Review of their Geological and Production Characteristics. IPTC 13156 (2009)
4. Kumar, S., Sahoo, M., Chakrabarti, S.K.: Geomechanical restoration based fracture modelling of mumbai high unconventional basement reservoir. In: SPG India 11th International Biennial Conference & Exposition (2015)
5. Misra, A.A., Bhattacharya, G., Mukherjee, S., Bose, N.: Near N-S paleo-extension in the western Deccan region in India: does it link strike-slip tectonics with India-Seychelles rifting? Int. J. Earth Sci. **103**, 1645–1680 (2014)
6. Mukherjee, S.K., Rao, S.C., Chitnis,S.N., Ray, B.B.: An integrated approach towards characterising metamorphic basement reservoirs: a case study from Mumbai high South Basement Reservoir. ONGC Bull. **51**(2), 62–77 (2016)

Characterizing Pandanallur Fractured Basement Reservoir: A G and G Enigma

S. K. Mukherjee, J. Sarkar, and S. B. Chitnis

Abstract

Fractured basement reservoirs are challenging to model due to the complexity of processes involved in the generation and preservation of the fracture network as well as the heterogeneity of the host rock. The Pandanallur basement, a predominantly granite gneiss fractured reservoir, exhibits wide scale vertical and lateral heterogeneity of fracture patterns and stress parameters, which has a bearing on migration and the accumulation of hydrocarbons and fluid flow potential. The best way to characterize such unconventional reservoirs is through an integrated approach combining fracture intensity and network modeling with inputs from petrophysical, kinematic and geo-mechanical analyses. The synergetic integration adapted for the Pandanallur basement reservoir has been able to characterize the potential flow contributing fractures and reduced economic risks by optimizing new well targets.

Keywords

Drilling induced tensile fractures • Borehole breakout
Ant track • Fracture intensity

1 Introduction

The NE-SW oriented Pandanallur structure, located on the northern flank of Kumbakonam-Madanam ridge, falls in the southern part of Ariyalur–Pondicherry sub-basin of the Cauvery Basin [2]. The prevailing in-situ stress direction in the Ariyalur-Pondicherry sub basin, as deduced from Drilling Induced Tensile Fractures (DITF) and bore hole breakouts in few wells drilled in basement, is found to be NW-SE to a WNW-ESE direction.

Analysis of hydrocarbon distribution in basement in Pandanallur area demonstrates a concentration of basement hydrocarbon occurrences in the flanks of prominent horsts, where there's a juxtaposition of source rock against basement highs, as well as high angle faults offset by cross faults which have created reservoir conditions for basement. Out of seven key wells drilled to basement, two wells viz., PN-A and C drilled on the crest of basement highs were dry (Fig. 1).

The discovery well PN-E is optimally placed at the flank of a horst in proximity to the source (Fig. 2). Two subsequent appraisal wells PN-F and G, the former drilled near the crest of basement high in blocks adjacent to PN-E, had hydrocarbon indications but did not flow.

To deal with the heterogeneity observed in drilled wells, an approach synergizing simultaneous analyses of seismic data, borehole images, sonic measurements, conventional log data, geochemical and geo-mechanical parameters was adopted to generate a robust fracture model of basement reservoir.

2 Methodology

The relative importance of fractures in controlling rock hydraulic properties depends on the fracture density and orientation, along with the hydraulic and mechanical characteristics of fractures themselves. It has been shown that most hydraulically conductive fractures are those that are critically-stressed in the current stress regime [1].

Conditioned and smoothened 3D pre-stack time migrated data (PSTM) was used to generate various horizon and volumetric attributes to define the geometry of the seismic discontinuities. Variance cube for edge detection was the input to generate the final ant track volume highlighting

S. K. Mukherjee · S. B. Chitnis
Oil and Natural Gas Corporation Limited, Centre of Delivery—Basement Exploration, WOB, Priyadarshini, Sion (E), Mumbai, 400022, India

J. Sarkar (✉)
Oil and Natural Gas Corporation Limited, Cauvery Basin, CMDA Tower-I, Egmore, Chennai, 600008, India
e-mail: sarkar_jayanta@ongc.co.in; jyntsrkr@gmail.com

F. Rossetti et al. (eds.), *The Structural Geology Contribution to the Africa-Eurasia Geology: Basement and Reservoir Structure, Ore Mineralisation and Tectonic Modelling*,
Advances in Science, Technology & Innovation, https://doi.org/10.1007/978-3-030-01455-1_29

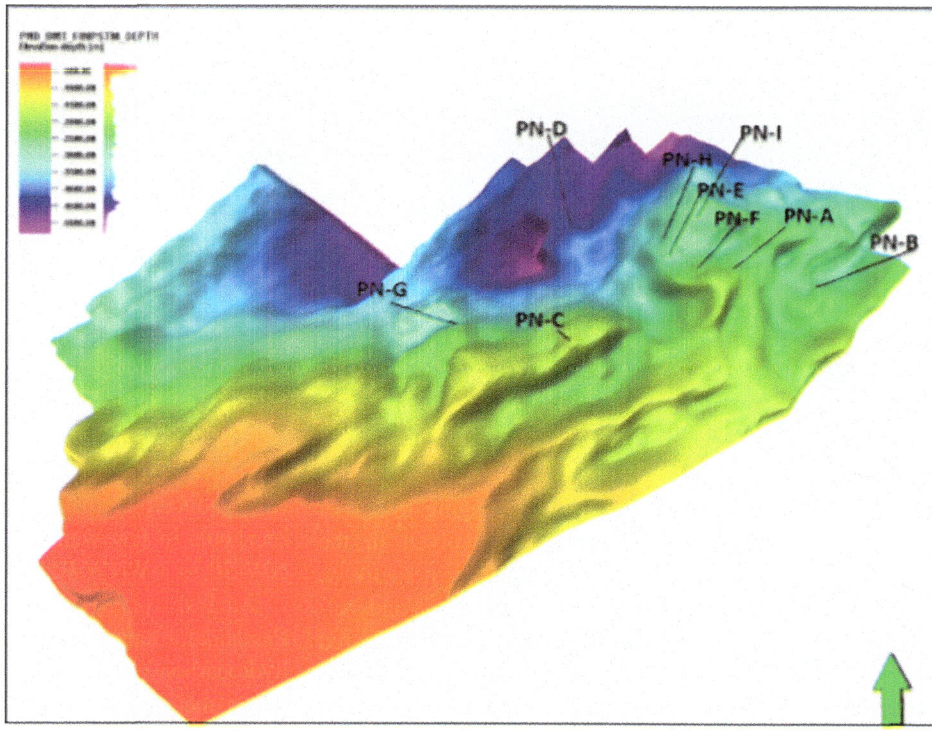

Fig. 1 3D depth relief of Pandanallur basement high displaying well locations

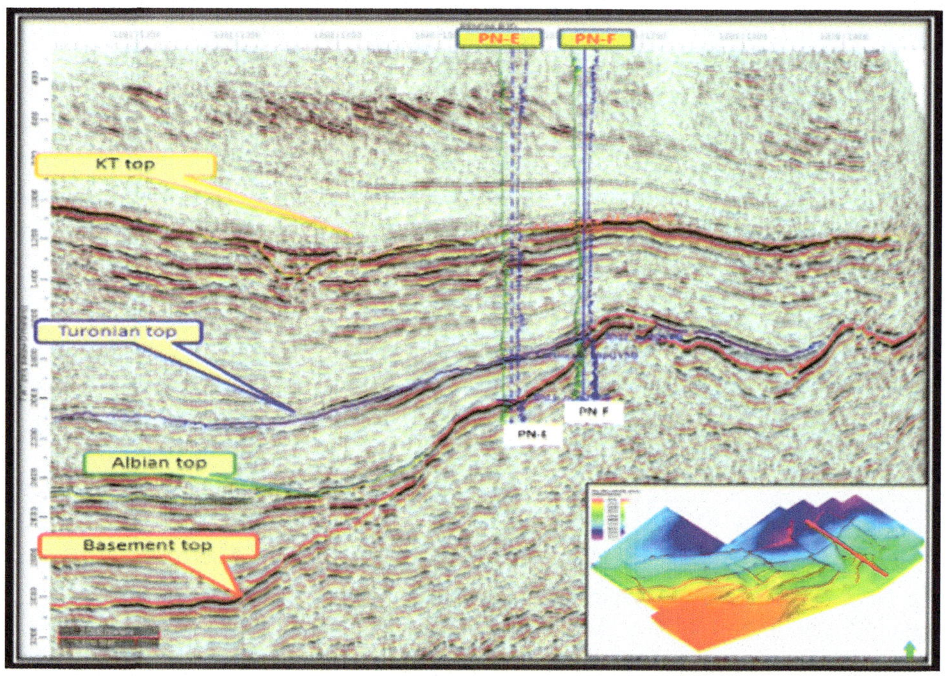

Fig. 2 Seismic arbitrary line through wells PN-E and PN-F

basement discontinuities. Analyses of image logs of basement intervals of three wells viz., PN-E, F and G, shear logs and other conventional petrophysical data were used to decipher the fracture drivers. Fracture intensity logs were generated using the fractural properties of the three wells. Artificial Neural Network (ANN) tool helped bring out the best correlatability between seismic attributes and petrophysical properties and identified the best fracture drivers for characterization [4]. The second step was property modelling with an upscaled intensity log. The final step was the integration of well data analysis and seismic attribute analysis to build fracture intensity model.

Morphotectonic analysis was carried out for Cauvery Basin to demarcate geomorphic highs indicating maximum intersection density of lineaments reactivated under present stress regime. The in-situ stress direction (SH_{max}) was deduced from drilling induced fractures. World stress map and focal plane solutions were found to be oriented in a NW-SE to ENE-WSW direction [3]. Geo-mechanical restoration was carried out to delineate areas of high strain accumulation.

Oil to oil and oil to source correlation using geochemical fingerprinting techniques for PN-E oil vis' a vis other basement oils in Ariyalur–Pondicherry, were carried out to understand the migration and accumulation pattern of hydrocarbons in Pandanallur basement reservoir [5].

3 Case Study and Discussion

Morphotectonic analysis have brought out that the E-W oriented faults are the most reactivated trends under present day stress regime. Geo-mechanical restoration studies predict an overall strike slip stress regime with the likelihood of critically stressed fractures being oriented N-S to NNW-SSE and in few cases almost E-W.

PN-G basement oil is similar to PN-E oil in terms of geochemical finger prints and API gravity, thus establishing the pool connectivity to same kitchen area [5].

FMI™ logs of the wells PN-E, F and G were analyzed to understand the nature of fractures, their distribution patterns and their orientation. In the well PN-E, the bottom interval is envisaged to have contributed to production along with a section just 100 m below basement top with dominant fracture strike direction oriented ENE-WSW. Bad hole conditions are evident on caliper log of this interval points to a high strain accumulation zone in this interval (expression of shear stress). Ant track section and fracture intensity models suggest good fracture intensity in these intervals (Fig. 3).

The FMI logs of well PN-F show the presence of dominantly NE-SW striking fractures with south easterly dips. Principal horizontal stress, as deduced from analyses of drilling induced tensile fractures (DITF) in wells PN-D and F, is oriented NE-SW. This is not in alignment with the regional stress maxima and it is possible that none of the fractures are in slip state (i.e., critically stressed fractures are absent; Fig. 4).

Fracture connectivity analysis on ant track volume of wells PN-E and F demonstrates a NNW-SSE to N-S connectivity of fractures encountered in PN-E to the kitchen in the north, whereas fractures are mostly oriented NE-SW in well PN-F.

In PN-G, fracture strike is dominantly NE-SW. A few sub-vertical fractures are oriented NNW-SSE, with maximum slip tendency (critically stressed) are seen near the basement top. This is in alignment with the maximum horizontal stress azimuth deduced from FMI™ and Sonic scanner analyses. This observation is also validated in ant

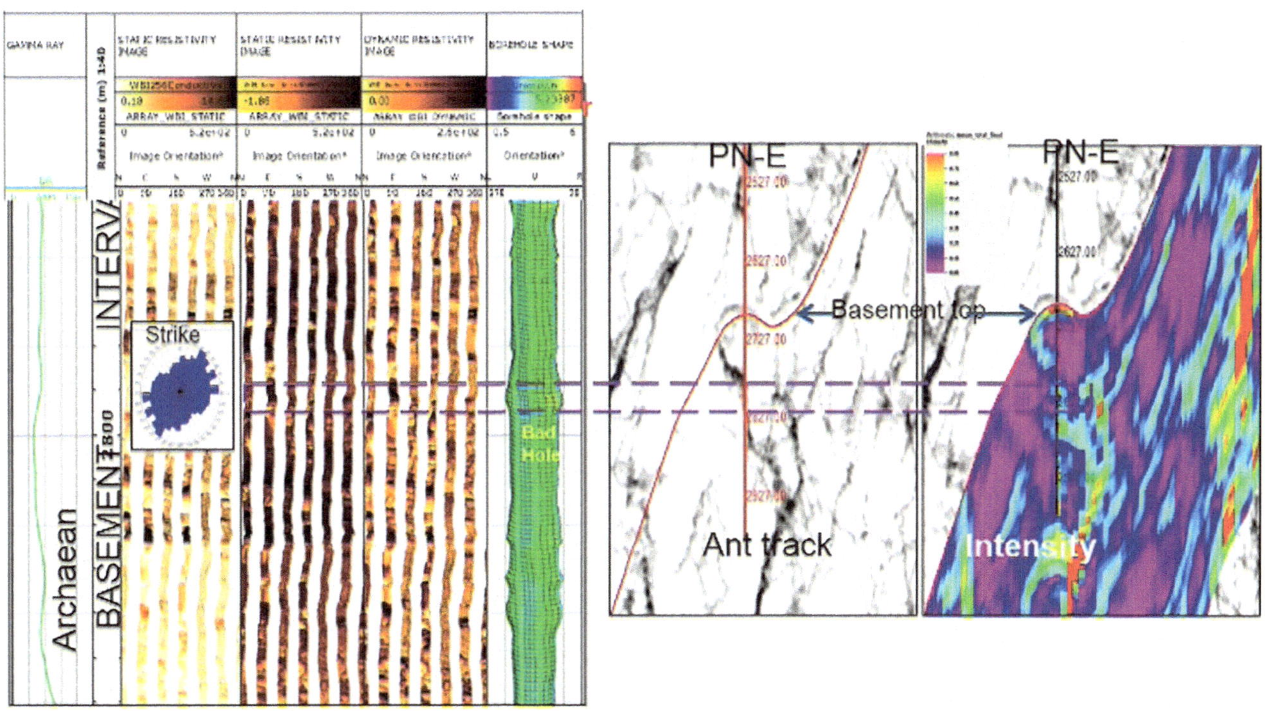

Fig. 3 Composite showing fractured interval in well PN-E (zone marked with blue dotted line showing calibration of fractured interval on FMI, ant track and intensity sections)

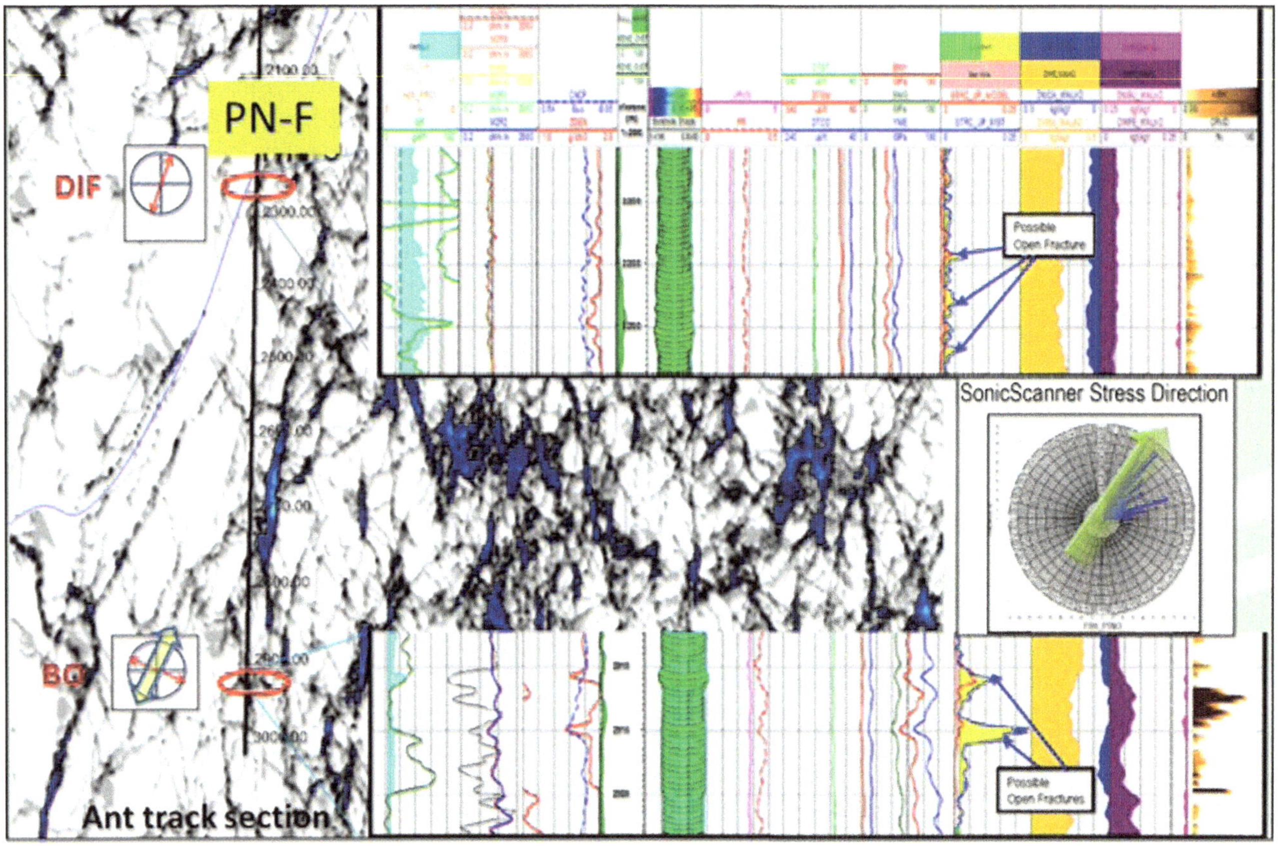

Fig. 4 Fracture calibration on ant track section through well PN-F using interpreted composite of ECS-Sonic Scanner log

track and intensity section drawn through the well. CMR™ log also indicates very minor free fluid and poor permeability throughout the Basement section.

Lateral variation of stress orientation observed in drilled wells may be attributed to the location of these wells in different fault blocks as well as proximity of a well to the fault. Vertical variation of stress seems to be due to rheological changes in basement lithology and/or presence of different adjoining faults at different depths, particularly shallow expressions of shears, which have affected the Cauvery basin at many places.

4 Conclusions

The current understanding of Pandanallur basement indicates the brecciated nature of the reservoir supported by fracture permeability. Accumulations are primarily controlled by faults connected to local sources. Short distance migration is the key in zeroing in on prospective locales as is evident from the absence of hydrocarbons in wells PN-A and B. First available flank structures against faults with significant damage zone seems to be the locales where hydrocarbon accumulation can be expected.

Fracture connectivity analysis is of prime importance in deciding future well placements as has been planned in prospects PN-H and PN-I. In the absence of PLT, it is not possible to assess the contribution of non-weathered fractured basement.

Static models alone cannot account for all basement heterogeneities. An approach integrating 3D fracture intensity model with well data is the best method to optimize well locations by predicting to a fair degree of accuracy the fracture intensity expected at the well bore.

Acknowledgements Authors express their sincere gratitude to ONGC for kind permission to publish this paper. Views expressed in the paper are those of authors only and not of ONGC.

References

1. Johri, M., Zoback, M.D., Hennings, P.: Observations of Fault Damage Zones at Reservoir Depths, pp. 11–446. ARMA (2011)
2. Kashyap, P., Haritha, M., Subrahmanyam, O., Vema, J., Ganesh, N., Ayyadurai, M., Josyulu, B.S.: Delineation of Basement fracture pattern in Madanam-Pandanallur area of Cauvery basin. In: 10th Biennial International Conference & Exposition. SPG India, Kochi (2013)

3. Mazumder, S., Tep, B., Pangtey, K.K.S.: A Remote Sensing and GIS Based Integrated Approach for Identifying Promising Areas for Basement Exploration in Cauvery Basin. SPG India, Jaipur (2015)
4. Mukherjee, S.K., Sarkar, J., Kadam Kanchan, N., Rao, S.C., Chitnis, S.N.: Granitic basement reservoir fracture modelling, challenges and its mitigation: a case study from Khoraghat field, Assam. In: 4th SPWLA India Symposium (2017)
5. Sivan,P., Nidhi: High resolution geochemical characterization of oils from Basement fields. Cauvery Basin, KDMIPE, ONGC Report unpublished (2017)

Petroleum Play Potential in the Thrust and Fold Belt Zone of the Offshore Timor-Tanimbar, Eastern Indonesia

Sugeng Sapto Surjono, Muhamad Rizki Asy'ari, and Arif Gunawan

Abstract

The collision zone in the outer Banda Arc has become one of the important petroleum exploration targets in eastern Indonesia region. There are some fields that produce hydrocarbon from collision zones in those area, such as the Bula and Oseil oil fields in Seram and the Tiaka Field in Eastern Sulawesi. The study aims to reveal the petroleum potential in thrust and the fold belt of the Timor-Tanimbar region as a product of the collision along the Banda Arc. Surface and subsurface data integrated with some previous researches are utilized in this study. The data include 2D seismic, isostatic residual anomaly gravity, and hydrocarbon indication from the result of coreseep geochemistry analysis. The integration of all the data lead to the building of the play concept in thrust and fold belt zone within the study area. The petroleum play concept in the thrust and fold belt zone of the study area is supported by proven petroleum elements developed since Mesozoic. The petroleum trap of thrust and fold system was generated by the collision of Australian continental plate and Banda Arc during Plio-Pleistocene. This process is also responsible for the increase of the sediment thickness, which triggered the generation and expulsion of hydrocarbon then followed by the migration towards the thrusted anticline traps to the south.

S. S. Surjono (✉) · M. R. Asy'ari
Geological Engineering, Universitas Gadjah Mada, Jl. Grafika No. 2, Kampus UGM, Yogyakarta, 55281, Indonesia
e-mail: sugengssurjono@ugm.ac.id

A. Gunawan
TGS-NOPEC Geophysical Company, Pondok Indah Office Tower, Jl. Sultan Iskandar Muda Kav V-TA, Jakarta, 12310, Indonesia

Keywords

Collision zone • Thrust and fold belt • Timor-Tanimbar Petroleum play

1 Introduction

The offshore of the Timor-Tanimbar region is one of the frontier exploration areas in Eastern Indonesia. This area is located in the collision zone between the Australian and Banda Arc continental plates. The region comprises islands of outer Banda Arc which is bounded by Timor Island to the west and Tanimbar Island to the east. The petroleum exploration of this area has become interesting due to the existence of some proven oil and gas fields such as Abadi, Bayu-Undan, Sunrise Trobadour, and Elang-Kakatua [1, 2] and challenging due to the limited data and the complexity of geological settings.

Collision zone particularly in thrust and fold belt zone become one of the important petroleum exploration targets in the world. Several proven oil and gas fields such as in Zagros (Iraq and Iran), East Venezuela, and Papua New Guinea [3] have produced a significant quantity of hydrocarbon in this zone. Also, there are some fields in Indonesia which produced hydrocarbon in the collision zone, such as Bula and Oseil oil fields in Seram and Tiaka Field in Eastern Sulawesi. The new data acquired around the study area is expected to answer the exploration challenge of new play in the thrust and fold belt zone of outer Banda Arc system.

2 Research Method

New 2D seismic data from TGS-NOPEC spec survey are interpreted to identify the markers and the presence of structures along the platform up to the thrust and fold belt zone. The second zone aforementioned has poor seismic data quality. This seismic data is not calibrated due to no well

F. Rossetti et al. (eds.), *The Structural Geology Contribution to the Africa-Eurasia Geology: Basement and Reservoir Structure, Ore Mineralisation and Tectonic Modelling*,
Advances in Science, Technology & Innovation, https://doi.org/10.1007/978-3-030-01455-1_30

data being available in this area. The isostatic residual anomaly map of gravity data is used to identify the distribution of thickness sediment layers around thrust and fold belt zone. Result of coreseep geochemistry analysis are also utilized to find out the distribution hydrocarbon traces occurred as seepages on the seafloor around the thrust and fold belt zone and surrounding area. The interpreted data is integrated to build the play concept within the study area.

3 Results

3.1 Seismic Interpretation

There are seven key markers identified based on the seismic characters and regional stratigraphy data from the seismic interpretation (Fig. 2). The markers are base Permian (red), top middle Jurassic (yellow), top late Jurassic (green), base Cretaceous (dark blue), base Tertiary (light blue), base Miocene (orange), and base Plio-Pleistocene (violet). All of them are continuously distributed from south to the north.

In the N-S section, the thrust and fold belt area is located in the northern part, while the platform area is in the southern part. Timor-Tanimbar Trough is encountered between those areas. A thin-skinned deformation has occurred near the trough with decollement surface is on the Tertiary section.

3.2 Isostatic Residual Gravity Anomaly

Generally, the study area has a low isostatic residual gravity anomaly (blue color) in the central area with the relative SW-NE trend (Fig. 1). The higher gravity anomaly (red color) is distributed widely in the northern and the southern part of the study area. The low gravity value occurs around the collision zone. The blue area indicates that the area is composed of thick sediments and potential to have greater source rock maturation.

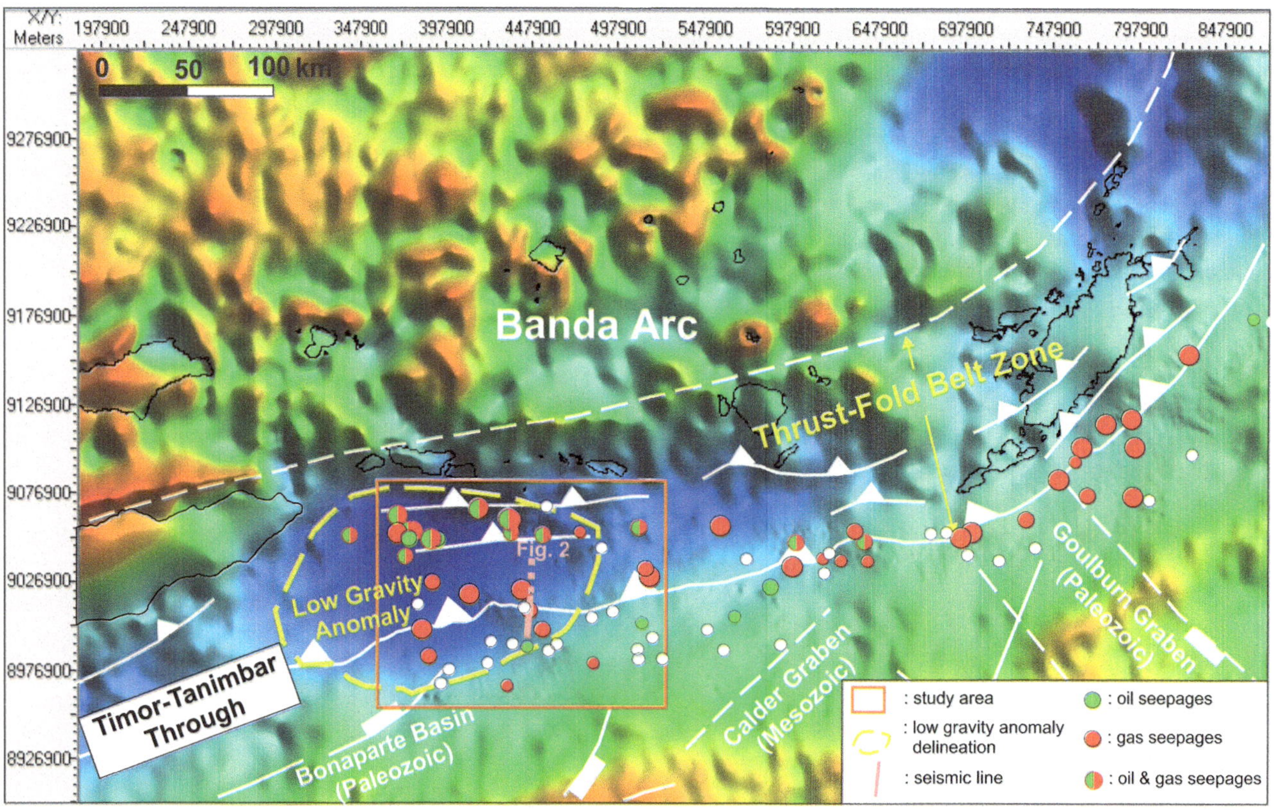

Fig. 1 The isostatic residual gravity map of Timor-Tanimbar area with the distribution of hydrocarbon seepages [4] and tectonic features [5]. The blue color indicates low gravity value representing thickest sediment in that area (also shown by seismic section in Fig. 2). Oil and gas seepages occurred mostly around this area

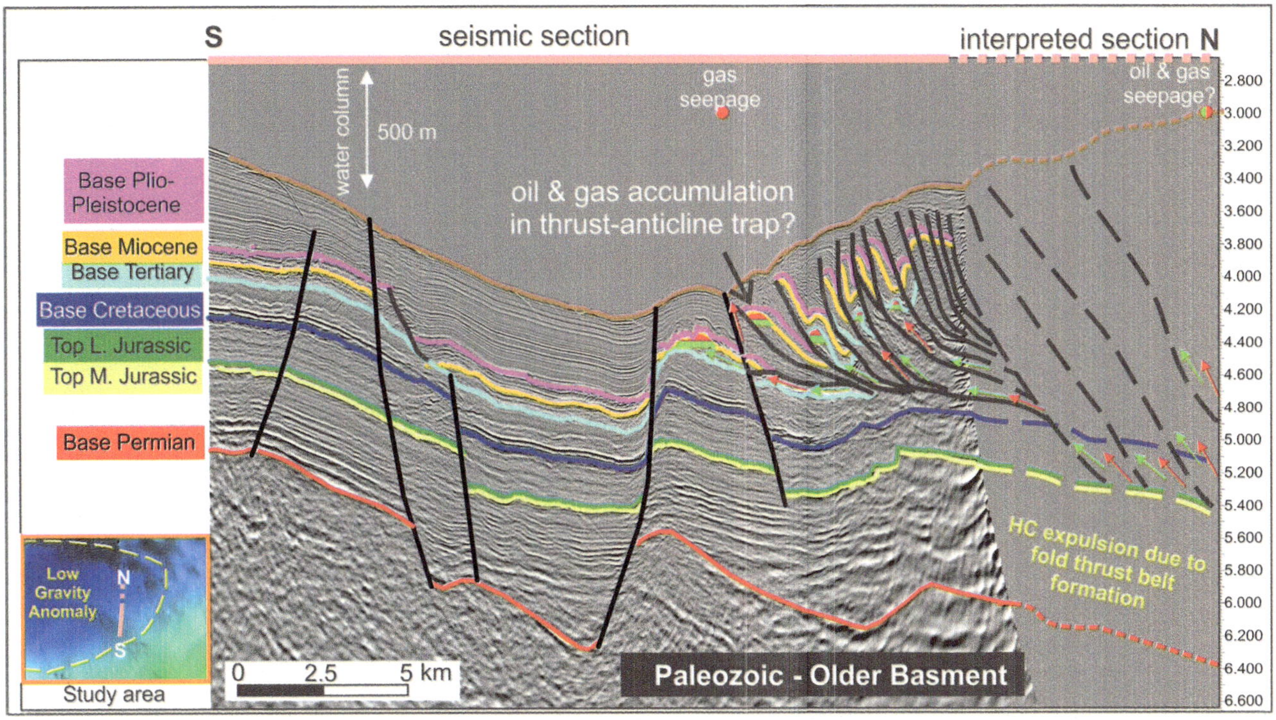

Fig. 2 The petroleum play concept in the study area. Thrusting causes the increase of sediments thickness and generation and expulsion of hydrocarbon from Jurassic source rocks interval. The hydrocarbon migrated through faults and accumulated in thrust-anticline structures and some others reached to the surface as seepages

3.3 Distribution of Hydrocarbon Indication from Coreseep Geochemistry Data

Results of coreseep geochemistry analysis from TGS-NOPEC indicates some thermogenic oil and gas seepages found in the study area [4]. The plotting of this data by integrating gravity anomaly and tectonic features shows that the hydrocarbon occurrence is distributed along thrust and fold belt zone. More specific, the greater oil and gas seepages were found in the low gravity anomaly zone.

4 Discussion

The petroleum play concept in the thrust and fold belt zone in the offshore Timor-Tanimbar area can be identified by the integration of new gravity and 2D seismic data of this area for the first time with hydrocarbon indication from coreseep data. Based on the gravity interpretation, the thickest sediment occurred in the western part of the study area due to the biggest magnitude of thrust and fold system compared to the eastern area. The source rocks from Jurassic interval are expected to be matured and has expelled some oil and gas in that zone. The evidence is proven by major occurrences of oil and gas traces. The expelled source rocks in the northern part migrates to the south through the thrust fault systems. The hydrocarbon then accumulates into thrusted anticlines in Tertiary and Cretaceous reservoir rocks (Fig. 2). Further, the improvement of seismic data in the thrust and fold belt zone is required in order to obtain a better understanding of potential trap in the study area.

5 Conclusions

This study has revealed a good petroleum potential in the thrust and fold belt zone in the offshore study area between Timor and Tanimbar region from its play concept for the first time. The generation and expulsion of hydrocarbon is triggered by the addition of sediments due to thrusting to the northern part of the thrust and fold belt zone and migrates to the thrusted anticline traps to the south.

References

1. Barber, P., Carter, P., Fraser, T., Baillie, P., Myers, K.: Paleozoic and Mesozoic petroleum systems in the Timor and Arafura Seas, Eastern Indonesia. In: Proceedings of the 29th Annual Convention & Exhibition, IPA03-G-169. Indonesian Petroleum Association, Jakarta (2003)

2. Surjono, S.S., Hidayat, R., Wagimin, N.: Triassic petroleum system as an alternative exploration concept in offshore western Timor Indonesia. J. Pet. Explor. Prod. Technol. **10**, 543 (2017)
3. Hussein, D., Collier, R., Lawrence, J.A., Rashid, F., Glover, P.W.J., Lorinczi, P., Baban, D.H.: Stratigraphic correlation and paleoenvironmental analysis of the hydrocarbon-bearing Early Miocene Euphrates and Jeribe formations in the Zagros folded-thrust belt. Arab. J. Geosci. **10**, 543 (2017)
4. TGS-NOPEC: Jamdena regional geochemical survey SGE program interpretive report, Technical Report #08-2032 (2008)
5. Charlton, T.R.: The petroleum potential of East Timor. Aust. Pet. Prod. Explor. Assoc. J. **42**, 351 (2002)

Origin and Preservation of Microbialite-Dominated Carbonate Reservoirs, Lower Cambrian Xiaoerbulake Formation, Tarim Basin (China)

Qingyu Huang, Suyun Hu, Wei Liu, Tongshan Wang, and Congsheng Bian

Abstract

Microbialite-dominated carbonate reservoirs of the Lower Cambrian Xiaoerbulake Formation show a huge petroleum exploration potential in the Tarim Basin, NW China. In this study, the sequence stratigraphic framework, sedimentary model and pore-system evolution were constructed to investigate the formation and preservation mechanisms of the age-old microbial carbonate reservoirs. Two 3-order sequences were divided in the Xiaoerbulake Formation. The good reservoirs were mainly developed in the microbial moundsand shallow water shoals of the upper sequence (SQ2). The origin of the microbial carbonate reservoirs is attributed to near-surface freshwater dissolution evidenced by non-fabric selective pores, meteoric cements and slightly negative $\delta^{13}C$ and $\delta^{18}O$ values. The facies succession also controls the development and the space texture of the reservoirs. For the preservation of the reservoirs, we argue that the retention diagenesis (early-stage dolomitization and methanogenesis), evaporite—carbonate cycles and relatively stable tectonic settings are important for pore maintenance during long-term burial.

Keywords

Microbial carbonate • Reservoirs • Origin • Preservation • Cambrian • Tarim basin

1 Introduction

Microbialite-dominated carbonate reservoirs of the Lower Cambrian Xiaoerbulake Formation show huge petroleum exploration potential in the Tarim Basin, NW China [1]. Previous studies focused on the sedimentary and reservoir characteristics of the microbial carbonates of the Xiaoerbulake Formation. However, the formation and preservation mechanism of the ancient microbial carbonate reservoirs were less investigated due to the complicated rock textures and the shortage of modern analogs. The main objectives of this paper, therefore, are to (1) investigate the sequence stratigraphy of the Xiaoerbulake Formation and the key facies that form the carbonate reservoirs including the lateral and vertical extent, and the sedimentary model; (2) identify the origin of the reservoirs dominated by microbial carbonates; (3) gain insight into what geological factors can inhibit the diagenetic processes and preserve the porosity and permeability in these age-old carbonate rocks. A better comprehension of the evolution of the microbial carbonate porosity will benefit to predict the spatial distribution of the good-quality reservoirs in the formation in this basin.

2 Methods

A total of 350 samples were selected from more than 230 m of two outcrop sections and 85 m of cores (5 wells). For each sample, a polished thin section, trace element composition and stable isotope analyses were performed. Seven samples were tested by laser micro zone isotope analysis for determining the isotope components of the individual cement phases in diagenetic investigation.

Q. Huang (✉) · S. Hu · W. Liu · T. Wang · C. Bian
Research Institute of Petroleum Exploration and Development, PetroChina, Beijing, 100083, China
e-mail: huangqingyu@petrochina.com.cn

F. Rossetti et al. (eds.), *The Structural Geology Contribution to the Africa-Eurasia Geology: Basement and Reservoir Structure, Ore Mineralisation and Tectonic Modelling*, Advances in Science, Technology & Innovation, https://doi.org/10.1007/978-3-030-01455-1_31

3 Results

3.1 Sequence Stratigraphy and Sedimentology

Based on the detailed research of the field sections, two 3-order sequences were divided in the Xiaoerbulake Formation (Fig. 1). The microbial mounds which compose of the main reservoirs in the formation mainly occur in the upper portions of the high system tract (HST) in each sequence (Fig. 1b). The sedimentary model of the microbial carbonates is the weakly rimmed platform (Fig. 1c). Because of the limited platform margin barriers, the microbial mounds and shoals can be extensively distributed in the intra-platform area, the lateral extension more than 100 km [2].

3.2 Characteristics of the Reservoirs

The most common pore types observed are microbial growth-framework pores, dissolution-enlarged interparticle pores and vugs. Secondary pores include intercrystal pores, dissolution-enlarged fractures and microfractures. The stratified microbialites in the upper part of the SQ1 are characterized by microbial growth-framework pores, while the microbial mound of the upper part of the SQ2are characterized by vugs, dissolution-enlarged growth-framework pores and interpartical pores. The upper part of the SQ2 has better porosity (1.08–10.80%, average 4.23%) than that of the SQ1 (0.09–2.06%, average 1.78%).

Unsmooth surface, erosional pits and millimeter- to centimeter-size vugs are observed on the upper part of the microbial mound in the Shiairike and Xiaoerbulake outcrop sections (Fig. 2a, b). In thin-section, the meniscus cements between the microbial grains and the isopachousbladed cements around the microbial clots and grains occur in the reservoirs (Fig. 2c, d). Meanwhile, the cathodoluminescence thin-sections show minor blotchy red or bright red zones withmoderate intensity(Fig. 2e, f). Additionally, the oxygen and carbon isotopic values of the reservoirs near the sequence boundary are slightly negative.

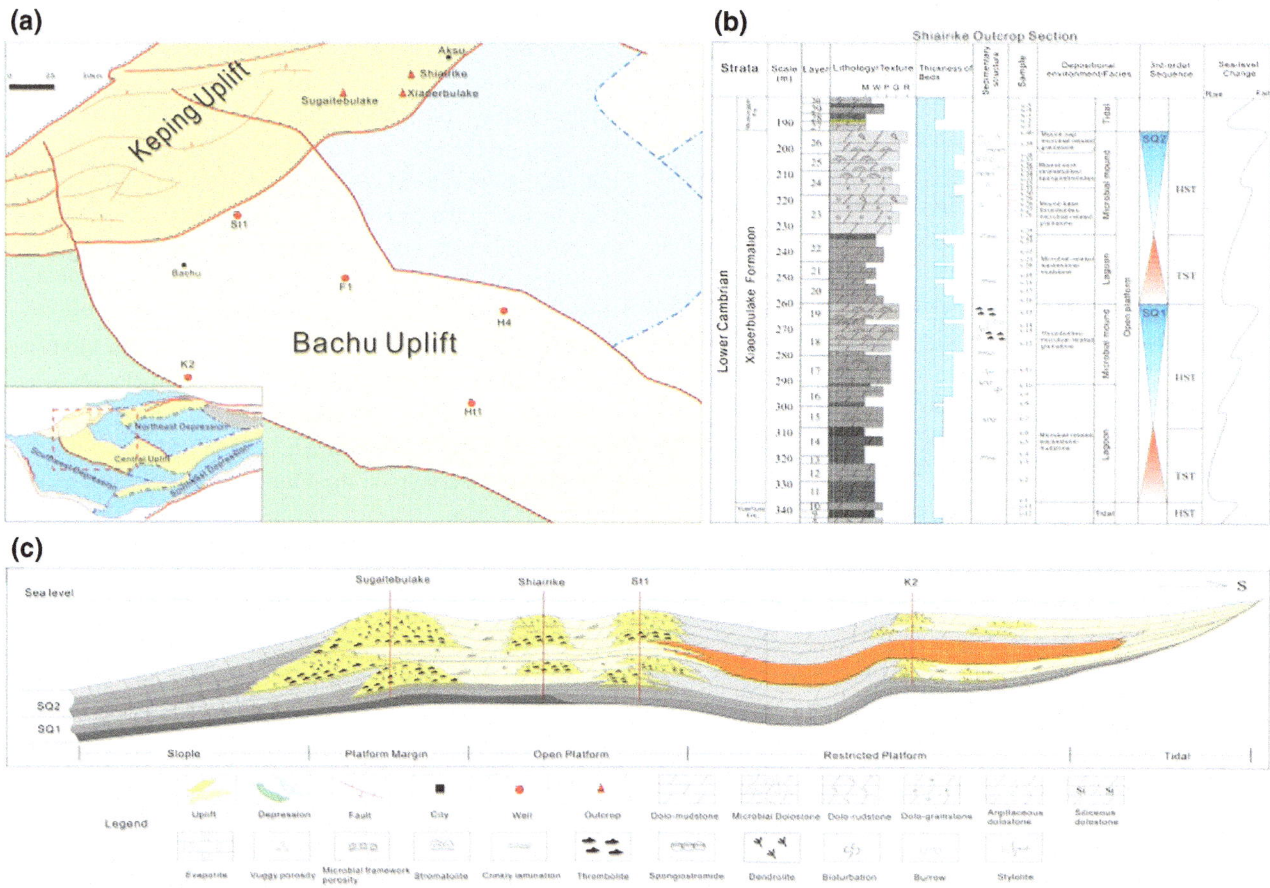

Fig. 1 **a** Schematic tectonic map of the northwesternTarim Basin, showing the position of the studied outcrops and wells. **b** Sequence stratigraphic section of the Xiaoerbulake Formation at the Shiairike area. **c** Sedimentary model of the Xiaoerbulake Formation at the northwestern Tarim Basin

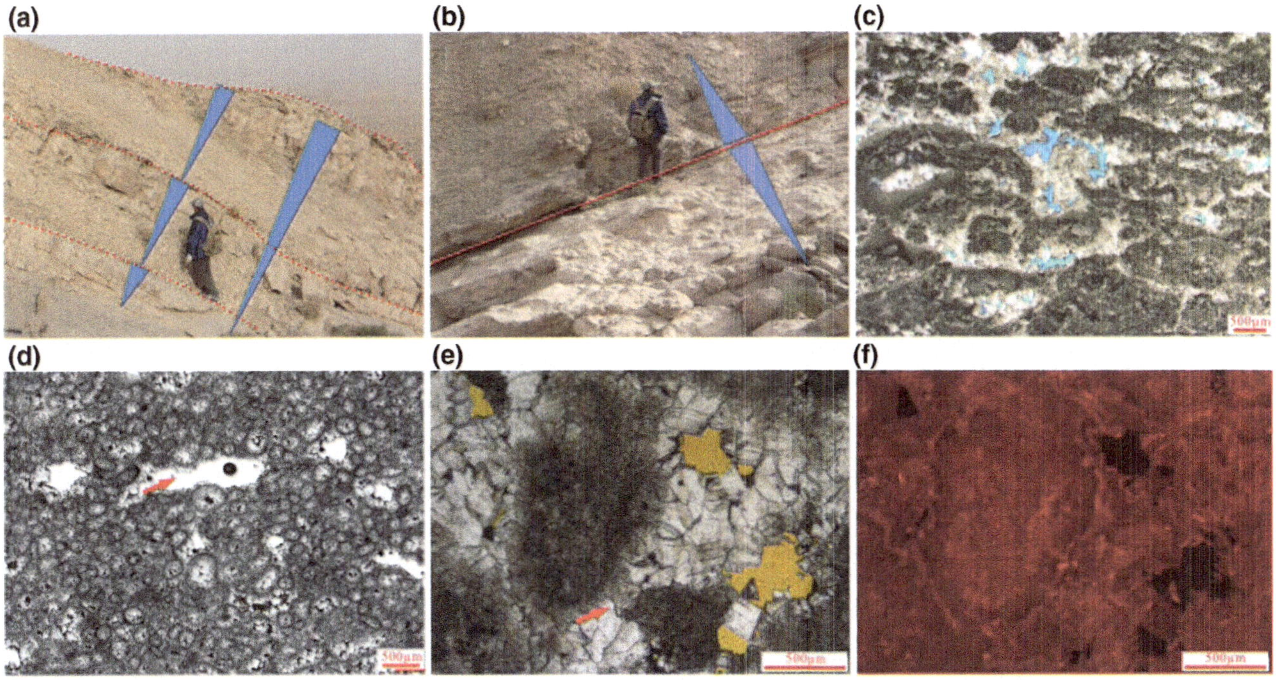

Fig. 2 **a** Microbial mound in the Shiairike outcrop section. **b** Sequence boundary on the top of the SQ2 in the Shiairike outcrop section. **c** Thrombolite dolostone with obvious clotted structure, blue part is microbial growth-framework pore. **d** Spnogiostromide dolostone (foam-like microbialite), red arrow shows the vuggy porosity. **e**, **f** Meniscus cement between the microbial-related grains, (**f**) is the cathodoluminescence figure

4 Discussion

The origin of the microbial carbonate reservoirs in the Xiaoerbulake Formation is attributed to the dissolution of the meteoric water resulted from the short-term subaerial exposure events, based on the petrographic and geochemistry data. In addition to the dissolution, facies succession also controls the development and the space texture of the reservoirs. The mound core has the best reservoir and the mound cap takes the second place, and the mound base is worst in the interior of the mound.

The preservation of the reservoirs is related to the retention diagenesis, evaporite—carbonate cycles and relatively stable tectonic settings during long-term burial. The dolomitization and methanogenesis in the early diagenetic stage are the main retention diagenesis. The early dolomitization can effectively resist the physical and chemical compaction due to the special crystal texture of the dolomite which can form more firm framework than the calcite [3]. The methanogenesis during early diagenesis produce microbial methane gas which can form a two-phrase system in the pore space to prevent pore water flow, reducing the formation of cements [4]. The evaporites on the top of the Xiaoerbulake Formation can slow down the diagenetic process because of their high heat conductivity during the burial. Meanwhile the stable tectonic setting forms a closed flow system and leads to less burial and/or late-stage hydrothermal fluids activities. Thus, the sub-salt microbial carbonate reservoirs can keep original fabrics and early formed porosity very well.

5 Conclusions

Two 3-order sequences were divided in the Xiaoerbulake Formation. The microbial mounds and shallow water shoals which primarily developed in the upper sequence (SQ2) have good porosity.

The origin of the microbial carbonate reservoirs is attributed to near-surface freshwater dissolution, and the facies succession also controls the development and the space texture of the reservoirs.

For the preservation of the reservoirs, we argue that the retention diagenesis, evaporite—carbonate cycles and relatively stable tectonic settings are important for pore maintenance during long-term burial.

References

1. Jinhu, D.U., Wenqing, P.A.N.: Accumulation conditions and play targets of oil and gas in the Cambrian subsalt dolomite, Tarim Basin, NW China. Pet. Explor. Dev. **43**(3), 327–339 (2016)
2. Huang, Q., Suyun, H., Pan, W., et al.: Sedimentary characteristics of intra-platform microbial mounds and their controlling effects on the development of reservoirs: a case study of the Lower Cambrian Xiaoerbulake Formation in the Keping-Bachu area, TarimBasin. Nat. Gas. Ind. **36**(6), 21–29 (2016)
3. Lucia, F.J.: Origin and petrophysics of dolostone pore space. In: Braithwaite, C.J.R., Rizzi, G., Darke, G. (eds.) The Geometry and Petrogenesis of Dolomite Hydrocarbon Reservoirs, vol. 235, pp. 141–155. Geological Society (London) Special Publication (2004)
4. Kenward, P.A., Goldstein, R.H., Brookfield, A.E., et al.: Model for how microbial methane generation can preserve early porosity in dolomite and limestone reservoirs. AAPG Bull. **96**(3), 399–413 (2012)

Insight into the Tectonics of the Kaboudia Area and Related Petroleum Systems, Eastern Tunisian Offshore (Tunisia)

Mohamed Arab, Mohamed Hassaim, Chokri El Maherssi, Karim Belabed, and François Roure

Abstract

The Kaboudia permit belongs to the Pelagian domain that has experienced extensional tectonics in the framework of the Tethys oceanic opening since Triassic. As the rest of the Northern African Mesozoic basins, peritethyan basins underwent a syn-rift opening during Late Triassic-Jurassic with coeval magmatic intrusions. This extensional period extended until Lower Cretaceous within the Tunisian basins. After Austrian uplift and erosional event, the former tectonic subsidence and extension was followed by a post-rift phase, which is marked by the deposition of a thick shale series of Fahdene during Albian-Cenomanian period. The Late Cretaceous to Eocene represents tectonically a quiet period that was characterized by the deposition of condensed sections and compressional periods such as, Santonian and Priabonian phases. Horst and grabens tectonics associated to a transtension occurred during Upper Miocene along the NW-SE extensional faults that are related to the Opening of the Tyrrhenian Sea and subsequent movement of Sicily with regard to Africa in the framework of the Ionian subduction. The Kaboudia permit conceals already oil accumulations within the Serdj and Allam Cretaceous carbonate reservoirs that are sourced from Lower Fahdene (Vraconian) source rock. After lab data, only few samples of such shale have shown a certain potential. However, Delta Log R method and seismic interpretation show source indications within the paleo-troughs around Mahdia structure and Kuriat furrow (near Taboulba). On the other hand, lab data shows also few samples from Allam and Serdj Aptian shale with higher TOC's and residual potential S2. In terms of petroleum system working, the Mouelha could charge Allam limestone reservoir in a perdescensum. However, the reservoir below might be charged only when the latter is comprised within the same structural closure of the structure (trap). In addition, the Serdj reservoir could be charged downward by Allam and Mouelha shale within the kitchen zones. The postmortem analysis of exploration results at Kaboudia and the surrounding blocks, revels that the main risk regarding the petroleum system is the source rock potential, its extension and at a lesser degree the seal capacity.

Keywords

Pelagian domain • Syn-rift • Transtension
Source rocks

1 Introduction

The Kaboudia permit belongs to the Tunisian offshore pelagean domain. It experienced a syn-rift stage during Late Jurassic to Lower Cretaceous then a post rift phase at Albian_Cenomanian times. The sedimentary series reached by wells, which are shale dominated, varies between 2500 and +4000 m. After a tectonic period of compressions and erosions a new period of extension governed by NW-SE faults occurred at Late Miocene. Regarding the uncertainties related to the tectonic evolution, the present work will come up with additional interpretations based on new data that will clear up the geological evolution of the Kaboudia and thus a whole eastern Tunisian offshore.

Before undertaking any petroleum system evaluation, we need to reconstruct the tectonics and the whole geological events such as rifting period and uplifts. The thermal maturity of the source rocks depends on the geothermal framework that is governed by these geological events. The study area comprises two main plays, Lower Cretaceous

M. Arab (✉) · M. Hassaim · C. El Maherssi
Numhyda.r.l. Sonatrach/Etap JV Company, Lac Leman, Leman Center Tower, Lac1, Tunis, Tunisia
e-mail: mohamed.arab76@gmail.com; arab@numhyd.com

K. Belabed
SIPEX/Sonatrach, Algiers, Algeria

F. Roure
Ifp-Energies Nouvelles, Avenue Bois Préau Rueil Malmaison, Paris, France

F. Rossetti et al. (eds.), *The Structural Geology Contribution to the Africa-Eurasia Geology: Basement and Reservoir Structure, Ore Mineralisation and Tectonic Modelling*,
Advances in Science, Technology & Innovation, https://doi.org/10.1007/978-3-030-01455-1_32

Serdj/Allam carbonates (carbonate platform), Miocene siliciclastics/limestones formations deposited in shorface and locally in deltaic system. It is mainly extensional domain with local inversions and erosions. The study area conceals Mahdia oil accumulation and hydrocarbon shows from Serdj and Allam fractured reservoirs.

So far, the relatively low prospectivity and modest exploration success is mainly due to low source rock potential, hydrocarbon preservation risk and the presence of CO_2 and H_2S. The present report is an overview of structural and tectonic evolution of the Kaboudia permit and its petroleum system analysis especially the source rock identification and the way the hydrocarbon charge may occur within the Lower to Middle Cretaceous petroleum system.

2 Settings

The Kaboudia offshore area lies in the Pelagian Sea (Fig. 1) between [34° 12′ 50″N–36° 10′ 50″N] and [11° 11′ 50″E–13° 11′ 50″E]. It covers about 32,800 km^2. It is located between the gulf of Hammamet to the north and the gulf of Gabes to the south. To the east it is limited by the Malta escarpment, which illustrates the transition towards the deeper zones of the Ionian Sea [3]. On the contrary to Cretaceous period, which is characterized by moderate tectonics, the Miocene is considered as highly dynamic period where occurred coevally extension and inversion with local erosions. The Gulf of Hammamet area lies in the Tunisian offshore, east of the Atlas front (Fig. 2). The Tunisian Atlas is a foreland folded basin that grew up during the Cenozoic collision [1, 2, 5]. Kaboudia area makes the junction between Hammamet Gulf and Gabes Gulf, and experienced several extensional events during Triassic-Jurassic and Lower Cretaceous period related to the Tethys rifting (Fig. 3). The Albian-Cenomanian represents the post-rift period where over 1500 m of shale are deposited by thermal subsidence [4]. Paleocene- Eocene records a decrease of subsidence marked by the deposition of a condensed series, with mean burial rate of 2 m/my. This period represents an uplift period, which terminates with important erosion at Kaboudia, at Priabonian time which is related to the tectonics of horsts and grabens. Furthermore, the Santonian and Late Burdigalian local erosions are induced as a result of a lateral blocking of an extension (Fig. 2). Later on, a tectonic of horst and grabens governed by NW-SE faults was

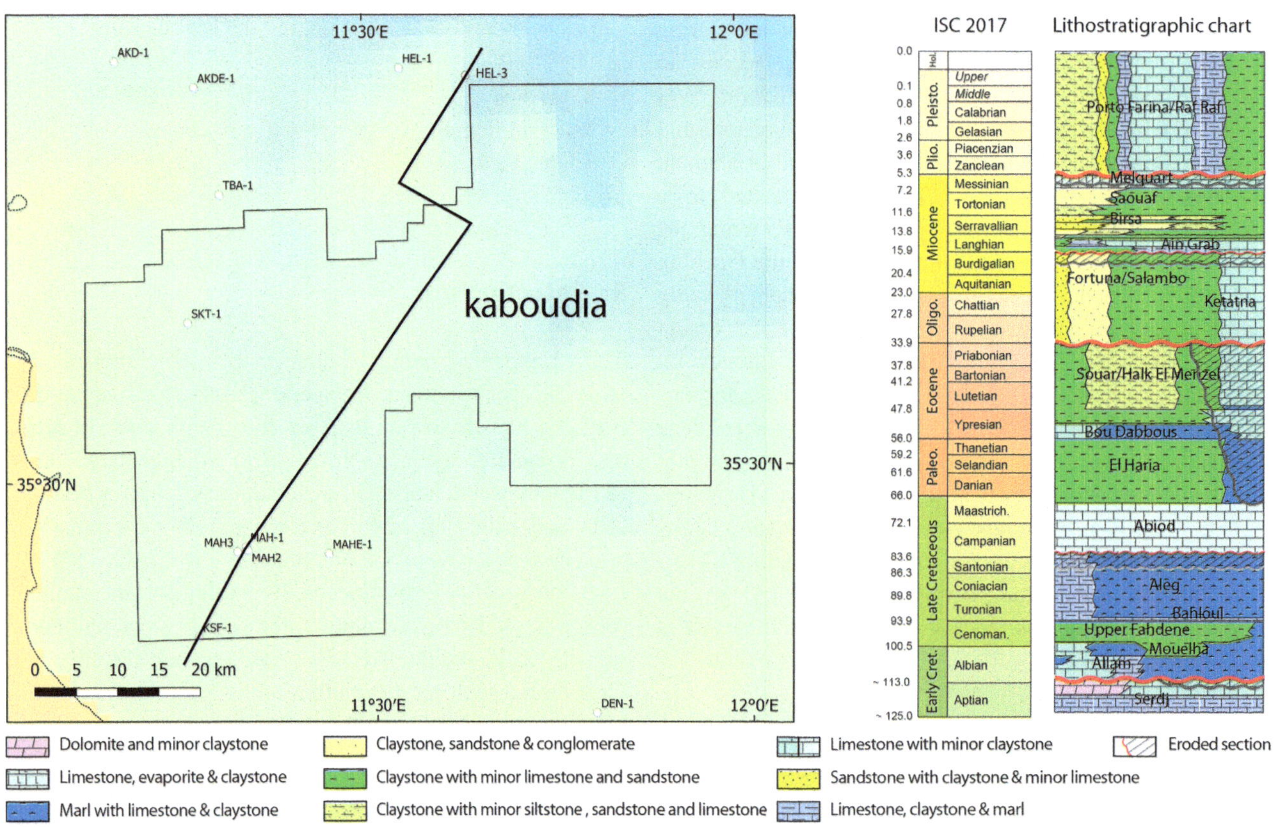

Fig. 1 Location map of the Kaboudia permit and the interpreted seismic profil, Eastern Tunisian offshore

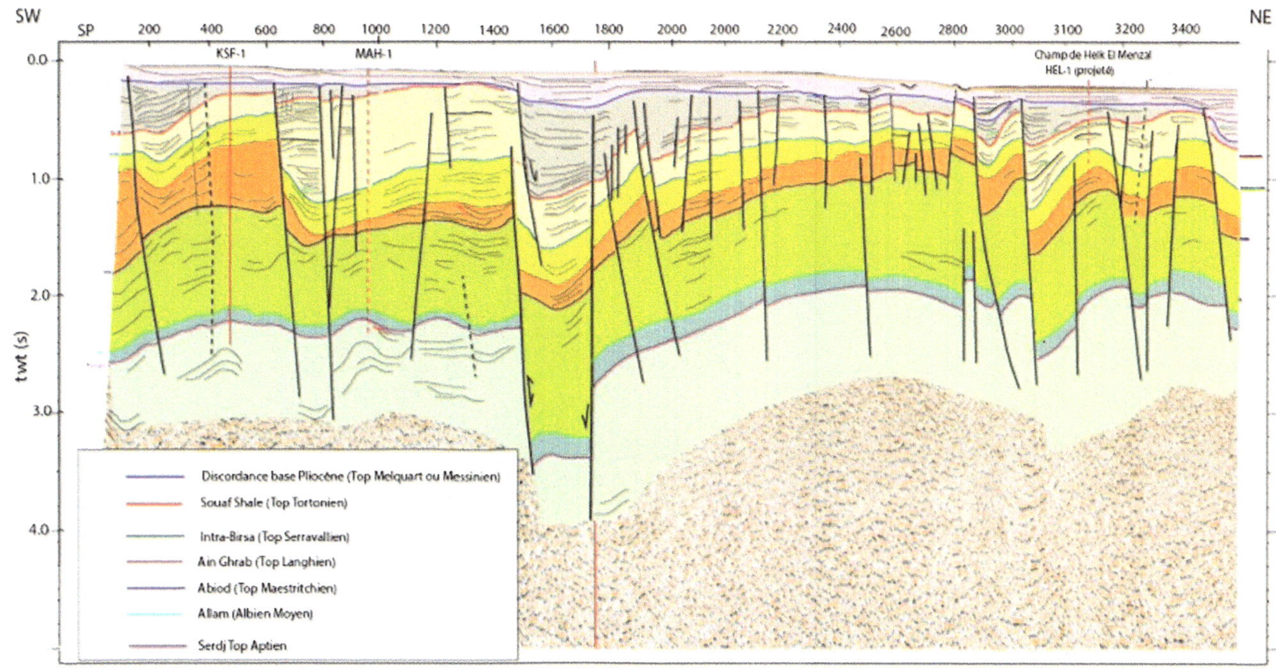

Fig. 2 NE-SW Seismic-stratigraphic section showing the structure and gross stratigraphic units of the Kaboudia permit, Tunisian offshore. Geoseismic section location (see Fig. 1)

triggered at Tortonian and extended regionally to Messinian period, where thick shale associated locally to evaporates and limestones were deposited within the grabens. At the End Messinian, an important erosion event occurred inducing important erosions, such event may be related partly the convergence between Africa and Europe (Fig. 2).

3 Results and Discussion

In addition to Allam shale, the Mouelha formation (equivalent formation in outcrops) of Late Albian age (Vraconian) is the main source rock, it shows weak organic richness and

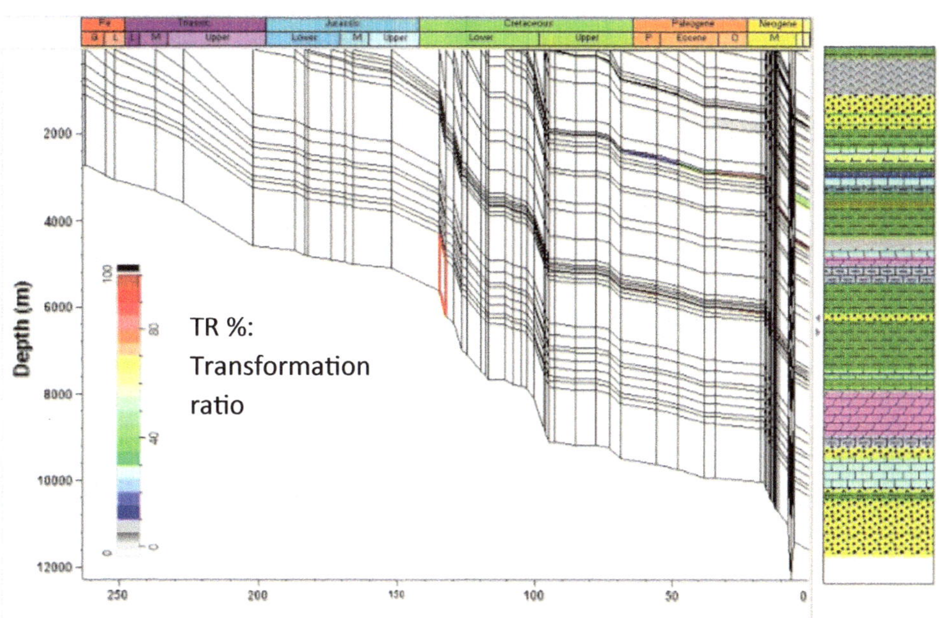

Fig. 3 Burial history and timing of hydrocarbon generation from Lower Fahdene or Mouelha source rock (Vraconian), virtual well from Northern Mahdia trough

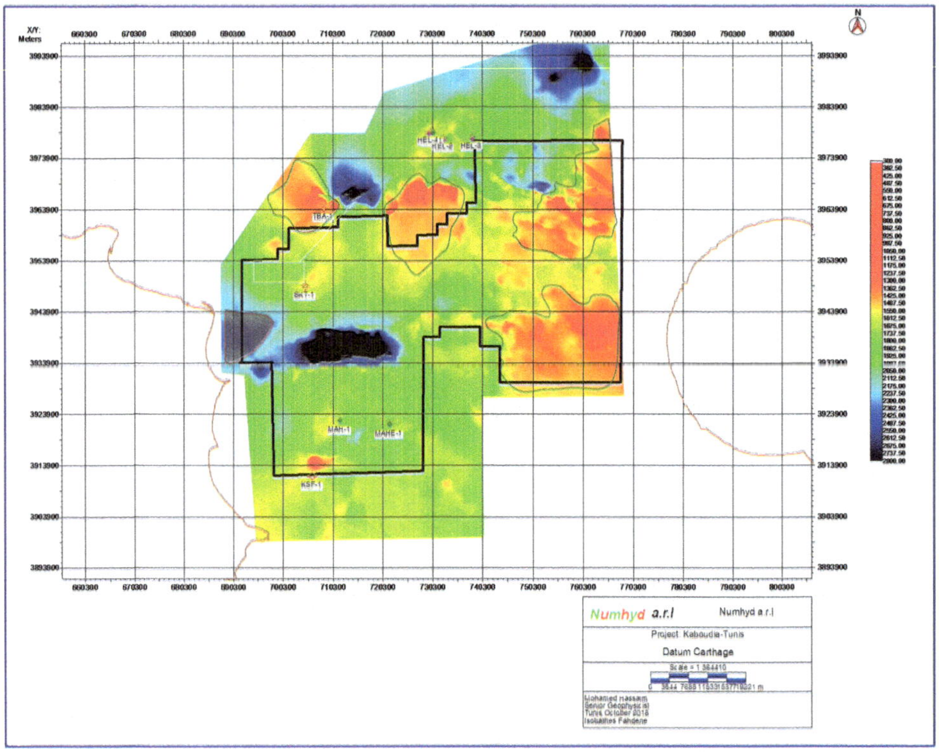

Fig. 4 Depth map, organic richness (residual TOC) and maturity map at the base of Lower Fahdene source rock (Mouelha)

little potential (Fig. 4) in the drilled structures, since it is possible that the latter represent paleoreliefs. Given the paleogeographic conditions of the period of deposition (Vraconian), the most organic rich and potential shale/calcareous formations are encountered in the onshore domain, such as Jehaf area and most probably within the paleo-grabens of Vraconian times at offshore domain, known as favored anoxic zones. The lab data (Leco and pyrolysis) conducted on Mouelha and Upper Fahdene shale samples depict generally low TOC (total organic carbon) values (0.4–1%) while the TOC calculated from logs by Delta log R method witnesses a presence of thin potential layers, like in TBA-1 and MAH-1. The low TOC values could be explained by the high sedimentation rate (>350 m/Ma, Fig. 3). The type of organic matter as interpreted from HI/Tmax and HI/OI is seemingly continental or mixed origin. However, the integration of optical observation of organofacies depicts rather marine type II, having just undergone an oxidation which caused the reduction of the initial potential. The average initial potential should be 600 mg/g TOC but it is reduced to less than 250 g HC/g TOC. In terms of maturity, Mouelha shale is currently in condensate to wet gas window(Ro averaging 1.6%, equivalent to a temperature of 150–160 °C) in depressive zones (furrows) and just in early oil generation (Ro = 0.65% and men T_{max} = 435–437 °C, Fig. 4) in the rest of Kaboudia permit. Within the troughs, the maturity is estimated after 1D geochemical models, depth maps (Fig. 4) and the observed API density of the accumulated oils. The source rock efficiency and kitchen zones extension (Fig. 4) which are the main sensitive parameter in the petroleum system working in Kaboudia permit, depends on the extension and duration of anoxic zones, maturity and the kinetics of the organic matter.

4 Conclusion

Kaboudia area as the rest of the eastern Tunisian offshore experienced two extensional events, at Jurassic-Lower Cretaceous and Late Miocene along NW-SE parallel faults, respectively subscribed in the framework of Thetis Ocean and the opening of the tectonics induced by the Calabre arc and extension of Tyrrenian Sea. Such Miocene neo-rifting caused an elevation of heat flow that is shown by high geothermal gradients. The presence of the discovered oil accumulation and oil shows recorded in the study area are related to a mature Mouelha source rock but also Aptian Serj and Allam shale (Albian) that would have been deposited in anoxic environment within the paleo-troughs such as, Kuriat near Taboulba and around Mahdia structure. Regarding the low potential of such source rocks and the short hydrocarbon migration distances or drainage along the low permeability Cretaceous reservoirs (Serj and Allam), small closed structures are privileged in terms of hydrocarbon charge and preservation. However the eastern frontier area of Kaboudia could be prospective essentially with the presence of deeper source rocks and/or the long migration paths along Miocene permeable reservoirs from the northwest of the permit. Besides, the northern part close to Helk El Menzal oil field is still the less risky for further exploration activity. The main geological risk is related to the source rock distribution.

References

1. Anderson, J.E.: The Neogene structural evolution of the western margin of the Pelagian platform, central Tunisia. J. Struct. Geol. **18**(6), 819–833 (1996)
2. Benaouali-Mebarek, N., Frizon de Lamotte, D., Roca, E., Bracene, R., Faure, J.L., Sassi, W., Roure, F.: Post-cretaceous kinematics of the atlas and tell systems in central Algeria: early foreland folding and subduction-related deformation.C. R. Geosci. **338**, 115–125 (2006)
3. Ben Brahim, G., Brahim, N., Turki, F.: Neogene tectonic evolution of the Gulf of Hammamet area, Northeast Tunisia offshore. J. Afr. Earth Sci. **81**, 16–27 (2013)
4. Grasso, M., Torelli, L., Mazzoldi, G.: Cretaceous-Palaeogene sedimentation patterns and structural evolution of the Tunisian shelf, offshore the Pelagian Islands (Central Mediterranean). Tectonophysics **315**, 235–250 (1999)
5. Piquet, A., Tricart, P., Guraud, R., Laville, E., Bouaziz, S., Amrhar, M., Ait, R.: The Mesozoic-Cenozoic Atlas belt (North Africa): an overview. Geodin. Acta **15**, 185–208 (2002)

Structural Styles, Petroleum Habitat and Traps in the Pelagian-Sirt Basins, Northern Africa: An Overview and Future Exploration Developments

Sami Khomsi, Francois Roure, Najoua Ben Brahim, Chokri Maherssi, Mohamed Arab, and Mannoubi Khelil

Abstract

The Pelagian-Sirt platform of Northern Africa is a structurally complex area at the triple junction of three major structural domains: the Alpine thrust wedge to the northwest, the Pantelleria-Linosa-Malta grabens system to the northeast and the Sirt basin to the southeast. The Pelagian domain encompasses the easternmost foreland basins of the Atlas in the west. This position combined with variations in the lithostratigraphic assemblages result in a complex geological evolution with the sur-imposition of several tectonic events controlling the individualization of petroleum systems with different structural traps. This work aims at (1) summarizing some new developments on the structural styles as well as the major tectono-sedimentary events as they arise from new collected subsurface data and field observations from both the offshore and onshore portions of the Pelagian platform, and (2) describing the structural style of major petroleum traps in the area, combining seismic interpretations and field observations.

Keywords

North Africa • Pelagian-Sirt domain • Structural styles Seismic • Regional sections • Petroleum traps

S. Khomsi (✉)
Geo-Exploration Department, Faculty of Earth Sciences, King Abdulaziz University, Jeddah, Saudi Arabia
e-mail: samykhomsi@yahoo.ca

F. Roure
Institut Français du pétrole, IFP énergies nouvelles, Paris, France

N. B. Brahim · M. Khelil
Laboratoire georessources, CERTE, Carthage University, Tunis, Tunisia

C. Maherssi · M. Arab
Exploration and Production Management Departments, NUMHYD a.r.l., Tunis, Tunisia

N. B. Brahim · M. Khelil
Faculté des sciences de Tunis, Université Tunis al Manar Tunis, Tunis, Tunisia

1 Introduction

The potential petroleum plays of the Pelagian-Sirt Domain [1] are tentatively classified according to their structural styles and major controlling structural features and tectonic events. The classifying elements are: age of the main deformation pulses, structural type features and trend, detachment levels, fracturation and migration pathways.

2 Classification of the Petroleum Traps

Seismic interpretations in the Pelagian-Sirt structures allow us to recognize five major types of traps:

- Contractional structures of different ages developed from different detachment levels. They were charged from the Cenomanian-Turonian blak shales during the flexural phase.
- Thrust structures (Late Miocene compressions) reactivating older inherited extensional structures and in some cases related to halokinesis of the Triassic salt [5].
- NE-SW Cretaceous normal faults cutting the pre-Triassic substratum in the offshore domain, controlling horst and grabens configuration. The hanging horst crests controlled the deposition of the Early Eocene Nummulitic limestones. These important reservoirs were later folded in large anticlinal structures. Some of the faults underwent strike slip reactivation during Late Miocene-Quaternary controlling secondary fracturation features.
- Miocene graben features especially in the northern part of the Pelagian block which recorded very high subsidence rates in some depocenters with cumulative thicknesses in many places reaching 3000 m. These thick siliciclastic series were folded and inverted during Late

F. Rossetti et al. (eds.), *The Structural Geology Contribution to the Africa-Eurasia Geology: Basement and Reservoir Structure, Ore Mineralisation and Tectonic Modelling*,
Advances in Science, Technology & Innovation, https://doi.org/10.1007/978-3-030-01455-1_33

Miocene-Quaternary, leading to individualization of turtle type anticlines and en echelon structures along major faults. The reservoirs were sourced by the Lower Miocene black shales.

– Diapiric and peri-diapiric structures anchored on deep seated faults and controlling the development of peri-diapiric structures on the flank of major diapirs (Fig. 1).

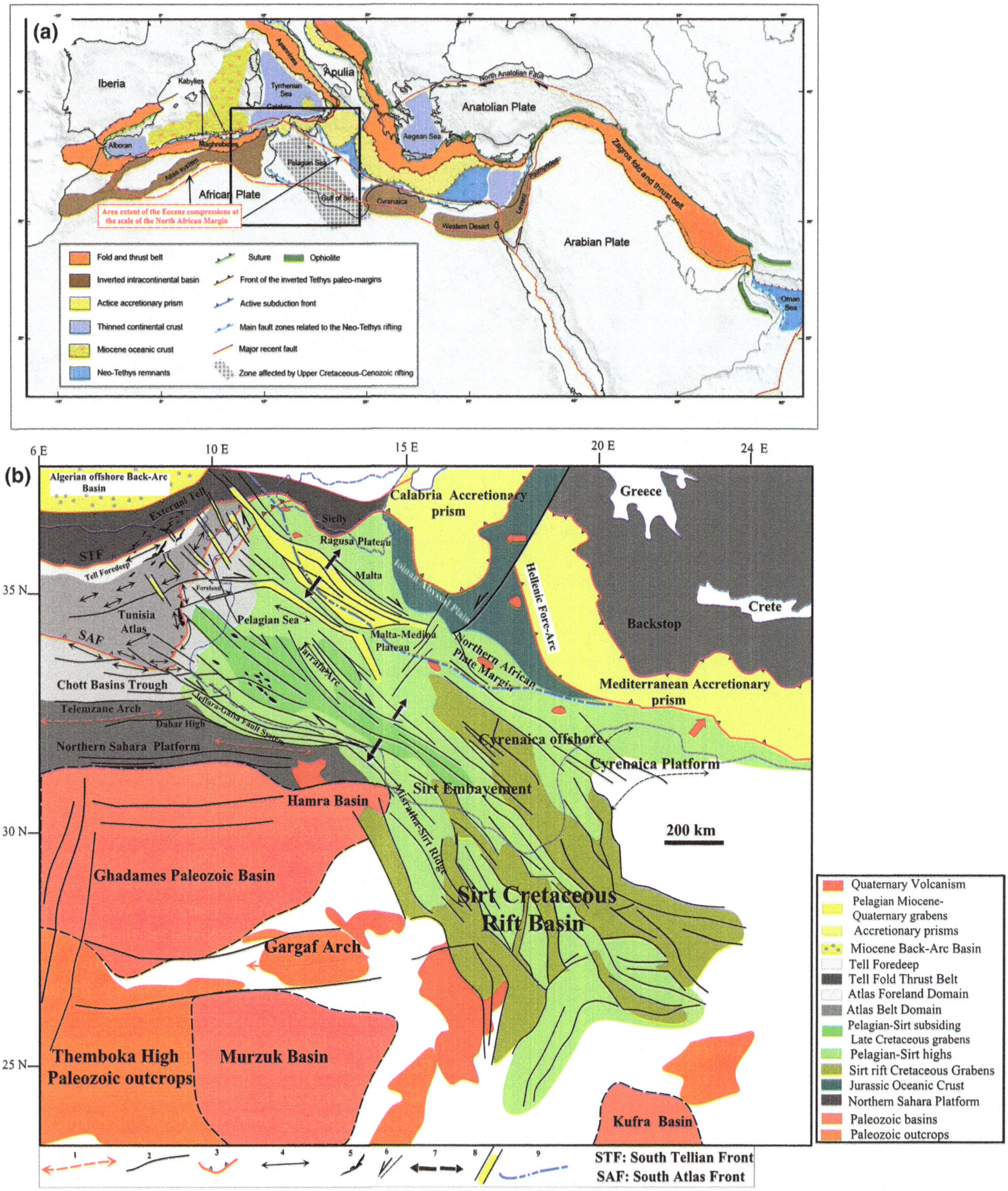

Fig. 1 Simplified structural map of the Eastern Maghreb and Sirt-Pelagian Domain (after 4)

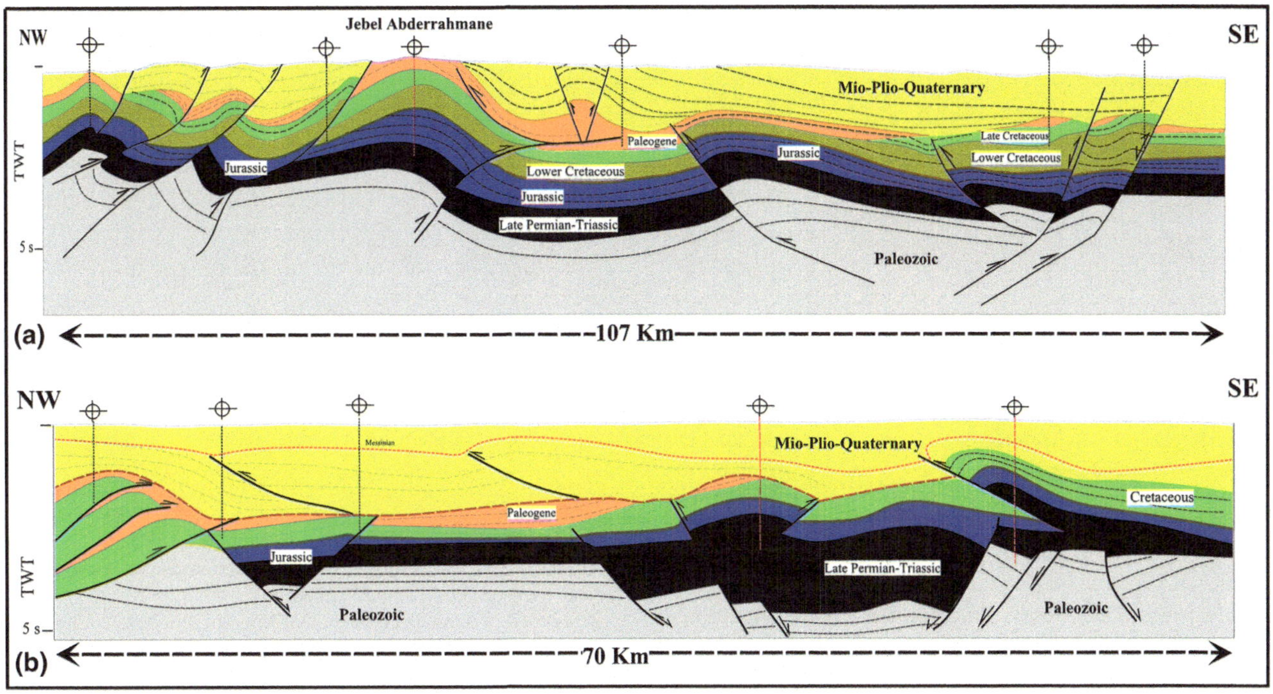

Fig. 2 Examples of structural interpretations and styles issued from seismic interpretations in the Sicily-Tunisia channel and Gulf of Hammamet (after 4 and 5)

3 Discussion

In the foreland basins of the Atlas, at the western termination of the Pelagian platform, the folds imaged by the seismic sections derive from positive Paleocene-Eocene inversions [2–5] and relate to faulted folds detached over the Triassic evaporites levels implying salt mobility and multiple 'détachement' levels. These faulted detachment folds exhibit two major geometrical patterns, i.e. either asymmetric or symmetric, depending on the relative dips of the forelimb and backlimb. The quasi-symmetric fold geometries are outlined by pop up configurations with relatively flat crest on which occur progressive onlaps of Oligo-Miocene series above Cretaceous series, open and rounded fold geometries and occurrence of footwall synclines. On the majority of folds imaged by seismic, it is worth noting that thrust faults terminate upwards within middle-Late Eocene seismic horizons and downward within the Triassic salt or the Lower Cretaceous shales. So, both symmetric and asymmetric faulted detachment folds are cored by Triassic salt layers. In this domain, the main reservoirs are represented by the Campanian-Maastrichtian limestones, which are involved in the folded structures and sourced by the Cenomano-Turonian black shales that attained maturity and oil window in coeval kitchens since Late Eocene and Oligocene with maximum sourcing during Early Miocene.

To the east, the structural traps are represented by symmetrical anticlines anchored on Jurassic- Cretaceous paleo-horst structures. This category of traps is widespread in the Karkena Archipelago and Gulf of Gabes structures, with reservoirs pertaining essentially to the Lower Eocene Nummulitic limestones, which have very good reservoir characteristics related to secondary fracturation.

Farther to the south-east, some petroleum traps are related to symmetrical to asymmetrical diapers involving Liassic evaporite, associated to Cretaceous normal faulting controlling the Sirt basin [4] (Figs. 2 and 3).

4 Conclusion and Perspectives

A set of unexplored or not-sufficiently explored deep structures are shown on the cross section. These structures can be very favourable petroleum targets taking into account the results of modelling and the lessons taken from other productive zones in the area such as Sidi El Kilani field. Especially closed anticlines exist and are not drilled. In other cases, structures associated to inversion tectonics and faulted zones are clearly shown on the seismic records. It seems that many of the wells drilled in the zone must be re-entered because the majority of old wells did not reach sufficient depths. Seismic interpretations allow us clearly to identify a good variety of potential structural plays ranging from fault

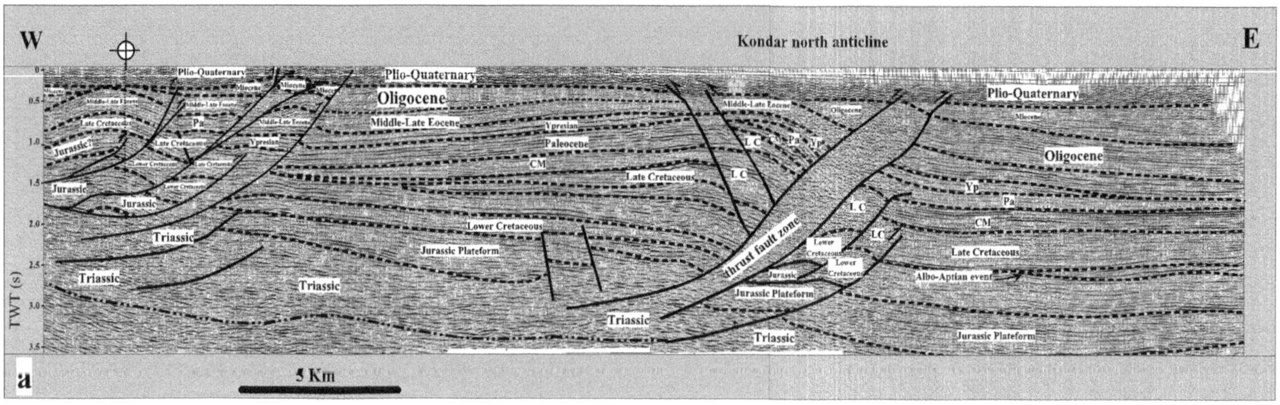

Fig. 3 Example of structural interpretation and style issued from seismic interpretations in Eastern Atlas foreland basins

related anticlines, ramps, fault propagation folds and passive roof duplexes, horst blocks, peri-diapiric structures, anticlines and en echelon fold anticlines which were identified in the subsurface. Deep reservoirs still also remain unexplored and must be taken in account.

Some examples will be presented and discussed in this presentation.

References

1. Roure, F., Casero, P., Addoum, B.: Alpine inversion of the North African margin and delamination of its continental lithosphere. Tectonics **31** (2012). https://doi.org/10.1029/2011TC002989
2. Benaouali-Mebarek, N., Frizon de Lamotte, D., et al.: Post-cretaceous kinematics of the Atlas and Tell systems in central Algeria: early foreland folding and subduction-related deformation. C. R. Geosci. **338**, 115–125 (2006)
3. Bracene, R., Frizon de Lamotte, D.: Origin of intraplate deformation in the Atlas system of western and central Algeria: from Jurassic rifting to cenozoic-quaternary inversions. Tectonophysics **357**, 207–226 (2002)
4. Khomsi, S., Frizon de Lamotte, D., Bedir, M., Echihi, O.: The Late Eocene and Late Miocene fronts of the Atlas belt in eastern Maghreb: integration in the geodynamic evolution of the Mediterranean Domain. Arab. J. Geosc. **9**(15), 125–142 (2016)
5. Casero, P., Roure, F.: Neogene deformations at the Sicilian-north African Plate boundary. In: Roure, F. (ed.) Peri-Tethyan Platforms. Editions Technip, Paris (1994)

Integrated Study of Bentiu and Aradeiba Reservoir Formations, Hamra East Oil Field—Muglad Rift Basin, Sudan

Tourkman Omayma, Cerepi Adrian, Mohamed Nuha, Zaroug Izzeldin, and Hussein Rashid

Abstract

The Muglad Rift Basin represents the largest of Sudan's interior basins with NW-SE orientation and up to 13 km of Cretaceous-Tertiary continental sediments. Bentiu and Aradeiba are the main reservoirs. This study deals with integrated multi-scale analysis using six sedimentological methods (conventional cores analysis and thin section description, well log interpretation, SEM, XRD, and two seismic sections). Nine siliciclastic lithofacies represent fluvial, deltaic and lacustrine depositional environments. On the field-scale, the formations reveal a variation in sandstone bodies, geometry, thickness, and architecture. On the macro scale, the reservoir quality varies laterally and vertically, controlled by depositional environments and a general fining upward trend with different degrees of connectivity. On a micro-scale, the reservoir quality is affected by diagenetic processes. Kaolinite is dominant as pore fill and coat. Average porosity and permeability in Bentiu are 23.8% and 369 mD, and 20.5% and 35.1 mD in Aradeiba.

Keywords

Integrated study • Bentiu and Aradeiba reservoirs
Muglad Basin • Sedimentological analysis

T. Omayma (✉) · C. Adrian
EA 4592 Géoressources et Environnement, ENSEGID-Bordeaux INP, 1, allée Daguin, Pessac, Cedex, 33607, France
e-mail: banohamo@gmail.com

T. Omayma · H. Rashid
Faculty of Petroleum Engineering and Technology, Sudan University of Science and Technology, Box: 72 Khartoum, Sudan

M. Nuha
Faculty of Petroleum and Minerals, Al Neelain University, Box: 12702 Khartoum, Sudan

Z. Izzeldin
El Rawat Operating Company, Box: 11778 Khartoum, Sudan

1 Introduction

The influential factors impacting on reservoir quality include a variety of facies associations, diagenesis and the depositional environment. Some authors have studied the sedimentology of reservoir formations in Muglad basin, such as Schull [1]. He linked the textural change and mineralogical component to reservoir quality. While Sayed [2] concluded that large-scale facies variations in the basin are primarily controlled by tectonic processes and structures. The aims of this study are: (i) to identify the lithofacies and electrofacies associations from cores and wire line logs; (ii) petrographical analysis to classify the sandstone types and integrate with SEM and XRD to investigate the diagenetic processes that effect the reservoir quality; (iii) to identify the depositional environment and construct a depositional model for the study area and (iv) to evaluate the reservoir quality of the Bentiu and Aradeiba reservoirs.

2 Geological Setting

The Muglad basin belongs to central African rift basins. It displays polyphase rifting. It is recognized as the largest basin among the Sudanese interior rift basins with NW-SE-orientation, extending across 120,000 km^2, up to 200 km wide, and over 800 km long (Fig. 1a). Locally it contains up to 13 km of Cretaceous-Tertiary sediments [3]. Three rifting phases recognized in Sudan are defined in Muglad Basin, resulting in the accumulation of 5400, 4200 and 5400 m of sediments for each phase, respectively. In the Muglad Basin, the sag phases are represented by lower Bentiu, lower Amal and Adok, which represent fluvial sands of sediment cycles F3, F2 and F1 respectively. The selected wells locate in Hamra East Oil Field (HAE oil field) at south of the Muglad Basin (Fig. 1b).

F. Rossetti et al. (eds.), *The Structural Geology Contribution to the Africa-Eurasia Geology: Basement and Reservoir Structure, Ore Mineralisation and Tectonic Modelling*,
Advances in Science, Technology & Innovation, https://doi.org/10.1007/978-3-030-01455-1_34

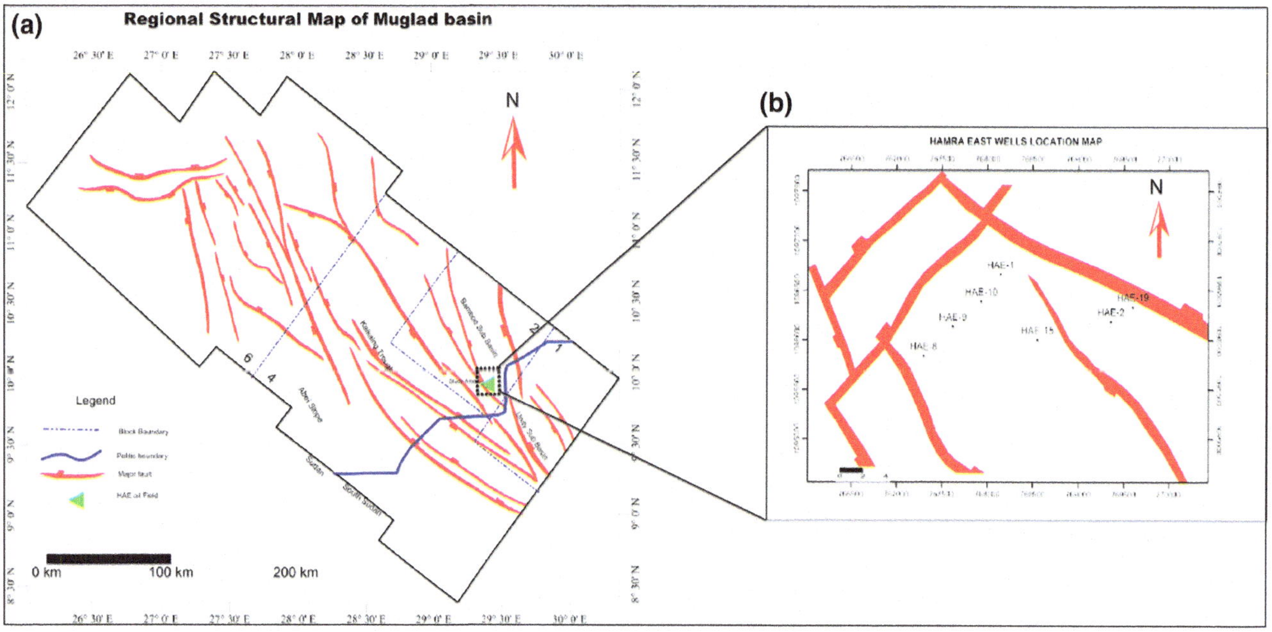

Fig. 1 **a** Regional structural map of Muglad Basin; **b** location map of studied wells within Hamra East Oil Field

3 Results

3.1 Lithofacies and Electrofacies Associations

Three lithofacies groups were classified as G1, G2 and G3 beside the electrofacies associations, both described in Table 1.

3.2 Sandstone Petrography

The mineral percentage counted under the microscope is illustrated in Fig. 2. The studied samples are classified (Fig. 3) according to the scheme of Pettijohn et al. [4] and Folk [5]. The results obtained from thin sections under OM, XRD and SEM helped to identify diagenetic processes effecting the reservoir quality. Some had positive effects like

Table 1 Facies and electrofacies classification

Facies group	Lithology description	Lithofacies	Sedimentary structures	Interpretation	Electrofacies association
G1	Medium to coarse grained sandstone	Lf1 Lf2 Lf3	Planar crossbedding Massive or faint lamination Through crossbedding	Fluvial channel Point and braid bars	Fining upwards sequence and box shape
G2	Very fine to fine-grained sandstone	Lf4 Lf2 Lf5	Fine lamination Massive or faint lamination Ripple marks	Creeves supply and levees	Fining upwards sequence and irregular shape
G3	Fine grained sediments Siltstone mudstone and shale	Lf6 Lf7 Lf8 Lf9	Fine lamination Siltstone Massive mudstone Bioturbated mudstone Laminated shale	Sand (channel-fill or crevasse splay) or lacustrine	Box shape

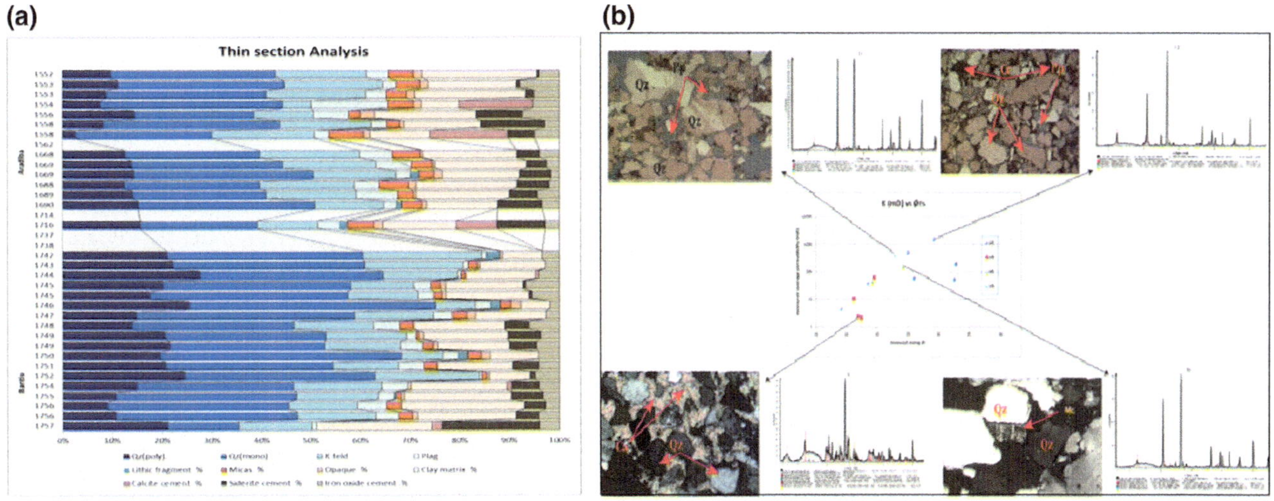

Fig. 2 **a** Mineral percentage counted under the microscope; **b** the relation between sedimentary/diagenetic facies and petrophysical properties such as porosity and permeability

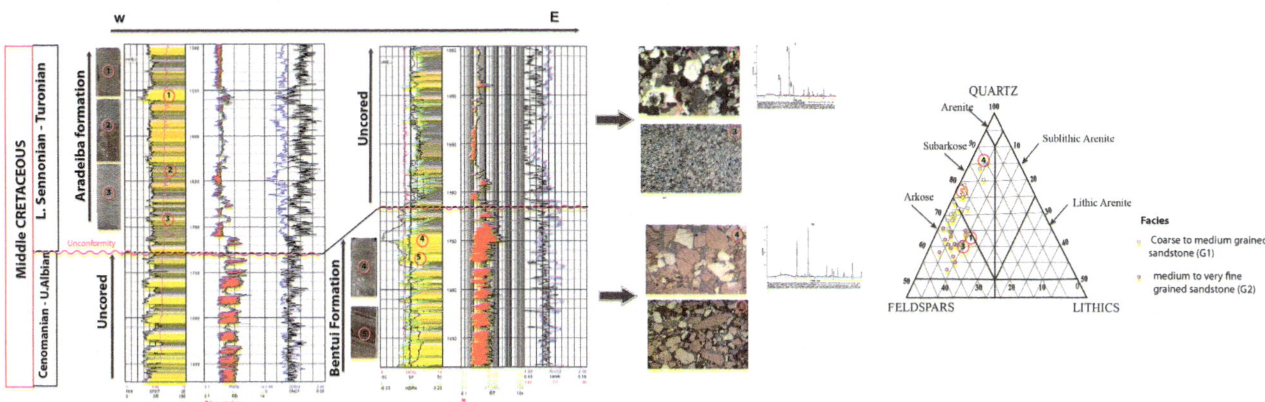

Fig. 3 Lateral and vertical facies distribution

dissolution or negative effects such as the precipitation of carbonate and kaolinite, and compaction (Fig. 3) (Table 2).

4 Discussion

Figure 3 shows vertical and lateral variations in facies association and stacking patterns in Bentiu and Aradeiba reservoirs within a fluvial/lacustrine environment. On macro-scale they are depositional environment, geometry and architecture. The study reveals fining upward trends with different degrees of connectivity. Bentiu shows well-connected sandstone bodies, while lower and upper Aradeiba which have moderate to lower connectivity of sandstone bodies. The Bentiu and Aradeiba formations consist of coarse to very fine-grained sandstone. Classified as arkoses and sub-arkoses they are mineralogically mature to sub-mature. On a micro-scale, diagenetic processes such as mineral precipitation, compaction and dissolution are main controls of the reservoir quality. The reservoir porosity and permeability are significantly affected by: kaolinite precipitation, clay matrix presence and pore-filling. This multi-scale analysis improves the understanding of the various facies associations, diagenesis and the depositional environment and consequently, of petroleum prospectivity in the study area.

Table 2 Hg-porosity (38 samples) and Air permeability (14 samples) for Bentiu and Aradeiba reservoir formations

Formation	Sub formation	K (mD)			Hg-porosity (%)		
		Min.	Max.	Average	Min.	Max.	Average
Aradeiba	Upper	2.3	59.9	31.1	16.9	33.7	23.1
		–	–	54.4	21.2	26.0	22.5
	Lower	38.7	46.3	42.5	14.0	19.5	17.8
		–	–	4.7	14.2	18.6	16.4
		–	–	–	11.6	24.3	18.0
Bentiu	Upper	2.1	1611.2	369.4	13.6	33.2	23.9

References

1. Schull, T.J.: Rift basins of interior Sudan: Petroleum exploration and discovery. Am. Assoc. Pet. Geol. Bull. **72**, 1128–1142, Tulsa
2. Sayed, A.M.I.: Sedimentology and reservoir geology of the middle-upper cretaceous strata in unity and heglig fields in SE Muglad Rift Basin, Sudan (2003)
3. McHarque, T.R., Heidrick, T.L., Livingston, J.E.: Tectonostratigraphic development of the interior Sudan rifts, central Africa. In: Ziegler, P.A. (ed.) Geodynamics of Rifting, Volume II. Case History Studies on Rifts: North and South America and Africa Tectonophysics, vol. 213, pp. 187–202 (1992)
4. Pettijohn, F.J., Potter, P.E., Siever, R.: Sand and Sandstone, 553 pp. Springer, New York (1987)
5. Folk, R.L.: Petrology of Sedimentary Rocks, 159 p. Hemphill, Austin, Texas (1974)

Organo-Geochemical, Petrography and Spectroscopy of Organic Matter Associated Ypresianphosphatic Series of the Maknassy-Mezzouna Basin Central Tunisia

Faten Jaballi, Mongi Felhi, Adnen Lafi, Kamel Zayani, and Ali Tlili

Abstract

The mineralogical and organo-geochemical analyses of borehole samples were carried out for the Ypresianphosphatic series of Jebel Jebbeus. This section corresponds to Chouabine Formation, which was deposited in Maknassy-Mezzouna Basin (central Tunisia) and is made up of three phosphatic layers alterned with dolomitic marls. Their thicknesses are 7, 2.3 and 6.1 m corresponding to the top, middle and the bottom layers. The aim of this study is to understand the nature, origin and preservation of organic matter (OM) within the phosphatic deposits of eastern basin. Three representative borehole sample of phosphate were investigated using Rock-Eval pyrolysis, X-Ray diffraction, Infrared analyses and the petrographic study of OM. This dataset shows that OM is immature. TOC contents slightly increases from the top to the bottom of the Chouabine Formation (0.33–0.51%), corresponding to increased contents of clay minerals content in the bottom of the formation. The clay fraction found in the middle and lower phosphate series is composed mainly of smectite. Petrography of the residual OM shows that the amorphous organic matter (AOM) is associated with land plant material. The orange AOM can be related to a high degree of aliphaticity, which is confirmed by predominatly of aliphatic bands of CH_2 and CH_3, likely by biochemical degradation by anerobic bacteria.

Keywords

Ypresianphosphatic series • Tunisia • Clay fraction
Amorphous organic matter • Infrared • Aliphaticity

1 Introduction

The Ypresianphosphatic successions in Tunisia are deposited in three basins, reported around Kasserine "Island": the north basin (El Kef basin), the south basin (Gafsa basin) and the eastern basin (Maknassy-Mezzouna basin) (Fig. 1). Geological survey established in the eastern basin shows that Ypresian phosphate series are constituted by rhythmic alternations of phosphatic and dolomitic marl layers. The mineralogy and geochemistry of phosphatic series have been widely studied. However, Ypresian organic matter is diverse in a compressive study. Several authors suggest that phosphorites deposits in the eastern basins correspond to small basins in a lacustrine environment [1]. According to [3, 6], marine phytoplankton is the major source of OM in south basin.

2 Materials and Methods

Borehole samples were sampled from syncline Jebbeus (Maknassy-Mezzouna basin). Each bulk rock sample was analysed by Rock-Eval pyrolysis. X-Ray diffraction was carried out on clay mineral. After the removal of carbonate and organic matter, the clay fraction prepared by centrifugation, air drying, ethylene glycol solvation and heating at 550 °C during 2 h. Kerogen was extracted from sediment, using chemical treatment to remove carbonates and silica. The residual organic matter was used to prepare palynological slides and KBr pellet for Infrared analyses. For the palynological analysis, the polynomorph assemblages were separated from Organic matter using $ZnCl_2$.

F. Jaballi · M. Felhi (✉) · A. Tlili
Georesources, Materials, Environment and Global Changes Laboratory, University of Sfax, 3018 Sfax, Tunisia
e-mail: mongi_felhi@yahoo.fr

A. Lafi
Department of Geological Engineering, National School of Engineers of Sfax, University of Sfax, Sfax, Tunisia

K. Zayani
Gafsa Phosphate Company, Research Center, 2130 Metlaoui, Tunisia

F. Rossetti et al. (eds.), *The Structural Geology Contribution to the Africa-Eurasia Geology: Basement and Reservoir Structure, Ore Mineralisation and Tectonic Modelling*,
Advances in Science, Technology & Innovation, https://doi.org/10.1007/978-3-030-01455-1_35

Fig. 1 Simplified geological map of the JbelJebbeus (Google Earth), showing the location of borehole log of Ypresienphosphatic series. Detailed lithological borehole log of phosphatic layer indicate the sample position

3 Results

3.1 Rock-Eval Pyrolysis Analysis

The residual petroleum potential of the rock S2 ranges from 0.48 to 0.74 mg HC/g rock. The value of TOC shows a relative increase from the top, middle and the lower phosphatic series, their values are 0.33, 0.47 and 0.51% respectively. The relatively important values of TOC were measured for borehole samples from the bottom (Fig. 2).

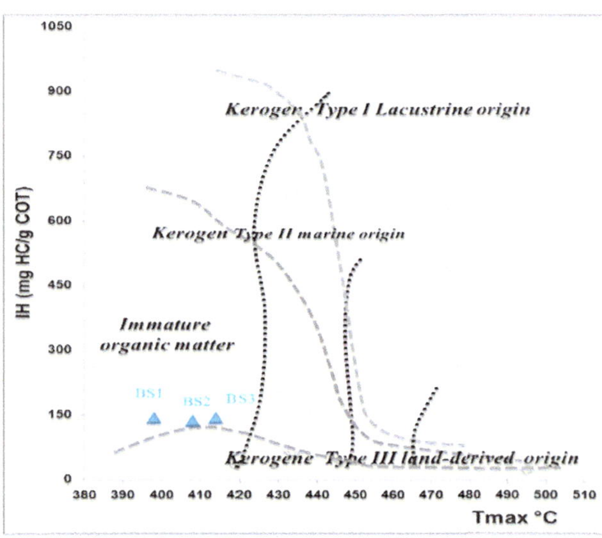

Fig. 2 IH-Tmax diagram showing types of OM in drilling sample in of phosphatic layers of the Jebbeus deposit

3.2 Oriented Preparations

The data obtained of X-Ray diffraction of clay minerals indicate that the principal clay minerals is smectite (~97% for the bottom and middle layers) with traces of kaolinite. The percentage of smectite values tends to decrease from 97% at the bottom to 60.5% at the top. The kaolinite mineral becomes abundant at the top of Chouabine Formation.

3.3 The Infrared Spectroscopy

The Infrared spectrum show that the maximum intensity of asymmetric and symmetric stretching vibrations of CH_2 are 2925 and 2855 cm^{-1}, respectively. The band detected about 1450 cm^{-1} is attributed to asymmetric deformation vibrations of the CH_2 and CH_3 groups. Thus, the infrared spectroscopy reveals the presence of symmetrical deformation vibration of CH_3 groups at 1376 cm^{-1} (Fig. 3).

3.4 Petrographic Study of Organic Matter

Petrographic observations of residual Kerogen reveals the presence of the AOM. The marine OM shows the coexistence of planktonic materiel (dinoflagellates, diatomes) and land plant material (pollen, spore). Observations on Transmitted light micrograph show the predominantly orange AOM with Brown and dark AOM.

4 Discussion

Rock-Eval pyrolysis data reveals that the analyzed samples have similar geochemical characteristics: TOC content and IH, Tmax parameters, which confirms that the same

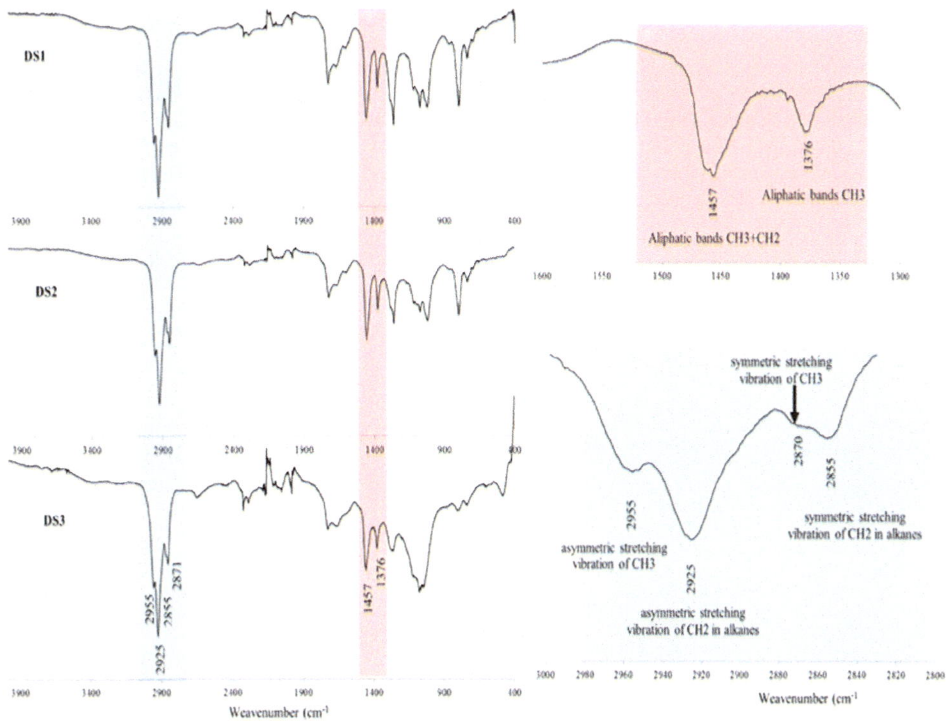

Fig. 3 Infrared band attributions in the three kerogen samples isolated from Phosphate layers and selected

conditions of deposit and preservation of the OM are associated to phosphorites. The low values of residual petroleum potential S2 show that the OM is immature indicating early diagenesis. Although, the low TOC contents can be related to relative oxicity of the depositional environment. This assumption agrees with previous studies of [2, 3]. The diagram IH versus Tmax indicates that oxidized OM of three samples is predominantly Type II, which has a phytoplanktonic marine origin with a contribution of land plant material. This result is confirmed by the study of residual of OM indicating, the presence pollen and spores. The organo-geochemical contents agree with the mineralogical composition of clay mineral, which shows a gradual increase of OM from the top, middle and the bottom sections that coincides with the occurrence of clay minerals in the bulk rock sample (0.5, 1.5 and 2.25%; respectively). The preservation of OM can be explained by the presence of clay minerals with high smectite content found in the lower part of phosphatic series. This assumption agrees with the previous studies of [4, 5], which indicate that OM is often more associated to the fine fraction. Analysis of organic matter by Infrared shows the predominantly of bands of CH_2 and CH_3 groups, which reflects the aliphatic character of kerogen [6]. The absorption bands of CH_2 groups are more intense than those of CH_3 groups, reflecting rather long aliphatic chains. This result is confirmed by the presence of characteristic bands at about 1457 cm^{-1} and at 1376 cm^{-1}. This assumption agrees with the previous studies of [7].

5 Conclusions

The organo-geochemical, petrographic and spectroscopy studies of OM associated with the Ypresianphosphatic series in borehole samples shows that TOC content increases at the bottom. This can be related to semi-restricted setting associated with suboxic-oxic conditions in lacustrine environments followed by diagenetic transformation and the lower plankton biomass, productivity. These samples are characterized by AOM Type II, which has a phytoplanktonic marine origin with a contribution of land plant material. The Infrared spectrum shows the aliphatic character of their residual OM. This agrees with the predominantly of aliphatic bands of CH_2 and CH_3 groups, which can be related to the biochemical degradation by anerobic bacteria.

References

1. Garnit, H., Bouhlel, S., Barca, D., Chtara, C.: Application of LA-ICP-MS to sedimentary phosphatic particles from Tunisian phosphorite deposits: Insights from trace elements and REE into paleo-depositional environments. Chem. Erde–Geochem. **72**, 127–139 (2012)
2. Tissot, B.P., Welte, T.H.: Petroleum Formation and Occurrence. Springer, New York, DH (1984)
3. Arfaoui, A., Montacer, M.: Geochemical characterization given by Rock-Eval parameters and n-alkanes distribution on Ypresian

organic matter at Jebel Chaker, Tunisia. Resource Geology **57**, 37–46 (2007)

4. Salamon, V., Derenne, S., Lallier-Vergès, E., Largeau, C., Beaudoin, B.: Protection of organic matter by mineral matrix in Cenomanien black shale. Org. Geochem. **31**, 463–474 (2000)
5. Tlili, A., Felhi, M., Fattah, N., Montacer, M.: Mineralogical and geochemical studies of Ypresianmarly clays and silica rocks of phosphatic series, Gafsa-Metlaoui basin, southwestern Tunisia implication for depositional environment. Geosci J **15**(1), 53–64 (2011)
6. Felhi, M., Tlili, A., Montacer, M.: Geochemistry, petrographic and spectroscopic studies of organic matter of clay associated kerogen of Ypresian series: Gafsa-Metlaoui phosphate basin, Tunisia. Resour. Geol. **59**, 428–436 (2008)
7. Landais, P., Rochdi, A.: In situ examination of coal macerals oxidation by micro-FT-ir, spectroscopy. Fuel **72**, 1393–1401 (1993)

Application of Seabed Core Geochemistry for Further Exploration in Banggai-Sula Area

Sugeng Sapto Surjono, Muhamad Rizki Asy'ari, and Arif Gunawan

Abstract

Banggai-Sula area is the complex tectonic region located in the east of Sulawesi. The recent basin, called Sula Basin, lies in the northern part of the Taliabu and Mangole island. It is part of frontier exploration area with no significant hydrocarbon discoveries. However, the adjacent basin, Banggai Basin is a mature basin with some fields such as Senoro, Donggi, Minahaki, Matindok, and Dongkala, Sukamaju, and Maleo Raja gas fields and Tiaka oil field. Previous drilled wells such as Loki-1 resulted in significant gas show while Alpha-1 indicated no show. Significant new directions for further exploration in Banggai-Sula area are required. By utilizing the new data, this study is aimed to approach this concern. Several surface and subsurface data are utilized in this study, including seabed core geochemistry, 2D seismic lines and the multibeam data. The geochemistry analysis result of seabed core data shows that the occurrence of hydrocarbon seeps mostly in the central part of the study area. The northern part area represents the most hydrocarbon seepages occurrence. The collision process leads to the generation and expulsion of hydrocarbon and provide suitable traps for hydrocarbon accumulation. This area is recommended for further exploration of Banggai-Sula region.

Keywords

Banggai-Sula • Seabed core geochemistry
Collision • Exploration recommendation

S. S. Surjono (✉) · M. R. Asy'ari
Geological Engineering Department, Universitas Gadjah Mada, Jl. Grafika no. 2, Yogyakarta, 55281, Indonesia
e-mail: sugengssurjono@ugm.ac.id

A. Gunawan
TGS-NOPEC Geophysical Company, Pondok Indah Office Tower 3, Jl. Sultan Iskandar Muda Kav V-TA, Jakarta, 12310, Indonesia

1 Introduction

Banggai-Sula is the complex tectonic region located in the east of Sulawesi. This area is regarded as amicrocontinental fragment of Australian and New Guinea continents due to its similar stratigraphical characteristics [1, 2]. The similarity is the Pre-Cretaceous stratigraphic units resting on Paleozoic granitic and metamorphic basement. The westward movement driven by the Sorong Fault implied to the collision of Banggai Sula Platform and East Sulawesi Ophiolitic Complexsince Plio-Pleistocene [1]. This process has formed Banggai and Sula Basins.

Sula Basin lies in the northern part of the Taliabu and Mangoleislands. It is part of frontier exploration area with no significant hydrocarbon discoveries. However, the adjacent basin, Banggai Basin is a mature basin with some proven fields such as Senoro, Donggi, Minahaki, Matindok, and Dongkala, Sukamaju, and Maleo Raja gas fields and Tiaka oil field [3]. Previous drilled wells in Banggai Sula area such as Loki-1 resulted in significant gas show while Alpha-1 indicated no show [4]. Significant new directions for further exploration in Banggai-Sula area are required by utilizing the new data of seabed core geochemistry interpretation, this study is aimed to approach this concern.

2 Research Method

Several surface and subsurface data are utilized in this study, including hydrocarbon seepage indication from seabed core geochemistry analysis, 2D seismic lines and the multibeam data. Seabed core data is analyzed geochemically by TGS NOPEC to identify possible oil and gas seepages occurred on the seafloor around the study area. The result is then used to find the distribution of the seepages based on geological condition. The 2D seismic lines are interpreted to identify the markers and structural configuration of the study area. Due to no well data available in this study area, the

F. Rossetti et al. (eds.), *The Structural Geology Contribution to the Africa-Eurasia Geology: Basement and Reservoir Structure, Ore Mineralisation and Tectonic Modelling*,
Advances in Science, Technology & Innovation, https://doi.org/10.1007/978-3-030-01455-1_36

markers referred to the previous study [5]. The multibeam data is utilized to identify some structural zones in relation with the occurrence of the seepages beneath the sea. All of data are synthesized to give a recommendation of future exploration target area in Banggai-Sula region.

3 Results

3.1 Seismic Interpretation

Several markers can be identified from the seismic interpretation, such as Paleozoic Basement, Jurassic Bobong and Buya Formations, Early Cretaceous Tanamu Formation, and Tertiary Salodik Formation. The North–South seismic section shows the passive margin of Mesozoic sequence at the south and collision complex at the north with some south-verging thrusts (Fig. 1). The collision complex could act as a trigger for hydrocarbon generation and expulsion in this area.

3.2 Seabed Core Geochemistry and Multibeam Analysis

Total 56 seabed core samples were analyzed in the laboratory. 26 samples have hydrocarbon shows while the other 30 samples have no significant show. The shows mostly occurred in the central part of the study area, from the north to the south. The oil seeps mostly occupy the southern and northern parts while the gas seeps in the central and northern parts of the area. Seabed cores with no hydrocarbon indication occurred in the northwest and the southeast. Based on multibeam data, the northern area is a collision complex zone which was identified by intensive thrusting shown by the map (Fig. 2).

4 Discussion

The geochemistry analysis result of seabed core data shows that the occurrence of hydrocarbon seeps mostly in the central part of the study area (Fig. 2). The northwest and southeast area have no hydrocarbon indication. Well Alpha-1 drilled in the offshore western part of the Taliabu island is a dry hole well. Well Loki-1 in the offshore northern part of Mangole island resulted a gas show.

Most of seabed core samples were taken in the northern area which is the collision complex. The data proved that the collision complex has a good hydrocarbon indication as well as a source maturation triggered by thrusting process. This area has become the important future target exploration. The oil accumulation could be found in a thrust-anticline traps of Mesozoic-Tertiary reservoirs (Fig. 1). However, the additional data are needed to obtain a higher certainty of geological condition in that area.

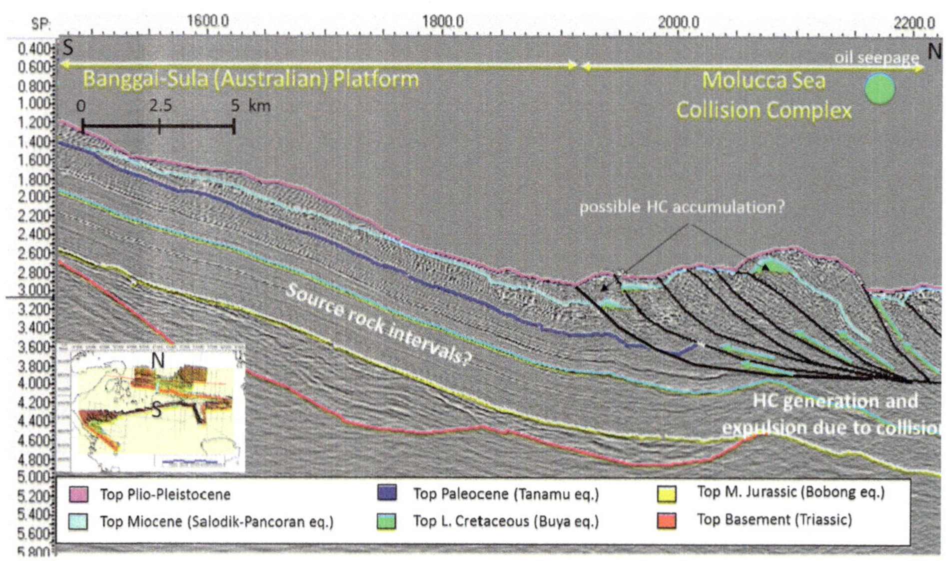

Fig. 1 N-S seismic cross section of Banggai Sula area. Collision causes the increase of sediments thickness and generation and expulsion of hydrocarbon from source rocks interval. The hydrocarbon migrated through faults and possibly accumulated in thrust-anticline structures and some others reached to the surface as seepages

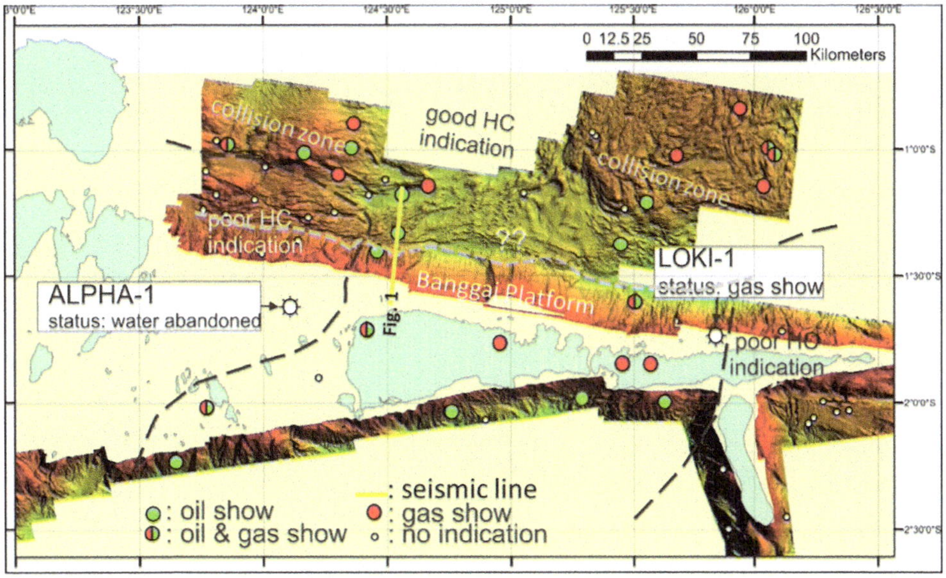

Fig. 2 The map showing the occurrence of hydrocarbon in the study area from seabed core and some onshore seepages from previous study [1, 4]. The hydrocarbon mostly occurred in the northern part of the study area

5 Conclusions

The hydrocarbon seepages in the seafloor mostly occurred in the central part of the study area. The northern part area represents the most hydrocarbon seepages occurrence. The collision process leads to the generation and expulsion of hydrocarbon and provide suitable traps for hydrocarbon accumulation. This area is recommended for further exploration of Banggai-Sula region.

References

1. Garrard, R.A., Supandjono, J.B.S: The geology of the Banggai-Sula microcontinent, Eastern Indonesia. In: Indonesian Petroleum Association, Proceedings of the 7th Annual Convention and Exhibition, IPA88-11.01. Jakarta (1988)
2. Surjono, S.S., WijayantiH, D.K.: Tectono-stratigraphic framework of Eastern Indonesia and its implication to petroleum systems. ASEAN Eng. J. Part C **1**, 110–124 (2011)
3. Satyana, A.H., Zaitun, S.: Origins of oils and gases at Banggai-Sula Microcontinent, Eastern Sulawesi-North Moluccas: constraints from biomarkers and isotope geochemistry-implications for further exploration of Cenozoic and Pre-Cenozoic objectives. In: Indonesian Petroleum Association, Proceedings of the 40th Annual Convention and Exhibition, IPA16-575-G. Jakarta (2016)
4. PND Data Availability (Daval): http://product.patranusa.com/daval. Last Accessed 30 Apr 2018
5. Watkinson, I.M., Hall, R., Ferdian, F.: Tectonic re-interpretation of the Banggai-Sula—Molucca Sea margin, Indonesia. In: Hall, R., Cottam, M.A., Wilson, M.E.J. (eds.) The SE Asian Gateway: History and Tectonics of the Australia-Asia Collision, Geological Society, London, Special Publication, vol. 355, pp. 203–224. The Geological Society of London (2011)

South Tunisia, Structures and Traps Evolution: A Review from a New 3D Mega-Merge Survey

Ferid Adouani, Ahmed Saadi, Francis Chevalier, and Noura Ayari

Abstract

Improving 3D seismic data by dedicated reprocessing and having access to regional mega-merge datasets is allowing the OMV exploration geologists access to a more detailed understanding of the structural evolution of the Ghadames Basin in the south of Tunisia, and to better characterize the associated structures. It also helps exploration activity to focus on remaining hydrocarbon potential in subtle traps, structural or stratigraphic.

Keywords

3D seismic • Mega-merge • Ghadames basin • South Tunisia • Extension • Compression • Strike-slip movements • Differential compaction

1 Introduction

The Ordovician and Silurian plays of the South Tunisian Ghadames Basin have been the focus of extensive investigation and exploration studies since the beginning of the exploration in this area in the 1960s. However, due to difficult access to the geometry of the subsurface, details about the traps' formation mechanisms and their development and evolution through time are still not documented in detail.

Even if the structural framework and chronology of tectonic events are well established at a regional scale and have been documented by several authors (e.g. Echikh [7], Bouaziz et al. [4], Galeazzi et al. [9]), the new available 2D and 3D seismic data are giving more information to improve the subsurface imaging and better evaluate subsurface structures and their geometries.

Moreover, the exploration strategy applied by most of the oil & gas companies has focused until now on trying to find oil or gas accumulations in structural traps—3-way- to 4-way dip closures—at Ordovician, Silurian and Triassic levels.

These traps are the results of a complex structural and sedimentological history of the region. A wide variety of structural patterns of different age, orientation and origin can be observed form a detailed analysis of the available seismic datasets.

2 Method and Database

This study aims to better constrain and understand the structural architectures and the detailed geodynamic evolution of Southern Tunisia area (Northern Ghadames Basin, Fig. 1), in order to better assess the kinematic of the structures, the impact on structural trap formation, and the hydrocarbon generation and migration.

OMV Tunisia is using a proprietary extensive regional 3D seismic data set (around 5600 km^2) issued from the reprocessing and merging of nine 3D sub-surveys. A large 2D seismic data set (>1000 km) tied to existing wells (Fig. 2), integrated with surface geology, gravity and magnetic data as well as seismic attributes were also included.

A detailed mapping of all possible extensive and compressive events has been conducted over the 3D seismic mega-merge, from the Top Basement to the Hercynian Unconformity. Main structures have been studied in detail in order to reconstruct their kinematic and the associated stress field. The chronology of the deformation was assessed when possible and summarized into a detailed timescale.

3 Results

Detailed structural mapping and fault/lineaments analysis of the Top Basement and the Cambro-Ordovician intervals showed the main NE-SW trend. On top of this, other directions such as N-S, NW-SE and E-W can be seen, but

F. Adouani (✉) · A. Saadi · F. Chevalier · N. Ayari
OMV (Tunesien) Production GmbH, Imp. du Lac Turkana, Les Berges du Lac, 1053 Tunis, Tunisia
e-mail: ferid.adouani@omv.com

F. Rossetti et al. (eds.), *The Structural Geology Contribution to the Africa-Eurasia Geology: Basement and Reservoir Structure, Ore Mineralisation and Tectonic Modelling*,
Advances in Science, Technology & Innovation, https://doi.org/10.1007/978-3-030-01455-1_37

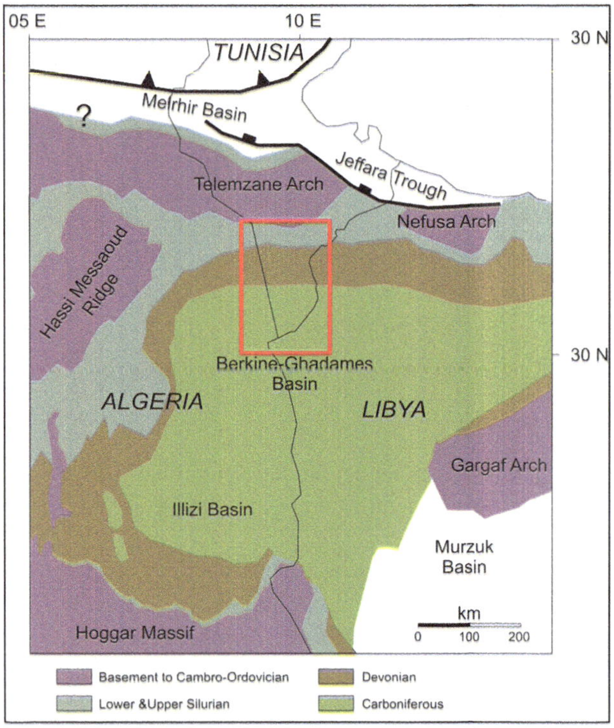

Fig. 1 Location map of the Northern Ghadames Basin and its regional structural context in North Africa (modified after [9]). The red rectangle shows the area of interest

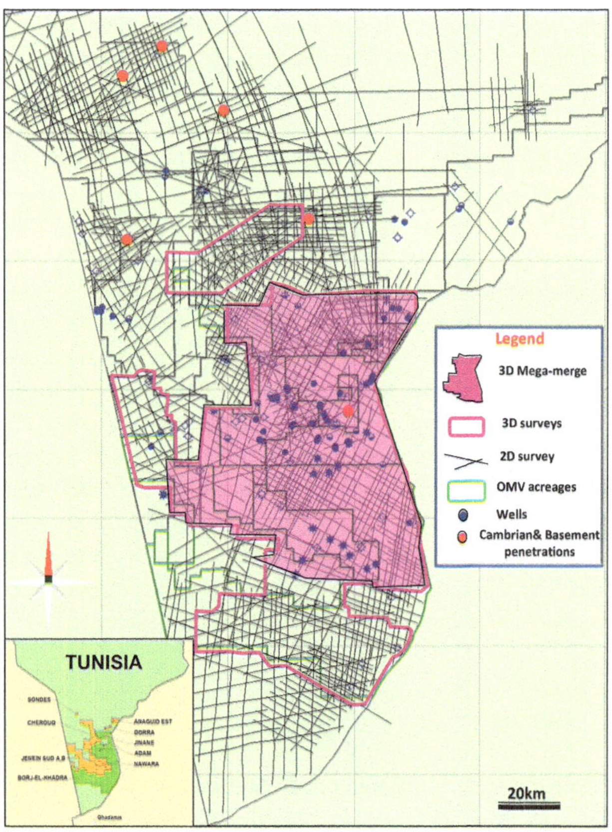

Fig. 2 OMV Tunisia seismic and well database used for regional studies

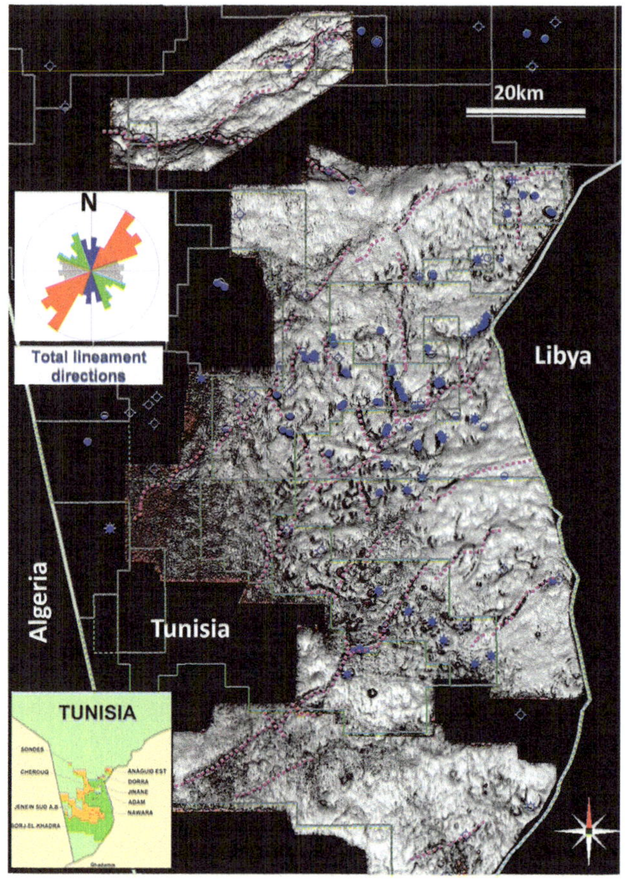

Fig. 3 Map of faults and lineaments (deduced from variance Attribute slice @ Ordovician level) associated with rose diagram

less expressed (Fig. 3). From literature, these faults have been initiated through the Proterozoic Pangea continent break-up and reactivated during the Pan-African orogeny [2–4, 7, 10, 11].

Extensional features are mainly visible in the Lower Cambrian series in the center of the study area (Adam/BEK concessions). These structures are materialized by normal faulting, block tilting, horst and graben development, basement erosion, and onlap geometries on basement escarpments.

Upper Cambrian to Ordovician period is dominated by stable subsidence over the whole area of interest, with minor reactivation of pre-existing fault system (differential compaction?) and low basin differentiation. No major plate movements are documented for the AOI, even if regional tectonic movements are evocated for the Southern Ghadames Basin in Algeria and Libya [5–8, 12, 14].

Silurian is complex to understand in terms of tectonics/structural evolution as many different features can be observed in the seismic, such as reverse faulting, compaction phenomenon, vertical faults most-likely related to strike-slip movements, etc. The amplitude of these movements is relatively low, and interpretation remains uncertain in some places. Assessing the correct stress field over the area is challenging in many ways. However, a slight compression phase can be deduced from most of the observations [12, 13, 15].

Devonian to Carboniferous section is again dominated by stable subsidence with small reactivation of pre-existing faults system mainly due to differential compaction.

The Upper Paleozoic (Carboniferous) is marked by regional uplift, erosion and basin-scale tilting during the deformation linked to the Hercynian orogeny caused by the collision between Laurasia and Gondwana [1, 2, 4, 7, 11, 14]. During this time, and towards the Northern edge of the Ghadames basin (example of the El Borma area), the preexisting faults system has been reactivated as transpressional structures with strike slip or reverse minor displacement, depending on their respective original orientations.

The recent tectonics events related to the Austrian (Aptian) and Alpine phases resulted in a gentle regional tilting of the basin with little remodeling of preexisting structural features and the reshaping of the actual basin architecture and traps.

4 Conclusions

From dedicated 3D seismic reprocessing and regional mega-merge, OMV is able to get access to a more detailed tectonic evolution at the scale of the Northern Ghadames Basin. Characteristics and the kinematic of structures, impact on sedimentation of reservoir and non-reservoirs layers, and migration of hydrocarbon can then be better assessed, in order to help exploration geologists get access to the remaining hydrocarbon resources in the South of Tunisia.

References

1. Acheche, M., et al.: Ghadames basin, Southern Tunisia: a reappraisal of Triassic reservoirs and future prospectivity. Am. Assoc. Petrol. Geol. Bull. **85**, 765–780 (2001)
2. Aissaoui, M.N., et al.: Petroleum assessment of Berkine-Ghadames Basin, southern Tunisia. AAPG Bull. **100**(03), 445–476 (2016)
3. Boote, D.R.D., et al.: Paleozoic petroleum systems of North Africa. In: Macgregor, D.S., Moody, R.T.J., Clark-Lowes, D.D. (eds.), Petroleum Geology of North Africa, vol. 132, pp. 7–68. Geological Society of London (1998)
4. Bouaziz, S. et al.: Tectonic evolution of the northern African margin in Tunisia from paleostress data and sedimentary record: Tectonophysics **357**, 227–253 (2002)
5. Craig, J., et al.: Structural styles and prospectivity in the Precambrian and palaeozoic hydrocarbon systems of North Africa. In: The Geology of East Libya, vol. IV, pp. 51–102. Earth Science Society of Libya, Tripoli (2008)
6. Dhaoui, M., et al.: Gravity analysis of the Precambrian basement topography associated with the northern boundary of Ghadames Basin (southern Tunisia). J. Appl. Geophys. **111**, 299–311 (2014)
7. Echikh, K.: Geology and hydrocarbon occurrences in the Ghadamis Basin, Algeria, Tunisia, Libya. Geol. Soc. Lond. Spec. Publ. **132**, 109–129 (1998)
8. Gabtni, H., et al.: Basement structure of southern Tunisia as determined from the analysis of gravity data: implications for petroleum exploration. Petrol. Geosci. **18**, 143–152 (2012)
9. Galeazzi, S., et al.: Regional geology and petroleum systems of the Illizi-Berkine area of the Algerian Saharan Platform: an overview. Mar. Petrol. Geol. **27**, 143–178 (2010)
10. Guiraud, R.: 1998: Mesozoic rifting and basin inversion along the northern African Tethyan margin: an overview. Geol. Soc. Lond. Spec. Publ. **132**(1), 217–229 (1998)
11. Klett, T.R.: Total petroleum systems of the Triassic/Ghadames Province, Algeria, Tunisia and Libya—The Tannezuft-Oued Mya, Tannezuft-Melrhir and Tannezuft-Ghadames. US Geol. Surv. Bull. **2202-C**, 118 p (2000)
12. Le Heron, D.P., Howard, J.: Evidence for Late Ordovician glaciation of Al Kufrah Basin, Libya. J. Afr. Earth Sci. **58**, 354–364 (2010)
13. Rusk, D.C.: Petroleum potential of the underexplored basin centers—A twenty-first-century challenge. AAPG Bull. **74**, 429–452 (2001)
14. Sikander, A.H.: Structural development, geology and hydrocarbon potential of the Ghadamis and Murzuq Basins. In: Second Symposium on the Sedimentary Basins of Libya, vol. 2, pp. 281–326 (2003)
15. Soua, M.: Paleozoic oil/gas shale reservoirs in southern Tunisia: an overview. J. Afr. Earth Sci. **100**, 450–492 (2014). https://doi.org/10.1016/j.jafrearsci

Integrated Petrophysical Study of Acacus Reservoir (South of Tunisia)

Garci Sana

Abstract

The Silurian is subdivided in the Tannezuft (shale dominated) and Acacus (sand dominated) formations. The Acacus itself is subdivided in Acacus A, B & C, of which the Acacus B includes more shale than the others. During Acacus A, a time span of ~1–1.5 million years (Gorstian & lower Ludfordian), sediments of ~250 m were deposited. Petrophysical properties are quite variable and inhomogeneous since the reservoir shows layer heterogeneity and complex mineralogy related to diagenesis process, mainly chloritization and siderite cementation that deacrease the quality of the reservoir. The water saturation estimation is a critical task, the presence of chlorite and ferruginous mineral affects the logs resistivity responses that appear to be too low, conducting to an overestimation of the water saturation. For this purpose we did this integrated petrophysical study on the acacus that include more than 25 wells and it began by: Stratigraphic correlation, petrophysical interpretation using IP software, then an advanced petrophysical interpretation using the mineral solver, comparaison between the two methods of interpretation. After doing this tasks we conclude that scale data, rock typing and NMR data can reduce the petrophysical interpretation uncertainty.

Keywords

Acacus • Petrophysical interpretation • Rock typing
Chlorite • Complex mineralogy mineral solver

1 Introduction

Southern Tunisia, the Lower and Upper Silurian Formations crop out mainly in the relatively stable Ghadames Basin and the tectonically active Jeffara Basin. The Silurian is subdivided in the Tannezuft (shale dominated) and Acacus (sand dominated) formations. The Acacus itself is subdivided in Acacus A, B and C, of which the Acacus B includes more shale than the others. During Acacus A, a time span of ~1–1.5 million years (Gorstian and lower Ludfordian), ~250 m sediments were deposited., the Acacus A is subdivided in units A1–A9 (with subunits) in South Tunisia. Acacus A can generally be depicted as a thin bed—low resistivity pay characterized by alternated sands and shale streaks.

2 Petrophysical Interpretation

Petrophysical properties are quite variable and inhomogeneous since the reservoir shows layer heterogeneity and complex mineralogy related to diagenesis process, mainly chloritization and siderite cementation that decrease the quality of the reservoir. Sometimes for modelling purpose, the heterogeneous physical properties can be replaced by a single representative value (e.g., use of representative density in Mukherjee 2017, 2018, in press-1, 2).

The water saturation estimation is a critical task, the presence of chlorite and ferruginous mineral affects the logs

G. Sana (✉)
ETAP: Entreprise Tunisienne D'activitès Pètrolières, Tunis, Tunisia
e-mail: sana.garci@etap.com.tn

F. Rossetti et al. (eds.), *The Structural Geology Contribution to the Africa-Eurasia Geology: Basement and Reservoir Structure, Ore Mineralisation and Tectonic Modelling*,
Advances in Science, Technology & Innovation, https://doi.org/10.1007/978-3-030-01455-1_38

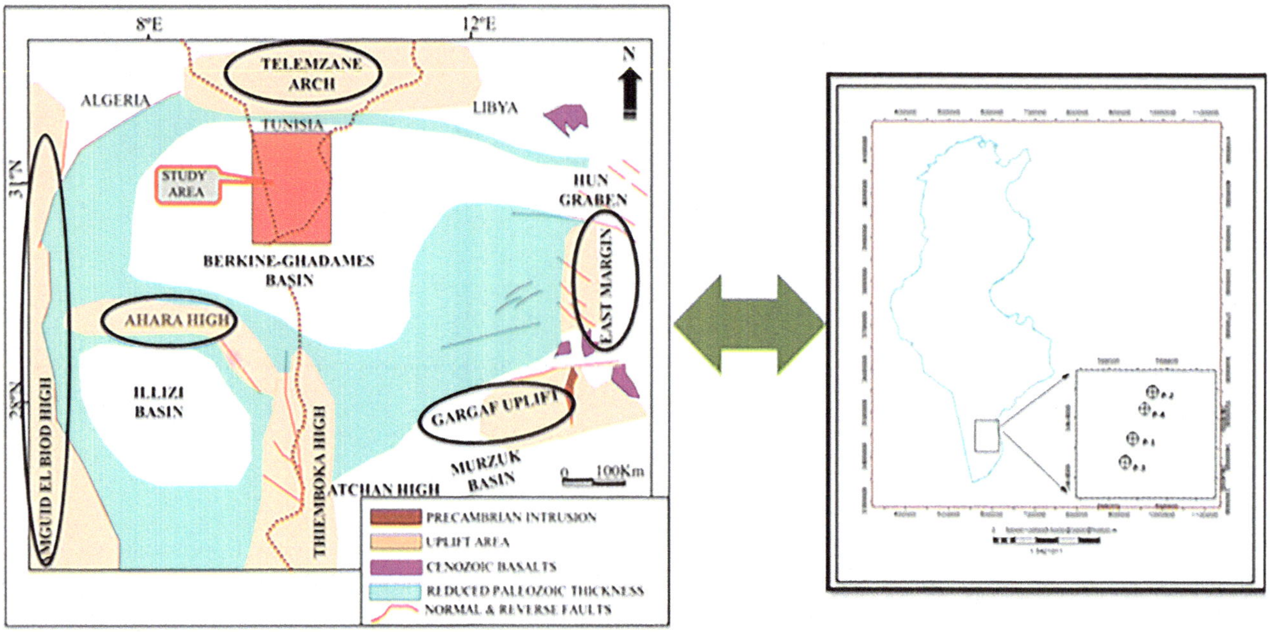

Fig. 1 Geographical and Geological location of the study area [2]

resistivity responses that appear to be too low, conducting to an overestimation of the water saturation.

For this purpose, we did this integrated petrophysical study on the acacus that includes more than 25 wells and it began by:

- Stratigraphic correlation,
- Petrophysical interpretation using IP software,
- Rock typing: It is essential to configure a petrophysical rock classification concept based on the acquisition of sufficient petrophysical data to properly understand the performance of the complex Silurian reservoir (Figs. 1 and 2).

3 Results

The hydraulic flow unit approach was used to identify reservoir rock quality and group the petrophysical data of the Acacus and Tannezuft reservoir sands into different rock types.

This method introduced by Amaefule et al. [1] is frequently used for characterizing complex variations in pore geometrical attributes and for accurate zoning of reservoirs into distinct units with similar fluid flow characteristics.

It involves a modified Kozeny-Carmen equation and the concept of mean hydraulic radius and is based on the

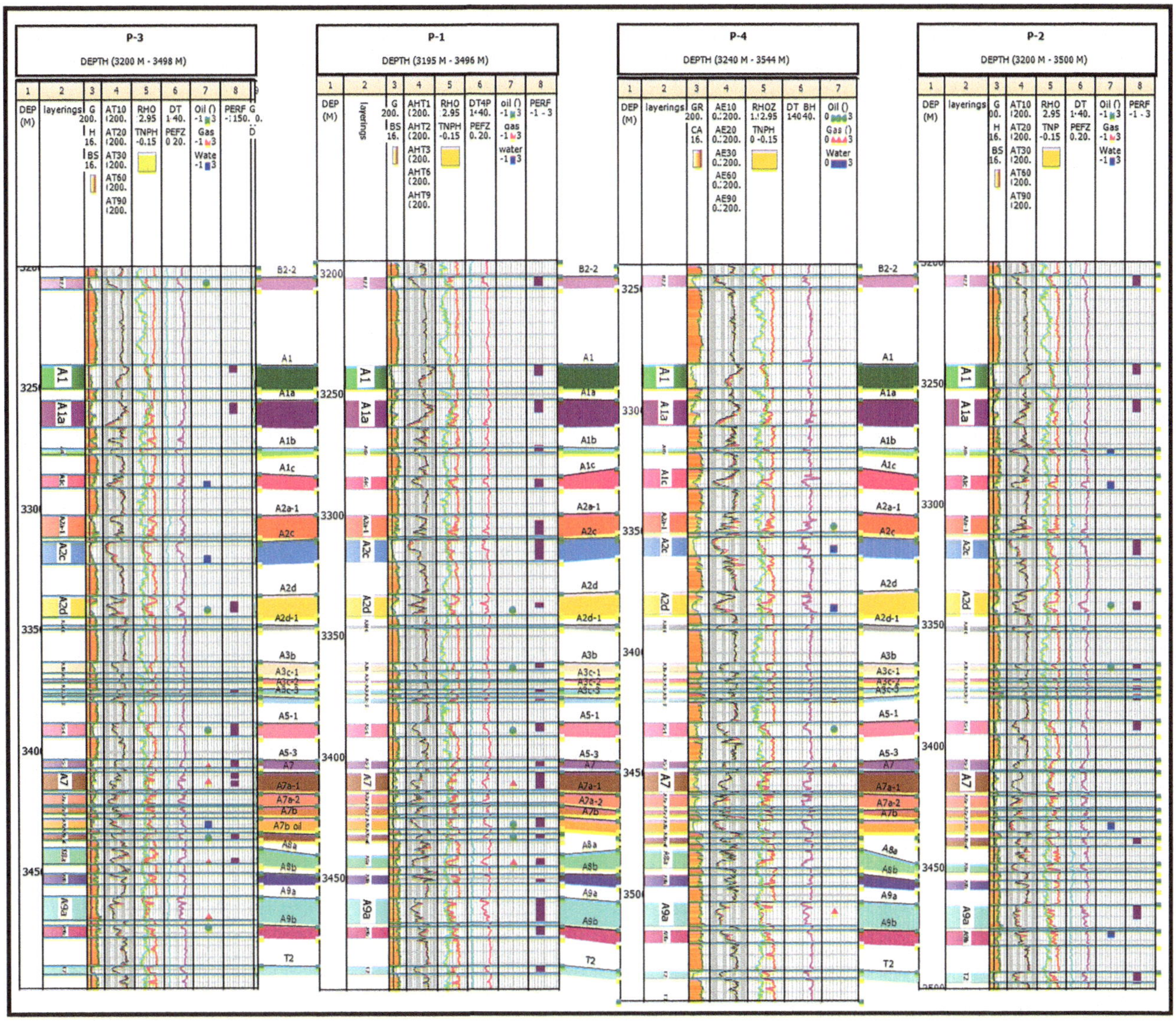

Fig. 2 Stratigraphic correlation

definition of three terms: the flow zone indicator (FZI), the reservoir quality index (RQI) and the normalized porosity index (PHIZ). The flow zone indicator (FZI) parameter depends on surface area and tortuosity variations and offers the combination of core porosity and permeability data through the following equation:

$$FZI = RQI/PHIZ = 0.0314\,\mathrm{sqrt}(k/phie)\,(1 - phie)/phie$$

RQI and PHIZ are defined as:

$$RQI = 0.0314 \times \mathrm{sqrt}(k/phie)$$
$$PHIZ = Phie/(1 - Phie)$$

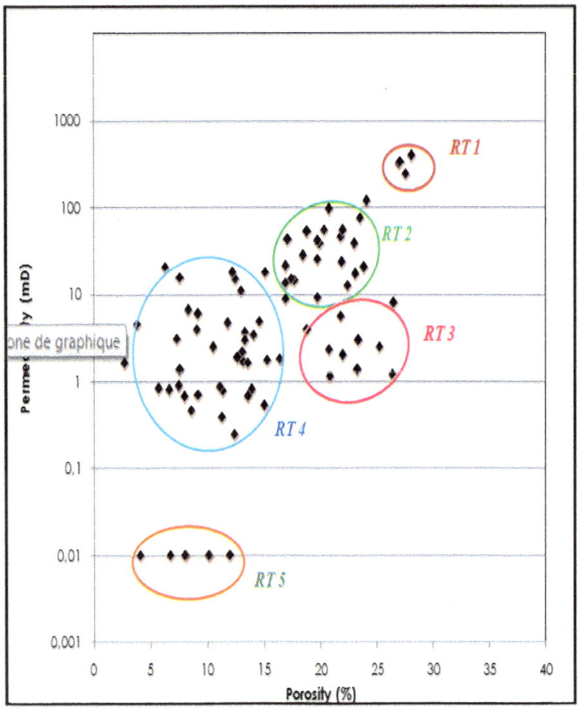

Rock-type	Porosité	Perméabilité
RT-1	25-30	100-500
RT-2	15-25	10-100
RT-3	17-27	1-10
RT-4	5-15	0.5-30
RT-5	4-12	0.001-0.01

Fig. 3 Rock typing results for reservoir zone (from level A1 (3250 mMD) to level A9b (3550 mMD)

where k is in mD, Phie in fraction, RQI and FZI in microns five rock types were identified within the Acacus and Tannezuft reservoir based on FZI.

Variations are shown in the permeability versus porosity plot. These hydraulic units were discriminated based on a statistical frequency distribution (Fig. 3).

4 Discussion

- Then an advanced petrophysical interpretation using the mineral solver probabilistic method.
- Comparison between the two methods of interpretation the probabilistic and the deterministic model.

5 Conclusions

After doing this petrophysical advanced petrophysical study, we conclude that a detailed mineralogical and sedimentological study based on the Aacacus existing core and the ECS run logging are recommended in order to generate a good mineralogical model.

The digital acoustic log (DSI) in combination with the imaging logs can provide the open fracture identification (stoneley waves) and the anisotropy of reservoir identifying the fracture plane azimuth, and could provide overburden geomechanic parameters.

The CMR logging is recommended in new drilled wells for free fluid localization and porosity and permeability calibration.

The imaging log is very recommended to avoid the faults and to localize the bars, channels and fractures.

References

1. Amaefule, J., Altunbay, M., Tiab, D., Kersey, D., Keelan, D.: Enhanced reservoir description using core and log data to identify hydraulic flow units and predict permeability in uncored intervals/wells. SPE **26436**, 205–220 (1993)
2. Aissaoui, N., Bedir, M., Gabtni, H.: Petroleum assessment of Berkine–Ghadames Basin, southern Tunisia. AAPG Bull. (2016)
3. Troudi, H.: Ordovician play in North Africa Sahara platform basins. Etap Memoir (2006)

Behavior of Evaporitic Material During the Structuration of the Eastern Saharian Atlas, Impact on Oil Prospects

Chacha Aziza, Zellouf Khemissi, and Belfar Farid

Abstract

The study's area is located at the Eastern part of the Oriental Saharan Atlas. This structure is characterized by a succession of anticlines and synclines oriented North East-South West to East North East-West South West, diapirs occupying the hearts of some folds, rifts globally oriented North West-South East and a dense network of accidents principally oriented, North West-South East, North East-South West, East West and North South; associated or affecting these structures. The terrains encountered in the area are mainly Meso-Cenozoic, characterized by marine facies, clay and carbonate interspersed occasionally by clay-sandstone deposits corresponding to the Neocomian, Barremian, Albian and Miocene. The Triassic basins in this region underwent an important evolution controlled by geodynamic factors that existed during Mesozoic. These factors are the causes of the remobilization of the Triassic evaporite material. The beginning of this movement took place during the Albo-Aptian. This period is characterized by a tilted blocks geometry, the mechanism that can be used for the development of diapirs of the Oriental Saharan Atlas which seems to be linked to the reactivation of an accident affecting the basement [1]. The interest in the study of diapirism is the result of hydrocarbon discoveries related to this movement (Gulf of Mexico, North Sea, Iran).

Keywords

Oriental Saharan atlas • Triassic basins
Diapirs • Evaporite material • Resedimented lenses
Tilted blocks • Diapirism • Hydrocarbon discoveries

1 Introduction

The study area (Fig. 1) is located East of the Eastern Saharan atlas. It extends from the meridian 6° 20′ west to the Tunisian border in the East, and from the southern edge of the Tellian Atlas in the North to the groove Aures Kef in the South.

This study focuses on the evaporite diapirism within the geodynamic and structural context of the Eastern Saharan Atlas, in particular through examples from the mountains of Mellègue. There the Triassic evaporite clay material is already described and recognized at the surface of Jebel Mesloula, Jebel Ouenza and Morsott. This study shows an interpretation of these Diapiric features both at the surface (on land) and on the surface (interpretation of seismic sections).

This study aims to determine the oil/hydrocarbon potential that may occur in the outskirts of some diapiric systems which remains so far unexplored. Creep evaporites rise in early bulges, which are interesting features for oil research. The history and the evolution of the basins thus allow us to assess the petroleum prospect of this region.

2 Materials and Methods

To carry out this work we have relied on:

- A bibliographical study on the geological and geodynamical evolution of the Eastern Saharan Atlas.

C. Aziza (✉) · Z. Khemissi
Department of Mining and Oil Deposits, Faculty of Hydrocarbons and Chemicals, University M'Hamed Bougara, Independence Avenue, 35000 Boumerdes, Algeria
e-mail: aziza.chacha3@gmail.com

Z. Khemissi
e-mail: khemissized@yahoo.fr

B. Farid
Exploration Division SONATRACH, Boumerdes, Algeria
e-mail: farid.belfar@yahoo.fr

F. Rossetti et al. (eds.), *The Structural Geology Contribution to the Africa-Eurasia Geology: Basement and Reservoir Structure, Ore Mineralisation and Tectonic Modelling*,
Advances in Science, Technology & Innovation, https://doi.org/10.1007/978-3-030-01455-1_39

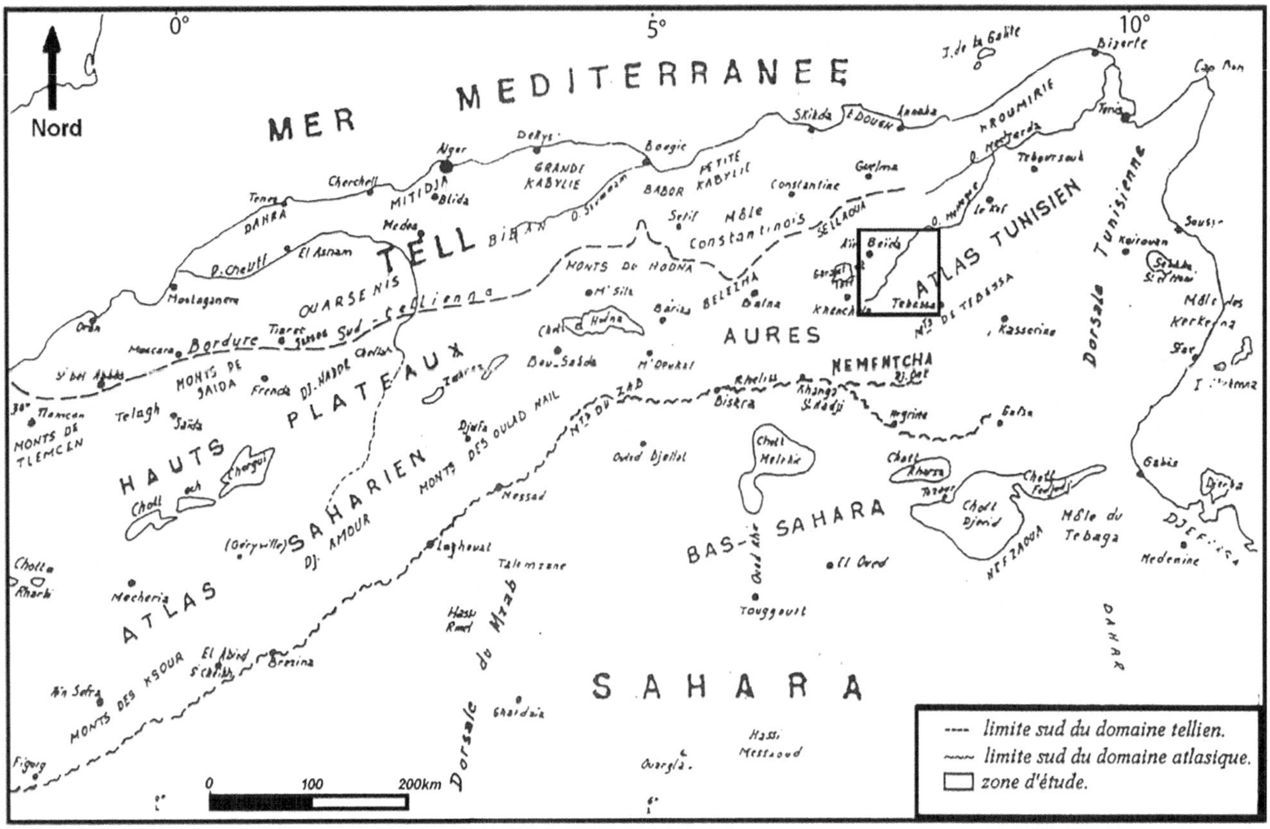

Fig. 1 Geographical location of the study area (in Guiraud, 1973)

- A stratigraphic study based on bibliographic data, drill data, core descriptions and correlations of the Mesozoic series in the basin.
- A structural analysis based on the interpretation of surface data (field measurements, geological maps, and LandSat images) and sub-surface (seismic sections and boreholes).
- A geodynamic study, which consists in reconstructing the different phases of mobilization of the Evaporitic material of the Triassic, and the opening modalities of the sub/basins of the Eastern Saharan Atlas, as well as their geometry.
- A summary of the petroleum aspect including the characteristics of the source rock, the nature of the reservoir rock and the different hydrocarbon traps suggested by the structural analysis.

In conclusion, we propose a geodynamic synthesis linking the diapirism of clay-evaporitic material and the geodynamical evolution of the South-East Constantine basin.

3 Results and Discussion

The structure of the study area is characterized by a succession of anticlines and synclines oriented NE-SW to ENE-WSW, diapirs occupying the hearts of some folds, with rifts globally oriented NW-SE and limited by inherited borders NE-SW faults that delimit tilted blocks from the Aptian and a dense network of accidents with the main following directions NW-SE (N 135–140) NE-SW (N 055), EW (N 080–100) and NS (No 170).

To try to find the relationship between the Triassic surface exposures and their subsurface expression we have used seismic reflection, indispensable tool in the exploration of the subsoil.

As shown in the SW-NE section passing through the rift Tebessa-Morsott (Fig. 2), we can observe:

- A central diapir that crosses the entire overlying sedimentary series (since the Jurassic) along two deep

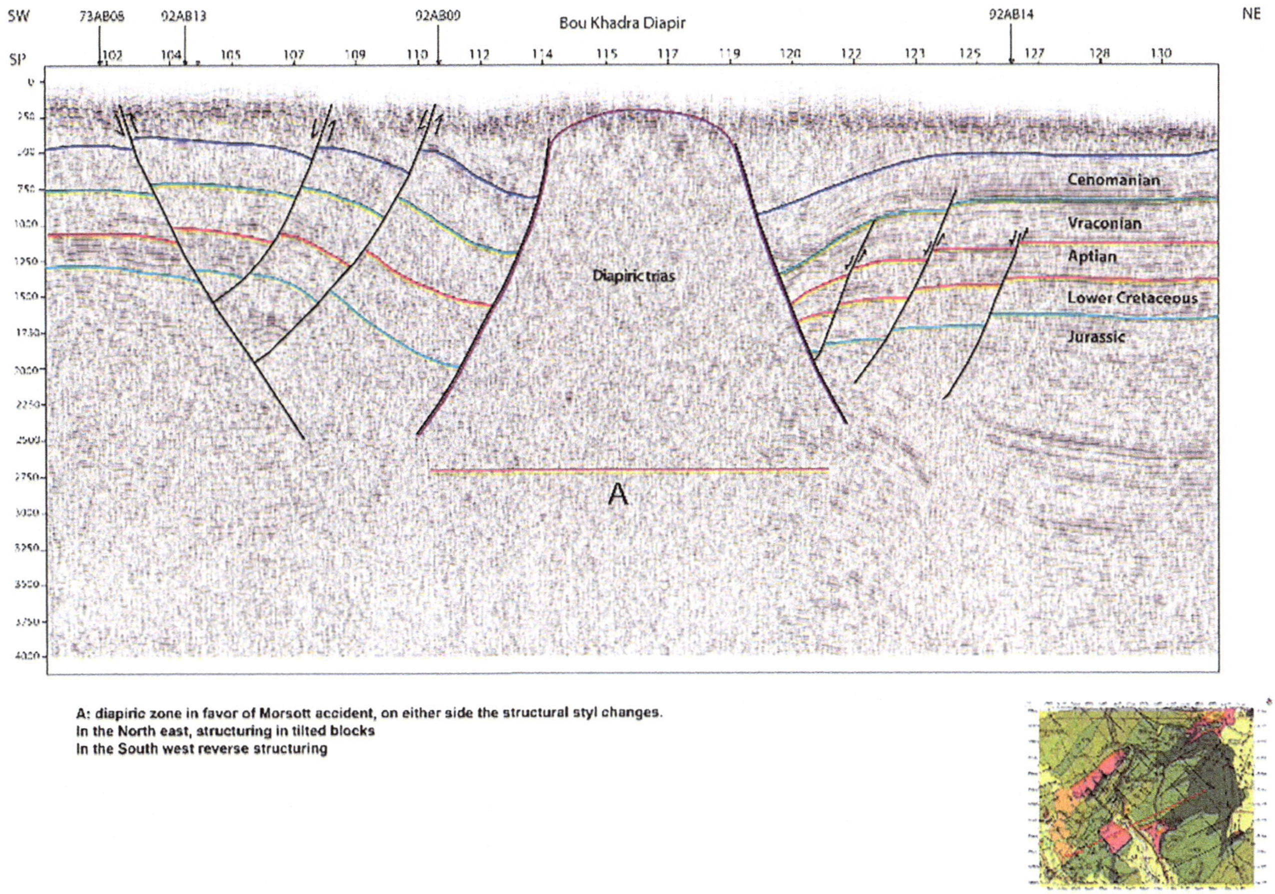

Fig. 2 The seismic section 76 AB 22

boundary faults and that reaches the surface forming the Bou Khadra diapir limited by two deep accidents direction SW-NE and NW-SE.

- In the South West sedimentary series are affected by a network of listric faults that operate in an opposite movement.
- Tilted blocks structures in the North East limited by normal faults (dated/active?) from the Jurassic to the Cretaceous period, and which reflect the structuration of Mesozoic basins during this period.

In the study area, the petroleum systems would be located in the Upper Albian, the Cenomanian and the Turonian. Only the Turonian is potentially favorable to the generation of hydrocarbons and has a state of maturation within the oil phase. The rocks of the Albian and Cenomanian show levels with rates appreciably TOC, and states of maturation passing of the oil window in the gas windows, but they have low petroleum potential.

The reservoir levels have locally significant petrophysical characteristics with high energy levels. By sedimentation areas these characteristics diminish and the low permeability values are a problem for the production of these tanks except in cases of fractures that can improve its qualities.

In the Ain Beida permit different types of traps (structural, stratigraphic, mixed and pitfalls associated with diapirs) exist, but their approach is often made difficult by the complex geology. Exploration of these traps requires detailed studies of structural geology and geophysics.

Early structures from Austrian and Emscherian phases are of great interest to the trapping of hydrocarbons generated from Cenomanian-Turonian and Vraconian source rocks. The search for these traps requires better quality/resolution of seismic data.

4 Conclusion

The analysis of seismic profiles calibrated to the well and field data, and mapping reveal a differentiated structure from the Triassic which are first associated to a fault network, oriented NE-SW, NS, EW and NW-SE limiting horst and grabens and drawing tilted blocks, and secondly associated with the rise of the upper Triassic evaporites, which is done by intrusions along these faults.

The Diapir activity began at least during the Aptian linked to the distention (tilted blocks) that characterized the Oriental Saharan Atlas during the Cretaceous [2]. The rise of the evaporite material from lithostatic anomalous zones or density is guided by the interference of two major structural directions: the first NE-SW is marked by Triassic extrusions. The second, NW-SE to WNW-ESE is materialized by collapsed ditches. This structural control is shown by geophysical data (seismic sections) that show lineaments and deep crashes in the same direction.

Most of the diapirs in the study area draw a morphological domes oriented NE-SW, flush as stretched elliptical body, forming a string of clay-carbonate Gypsum masses.

In the eastern Saharan Atlas the effects of multiphase mobilization of evaporitic material on the oil system are multiple:

- Initiation of early anticlinal structures (salt dome) in the Albo-Cenomanian and Aptian.
- Lateral Collapses in tilted blocks, which makes it possible to have, at the apices of the blocks, the development of high energy facies favoring the initial porous network. In the subsident zones, mother rocks have been developed (Upper Albian, Turonian and Cenomanian).
- Permeability barriers formed by evaporite promoting trapping on the flanks of diapirs.

References

1. Vendeville, B.: Field of faults and tectonic extension: experimental modeling. Thesis of science, 380 p., Université de Rennes (1987)
2. Vila, J.M.: The Alpine chain in eastern Algeria and the borders Algéro-tunisiens. Bachelor thesis science, 2 t., 665 p., Paris VI (1980)

The State of the Nearshore Bottom as an Index of the Shore State, South Baltic Coast Examples

Kazimierz Szefler, Radosław Wróblewski, Janusz Dworniczak, and Stanisław Rudowski

Abstract

The aim of this work is to present the results of seabed surveys in the deeper nearshore zone, which indicate an intense transport of sandy bedload into the sea. This translates into a negative bottom sediment balance and erosive character of the shore incongruent with the intense protective actions including a lot of beach nourishment. The phenomenon of irrevocable transport of sandy bedload into the sea has not been sufficiently accounted for in the studies on the shore protection so far. This might be due to the fact that an assessment of bedload transport requires adequate knowledge of the character and state of the bottom in the deeper nearshore zone. Other basic requirements include multibeam echo-sounder, side-scan sonar measurements and seismic profiling supplemented with other methods such as collection of grab and core samples, underwater TV, etc.). The paper provides examples of the assessment of nearshore bottom state conducted in the vicinity of Ustka (where the dune shore is typical of the Polish coast) and near Kołobrzeg (where rocky outcrops occur on the nearshore bottom and which is a unique coastal form in this area). The state of the nearshore bottom in the deeper nearshore zone may constitute a relative state index in the assessment of the long-term state of the whole coastal zone, while the nearshore bottom of the shallower nearshore zone (including the sandbars zone) serves as a temporary state index. These results are important in the assessment of all the shores formed by wave activity.

Keywords

Bedload transport out to sea • Beach nourishment
Non-invasive seabed surveys • DTM • Baltic Sea
Poland

K. Szefler (✉) · R. Wróblewski · J. Dworniczak · S. Rudowski
Maritime Institute in Gdańsk, Długi Targ 41/42, 80-830 Gdańsk, Poland
e-mail: kaszef@im.gda.pl

1 Introduction

The Polish coast of the Baltic Sea has been formed as a result of the post-glacial transgression of the sea over the past several thousand years [6]. The entire coast is subject to progressive abrasion including the changes in the location of the shoreline by 100 metres in 100 years. It includes mostly sandy dune shores with moraine cliff coast sections. Tides do not occur here, yet the changes in the sea level associated with storms are even up to 2–3 m [6]. Although direct onshore and offshore surveys have been conducted in the area of the Polish coast for years, they were usually limited to its onshore part (cliffs, dunes, beach) and the sandbars zone. The nearshore surveys are conducted mainly along profiles perpendicular to the shore, delineated to the depth of approximately 5 m.

The term "nearshore" describes here a zone affected by waves with unclosed orbits resulting from the transformations that open sea waves undergo when approaching a shallower nearshore bottom. A non-linear mass transport (and not linear currents) occurs here, i.e. a consequential transport of water and sediment masses which is variable in time, space and direction. The seaward storm surge boundary of the nearshore zone is located at the depth of approximately 11 m (in the area in question). Two parts have been differentiated within this zone. The shallower nearshore zone extends to the depth of 5–6 m and includes the sandbars zone. The surveys of the shore state have been conducted so far almost exclusively in the area of the shallower nearshore. Nowadays, it is possible to perform in-depth surveys of the deeper nearshore zone and provide a cartographic description of results (DTM). Two examples of such studies are presented herein, which indicate how useful and beneficial the deeper nearshore surveys are in the assessment of all kinds of shore shaped by wave activity.

F. Rossetti et al. (eds.), *The Structural Geology Contribution to the Africa-Eurasia Geology: Basement and Reservoir Structure, Ore Mineralisation and Tectonic Modelling*,
Advances in Science, Technology & Innovation, https://doi.org/10.1007/978-3-030-01455-1_40

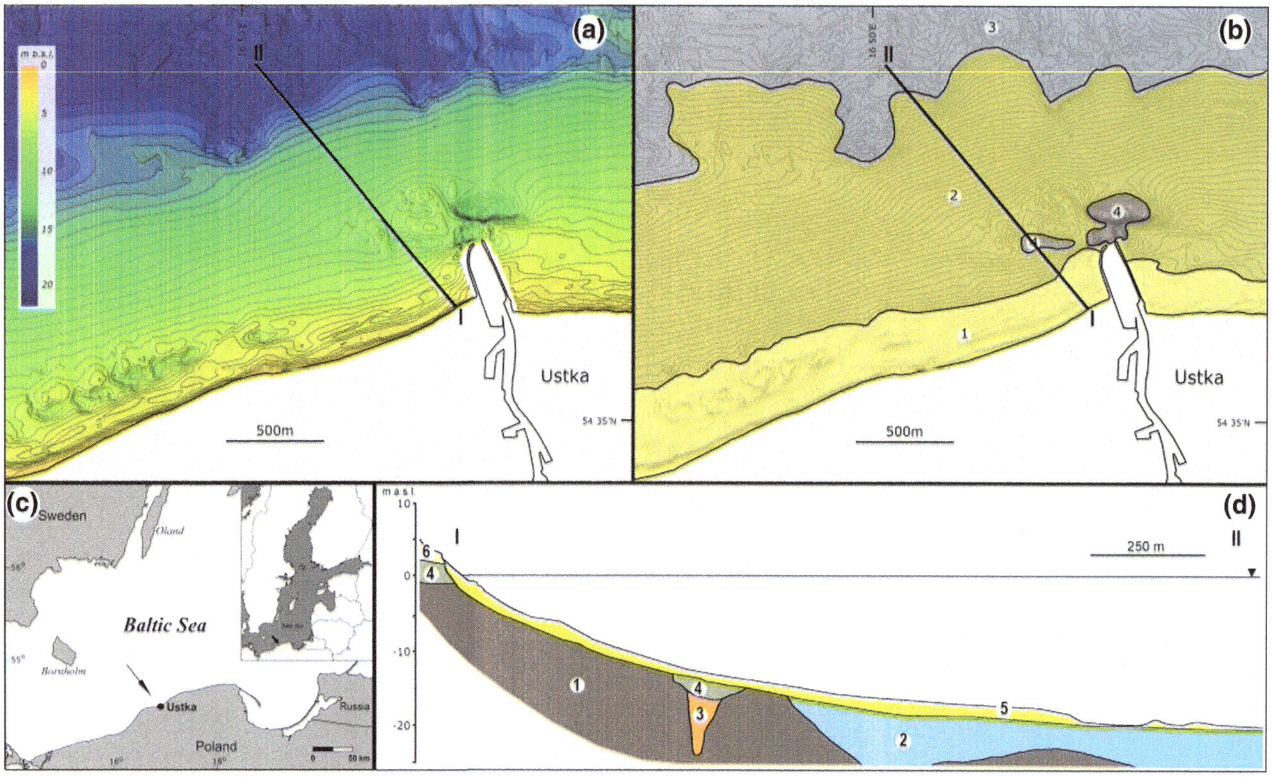

Fig. 1 The Ustka area: A—bathymetric map; B—geomorphological map: 1—sandbar zone, 2—shore slope, 3—open sea seabed, 4—anthropogenic incisions; C—locational map; D—geological cross-section (I–II): 1—till, 2—clay and silt, 3—sand and gravel, 4—riverine sand, 5—marine sand, 6—aeolian sand

2 Materials and Methods

The study is based on the results of surveys of the Maritime Institute in Gdańsk conducted as part of different tasks performed within the area of the Polish coast [2, 5]. Mainly, measurements with a multibeam echosounder and a side-scan sonar were applied. The precise location of the measurement pointing to the seabed was ensured by the DGPS system supported by the adequate navigation systems. The collected measurement data allowed for the drawing of a DTM of the seabed relief. The character of the seabed surface was determined using a side-scan sonar and underwater TV, while its lithology and structure were determined using the seismic profiling developed based on the results of the scoop and core samples analyses.

3 Results

The survey results are presented in the form of two examples from the Polish coast of the Baltic Sea. The first example comes from the vicinity of Ustka and is associated with a sandy dune coast [5]. An intense bedload transport out to sea occurs here (Fig. 1), which forms a vast underwater shore slope extending outside the sandbars zone to the depth of 15–16 m, i.e. into the open sea bottom. It consists of medium- and coarse-grained sand with interbedding of massive multi-grained sediments. This sediment is associated mainly with a rapid deposition caused by the turbulent seaward "rip current". [1].

In the second example, on a rocky seabed in the near-shore zone [3, 4], Jurassic sandstone and mudstone are

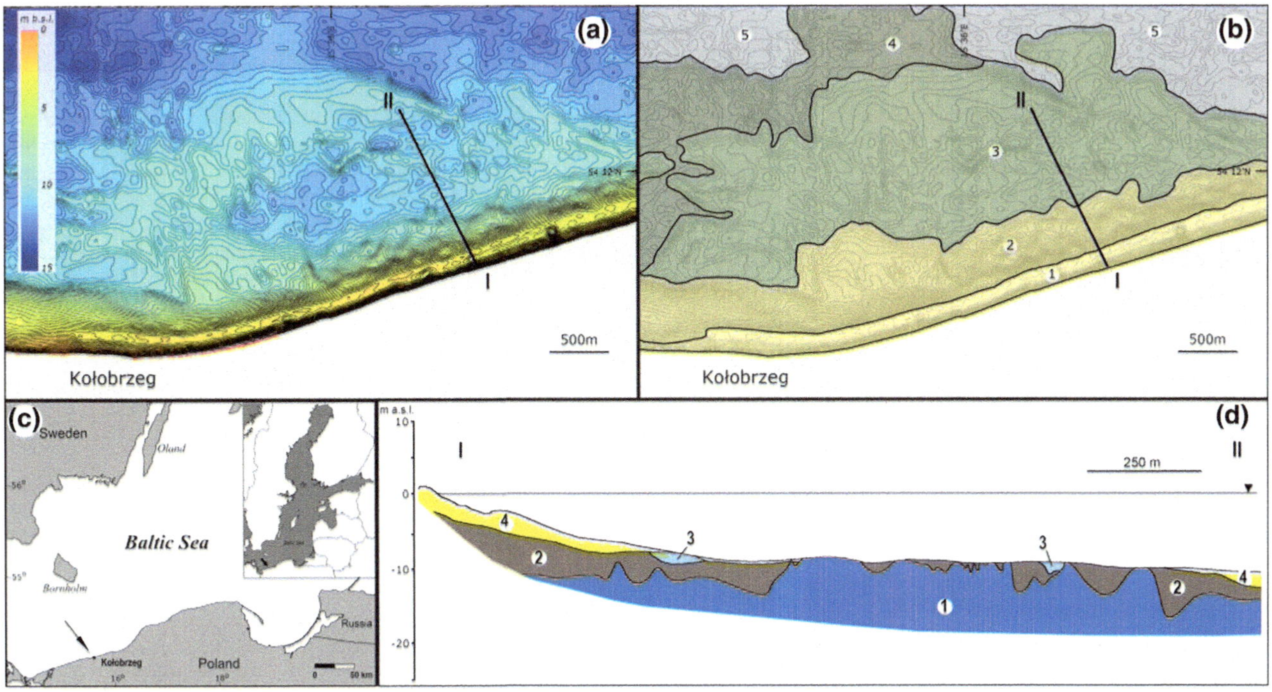

Fig. 2 The Kołobrzeg area: A—bathymetric map; B—geomorphological map: 1—shore terrace with sandbars, 2, 3, 4—terrace levels of the rocky seabed with a variable marine sand cover, 5—open sea seabed with sand deposition; C—locational map; D—geological cross-section (I–II): 1—sandstone and mudstone, 2—till, 3—mud, 4—marine sand

revealed from under a thin layer of modern marine sand over vast areas (Fig. 2). The seabed here is formed by several layers of erosive terraces. On the highest one, there spreads a narrow strip of poorly developed sandbanks with a thin strip of bedload deposited at the foot of this terrace. The entire bottom surface in the deeper nearshore zone is characterized by a repeated transport of recent marine sand in various directions. The bedload runoff reaches the bottom of the open sea to the depth of more than 11 m (at the distance of over 2500 m from the shore), where a relatively constant deposition of transported bedload takes place.

4 Discussion and Conclusions

The paper provided two examples presenting different bottom images in the area of the deeper nearshore zone. In the vicinity of Ustka (Fig. 1), there is a shore slope extending to the depth of approximately 16 m. Its character indicates an intense deposition of sediments from the shallower parts of the bottom, i.e. associated with the bedload transport out to sea, even outside the nearshore zone (the storm surge boundary of the nearshore zone is located here at the depth of *ca.* 11–12 m). The nearshore zone in the area of Kołobrzeg (Fig. 2) is relatively shallow (with a depth of up to 10 m) and large (over 2500 m). The character of the nearshore bottom indicated a multi-directional repeated redeposition of sand forming a changeable cover with numerous outcrops of solid Jurassic rock with an erosive top. This indicates an intense bottom erosion and seaward bedload transport during storms. What characterises both these examples, is the irrevocable seaward transport of bedload. Thus, the state of the deeper nearshore zone proves the shore erosion.

The presented survey results indicate that bedload transport out to sea is an important issue which should be taken into account when assessing the state of the shores. Additionally, an effort could be made to restrain the seaward bedload transport (especially in the context of sand resources degradation becoming a global issue) and discuss how to do it in relation to the regional coast conditions.

The numerous shore protection methods and techniques used thus far can hardly be assessed as satisfactory despite the long years of efforts. This is largely due to the hitherto lack of proper recognition of the deeper nearshore bottom [2, 5].

Acknowledgements The authors wish to thank all the reviewers for their constructive remarks, which helped us express our ideas more precisely and improve the article's quality.

References

1. Gruszczyński, M., Rudowski, S., Semil, J., Słomiński, J., Zrobek, J.: Rip currents as a geological tool. Sedimentology **40**(2), 217–236 (1993)
2. Nowak, J., Rudowski, S., Wróblewski, R., Sitkiewicz, P., Lisimenka, A., Gajewski, Ł.: Application of a multibeam echosounder for the digital imaging of the bottom relief of seabeds and rivers. Bull. Marit. Inst. Gdańsk **30**(1), 86–89 (2015)
3. Rudowski, S., Gajewski, J., Hac, B., Makurat, K., Nowak, J., Szefler, K.: The role of an adequate determination of the state of nearshore bottom bed load as a scaling factor controlling the state and the changes of the shore in the Kołobrzeg region. In: Proceedings of the 11th International Coastal Symposium, Journal of Coastal Research, Special Issue 64, pp. 816–819 (2011)
4. Rudowski, S., Sitkiewicz, P., Wróblewski, P., Makurat, K.: Solid rocks on the nearshore seabed—the distribution and potential impact on coastal processes in the Kołobrzeg region, the Southern Baltic. Oceanol. Hydrobiol. Stud. **46**(1), 62–73 (2017)
5. Szefler, K., Gajewski, Ł., Hac, B., Rudowski, S., Wróblewski, R.: The Technology of the seabottom surveys for the monitoring of the coastal area. In: Sana, A., Hewawasam, L., Waris, M.B. (eds.) Proceedings of the Fifth International Conference on Estuaries and Coasts (ICEC 2015), Conference 2015, pp. 175–178 (2015)
6. Uścinowicz, S.: The Baltic Sea continental shelf. In: Chiocci, F., Chivas, A. (eds.) Continental Shelves of the World: Their Evolution During the Last Glacio-Eustatic Cycle, pp. 1–35. The Geological Society of London (2014)

Regional Basement Interpretation and Its Impact on Ordovician and Silurian Structures (Ghadames Basin)

Dalila Samaali, Sarah Boudriga, Walid Mahmoudi, and Ahmed Nasri

Abstract

Seismic interpretation of southern Tunisia (Ghadames basin) was carried out to identify the relationship between the geodynamic context and the petroleum play of the basin. This study encompasses the elaboration of regional structural maps of the different hydrocarbon bearing reservoirs (Paleozoic). These maps indicate that in Southern Tunisia, Ghadames basin is marked by master NE-SW trending faults interference with E-W and N-S direction. Advanced seismic interpretation (seismic attributes) was applied to evince that the reactivation of the ancient faults (inherited from the basement) played an important role in the structuration of the Paleozoic series. In fact, by using seismic attributes, we can prove that almost all faults affecting the Paleozoic are originated from the basement. We utilize seismic attributes to shed the light on the idea that high number of producing field of Ordovician and Silurian "which are traditionally admitted as purely stratigraphic traps" occur along faults alignment. This explains that these fields are in part the result of a specific depositional environment and in part the result of the basement involvement in the formation of these petroleum fields. This study illustrates that advanced seismic interpretation can provide additional understanding and a new concept about the prospectivity of southern Tunisia.

Keywords

Basement • Fault interpretation • Regional seismic interpretation • Ordovician and Silurian Gravity • Seismic attributes • Variance Edge detection • Gravity

1 Introduction

The Ghadames Basin is a large sedimentary basin encompassing over 350,000 km^2 located in southern Tunisia, eastern Algeria and northwestern Libya [1, 2]. The basin is produced mainly by Paleozoic reservoirs (Silurian and Ordovician).

While it has previously been considered to be an intracratonic sag basin [1], recent detailed studies of the structures within the basin suggest that it has a more complex geological history involving a series of compressional and extensional tectonic events [2].

The Silurian four-ways dip structures are usually identified from the seismic shifting of the Ordovician horizon and had never been interpreted as distinct marker (below seismic resolution). So that the origin and hydrocarbon path way of these structures are poorly understood. Silurian traps are mainly attributed to stratigraphic traps. In this study we believe that these structures are somehow the results of the reactivation of the inherited basement faults.

Besides, except few conventionally producing field, the majority of Ordovician reservoirs encountered by wells are tight sandstones. In some areas, the natural flow production is assured where the fracture networks is well developed.

Indeed, the architecture of the reservoirs in the main fields in Ghadames basin is the result of the reactivation of the deeply-rooted basement faults. So we are confident that a good understanding of the geodynamic and structural configuration of the basement is a master key to achieving a reliable petroleum characterization of the basin.

D. Samaali (✉) · S. Boudriga · W. Mahmoudi
Entreprise Tunisienne d'activité pétrolière ETAP, Tunis, Tunisia
e-mail: dalila.samaali@etap.com.tn

S. Boudriga
e-mail: sarah.boudrigua@etap.com.tn

W. Mahmoudi
e-mail: walid.mahmoudi@etap.com.tn

A. Nasri
Mazarine Energy, Tunis, Tunisia
e-mail: a.nasri@mazarine-energy.com

F. Rossetti et al. (eds.), *The Structural Geology Contribution to the Africa-Eurasia Geology: Basement and Reservoir Structure, Ore Mineralisation and Tectonic Modelling*,
Advances in Science, Technology & Innovation, https://doi.org/10.1007/978-3-030-01455-1_41

2 Materials and Methods

To elaborate our task the geological history, stratigraphic context and gravity analysis [3] were used as a starting point to understand the structural model and to better identify the main lineaments directions.

We used all available wells data (>200) where 46 wells reach the Ordovician and only two (Oued Zar and Sanghar-1) reach the Pre-Cambrian (Fig. 1). We also used ten 3D seismic cubes that cover almost 80% of the studied area.

In this approach, we applied many efficient geophysical techniques such as advanced surface and volume attributes analysis (Fig. 2) to define the major fault network that affect the Ordovician and Silurian reservoirs.

Ordovician natural production sweet spots (Oued Zar, Durra, Chaabet El Markanti …) are located very close to the regional basement faults (Fig. 2). The reactivation of the basement faults are the primordial factor not only for the development of good fractures networks but also for the hydrocarbon migration route from source (hot shale) rock to reservoirs.

Moreover the Ordovician reservoir thickness, the net sand and petrophysical characteristics (porosity-permeability) are mainly controlled by the depositional environment fairways and trends which are also structurally-related to the basement structuration.

The identified Accacus traps 'distribution in the studied area are well fitting the direction of the major basement lineament (Fig. 3).

3 Results

Seismic interpretation leads to the following interpretations:

- Fault density is higher in the center of the area.
- The evidence of NE-SW master faults interfering with E-W and N-S fault direction.
- The western and southern parts seem to be more stable and less structured.

4 Discussion

The integration of the available data sets over Ghadames basin has demonstrated that it consists of a complex sag basin with a series of structural highs and lows.

This basin architecture is controlled by the deeply rooted Pre-Cambrian faults.

The hydrocarbon expelled from the Silurian hot shale source rock migrated to both Ordovician (Jeffara, Bir ben

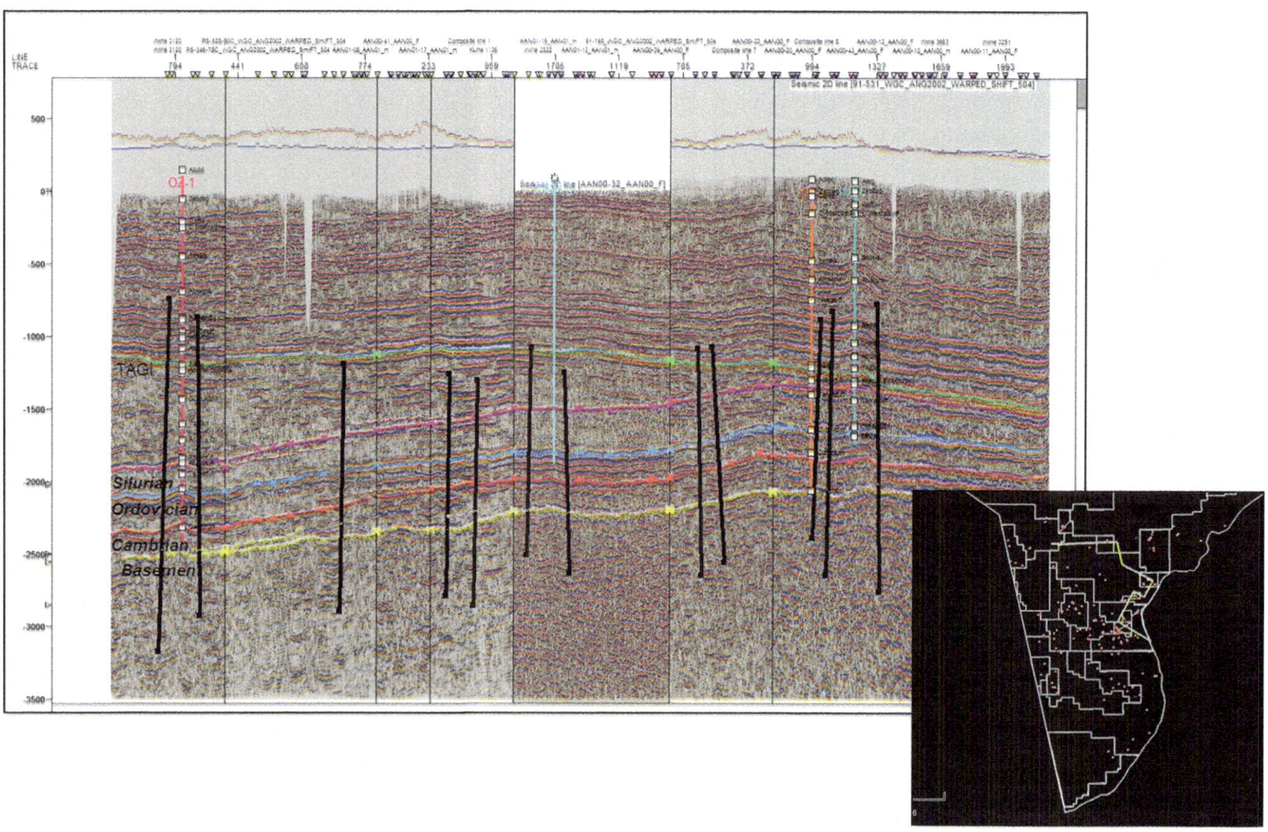

Fig. 1 Regional seismic interpretation section crossing SN-1 and OZ-1 wells

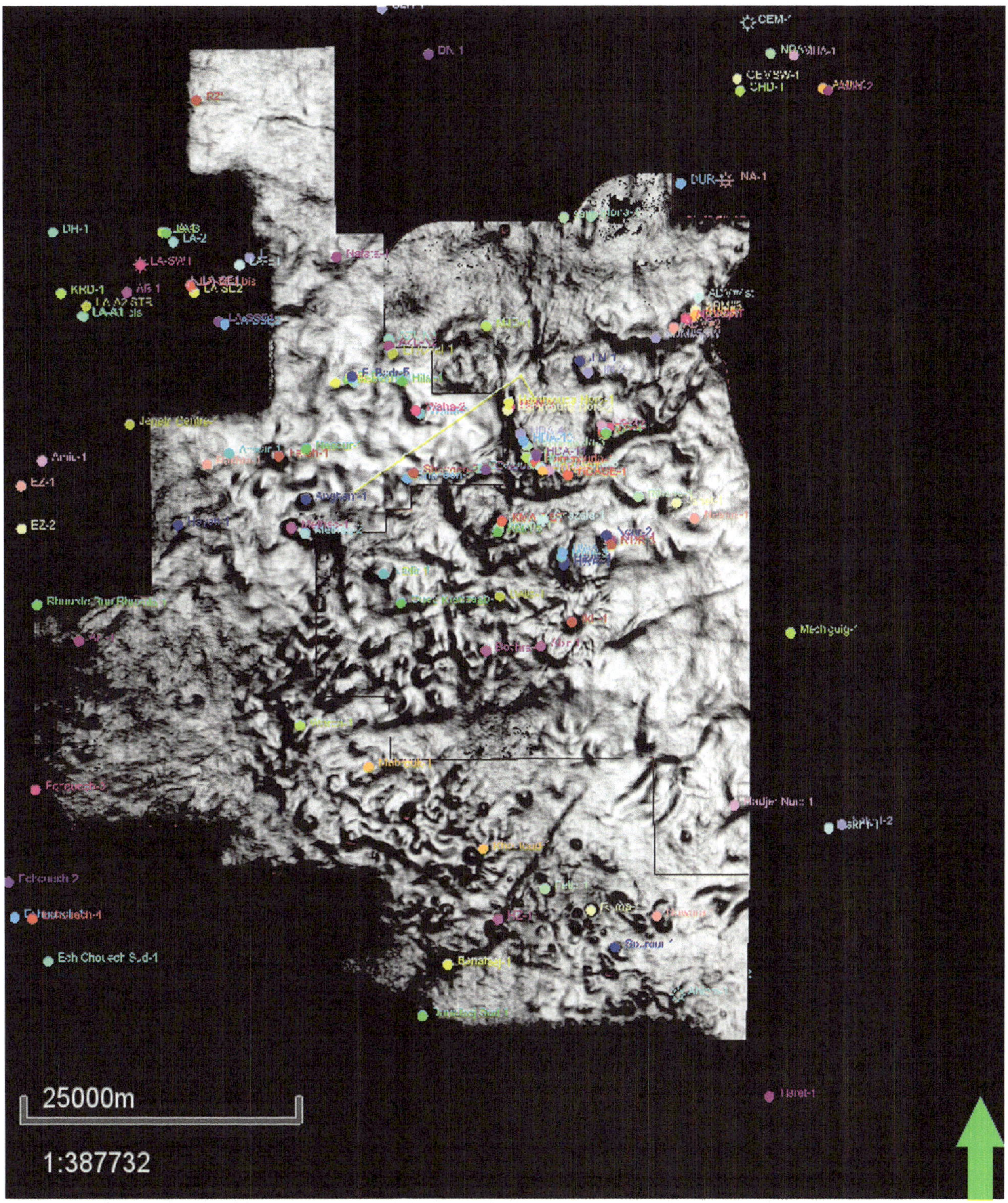

Fig. 2 Variance attribute within Ordovician reservoir

tartar, and Kasbah Leguine) and Silurian (Accacus A) reservoirs via this deep seated Pre-Cambrian faults system.

Ordovician reservoirs (fractures networks) and Silurian Accaus A shaling-out zones are widely guided by the reactivation these lineaments within different successive tectonic phases.

The 46 wells that reaching the Ordovician level have shown encouraging results with high gas readings while

The Lithostratigraphic Analysis and Petrophysical Characterization of the Campanian-Maastrichtian Formation in the Pelagian Basin (Northeast Tunisia)

Wafa Abdelkhalek, Fetheddine Melki, Klaus Bauer, Michael Weber, and Abdelhamid Ben Salem

Abstract

The Campanian-Maastrichtian formation was studied and defined in a Pelagian basin, mainly, in the Sidi El Kilani oil field. The carbonates of the Abiod formation are considered as a potential reservoir for several years. The present study aims to characterize the lateral variation in thickness and lithology of the Abiod reservoir based on borehole lithology logs and electrical well logging. This chalky carbonate shows a relevant porosity and permeability variance. Furthermore, the subsurface data have allowed recognizing the internal architecture of Pelagian Basin which is a stable domain, but also a rather a complex structural domain. The sedimentary facies of the Abiod formation (Campanian-Maastrichtian) have been guided by the existence of deep faults and older structures which evolved into major E-W and N-S faults and characterize the geodynamic structures and tectonic framework. The structural framework based on the major structures controls the existence of the deep reservoir and its distribution in the basin. Throughout the reservoir of the Abiod formation, areas are widespread where there is relevant thickness and relevant porosity, particularly along the major faults corridors and in the summit parts of the main anticlines.

Keywords

Upper cretaceous • Subsidence • Limestone reservoir Pelagian basin

W. Abdelkhalek (✉) · F. Melki
Faculty of Science of Tunis, University Campus, 2092 Manar, Tunisia
e-mail: w.abdelkhalek@yahoo.fr

K. Bauer · M. Weber
Research Center for Geoscience, Telegrafenberg, 14475 Potsdam, Germany

A. B. Salem
Entreprise Tunisienne d'Activité pétrolière, 54, Avenue Mohamed V, 1002 Tunis, Tunisia

1 Introduction

The Pelagian Basin is located in northeast Tunisia and has been an area of significant interest in the past few years. The Tunisian Sidi El Kilani oil field was discovered by KUF-PEC, a Tunisian company, in April 1989. Different sets of data helped identify an extensive Cretaceous phase. These Alpine and Atlassic phases decrease from the West to the East, where strongly folded and faulted zones can be observed. The carbonates of the Abiod formation show a relevant porosity value of around 25%. Also, a preferential direction was observed where the reservoir reached the maximum thickness in the Basin. In particular, the fold registered horst anticlines in wells W9 and W11. Finally, during the Upper Cretaceous, the E-W and N-S tectonic corridors of the Sahel were reactivated in a transtensive, transpressive dextral and sinistral sliding system [2], generating zones of apertures in grabens and depocentre subsidence neighboring zones.

2 Geological Settings

The present research is localized in the Sahel Foreland basin characterized by geological outcrops ranging from Miocene to Quaternary (Fig. 1). The Sahel is a stable domain belonging to Eastern Tunisia. The surface appears slightly tectonized because the structure is almost completely buried under Mio-Pliocene sediments. Its morphology shows a gradient of up to 200 m, including hills, plains and depressions. However, the tectonic framework becomes more complex in the deeper parts of the basin including sets of horst and graben.

F. Rossetti et al. (eds.), *The Structural Geology Contribution to the Africa-Eurasia Geology: Basement and Reservoir Structure, Ore Mineralisation and Tectonic Modelling*,
Advances in Science, Technology & Innovation, https://doi.org/10.1007/978-3-030-01455-1_42

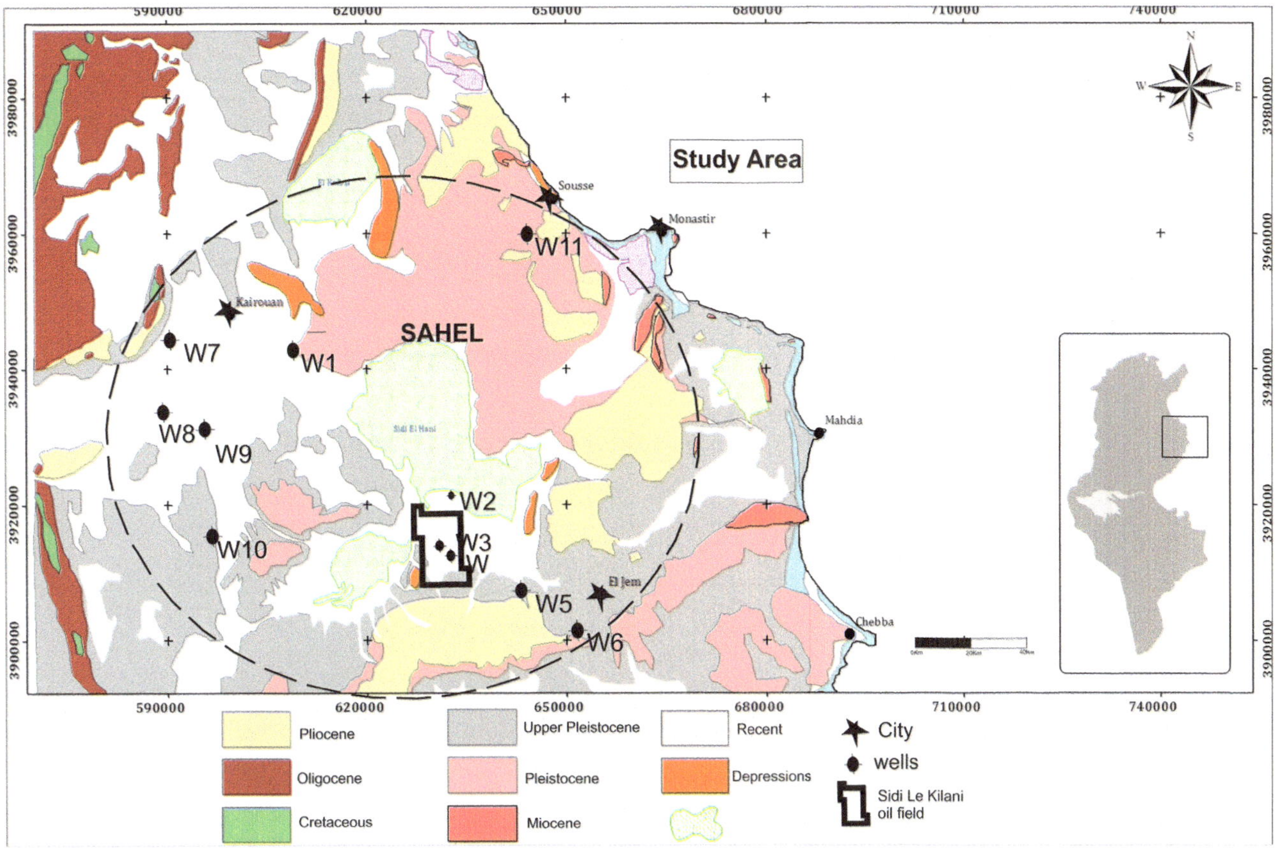

Fig. 1 Simplified geological and well location map of the study area (Modified from Tunisian geological map 1/500,000)

3 Materials and Methods

The Campanian-Maastrichtian deposits facies and structure have been highlighted by interpreting 2D seismic reflection profiles and wireline logging of petroleum wells. The data contains 33 seismic lines obtained from the Tunisian Petroleum Company (ETAP). The data was processed using the "Petrel" software. A compositional analysis has been used for the Campanian-Maastrichtian interval of the Abiod formation in eleven wells applying the Gamma-ray, density and neutron logs to solve the propositions of limestone, dolomite, and the porosity factor. The wire-logging requires a subdivision of its precise representative facies of wells analyzed and implanted in the study area. However, the facies of the Abiod formation were predicted to divide this reservoir into four units. The choice of wells was made according to the presence or absence of the Abiod formation.

The analysis of logging data was supported by the application of interactive petrophysics clustering techniques. For selected boreholes, different logs were combined to form data patterns. Unsupervised learning was applied to identify clusters with characteristic petrophysical signatures. The different clusters were interpreted as lithological units [1].

4 Result and Discussion

The logging and lithological study of the Abiod reservoir reveals that the vertical succession of facies can be summarized into four units (Fig. 2), of which two units represent two calcareous bars of the Abiod formation, composed essentially of chalky carbonates.

Moreover, a strong subsidence of all the Upper Cretaceous is accentuated for the Abiod formation at the level of the permit of Sidi El Kilani (black box in Fig. 1) and is also an improvement on the Abiod reservoir through fracturing (decrease in porosity). However, the facies of the Abiod formation were predicted to divide this reservoir into four units U1, U2, U3, and U4. The sediments of the Campanian-Maastrichtian age are chalky clay-limestone alternations with planktonic microfauna. A preferential direction of the NW-SE basin is testified by the wells having a medium porosity and enrichment in clays (Fig. 3). On the other hand, towards the western side, at the level of the paleo-high, one meets a fluctuating argillosity and maximum porosity. The tectonic movements of compressions and distensions are accompanied by an obvious lateral component which will play an important role in the structuring of the Sahel basin during the Cenozoic.

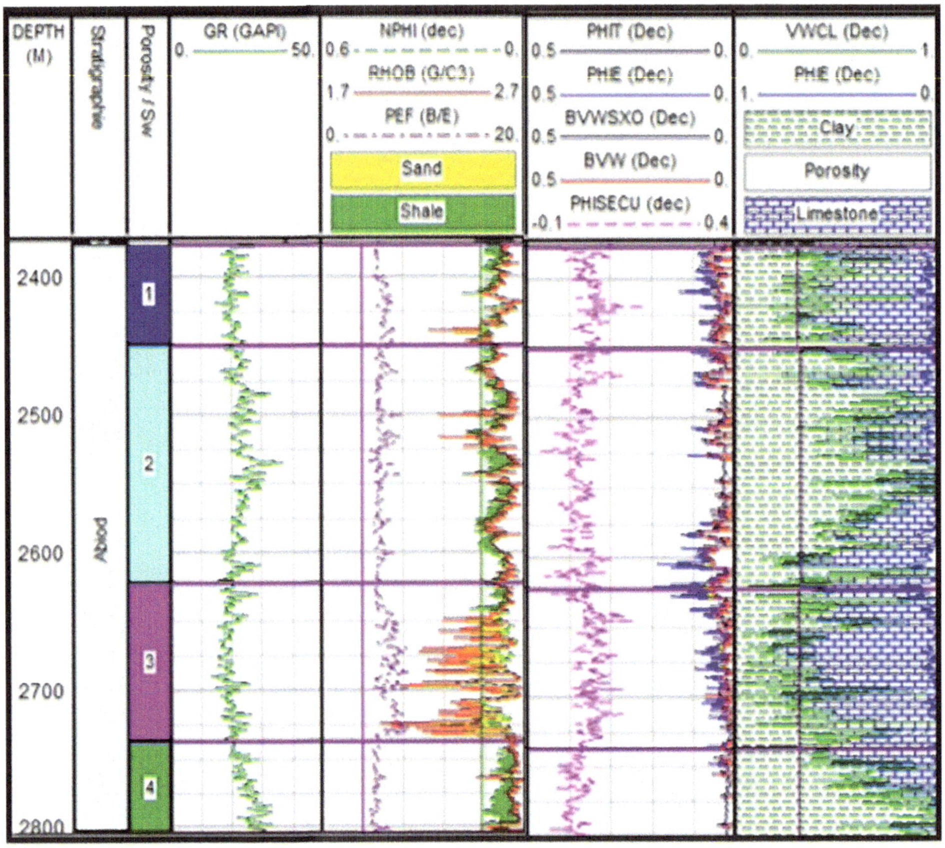

Fig. 2 Stratigraphic analysis and petrophysical characterization

The isopach map presented in Fig. 3 shows that the reservoir units have a maximum thickness and a preferential direction with high clay content values on either side. The volume of clay increases outside in wells W5, W7, W8, W9, W10 and W11 to reach 30%. Thus, this reservoir shows weak porosity values at this dominant direction. At wells W1, W2, W3, W4, and W6 it reaches in average value of 15%. On the other hand, in the summit parts of the main anticlines at wells W7 and W10 we observe increased porosity at 25%. Also, the porosity increases at wells W7, W8, and W10 to reach a relevant value of about 20%.

In this paper, primarily, the volume of shale and porosity are estimated from well log responses and cross-plot. The determination of this succession due to differences in log characteristics in Fig. 2. The lithology cross plot was derived from each unit in the formation (Table 1). This article indicates that petrophysics characteristics and seismic interpretation are in accordance.

5 Conclusion

The compressional and extensional tectonics have played an important role in the configuration of the Pelagian Basin during the Cenozoic. The sediment of Campanian-Maastrichtian age are chalky clay-limestones alternations with planktonic microfauna [3]. These facies are marked on several logging recordings, especially at the level of the Gamma-Ray and neuron-density curves. A preferential direction of the NW-SE basin testified to the wells having medium porosity and an enrichment in clay. However, in the western side, a relevant porosity and decrease of argillosity values was registered. This heterogeneity indicates a structural disturbance that is probably induced by the Triassic rise. It perhaps provides an opportunity for future research development.

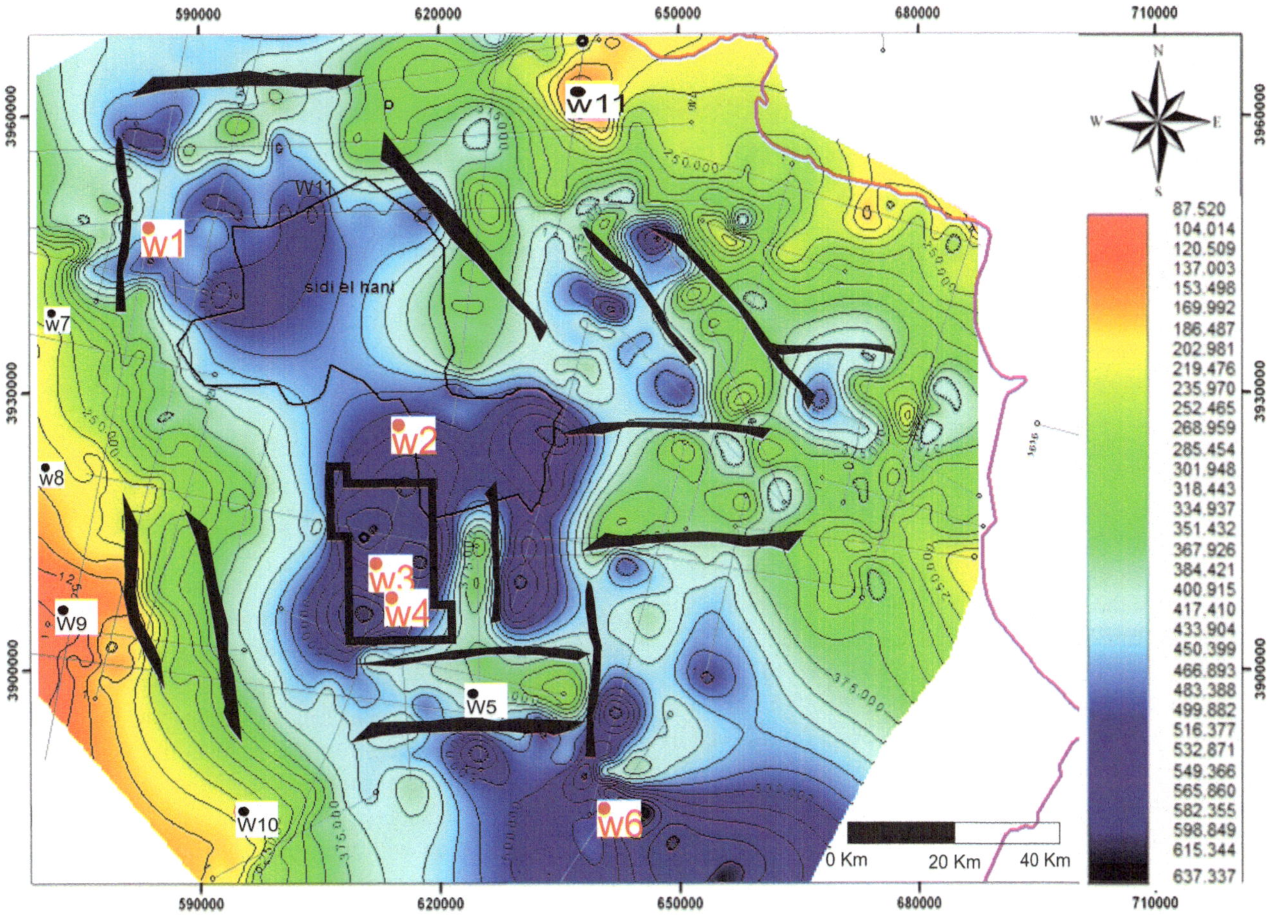

Fig. 3 Isopach map of the Cretaceous Abiod formation. Black boundary indicate faults

Table 1 The reservoir summary of the reservoir units

Zn	Zone name	Top	Bottom	Gross	Net	N/G	Av Phi	Av Sw	Phi * H	Phi So * H
1	U1	2376.00	2449.00	73.00	8.54	0.117	0.096	0.471	0.82	0.43
2	U2	2449.00	2622.00	173.00	27.88	0.161	0.101	0.651	2.83	0.99
3	U3	2622.00	2737.00	115.00	9.45	0.082	0.107	0.373	1.02	0.64
4	U4	2737.00	2804.00	67.00	1.05	0.016	0.107	0.375	0.11	0.07
All zones		**2376.00**	**2804.00**	**428.00**	**46.94**	**0.110**	**0.1012**	**0.554**	**4.77**	**2.13**

References

1. Bauer, K., Kulenkampff, J., Henninges, J., Spangenberg, E.: Lithological control on gas hydrate saturation as revealed by signal classification of NMR logging data. J. Geophys. Res. Solid Earth **120**, 6001–9017, Germany (2015)
2. Bedir, M., Arbi, A., Khomsi, S., Houatmia, F., Aissaoui, M.: Seismic tectono-stratigraphiy of fluvio deltaic to deep marine Miocene silicolastic hydrocarbon reservoir systems in the Gulf of Hammamet, northeastern Tunisia. Arab. J. Geosci. (2016)
3. Negra, M., M'rabet, A.: The Abiod facies repartition and reservoir aspect. In: 4th Proceeding of the Tunisian Petroleun Exploration Conference, pp. 495–507. ETAP, Tunis (1994)

Fracture Properties in Naturally Fractured Carbonate Reservoir (Gourigueur Eocene Syncline, North-Eastern Algeria)

Saad Hireche, Azzedine Bouzenoune, and Djelel Bouti

Abstract

The Gourigueur fractured Eocene carbonate rocks constitute one of the groundwater aquifers in the Tebessa region (north-eastern Algeria). The present work aims to characterize the fractures of this aquifer from a multiscale analysis. The Gourigueur syncline oriented NE-SW, extends over twenty kilometres from SW to NE and about five kilometres from NW to SE. The fracture orientations measured from the outcrops have allowed us to identify two of the most dominant fracture sets (d1 and d2) which are respectively a longitudinal set, which is normal to the fold axis, and a transverse set, much less abundant, which is oblique to the trend of the structure. Their mean orientations are respectively NNE-SSW and NNW-SSE (Hireche in Analyse multi-scalaire de la fracturation du réservoir carbonaté éocène du synclinal de Gourigueur (Région de Tébessa, Algérie Nord orientale), [1]). The microfaciologic analysis allowed us to show that the porosity of these Eocene fractured carbonate rocks is mainly controlled by the fracture porosity; the matrix porosity is very low or even null for some facies (Hireche in Analyse multi-scalaire de la fracturation du réservoir carbonaté éocène du synclinal de Gourigueur (Région de Tébessa, Algérie Nord orientale), [1]).

Keywords

Fracture • Reservoir • Carbonate • Eocene
Algeria

1 Introduction

The Gourigueur fractured Eocene carbonate rocks constitute one of the groundwater aquifers in the Tebessa region (north-eastern Algeria) [2]. The present work aims to characterize the fractures of this aquifer from a multiscale analysis. This study describes a detailed analysis of fracture parameters such as fracture lengths, orientations and densities. It combines the data from the satellite images, available geological maps, outcrop studies and microscopic analysis to characterize the geometry and distribution of the fracture networks in relation with the syncline structure. A comparison of the results based on levels of analysis has been conducted to assess the relevance of methodological various tools used in the treatment of this type of problem.

2 Méthodology

Satellite images of "Google earth" clearly show the fracture networks of the Gourigueur massif (Fig. 1). These images allowed us to develop fracture maps used as a base document for the analysis of these fractures using "One-dimensional line sampling" (scanlines), and a "window sampling" (rectangular area) [3]. An outcrop study has been undertaken around the Gourigueur Syncline, were fracture data have been collected using scanlines and rectangular areas called the inventory method on bedding surfaces in the limbs and hinge of the syncline [4]. Each scanline was marked with measuring tape and was around 15 m long. The measurement areas were around 12 m^2. Data sampled from each scanline include: fracture spacing, distance along the scanline, fracture strike and dip, fracture length, fracture aperture. Fractures are divided into sets and populations based on their orientation with respect to the geometry of Gourigueur Syncline. Informations provided by outcrop study allow a statistical analysis of the fracture

S. Hireche (✉) · A. Bouzenoune · D. Bouti
Laboratory of Geological Engineering (LGG), University Mohamed Seddik Benyahia, Jijel, Algeria
e-mail: saadhireche@yahoo.fr

F. Rossetti et al. (eds.), *The Structural Geology Contribution to the Africa-Eurasia Geology: Basement and Reservoir Structure, Ore Mineralisation and Tectonic Modelling*,
Advances in Science, Technology & Innovation, https://doi.org/10.1007/978-3-030-01455-1_43

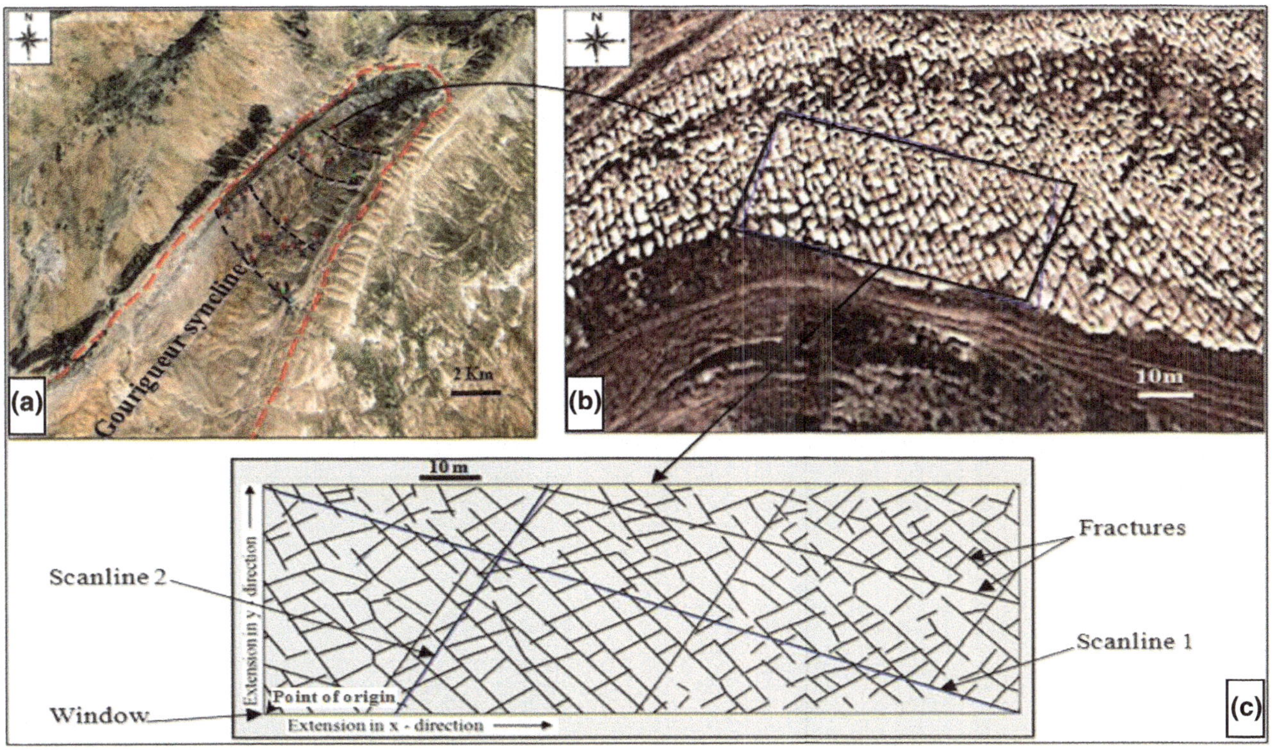

Fig. 1 Extraction and development of a fracturing map from images "Google earth". **a** Overview of allure of Gourigueur syncline. **b** Location, geometry and dimensions of "windows" used. **c** Fracturing Map extracted from the image "Google earth" and situation of "scanlines" used

orientation, fracture spacing distribution, frequency, intensity and density. Dual aim of this study includes microfaciological textures, diagenetic characterization and qualitative evaluation of the porosity of Eocene carbonate rocks of the Gourigueur Syncline; forty samples were collected at different localities. The spatial distribution of these localities was made in such a way as to follow the variations of the facies and porosity depending on the structural position in the syncline; that is, the forelimb, backlimb and hinge.

3 Microfaciological Textures, Diagenetic Characterization and Porosity Evaluation

The majority of microfacies identified from the study are characterized by packestone type of microfaciologic textures reflecting a degree of compaction developing "pseudo-stylolites" at the contacts between the particles. It reflects a landfill that firstly, contributed to the reduction of the original rock porosity and on other hand is behind other diagenetic processes (chemical precipitation, recrystallization …) (Fig. 2). In light of this microfaciologic and diagenetic study, it appears that these rocks are, at this microscopic scale, characterized by low matrix porosity (intra- and inter-particle).

4 Characterization of Fracture Networks

The Gourigueur syncline is oriented in an NE-SW direction, it extends over twenty kilometres from SW to NE and about five kilometres from NW to SE. The fold axis trends N34E and plunges 4° toward SW. At its north-eastern end, it has a closing conical peri-termination.

The orientations of fracture sets at different outcrop locations of the syncline were recorded and subdivided into four main populations. The two most dominating sets are striking NE–SW and NW–SE; that is respectively, a longitudinal set (d1), which is sub-parallel to the fold axis with a mean direction of N43°E 77NW and a transverse set, which is perpendicular to the fold axis with a mean direction of N132°E 81SW. In addition, a population of a third (d3) and fourth set (d4) of both NNE-SSW and NNW-SSE striking fractures are much less abundant. They are oriented oblique to the axis of the syncline. Overall, the fractures are joints, and offer near perpendicular orientation to the bedding.

Table 1 summarizes the main characteristics of fractures such as the fracture spacing distribution calculated using the coefficient of variation, which shows that distribution is moderately regular. Overall fractures have a high frequency in the hinge rather than in both limbs. It also appears that the bed thickness and the bedding dip have no significant effect

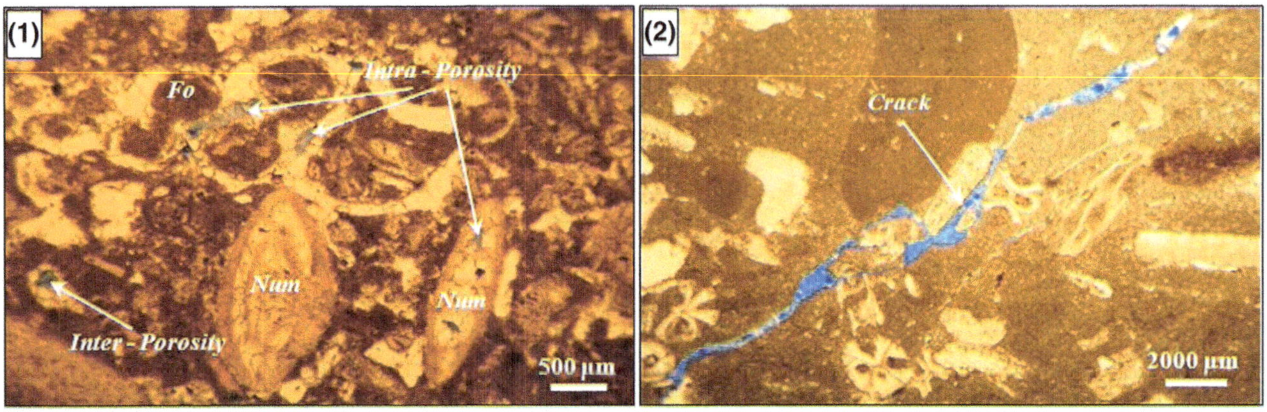

Fig. 2 Photos illustrating examples of porosities affecting Eocene carbonate rocks of the Gourigueur Syncline. Photo 1—matrix porosity intra-and interparticle (blue color). Particles are represented by nummulites (Num) and foraminifera (Fo). Photo 2—microfissure filled in with residual fissure porosity (blue color)

Table 1 Fracture characteristics

	d1	d2
Direction	N43E77NW	N132E81SW
Coefficient of variation	0.24–0.77	0.18–1.27
Frequency (mean values)	Hinge: 1.56 NW Limb: 1.27 SE Limb: 1.43	Hinge: 1.63 NW Limb: 1.3 SE Limb: 1.05
Intensity (mean values)	0.25–7.77	0.15–5.08
Areal density (mean values)	Hinge: 1.02 SE Limb: 0.84 NW Limb: 0.97	Hinge: 1.16 SE Limb: 0.88 NW Limb: 0.82
Connectivity	0.14–4.38	

on the fracture frequency. Based on the intensity, the induced fracturing of the two dominating sets can be classified as "strong fracturing. The areal fracture density is generally highest at the hinge than in the two limbs, is controlled by the structural position (hinge and limbs). The Studied fractures are typically "conjugated" and shows a more developed connectivity in the limbs.

Our comparative analysis of outcrop and satellite images results show some similarities, in particular In terms of the fracture sets, their orientation, density, both outcrop and satellite images data demonstrate a well-connected open fracture network.

5 Conclusion

In conclusion, it appears that the Gourigueur Eocene carbonate aquifer is characterized by very low matrix porosity. Fracturing is the main parameter controlling the porosity and permeability and therefore the hydrodynamic properties and the development of karst in this reservoir.

References

1. Hireche, S.: Analyse multi-scalaire de la fracturation du réservoir carbonaté éocène du synclinal de Gourigueur (Région de Tébessa, Algérie Nord orientale). Master thesis, Univ. Mohamed Seddik Benyahia, Jijel, Algérie, 140p (2014)
2. Baali, F.: Contribution à l'étude hydrogéologique, hydrochimique et vulnérabilité à la pollution d'un système aquifère karstique en zone semi-aride. Cas du plateau de Cheria (NE algérien). PhD thesis, Université Badji Mokhtar - Annaba, Algérie, 177p (2007)
3. Zeeb, C., Gomez-Rivas, E., Bons, P.D., Virgo, S., Blum, P.: Fracture network evaluation program (FraNEP): software for analysing 2D fracture trace-line maps. Comput. Geosci. 11–22 (2013)
4. Ruhland, M.: Méthode d'étude de la fracturation naturelle des roches associées à divers modèles structuraux, Bull. Sci. Géol. 26. Strasbourg. 91–113 (1973)

Part V

Mediterranean Tectonics

Polyphasic Tectonic on the Deformation Zone in the Northern Foothill of the Blida Atlas, the Cross Section of Hammam Melouane, Algeria

Seifeddine Adjiri, Mohamed Naak, and Abdelkrim Yelles-Chaouche

Abstract

The Mitidja Basin is considered as an extensional tectonic structure created in the early Miocene, located in the north of Algeria. It covers an old transition zone in the tertiary collision between the "Internal and External Domain" of the central Maghreb. The Mitidja—Tell Atlas limit is a major geological contact that coincides with a slope and trench break separating the Cretaceous Tell Atlas, from the Neogene and Quaternary Mitidja plain. Hammam Melouane is one of the best sites that summarize the structural geology of this contrasting area Mitidja-Tell. Our observations on this site, supported by a detailed cartography, as well as a study of the kilometric folds and fractures that we interpret as deformations induced by a dextral compression sliding phase having participated in the uplift of the Atlas and in the collapse-subsidence of the Mitidjian basin during the Miocene. The structural analysis revealed a polyphasic tectonic beginning with an extensional phase appropriate to the upper Burdigalian, with a north-south direction and a post-Burdigalian tectonic inversion phase, expressed by a combination of a strike-slip and thrust movements (transpressive), a feature that marks a certain degree in the intensity of the deformations in the "post-nappe" tectonic.

Keywords

Mitidja • Tell Atlas • Post-nappe tectonic Fracking • Burdigalian

1 Introduction

The Mitidja Basin is a post-orogenic structure resulting from early Miocene (Upper Burdigalian) extensional tectonic. This post-orogenic tectonic follows the last stage of the Alpine collision that formed the Maghreb chain, where the alpine orogen stretch locally to create longitudinal depressed structures, considered as sub-parallel grabens to the Mediterranean chain and coastline. This orientation is compatible with a submeridian extensional stress.

The Mitidja Basin-Tell Atlas are two structures oriented substantially East-West, are separated by a sudden slope break, which corresponds to a geomorphological transition zone, representative of an important tectono-magmato-sedimentary model. Hammam Melouane is one of the best sites that summarize the structural geology of this contrasting area Mitidja-Tell, which highlights several phases of tectonic deformation, from an extensional phase to the tectonic inversion. In the field, this polyphasic tectonic is recorded in the Neogene formations [1], much more in the upper Burdigalian limestone which showed a very complex model of faults and folds structures.

The present study contributes to the cartography, structural and microtectonic analysis of the geological structures that characterize the Neogene deformation pattern of the Hammam Melouane cross-section, in order to understand the tectonic evolution of the Mitidja basin in the Mediterranean geodynamic context.

S. Adjiri (✉) · A. Yelles-Chaouche
Centre de Recherche en Astronomie Astrophysique et Géophysique (CRAAG), Bouzareah, Algeria
e-mail: adjiriseifeddine.geo@gmail.com

A. Yelles-Chaouche
e-mail: a.yelles@craag.dz

S. Adjiri · M. Naak
Université des Sciences et de la Technologie Houari Boumediene (U.S.T.H.B), Bab Ezzouar, Algeria
e-mail: naakmohamed@yahoo.fr

F. Rossetti et al. (eds.), *The Structural Geology Contribution to the Africa-Eurasia Geology: Basement and Reservoir Structure, Ore Mineralisation and Tectonic Modelling*, Advances in Science, Technology & Innovation, https://doi.org/10.1007/978-3-030-01455-1_44

2 Geological Settings

The Mitidja basin belongs to the north-central part of Algeria (Fig. 1), located between the longitudes 2.00° E and 3.80° E and the latitudes 36.20° N and 37.00° N. It is bordered to the North by the Mediterranean Sea and to the East by the great Kabylie, to the west by the Menaceur basin, to the south by the Atlas of Blida, Bou Maad. It consists of a vast depression covered by quaternary alluvial deposits [5, 6].

Our study area (Fig. 2) is located between the village of Bougara in the North and Hammam Melouane in the South. This area is a major deformation zone between the neogenic Mitidja Basin in the north, and the Blida Atlas "Schistosity Massifs" [3] in the south.

The area of Hammam Melouane consists essentially of Neogene formations (Upper Burdigalian and Langhian-Serravalian; [4] which correspond to the Miocene post-nappe discordant with the Upper Cretaceous (Senonian). We distinguish from bottom to top: (i) calcareous marl formations (upper Cretaceous). (ii) Discordant red conglomerate. (iii) A massive limestone formation (upper Burdigalian) straightened vertically. (iv) Marl sandstone and sand which constitute the Langhian-Serravalian series.

Volcanic rocks outcrop in the foothill of the Blida Atlas (Fig. 2), it is interstratified in the upper Burdigalian-Serravalian formations, already described by [2, 4, 5] (Glangeaud 1952). The microscopy analysis reveals two types of rocks: Andesite and Dacite.

3 Analysis of the Tectonic Elements of the Site

Kilometer structures mark this studied area, they are folds in the Neogene with a particular appearance already described by Bonneton [4], as well as escarpments and thrusts:

A. **The syncline of Rocher des pigeons**: (Fig. 3a) It is a marl heart syncline (Langhian-Serravalian) with limestone flanks (Upper Burdigalian), it has an axis oriented E–W, that plunge about 70° E. This fold is associated with a sinistral strike-slip fault oriented N40°; it focuses on Burdigalian deformation (Fig. 2).
B. **Escarpments**: (Fig. 3b) they are oriented NE–SW, NNE–SSW and E–W. These escarpments announce a pronounced northern vergence thrust in the Miocene post-nappe formations. One of these escarpments, where the outcropping of the limestone Miocene is brecciated, showed two major phases of deformation underlined by: Normal faults with 50° NW dip, and Reverse faults with 20° SE dip.
C. **Thrusts**: there is a northern vergence that is observed in the southern flank of the syncline 'Rocher des Pigeons' (Fig. 3c) where the vertical Burdigalian limestones rest on the Langhian-Serravalian marls through a reverse fault. Near the village of Hammam Melouane, we found a contact zone, the continental red Miocene "post-nappe", overlain by the gray schists of the

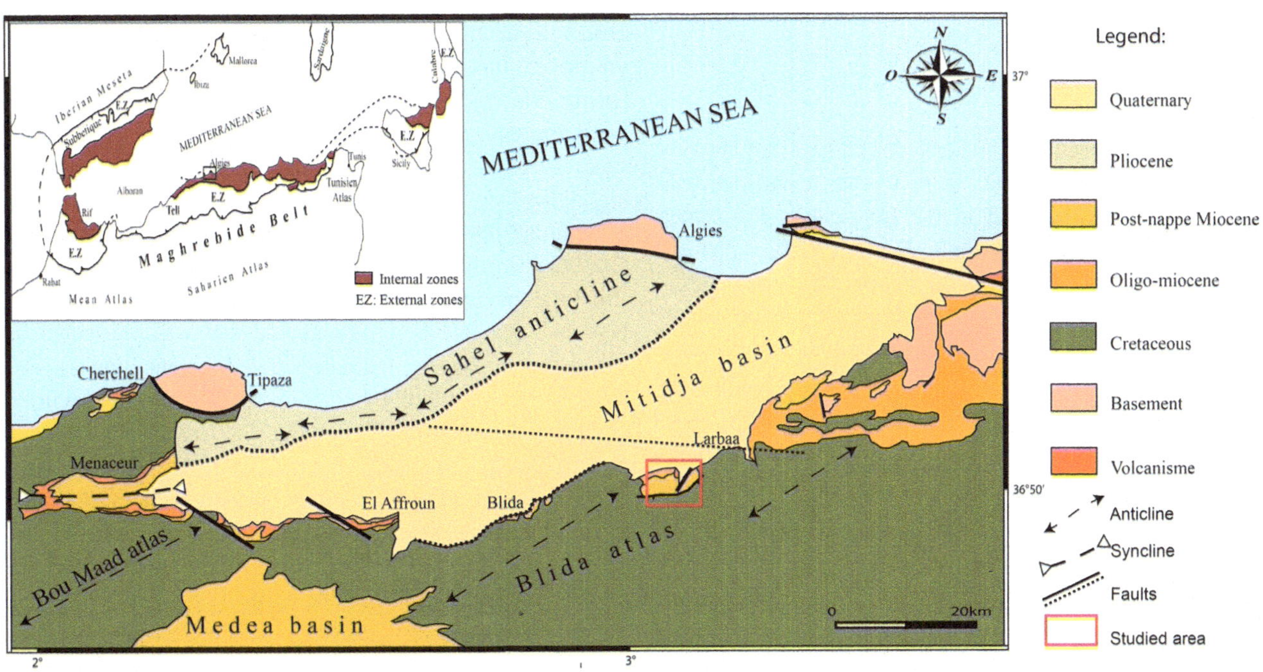

Fig. 1 Geological sketch map of the Mitidja Bassin and its surroundings regions

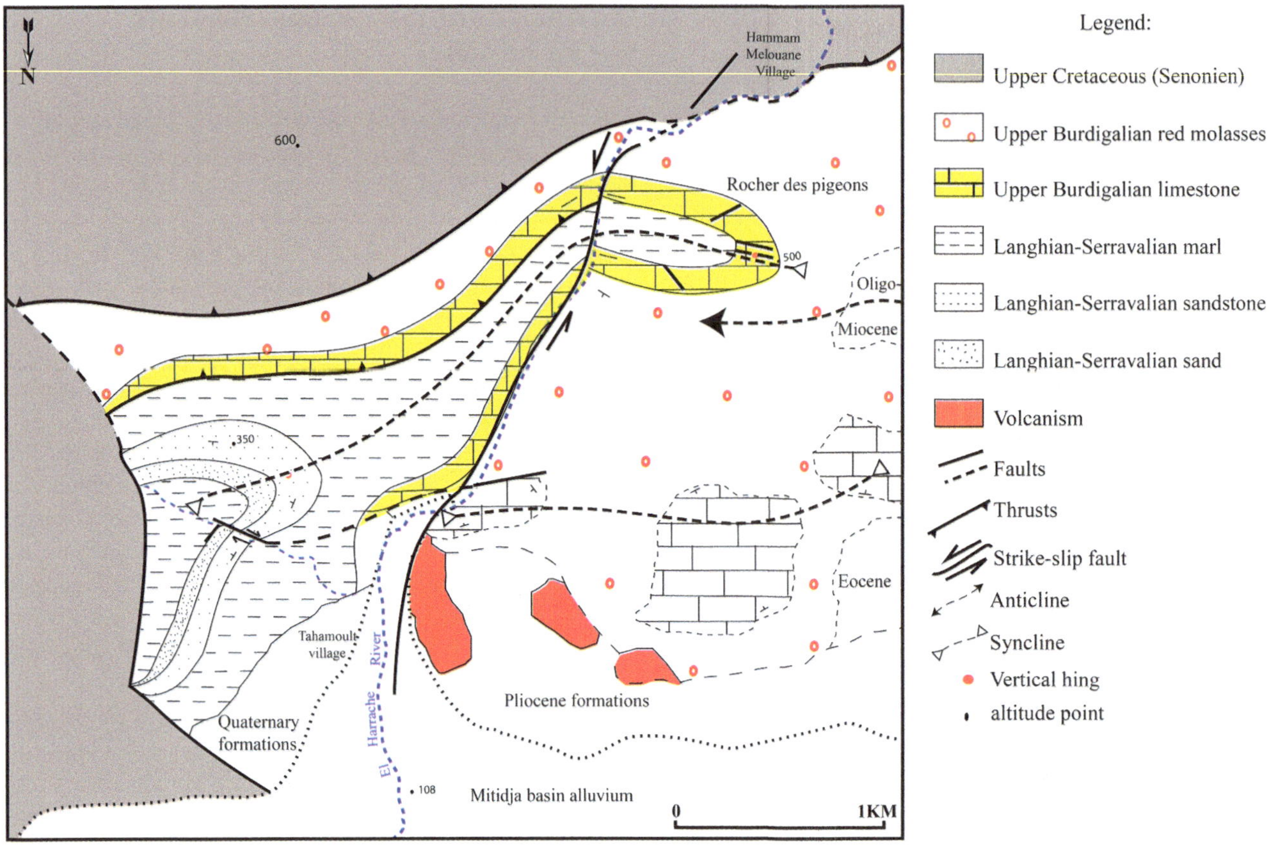

Fig. 2 Structural model of the Hammam Melouane region

Cretaceous (Senonian) and we can distinguish an angular discordance (Fig. 3d).

4 The Brittle Tectonics in the Study Area

The flexible deformation is associated with several fault system, the analysis at the thin section scale, clearly reveals extensional regime recorded in the Burdigalian limestones which is underlined by normal faults (Fig. 4a); this first phase is linked to the opening of the Mitidja basin. The same Burdigalian limestones shows another brittle deformation which corresponds to an inversion phase. It is highlighted by a reverse faults (Fig. 4b), thrusts (Fig. 4c) and strike-slip faults (Fig. 4d).

5 Discussion

The analysis of the different deformations in the Neogene of the study area appears to be part of the post-nappe tectonics. It has been possible to demonstrate the existence of a characteristic polyphasic tectonic that begins in an extensional context linked to the opening of the basin. We try to explain this relationship with the normal faults observed in the burdigalian limestones, as well as the volcanism that is localized in the southern edge of the basin. This extensional regime is taken up by other phases of deformation of compressive nature that allows the success of the Burdigalian-Serravalian organization; it is highlighted by kilometric folds deforming the Burdigalian formations, reverse, thrust and strike-slip faults. The relation between these tectonics elements is summarized in the structural model of the region (Fig. 2). The "Rocher des Pigeons" syncline is a vertical fold that was reoriented by a sinistral strike-slip fault (illustrated by the flanks), this tectonic is controlled by a transpression regime occurred according to us, in the Tortonien. The Cretaceous-Miocene limit showed a structural contrast between the red continental conglomerate "post-nappe", overlapped by dark and folded schists of the Upper Cretaceous. This clear difference in the style of deformation leaves no doubt of the presence of an angular discordance.

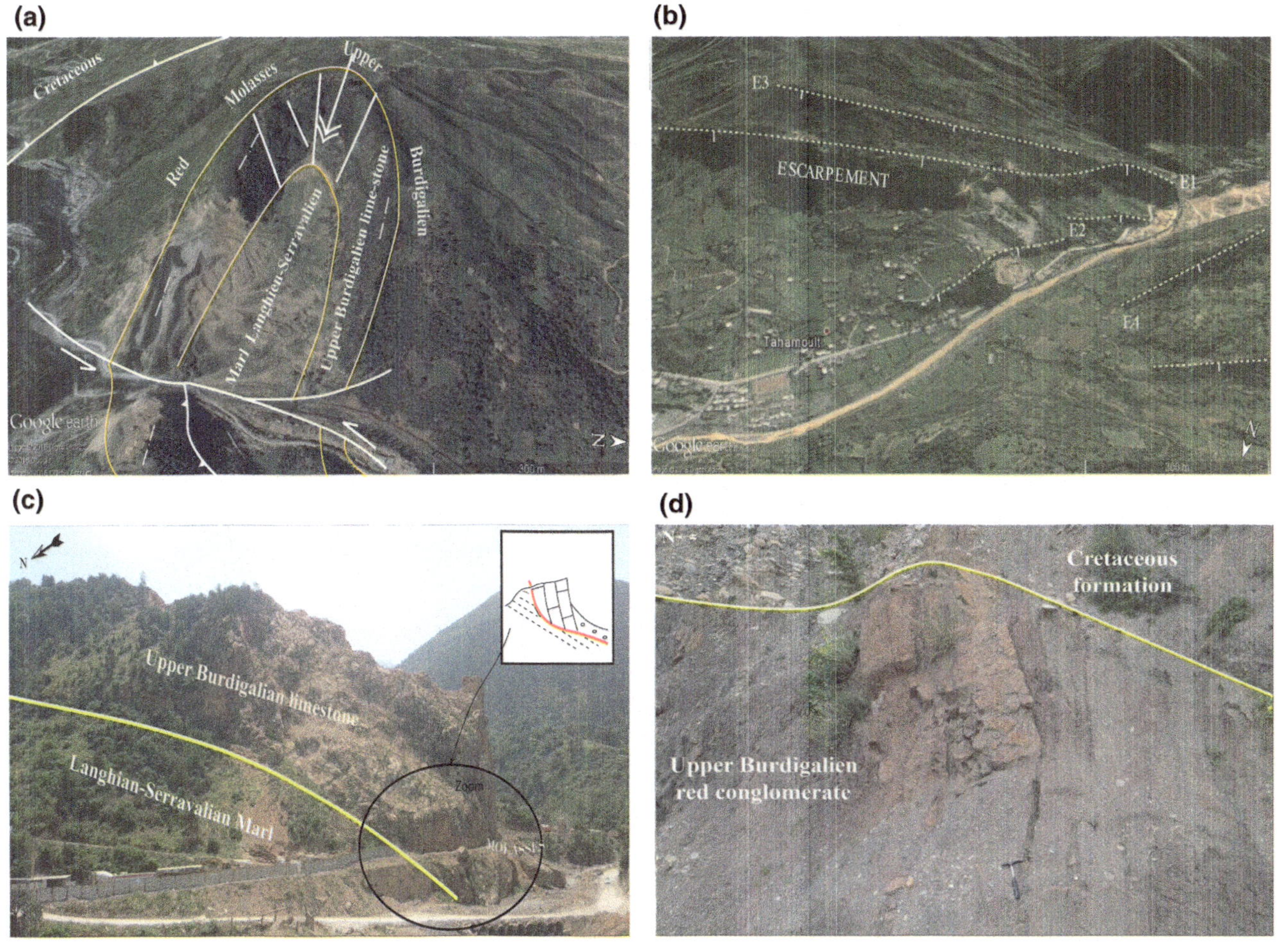

Fig. 3 **a** Interpretation of the Rocher des pigeons syncline; **b** Escarpments observed in the studied area; **c** Upper Burdigalian limestone thrust the Langhian-Serravalian marl; **d** Cretaceous-Miocene contact

6 Conclusion

The geological limit Mitidja Basin-Tell Atlas is a mobile zone. It shows a clearly deformational contrast between the discordant Neogene of the Mitidja Basin and the Cretaceous of the Blida Atlas. Hammam Melouane region is located at the heart of this transition zone. It is a very complex area marked by several tectonic phases, recorded in the Neogene formations. We have demonstrated for the first time an extensional regime in the marine limestones of the Burdigalian, which is the first phase linked to the opening of the Mitidja Basin. This phase is taken up by other phases with a compressive regime, underlined in the first step by a flexible deformation marked with kilometric folds (Rocher des Pigeons syncline). Subsequently, this fold is associated with strike-slip faults in a sinistral movement, which provokes the reorientation of the axe and the flanks of the syncline. This tectonic is controlled by a transpression regime that enables the success of the Burdigalian-Serravalian organization; Tortonian in our opinion. Therefore the compressive regime put the Cretaceous formations of the Blida Atlas in contact with the Neogene of the Mitidja Basin, highlighted with a thrust faults. All this tectonic train is recorded in a post-nappe context characterizing this southern edge of the Mitidja Basin. However, the superposition of the different phases needs more precision and analyses along the geological limit Mitidja Basin-Tell Atlas, to better understand the relationship between these two structures.

Fig. 4 **a** Normal fault in the Burdigalian limestone; **b** Reverse fault; **c** Thrust movement registered in the Cretaceous; **d** Sinistral strike-slip fault with tectonic markers in the Neogene formation (Sandstone)

References

1. Adjiri, S., Naak, M.: Géologie de la zone de fractures sud mitidjienne sur la transversale de hammam Melouane, Conférence, CMGA 7 – Alger, 20–22 Février (2018)
2. Ait Hamou, F.: Etude pétrologique et géochimique du volcanisme d'âge Miocène de la région de Hadjout (Ouest Algérois). Thèse de Magister en Géologie, spécialité Pétrologie-structurologie. USTHB, IST, Alger, Algérie (1987)
3. Blès, J.L.: Etude tectonique et microtectonique d'un massif autochtone tellien et de sa couverture de nappes: le massif de Blida (Algérie du Nord), Bull. Soc. Géol. Fr., (xiii), n° 5–6, pp. 498–511 (1971)
4. Bonneton, J.R.: Géologie de la zone de contact entre la Mitidja et l'Atlas de Blida au sud d'Alger, Thèse de Doctorat, 3ème cycle. Univ. Pierre et Marie Curie, Paris (1977)
5. Glangeaud, L.: Etude géologique de la région littorale de la province d'Alger, Thèse sc. Paris et bull. serv. Carte géol. Alger, 2eme série, strat, n° 8 (1932)
6. Glangeaud, L., Aymé, A., Caire, A., Mattauer, M., Muraour, P.: Histoire géologique de la province d'Alger, XIXe Congr. Géol. Inter., Alger, Monogr. Région., 1ere sér., Algérie, n° 22 (1952)

Late Hercynian and Alpine Deformation in Sidi Abdelaziz Area Small Kabylie—Algeria

Nacer-eddine Bouzekria

Abstract

The region of Sidi Abdellaziz is located in the center of the massif of small Kabylie (Algeria), it is part of the basement of the internal domain of the Maghrebids chain. The petrographic study of metamorphic rocks in this region has made it possible to highlight the upper sets of the Kabyle basement constituted by metapelitic series intruded by deformed granite. Structural analysis has shown that this granite is syntectonic and has been emplaced in a strike-slip reverse fault toward the East. This granite was later affected by a second N–S directional deformation, responsible for the formation of shear bands. The statistical analysis of the poles of the schistosity planes and the lineation in the inclosing The statistical analysis of the poles of the schistosity planes and the lineation in the metapelitic rocks gives a ESE–WSW direction, parallel to the emplacement of the granite body. The statistical study of fold axes directions and fault planes in the granite body and its metapelitic inclosing rocks reveals the existence of at least two compression phases, the first is NW–SE direction and the second is E–W. The sigmoidal elements derived from the Djennah shales indicate the existence of a shear zone towards the NW compatible with the structural elements related to the emplacement of the granite body. The region recorded a reverse-thrust materialized by lying down folds and reverse thrusting faults toward the NW in the metapelites of Sidi Abdellaziz, this movement is confirmed by the folding of the abnormal contacts separating the various units dipping towards the South on the surface and to the north at depth. These structural elements suggest of a polyphase deformation of the basement, it records the Hercynian and Alpine movements. This type of structure was highlighted in the East and West of the basement of Small Kabylie (Peucat et al. 1994).

Keywords

Deformation • Hercynian • Metapelite • Shear zone Lineation • Schistosity

1 Introduction

The studied region is located in the NE of Algeria, 40 km to the E of Jijel. It is located in the center of the small Kabylie basement which is part of the inner zone of the orogenic segment of the Maghrebids chains [4] (Fig. 1).

The internal area of the Maghrebids chain consists of two sets topped by satin shale Paleozoic age [3] and intruded by leucoganite.

The purpose of this work is to highlight the different directions of deformation that affected the central area of small Kabylie basement. This study is based on field observations and the microtectonic study.

2 Definition of Different Tectonic Unit

The central part of the small Kabylie basement outcrops from the west of the Sidi Abdellaziz city to Djennah region. This region consists of two sets separated by an abnormal contact of NE–SW direction (Fig. 2).

1. Lower sets: It is constituted by micaschists with Andalusite pseudomorphosed in kyanite; it is a deep slide basement that is outcropping in Machetat l'Arbaa, it appears at the abnormal contact that separates the unit of Sidi Abdellziz and Djennah, it was carried with the granitic intrusion of Bouyoucef.
2. Upper sets: It consists of two units, Sidi Abdellaziz unit and Djennah unit.
   - The unit of Sidi Abdellaziz outcrops to the NW of the studied region; it is represented by a metapelitic series with andalusite and garnet. The upper terms show intercalations of blue marble lenses.

N. Bouzekria (✉)
Ecole Normale Supérieure, Bp: 92 Kouba, Alger, Algérie
e-mail: bouzekria@ens-kouba.dz

F. Rossetti et al. (eds.), *The Structural Geology Contribution to the Africa-Eurasia Geology: Basement and Reservoir Structure, Ore Mineralisation and Tectonic Modelling*,
Advances in Science, Technology & Innovation, https://doi.org/10.1007/978-3-030-01455-1_45

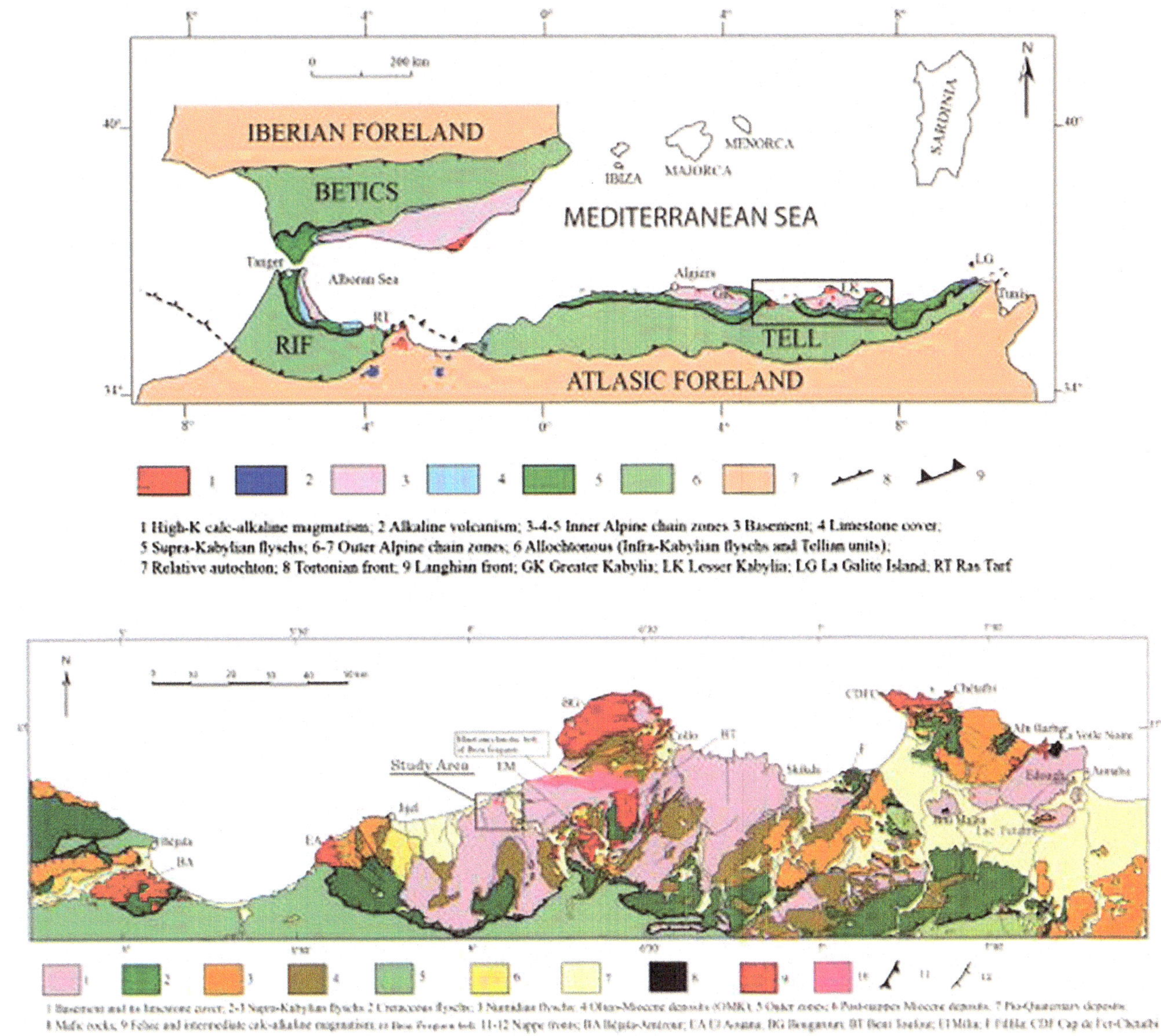

Fig. 1 Geological situation of small Kabylie in the Maghrebide chain (from Durand Delga [4] modified by Mahdjoub et al. [5])

The unit of Djennah is located at the NE of the studied region. It begins with an eyed porphyroid, and black quartzitic shales in which are interspersed with blue marble lenses; The series continues with shales with retromorphic garnet and ends with satin shales attributed to the Paleozoic [3].

3 Structural Study

The granite shows a N080 direction sub-horizontal elongation of quarto-feldspathic aggregates. These aggregates shows a continuous deformation, with flow and shear direction towards the E, this direction is compatible with the structural elements related to the emplacement of the granite body during the deformation D1 (Fig. 3) [5].

The second deformation affects the granite body it is crossed by three shear zone dipping to the S and carry a lineation of the same direction. This deformation (D2) is discontinuous, it is located mainly in the mylonitized bands [2], the deformations D1 and D2 have been dated Hecynians and tadi-Hercynians.

Stereographic projection of the poles of schistosity planes and the folding directions in the Djennah shale and in the Sidi Abdellaziz metapelites shows two directions of strain: NW–SE and N–S direction (Fig. 4). The projection of the poles of the foliation planes and the mineral lineation directions of

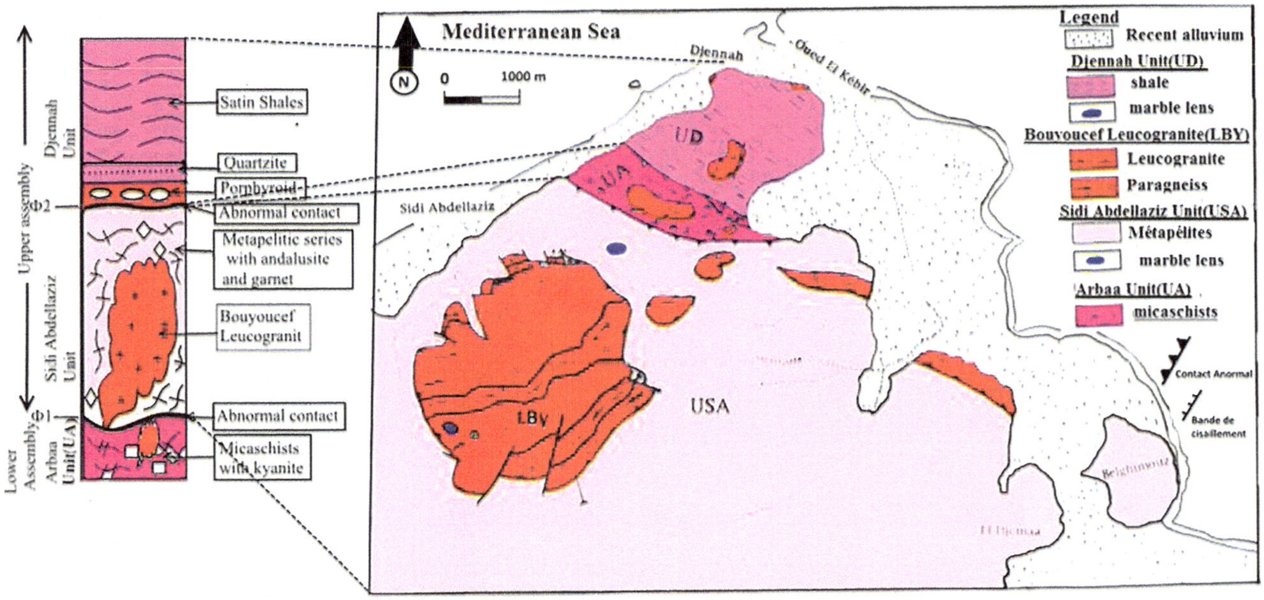

Fig. 2 Tectonic map (e: 1/50,000) and lithological column of different facies of Sidi Abdellaziz Djennah region

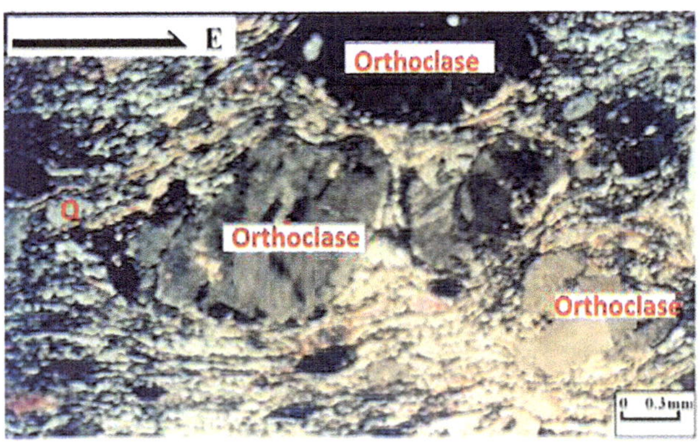

Fig. 3 Thin section showing the direction of deformation in the Quartzo-feldspathic elements from granites of Bouyoucef

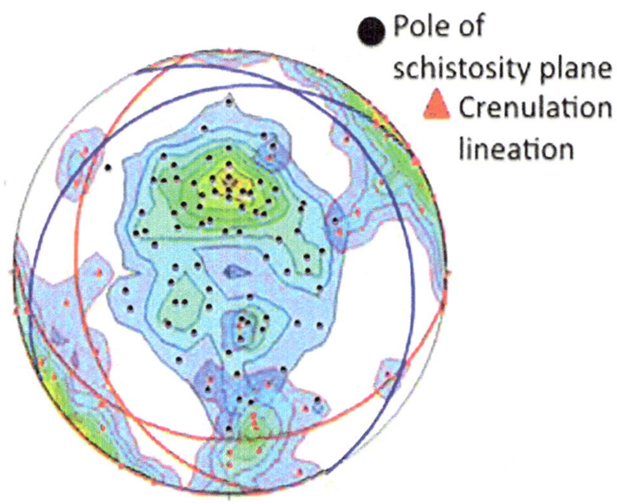

Fig. 4 Stereographic projection of the poles of the schistosity planes and fold axes in Sidi Abdellaziz metapelites

Bouyoucef granite reveals two directions of lineations which correspond to two deformation stresses, the first one is of E–W direction and the second of N–S direction (Fig. 5).

The Djennah region is affected by flat accidents, the deformation expressed by the sigmoidal elements gives a direction of SE shear to NW and the quartz strands affected by the conjugate faults show a direction of shortening N–S.

Reverse faults dipping towards the SE and folds lying down to the NW, this deformation has been attributed to late Alpine movements (Fig. 6).

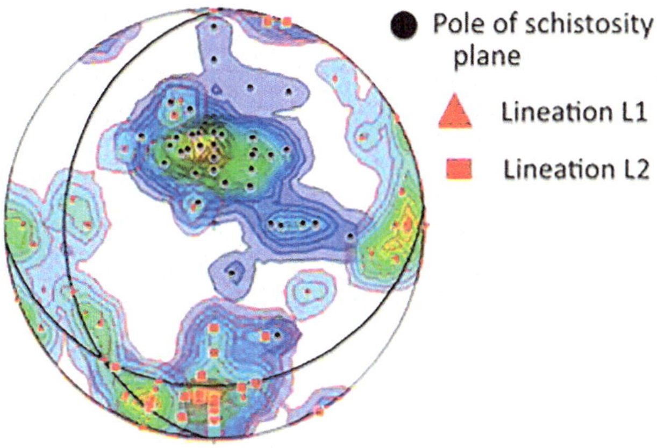

Fig. 5 Stereographic projection of the poles of the foliation planes and lineation in Bouyoucef granite

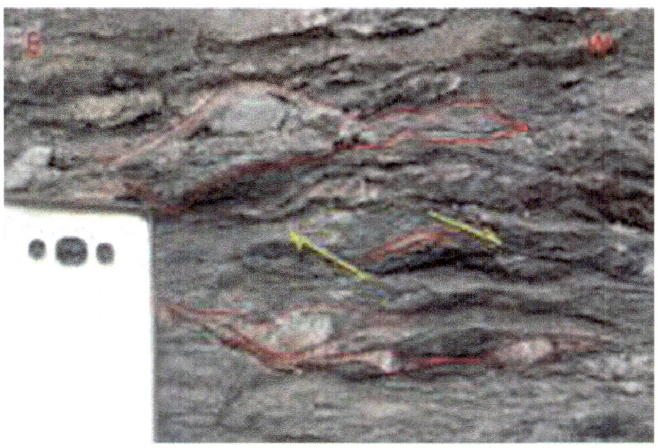

Fig. 6 Photo of sigmoidal elements in flat fault of the Djennah shales showing the shear direction towards the NW

4 Conclusion

Previous studies in the massifs of Grande and small Kabylie reveals the existence of several deformation phases during the Hercynian and Alpine orogeny. In Grater Kabylie, Saadallah et al. (1996) obtained ages around 25 Ma by the 40Ar/39Ar method on the granites of Sidi Ali Bounab.

In small Kabylie, Peucat et al. (1977) give mixed ages between Hercynian events 273 $\pm$ 6 Ma on U/Pb and Alpins (40–25) on Pb/Sr obtained on the Beni Ferguen granites located on the E of small Kabylie massif. The Bouyoucef (Sidi Abdellaziz region) and Beni Ferguen granites are located on a NE–SW fault.

The structural analysis of the Bouyoucef granite and its metapelitic inclosing rocks suggests several phases of deformation, the detail of this analysis allowed us to highlight Hercynian and Alpine events:

An E–W strike slip overlap synchronous with the emplacement of the granite, it followed by an N–S bastomylonitization, these two directions were respectively dated Hercynian and late Hercynian [6].

The second category of deformation is related to the behavior of small Kabylie basement during the alpine orogeny, it was affected by several phases of deformation. The first one is NW–SE direction, it is a compression phase known throughout the entire chain of Maghreb, it is responsible for the thickening of the basement and its shelling.

The second deformation is SE–NW directional, it appears at the contact between the granite and its metapelitic inclosing rocks and in the inclosing rocks also as sigmoidal elements indicating a flat shear thrust towards the NW, it demonstrate a post-thickening readjustment, The third deformation shows a backthrust towards the NW, and is classified as the late phases of the alpine orogeny [1].

References

1. Belhai, D., Merle, O., Saadallah, A.: Transpression dextre à l'Eocène supérieur dans la chaine des Maghrébides (Massif du Cheoua, Algérie).C. R. Acad. Sci. Paris, t, 310; série II, 795–800 (1990)
2. Bouzekria, N.: Etude géologique et structurale de la région d'El Anceur à Sidi Abdellaziz (Petite Kabylie - Algérie) Thèse magister; IST; USTHB; Algérie; 140p (1997)
3. Djellit, H.: Evolution Tectono-métamorphique du socle Kabyle et polarité de mise en place des nappes de flyschs en Petite Kabylie occidentale. Thèse de Doctorat. Univ. Paris XI. 205p (1987)
4. Durand Delga, M.: Mise au point sur la structure NE de la Berberie. Pub. Serv. Géol. Algérie. Nlle Série. Bull n° 39, 89–131 (1969)
5. Mahdjoub, Y., Choukroune, P., Kienast, J.R.: Kinematics of a complex Alpine segment superimposed tectonics and metamorphic events in the Petite kabylie Massif (Northern Algeria). Bull soc geol. France **168**, 649–661 (1997)
6. Peucat, J., Mahdjoub, Y., Drareni, A.: U-Pb and Rb-Sr geochronological evidence for late Hercynian tectonic and Alpine overthrusting in Kabylian metamorphic basement massifs (northeastern Algeria). Tectonophysics. **258**, 195–213 (1996)

Extensional Attenuation of Foreland Thrust Belts in the Western Mediterranean

Guillermo Booth-Rea, Seiffedine Gaidi, Lluis Moragues, Fetheddine Melki, Jose Miguel Azañón, Wissem Marzougui, Jorge Pedro Galvé, and Vicente Perez-Peña

Abstract

Extensional denudation in the western Mediterranean formed deep basins in the core of subduction-related orogenic arcs. It has contributed to the exhumation of metamorphic rocks in the hinterland of the Betics, Apennines, Rif and Tell. However, fewer cases have been described as extensional denudation affecting Foreland Thrust Belts (FTB). Here we describe the extensional collapse of FTBs around the western Mediterranean in response to mantle subduction dynamics in a setting of NW-SE convergence between Africa and Eurasia. The extension of FTBs has occurred at the edges of migrating subduction systems like the Gibraltar, Calabrian and Tunisian Tell-Atlas arcs when lithospheric tearing has propagated under the foreland region. Inboard of Subduction Transfer Edge Propagator (STEP) boundaries continental crust can undergo peeling of its mantle lithosphere driving extension, Si-K-rich magmatism and topographic uplift. This tectonic mechanism has been described as edge delamination in the Betics and Rif. We describe examples of extended FTB's in the Betics, Rif and Tell and their relation to mantle driven tectonic mechanisms.

Keywords

Extension • Foreland thrust belt • Western mediterranean
Slab roll-back • Slab tearing

G. Booth-Rea (✉) · L. Moragues · J. M. Azañón · J. P. Galvé
V. Perez-Peña
Department of Geodynamics, University of Granada, Granada, Spain
e-mail: gbooth@go.ugr.es

S. Gaidi · F. Melki
Department of Earth Sciences, Tunis El Manar University, Tunis, Tunisia

W. Marzougui
Office National de Mines, Tunis, Tunisia

1 Introduction

The extension in the western Mediterranean orogens has been related to slab roll back, tearing, detachment or delamination [e.g. 1, 2], or simply to the collapse of overthickened orogenic wedges [3]. This process has led to the development of extensional basins floored by thinned continental rocks, oceanic crust or even exhumed mantle in the core of the main orogenic belts of the western Mediterranean during the Miocene to Quaternary [e.g. 4–6]. Although extension is generally described to have affected the Internal zones of the orogenic wedges, recent work shows that locally the external foreland thrust belts (FTBs) are also thinned in the Apennines, Betics, Rif, Tunisian Tell and Sicily [e.g. 7–13], Fig. 1. We review and show several examples of extended FTBs in the western Mediterranean that suggest that this is a more common feature than previously thought. A part of this extension may be related to dynamics of the accretionary wedge [12]. However, we argue that another great part of the extension of FTB's in the western Mediterranean was driven by slab derived tectonic mechanisms like slab tearing, roll-back or peeling back during subduction.

2 Tectonic Setting

The western Mediterranean has accommodated NW-SE plate convergence between Africa and Eurasia since the late Cretaceous [14]. However, the main geological features of the region have developed in relation to lithospheric-scale related tectonic mechanisms like slab roll-back, detachment, tearing or edge delamination. This has produced material flow in directions oblique to the NW-SE direction of convergence between the plates, for example, westwards in the Gibraltar arc or eastwards in the Calabrian arc [1, 2]. Furthermore, the outward migration of these arcs, driven by mantle tectonic mechanisms, has been followed by

F. Rossetti et al. (eds.), *The Structural Geology Contribution to the Africa-Eurasia Geology: Basement and Reservoir Structure, Ore Mineralisation and Tectonic Modelling*,
Advances in Science, Technology & Innovation, https://doi.org/10.1007/978-3-030-01455-1_46

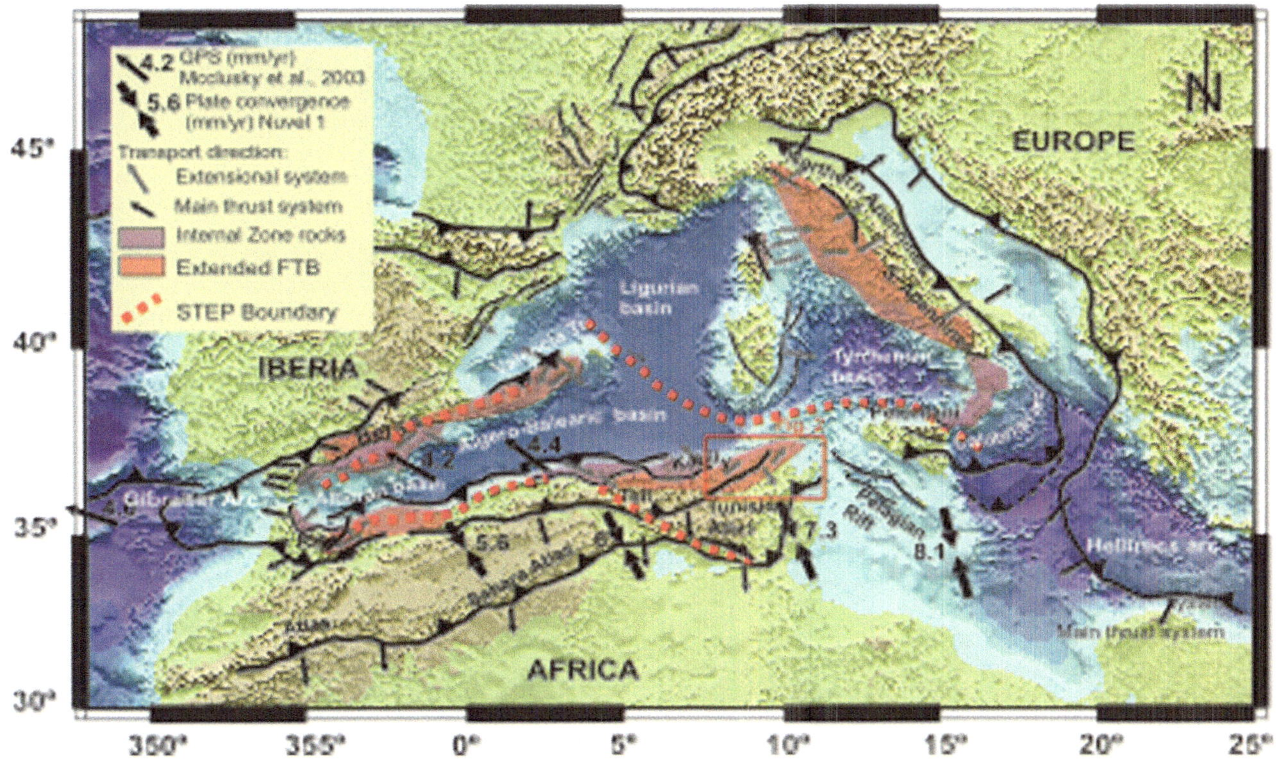

Fig. 1 Alpine tectonic domains of the western Mediterranean. Notice extended FTB's

extensional collapse of the internal parts of the orogens, producing the Alboran, Algero-Balearic, Ligurian and Thyrrenian basins from west to east between the Oligocene and the Quaternary. The migration of these arcs has been possible thanks to the opening of slab windows [2], defined as Subduction Transfer Edge Propagators by Govers and Wortel [15]. Finally, once the effects of slab retreat and tearing have migrated away, many extended regions of the western Mediterranean inverted due to shortening under Africa-Eurasia NW-SE plate convergence.

3 Results

Extensional denudation of FTBs occurred in the Betics, from Mallorca to the Central part of the Betics [8, 10, 12, 16], in the Rif [9] and in the Tell in Tunisia [13]. Furthermore, extension affecting the external thrust sheets has also been described in the Apennines and in Sicily [7, 11]. Both in the Betics and Rif there is a good control on lithospheric structure thanks to recent receiver function data that show lithospheric and crustal steps underneath the northern and southern branches of the Gibraltar arc [17]. *N* In these steps crustal thickness can increase from 32 to 55 km in a few km. For example, the Eastern Rif coincides with large strike-slip faults like the sinistral Nekor fault, interpreted as a STEP boundary [17]. Furthermore, outboard of the Rifean STEP boundary elongated extensional domes that exhume middle crust occur [9], equivalent to ones in the Betics. On the Betic side, lithospheric and crustal steps are noticeable under the FTB in the Eastern Betics [17], coinciding with the dextral Crevillente fault that worked as a STEP boundary during the Tortonian [18]. Extensional depocenters developed inboard of this STEP boundary straddling the contractive or transcurrent boundary between Internal and External zones [10]. Extensions bounded by strike-slip faults also occur in Mallorca [16]. A similar evolution with extension and related magmatism is observed in the Tunisian Tell-Atlas FTB that is also underlain by a slab body. Here the STEP boundary is represented at the surface by dextral structures like the Gafsa fault [13].

4 Discussion and Conclusions

The tearing of the subducting lithospheric mantle at STEP boundaries resulted in complex patterns of deformation in time and space where contractive deformation is followed by strike-slip and later by extensional denudation [e.g. 9, 13, 17]. STEP boundaries have sometimes propagated under the continental lithosphere reaching the external Foreland Thrust Belts, producing strike-slip deformation followed by extension of the FTB's in the Rif [9], the Betics [17, 18], northern Tunisia [13], the Appennines [7] and in Sicily [11] (Fig. 1).

Extensional collapse of the external FTB of the western Mediterranean is a widespread phenomenon that occurred at the edges of the migrating Gibraltar, Calabrian and Tunisian arcs since the Middle Miocene. The extension was driven by the propagation of STEP boundaries under the continental lithosphere underlying the FTB's. This mode of mountain building is characterized by efficient denudation mechanisms, namely the extension and topographic uplift, which propagate behind the initial contractive or transpressive orogenic wedges. Furthermore, the final result is that the orogen is stripped both of its upper crust, by extension and erosion, and of its lithospheric mantle by tearing and peeling back, inboard of STEP boundaries.

References

1. Lonergan, L., White, N.: Origin of the Betic-Rif mountain belt. Tectonics **16**(3), 504–522 (1997)
2. Faccenna, C., Piromallo, C., Crespo-Blanc, A., Jolivet, L., Rossetti, F.: Lateral slab deformation and the origin of the western Mediterranean arcs. Tectonics **23**(TC1012) (2004). https://doi.org/10.1029/2002tc001488
3. Platt, J.P., Vissers, R.L.M.: Extensional collapse of thickened continental lithosphere: a working hypothesis for the Alboran Sea and Gibraltar Arc. Geology **17**, 540–543 (1989)
4. Booth-Rea, G., Ranero, C., Martínez-Martínez, J.M., Grevemeyer, I.: Crustal types and Tertiary tectonic evolution of the Alborán sea, western Mediterranean. G-Cubed **8**, Q10004 (2007). https://doi.org/10.1029/2007GC001661
5. Prada, M., Sallares, V., Ranero, C.R., Vendrell, M.G., Grevemeyer, I., Zitellini, N., Franco, R.: Seismic structure of the Central Tyrrhenian basin: geophysical constraints on the nature of the main crustal domains. J. Geophys. Res.: Solid Earth **119**(1), 52–70 (2014)
6. Arab, M., Rabineau, M., Deverchere, J., Bracene, R., Belhai, D., Roure, F., Marok, A., Bouyahiaoui, B., Granjeon, D., Andriessen, P., Sage, F.: Tectonostratigraphic evolution of the eastern Algerian margin and basin from seismic data and onshore-offshore correlation. Mar. Pet. Geol. **77**, 1355–1375 (2016)
7. Carmignani, L., Kligfield, R.: Crustal extension in the Northern Apennines: the transition from compression to extension in the Alpi Apuane core complex. Tectonics **9**(6), 1275–1303 (1990)
8. Balanya, J.C., Crespo-Blanc, A., Azpiroz, M.D., Exposito, I., Lujan, M.: Structural trend line pattern and strain partitioning around the Gibraltar Arc accretionary wedge: insights as to the mode of orogenic arc building. Tectonics **26**(2) (2007)
9. Booth-Rea, G., Jabaloy-Sanchez, A., Azdimousa, A., Asebriy, L., Vazquéz-Vílchez, M., Martínez-Martínez, J.M.: Upper-crustal extension during oblique collision: the Temsamane extensional detachment (eastern Rif, Morocco). Terra Nova **24**, 505–512 (2012)
10. Rodríguez-Fernández, J., Azor, A., Miguel Azañón, J.: The Betic Intramontane Basins (SE Spain): Stratigraphy, Subsidence, and Tectonic History, pp. 461–479. Tectonics of Sedimentary Basins. Wiley, New York (2011)
11. Barreca, G., Scarfì, L., Cannavò, F., Koulakov, I., Monaco, C.: New structural and seismological evidence and interpretation of a lithospheric-scale shear zone at the southern edge of the Ionian subduction system (central, Eastern Sicily, Italy). Tectonics **35**(6), 1489–1505 (2016)
12. Jimenez-Bonilla, A., Torvela, T., Balanya, J.C., Exposito, I., Diaz-Azpiroz, M.: Changes in dip and frictional properties of the basal detachment controlling orogenic wedge propagation and frontal collapse: the external central Betics case. Tectonics **35**(12), 3028–3049 (2016)
13. Booth-Rea, G., Gaidi, S., Melki, F., Marzougui, W., Azañón, J.M., Zargouni, F., Galvé, J.P., Pérez-Peña, J.V.: Late Miocene extensional collapse of Northern Tunisia. Tectonics **37**, 1626–1647 (2018). https://doi.org/10.1029/2017TC004846
14. Dewey, J.F., Helman, M.L., Turco, E., Hutton, D.H.W., Knott, S. D.: Kinematics of the western Mediterranean. In: Coward, M.P., Dietrich, D., Park, R.G. (eds.) Alpine Tectonics, pp. 265–283. Special Publication Geological Society of London, London (1989)
15. Govers, R., Wortel, M.J.R.: Lithosphere tearing at STEP faults: response to edges of subduction zones. Earth Planet. Sci. Lett. **236** (1–2), 505–523 (2005)
16. Booth-Rea, G., Azañón, J.M., Roldán, F.J., Moragues, L., Pérez-Peña, V., Mateos, R.M.: WSW-ENE extension in Mallorca, key for integrating the Balearic Promontory in the Miocene evolution of the western Mediterranean. GeoTemas **16**(1), 81–84 (2016)
17. Mancilla, F.L., Booth-Rea, G., Stich, D., Pérez-Peña, J.V., Morales, J., Azañón, J.M., Martin, R., Giaconia, F.: Slab rupture and delamination under the Betics and Rif constrained from receiver functions. Tectonophysics **663**, 225–237 (2015)
18. Pérez-Valera, L.A., Rosenbaum, G., Sanchez-Gomez, M., Azor, A., Fernandez-Soler, J.M., Pérez-Valera, F., Vasconcelos, P.M.: Age distribution of lamproites along the Socovos Fault (southern Spain) and lithospheric scale tearing. Lithos **180**, 252–263 (2013)

Genetically Linked Sedimentary Basins to Define a Kinematic Model of the Central Mediterranean Extension

Alfonsa Milia and Maurizio Maria Torrente

Abstract

The Tyrrhenian Sea (Central Mediterranean) is an extensional area that formed during the collision between Africa and Eurasia. Through the analysis of well logs and seismic profiles we reconstructed the stratigraphic successions and the structural style of the Tyrrhenian sedimentary basins. The sequence stratigraphy interpretation of well logs permitted us to recognize third and fourth order depositional sequences and the stratigraphic signatures of the rift stages. The study area features several sediment overfilled/balanced infill basins that resulted from the superposition of several rifting events characterized by high rates of basin subsidence. These outcomes led us to propose five rifting episodes occurred from Miocene to Quaternary affecting the upper plate. The suggested geodynamic scenario takes into account a progressive rupture of the slab and formation of sedimentary basins in the upper plate.

Keywords

Basin analysis • Extensional modes • Triangular backarc basin • Forearc extension • Central mediterranean

1 Introduction

According to the classical literature, extensional areas are characterized by early-, syn-, climax and post-rift phases in response to the changes of the rate of subsidence. However recent papers documented a more complex evolution of rifts, due to the superposition of basins that grow in different structural conditions [e.g. 1] and characterized by a change in the direction of extension. Within this frame the stratigraphic succession of a rift basin results the product of a complex interplay between tectonics and sedimentation that is not yet completely understood [e.g. 2]. To study the evolution of a rift zone, it is fundamental to carry out the stratigraphic analysis of ancient and active rift basins. This analysis indicates that the spatial distribution and architecture of depositional systems controlled by the fault zones are influenced by the development of normal faults, their physiographic expressions, and variations in fault slip rates. Extensional sedimentary basins can be classified in terms of stratigraphy, age of formation, basin architecture, and bounding fault's geometry. It was proposed that continental extension has two end-members styles: rift settings and highly extended terrains, resulting in high-angle rift basins and supra-detachment basins, respectively [3].

Backarc rift basins are characterized by an underlying slab. The mode of extension and the stratigraphic signature of backarc basins are to a large extent controlled by the interaction of the upper plate with the slab [4] and bibliography therein. The interpretation of seismic reflection profiles calibrated by deep wells, by cores and dredges data and constrained by onshore geology, permitted us to reconstruct the tectono-stratigraphic evolution of the sedimentary basins linked to the evolution of the Tyrrhenian Sea backarc basin [5–8]. Our results show that extensional sedimentary basin formed since the late Oligocene. The reconstruction of genetically linked sedimentary basin documents the complex evolution over time of a polyphased backarc basin. The discrepancy between the current tectonic and geodynamic models of extensional deformation and the extensional features of the Tyrrhenian Sea led us to hypothesize a structural control on the backarc evolution by inherited geologic features of the subducting plate.

A. Milia (✉)
ISMAR, CNR, Naples, 80123, Italy
e-mail: alfonsa.milia@cnr.it

M. M. Torrente
DST, Università del Sannio, Via Portarsa 11,
82100 Benevento, Italy

F. Rossetti et al. (eds.), *The Structural Geology Contribution to the Africa-Eurasia Geology: Basement and Reservoir Structure, Ore Mineralisation and Tectonic Modelling*,
Advances in Science, Technology & Innovation, https://doi.org/10.1007/978-3-030-01455-1_47

2 Results

We interpreted seismic profiles, wells, and outcrops and recognized several periods of syn-rift affecting the basins over time. The analysis of the sedimentary successions led us to recognize third order depositional sequences, reconstruct the fault pattern, and furnish a detailed evolution of the sedimentary basins since the late Oligocene. During the middle Miocene, sedimentary basins architecture and fault pattern show a system of approximately N-S oriented normal faults and approximately E-W transform faults that were active since the late Oligocene. Two periods of extensional/transtensional tectonics (late Oligocene-Lower Burdigalian and Upper Langhian-Tortonian) were separated by a compressional event (late Burdigalian-Lower Langhian) [8]. These basins developed on the western Tyrrhenian margin, in the southern Apennine and onshore and offshore Calabria. Successively, during the Messinian, rift basins were active on the western Tyrrhenian margin and a supra-detachment basin formed on the Ionian margin of Calabria. In particular, the Stilo rift (off Calabria) features a Serravallian-Tortonian succession, bounded by a high-angle normal fault, and an adjacent supradetachment basin, Messinian in age, bounded by a low-angle normal fault. According to [9] the kinematics of rifting were characterized by the abandonment of the high-angle fault before the onset of low-angle normal fault activity. The Pliocene epoch was characterized by the opening of the triangle-shaped Vavilov depression, a fossil backarc basin, which tectonic evolution was investigated in both apex and bathyal parts. The data interpretation revealed that the apex basin corresponds to a sediment-balanced basin, recording several episodes of basin evolution. In contrast, the distal basin is an underfilled basin, characterized by localized volcanic activity and thin sedimentary succession covering an exhumed mantle. In summary the basin architecture of the Vavilov rift zone features rift and supra-detachment basins [7].

Throughout the course of the Lower Pleistocene episode, rifting spread over a wide area of the Tyrrhenian region and was characterized by three extension directions [2]: NE trending in the Campania Margin, SE-trending in the Marsili Basin and N-S in the Paola Basin. On the Eastern Tyrrhenian margin a triangular basin formed in the northern Gaeta Bay bounded to the south by a transform fault that transferred the extension in the Campania Plain. The stratigraphic signature of the Campania Plain basin documents a very high subsidence rate during this rifting episode. The Middle Pleistocene rifting episode featured a 90° rotation of the extension direction (NW-SE) with the morphology of the margin being profoundly influenced by NE-trending structural highs and half grabens transversal to the Apennine chain. During the Middle Pleistocene the subsidence rate of the Campania Plain reached the maximum value. The Middle Pleistocene dramatic change of direction of extension (from NE-SW to NW-SE) in the Tyrrhenian region can be linked to the change of direction of the tectonic transport (from NE to SE) of the southern Apennines thrust belt [2].

Two important events are fundamental in the evolution of the Tyrrhenian Sea [2]: (i) the skip of the extensional axial zone toward East in the Lower Pleistocene; (ii) the abrupt change of direction seen in the extension in the Middle Pleistocene.

3 Discussion and Conclusions

The reconstruction of the Paleo-Tyrrhenian evolution during the Miocene of the Central Mediterranean is based on the stratigraphic and tectonic constraints together with the age migration of the depocenters. Taking into account the position of the volcanic arc (Sardinia), the coeval extensional/transtensional sedimentary basins developed in a forearc position before the Tyrrhenian backarc opening. The Stilo rift is a rare example where the sedimentary basin is completely preserved. It represents the development of two end-member basins (from Serravallian-Tortonian rift to Messinian supra-detachment basin) during a continued extension with an abrupt change in style linked to a marked increase in extension rate.

The Pliocene rifting stage was characterized by the opening of a triangular basin. We documented that the rifting of the Vavilov triangular basin was synchronous with the apex to distal regions around a single Euler pole located in Latium, between 5.1 and 1.8 Ma. Besides the kinematic evolution of the Vavilov basin featured two stages: initial pure shear mode (5.1–4.0 Ma), that produced high-angle normal faults and syn-sedimentary wedges, followed by simple shear mode (4.0–1.8 Ma), that caused supra-detachment basins in the apex region. The final stage of extension in the distal region led to: (i) complete embrittlement of the crust; (ii) direct continuation of crustal faults to upper mantle depth; (iii) serpentinization and mantle exhumation [7].

The polyphase Tyrrhenian rifting features: early distributed continental deformation (Serravallian-Messinian), strain localization on the Vavilov bathyal basin (detachment faults that lead to mantle exhumation) (lower Pliocene) and the skip of the extension zone to the eastern Tyrrhenian margin (Campania Margin) during Quaternary times [2]. The proposed structural evolution maintains that regional stresses directions did not remain constant during Quaternary continental rifting but it recorded a change of 90° in the direction of extension.

The peculiar tectonic evolution of the extensional events that affected the Central Mediterranean, and its discrepancy with the current geodynamic models [e.g. 1] led us to hypothesize a genetic link between the deformation of the upper plate and the deformation of the slab.

The geologic setting of the Mediterranean region resulted from the consumption of the oceanic lithosphere between Africa and Europe and the extension in the lithosphere above the subduction zone. Based on tomographic images, seismological data, petrology of volcanic rocks and analog models, several authors reconstructed a complex geometry of the subducting slab bounded by a couple of STEP faults sensu [10]. The southern edge of the Ionian slab is currently imaged near Sicily, whereas the northern edge of the Ionian slab is poorly constrained. We propose a possible geodynamic scenario characterized by a subducting slab affected by progressive STEP propagation. When the eastward trench migration reached the continental lithosphere in the northern part of the margin, it stopped or significantly slowed down as the oceanic subduction continued in the southern part of the margin [2, 6].

References

1. Lavier, L.L., Manatschal, G.: A mechanism to thin the continental lithosphere at magma-poor margins. Nature **440**, 324–328 (2006)
2. Milia, A., Torrente, M.M.: Tectono-stratigraphic signature of a rapid multistage subsiding rift basin in the Tyrrhenian-Apennine hinge zone (Italy): a possible interaction of upper plate with subducting slab. J. Geodyn. **86**, 42–60 (2015)
3. Friedman, S.J., Burbank, D.W.: Rift basins and supradetachment basins: intracontinental extensional end-members. Basin Res. **7**, 109–127 (1995)
4. Cloetingh, S., Burov, E., Matenco, L., Beekman, F., Roure, F., Ziegler, P.A.: The Moho in extensional tectonic settings: insights from thermo-mechanical models. Tectonophysics **609**, 558–604 (2013)
5. Torrente, M.M., Milia, A.: Volcanism and faulting of the Campania margin (Eastern Tyrrhenian Sea, Italy): a three-dimensional visualization of a new volcanic field off Campi Flegrei. Bull. Volc. **75**(6), 1–13 (2013)
6. Milia, A., Iannace, P., Tesauro, M., Torrente, M.M.: Upper plate deformation as marker for the Northern STEP fault of the Ionian slab (Tyrrhenian Sea, central Mediterranean). Tectonophysics **710–711**, 127–148 (2017)
7. Milia, A., Torrente, M.M., Tesauro, M.: From stretching to mantle exhumation in a triangular backarc basin (Vavilov basin, Tyrrhenian Sea, western Mediterranean). Tectonophysics **710–711**, 108–126 (2017)
8. Milia, A., Valente, A., Cavuoto, G., Torrente, M.M.: Miocene progressive forearc extension in the Central Mediterranean. Tectonophysics **710–711**, 232–248 (2017)
9. Milia, A., Torrente, M.M.: Rift and supradetachment basins during extension: insight from the Tyrrhenian rift. J. Geol. Soc. **72**, 5–8 (2015)
10. Govers, R., Wortel, M.J.R.: Lithosphere tearing at STEP faults: response to edges of subduction zones. Earth Planet. Sci. Lett. **236**, 505–523 (2005)

Multidisciplinary Study of the Central High Atlas Diapiric Province in Morocco: Results from Analogue and Numerical Models

Jaume Vergés, Mar Moragas, and Jonas Ruh

Abstract

We present field data on good-quality outcrops of Early and Middle Jurassic widespread halokinetic depositional sequences in the Central High Atlas of Morocco defining large salt-related minibasins limited by a polygonal network of salt walls. Inversion tectonics mostly inverted and rejuvenated diapiric structures.

Keywords

Diapirism • Salt-related rift basins • Inversion tectonics
Subsidence analysis • Central high atlas
North Africa

1 Introduction

The present configuration of the Central High Atlas of Morocco evolved as a Jurassic salt-related rift basin inverted during the Alpine orogeny. An intricate polygonal network of salt walls separating deep minibasins, that sunk in salt-bearing Upper Triassic deposits, characterized the rift basin. Field data on both structural geology and sedimentology of halokinetic sequences combined in maps, balanced and restored cross-sections, thermal modelling and subsidence analysis, as well as analogue and numerical modelling, diagenetic studies on Jurassic-aged diapiric structures and minibasins to provide an integrated and new interpretation of Central High Atlas (e.g., [4]).

J. Vergés (✉) · M. Moragas · J. Ruh
Group of Dynamics of the Lithosphere (GDL), Institute of Earth Sciences Jaume Almera, ICTJA, CSIC, Lluis Sole i Sabaris s/n, 08028 Barcelona, Spain
e-mail: jverges@ictja.csic.es

2 Geological Setting

The subsidence evolution of the basin from Early to Middle Jurassic time was characterized by a Sinemurian-Pliensbachian rifting followed by long post rift phase. In the central part of the basin, the early Jurassic extension triggered salt-tectonic activity that spanned from Pliensbachian to Callovian times. The initial reactive-active diapir phase linked to the extension phase after 199 Ma was followed by a long period of passive diapiric growth that spanned roughly up to 167 Ma. The passive diapirism caused the development of extremely well exposed halokinetic geometries including bed thinning, onlaps and truncations that comprise composite stacks of halokinetic wedges and hook sequences flanking the diapir ridges (salt walls) (Fig. 1).

3 Combined Results from Field Data and Analogue Modeling

Analogue modelling applied to salt tectonics have proven to be an essential tool for the study and analysis of the mechanisms triggering the onset of diapirs and the evolution of diapiric structures and minibasins. Analogue modelling has also been applied to analyse the impact of the progradation of sedimentary systems above a ductile layer, upper Triassic salt-bearing rocks in nature, representing the source of diapirs.

Additionally, we present two models with increasing amounts of shortening to reproduce inversion tectonics during the Late Cretaceous–Tertiary Alpine compression (6% and 10%). These models show that shortening primarily affected the post-diapiric layers in the vicinities of the salt structures whereas very little deformation occurs away from diapirs (Fig. 2).

F. Rossetti et al. (eds.), *The Structural Geology Contribution to the Africa-Eurasia Geology: Basement and Reservoir Structure, Ore Mineralisation and Tectonic Modelling*,
Advances in Science, Technology & Innovation, https://doi.org/10.1007/978-3-030-01455-1_48

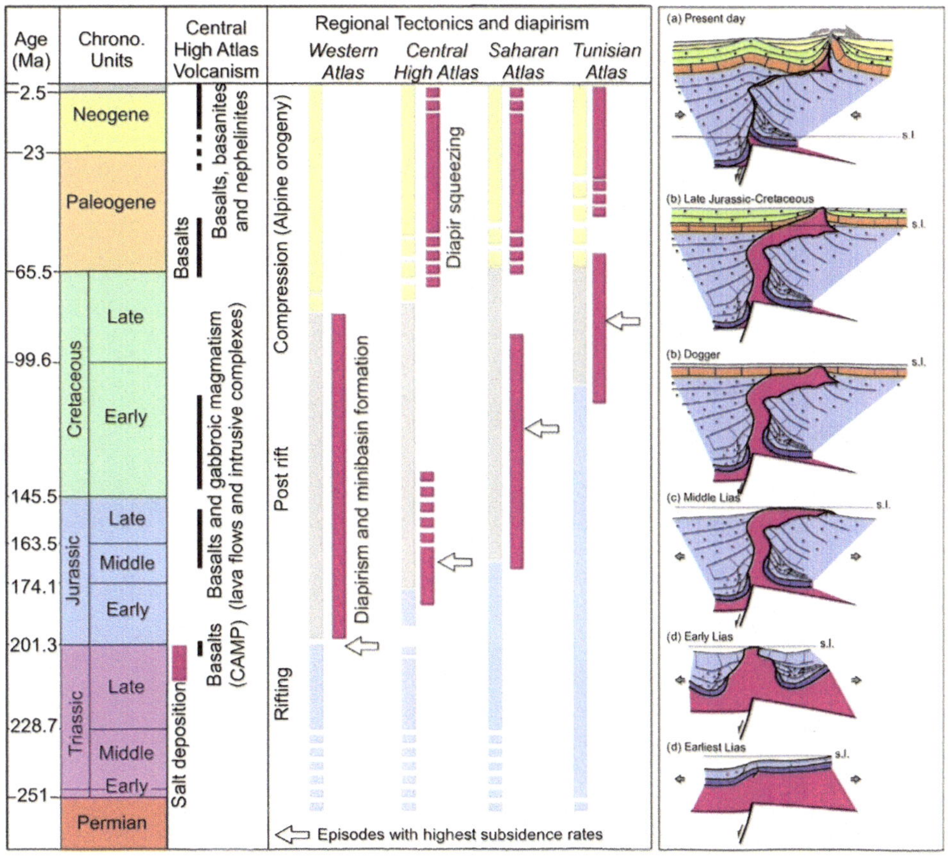

Fig. 1 Chart showing the tectonic and diapiric evolution during the Mesozoic extension and the Late Cretaceous–Tertiary compressive inversion of the Atlas system (Reproduced from [5])

4 Discussion and Conclusions

Subsidence analyses from the rift axis of the Central High Atlas Jurassic rift basin, where diapiric salt ridges and minibasins were developed, showed subsidence evolution characterised by high subsidence rates (tectonic subsidence up to 0.2 mm yr^{-1} and total subsidence up to 1 mm yr^{-1}), spatial and temporal migration of subsiding depocentres, and anomalous subsidence amounts during post-rift phase.

By using analogue and numerical modelling we established that, during the rifting phase, the tectonic subsidence related to salt withdrawal could be around 25% whereas during the post-rift phase this percentage increases up 65 to 80%. These high amounts of rate of bulk tectonic subsidence, which are partly attributed to the sunk of the halokinetic beds in the ductile layer, should be taken into account when calculating subsidence analysis since they should not be directly attributed to purely tectonic processes.

At the present-day, good-quality outcrops of Jurassic halokinetic strata in the Central High Atlas represent excellent analogues of the Triassic-Jurassic successions in other diapiric provinces of the Atlas system localized under a thick Mesozoic and Cenozoic sedimentary pile and only are revealed in seismic lines in Atlantic Morocco and Atlas in Algeria and Tunisia (as discussed in [1, 3]).

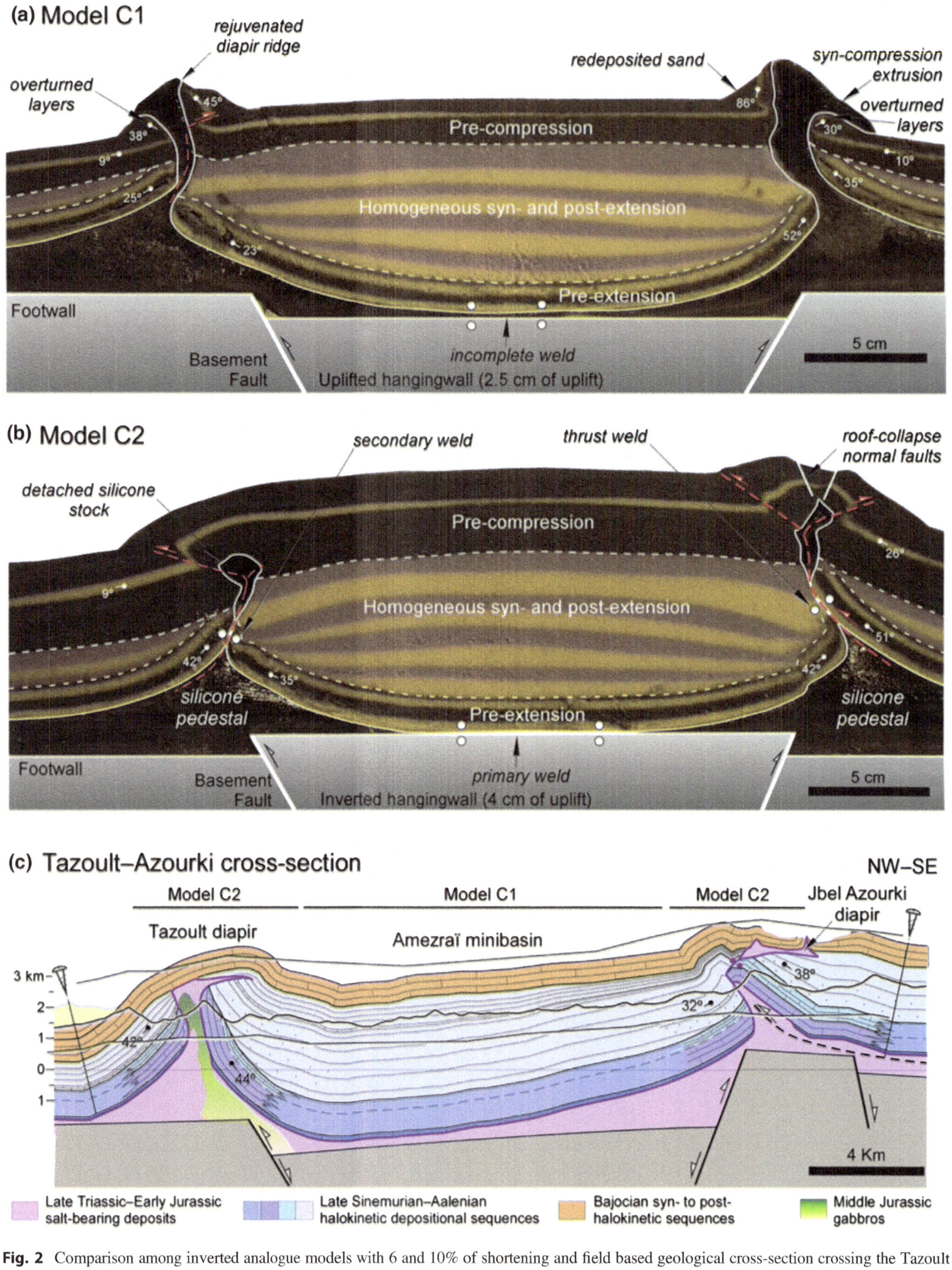

Fig. 2 Comparison among inverted analogue models with 6 and 10% of shortening and field based geological cross-section crossing the Tazoult and the Jbel Azourki salt walls (Reproduced from [2])

Acknowledgements We would like to thank all our co-authors for the good teamwork carried out during the scientific collaboration project between Statoil (Norway), the University of Bordeaux (France) and the Institute of Earth Sciences Jaume Almera of CSIC in Barcelona.

References

1. Martín-Martín, J.D., Vergés, J., Saura, E., Moragas, M., Messager, G., Baqués, V., Razin, P., Grélaud, C., Malaval, M., Joussiaume, R., Casciello, E., Cruz-Orosa, I., Hunt, D.W.: Diapiric growth within an Early Jurassic rift basin: the Tazoult salt wall (central High Atlas, Morocco). Tectonics **36**, 2–32 (2017). https://doi.org/10.1002/2016TC004300
2. Moragas, M., Vergés, J., Nalpas, T., Saura, E., Martín-Martín, J.D., Messager, G., Hunt, D.W.: The impact of syn- and post-extension prograding sedimentation on the development of salt-related rift basins and their inversion: Clues from analogue modelling. Mar. Pet. Geol. **88**, 985–1003 (2017). https://doi.org/10.1016/j.marpetgeo.2017.10.001
3. Moragas, M., Vergés, J., Saura, E., Martín-Martín, J.-D., Messager, G., Merino-Tomé, Ó., Suárez-Ruiz, I., Razin, P., Grélaud, C., Malaval, M., Joussiaume, R., Hunt, D.W.: Jurassic rifting to post-rift subsidence analysis in the Central High Atlas and its relation to salt diapirism. Basin Res. 1–27 (2016). https://doi.org/10.1111/bre.12223
4. Saura, E., Vergés, J., Martín-Martín, J.D., Messager, G., Moragas, M., Razin, P., Grelaud, C., Joussiaume, R., Malaval, M., Homke, S., Hunt, D.W.: Syn- to post-rift diapirism and minibasins of the Central High Atlas (Morocco): the changing face of a mountain belt. J. Geol. Soc. London **171**, 97–105 (2014). https://doi.org/10.1144/jgs2013-079
5. Vergés, J., Moragas, M., Martín-Martín, J.D., Saura, E., Casciello, E., Razin, P., Grelaud, C., Malaval, M., Joussiame, R., Messager, G., Sharp, I., Hunt, D.W.: Salt tectonics in the Atlas mountains of Morocco. In: Soto, J.I., Flinch, J.F., Tari, G. (eds.) Permo-Triassic Salt Provinces of Europe, North Africa and the Central Atlantic: Tectonics and Hydrocarbon Potential, pp. 563–579. Elsevier, Amsterdam (2017)

Analogical Compression Modeling of the Tunisian Atlas in the Mediterranean Context

Rabeb Dhifaoui, Pierre Strzerzynski, Régis Mourgues, Adel Rigane, and Claude Gourmelen

Abstract

Tunisia is divided into two major structural domains: the Tellian domain in the north and the Tunisian Atlas further south, both of which belong to the outer zones. In the outer zones, the deformation, dated from the Tortonian to the actual, is characterized by the coexistence of compressive, strike-slip and extensive structures. NE-SW oriented folds and thrusts affect the Mesozoic cover. These structures are more distant towards the south where they interact with two conjugated strike-slip corridors: the fault of Negrine-Tozeur and the N-S axis. The objective of this study is to replace this association of structures in a broader geological context and provide better constraints to deep-seated structures. For this, an analog modeling approach has been adopted. Several experiments were carried out. The results showed that the presence of a level of detachment at the base of the sedimentary series is necessary to correctly represent the compressive structures in central Atlas and extensive in the Tell and the north of the Atlas. Similarly, a free edge perpendicular to the direction of compression allows the formation of strike-slip and conjugate structures similar to the fault of Negrine-Tozeur and the N-S axis.

Keywords

Coexistence • Compression • Analogical modeling Tunisian Atlas • Strike-slip corridors

R. Dhifaoui · A. Rigane
Faculty of Sciences of Sfax, Laboratory: GEOGLOB (LR12E523), University of Sfax, Sfax, Tunisia

R. Dhifaoui · P. Strzerzynski (✉) · R. Mourgues · C. Gourmelen
Faculty of Sciences of Le Mans, Laboratory of LPG, University of Mans, UMR, Le Mans, 6112, France
e-mail: pierre.strzerzynski@univ-lemans.fr

1 Introduction

Tunisia lies at the eastern end of the Maghreb Alpine chain. Two oceanic areas border Tunisia: the Algero-Provencal basin opened in a back arc position during the Miocene in the north and the Ionian Sea, a relic of the Tethyan oceanic domain now subducting below the Calabria and the Mediterranean ridge in the East (Fig. 1).

Tunisia is divided into two major structural domains which belong to the outer zones: the Tellian domain in the north and the Tunisian Atlas further south. The Tellian and the northern part of the Atlas domains are associated with the highest relief culminating to 1500 m and topography progressively declines to the south. Tellian and Atlas domains consist of a Mesozoic to Cenozoic sedimentary cover located at the top of a basement of continental origin. The depth of the moho is 25–30 km in Tunisian atlas [1, 2].

In Tunisia, as in the rest of the Maghreb Alpine chain, Triassic evaporite forms a low strength layer (Fig. 2) located at the basement cover interface [3].

A complex tectonic and sedimentary history where extension, strike slip and shortening phases successively interplay with salt structure predate the formation of the Maghreb alpine. In the external zones, the deformation, dated from the Tortonian to the present, is characterized by simultaneous association of compressive, strike-slip and extensive structures. NE-SW oriented folds and thrusts affect the Mesozoic cover. Particularly abundant in the Tell and the North of the Atlas, these structures are scarcer towards the south where they interact with two conjugated strike-slip corridors: the Negrine-Tozeur and the N-S axis fault zones [4]. Structures compatible with directions of NE-SW then NE-SW extension direction post date the main shortening deformation in the Tell and Northern Atlas zones [5, 6].

The objective of this study is to replace this association of structures in a broader geological context and provide better constraints to deep-seated structures. For this, an analog modeling approach has been adopted.

F. Rossetti et al. (eds.), *The Structural Geology Contribution to the Africa-Eurasia Geology: Basement and Reservoir Structure, Ore Mineralisation and Tectonic Modelling*,
Advances in Science, Technology & Innovation, https://doi.org/10.1007/978-3-030-01455-1_49

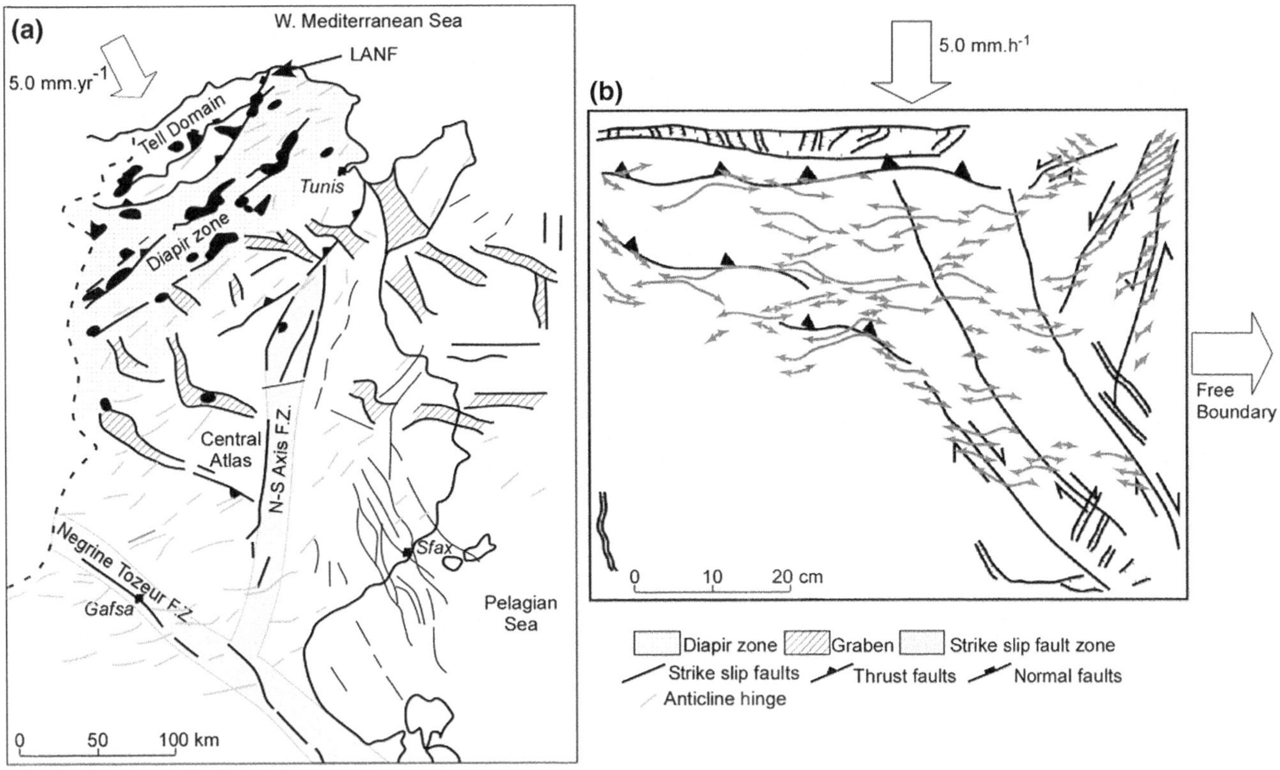

Fig. 1 **a** Structural scheme of Tunisia (modified after [4]). LANF: Low angle normal fault after [6]. **b** Interpretative sketch of the M6 experiment

Fig. 2 Strength profiles: (1) Strength profile for continental crust in the Tunisia. (2) Strength profile for experimental model

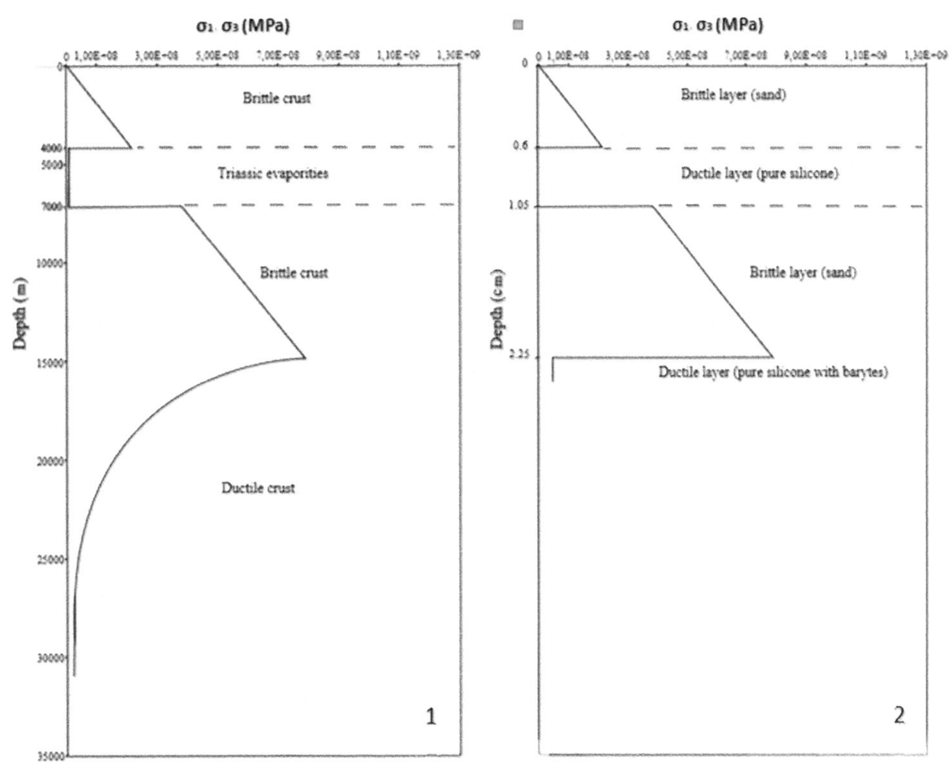

2 Method and Scaling

The experimental device includes a 60/60 cm PVC box. It is subjected to shortening simulating the convergence between the Eurasia and Africa plates and extensions in a direction, orthogonal to the compression, which represents a free edge located at the level of the Mediterranean ridge. This box is filled with layers of deformable materials: sand for brittle levels, upper crust and sedimentary cover and silicone to represent the ductile levels such as Triassic evaporates.

The models are made with similar materials respecting the principle of physical similarity between the model and its prototype in terms of rheologies, dimensions and boundary conditions. This approach is based on the Davy and Cobbold [7] procedure to calculate resistance profiles for the natural example and the corresponding experimental model (Fig. 2) by handling brittle/ductile materials to undergo the prototype to gravity and tectonic forces. Also, parameters that control the rheological profile of nature are: the petrology of rock, the heat flux to the surface, moho depth and speed of deformation.

3 Results

Several experiments were carried out by changing two parameters. First, experiments have been done assuming the presence or absence of a ductile layer located at the basement cover interface and simulating the role of Triassic evaporates. Second, we conducted an experiment at a lateral free boundary representing the influence of the Tethyan oceanic subduction below the Mediterranean ridge. Each experiment is subjected to a constant rate of shortening resulting in the formation of a sand wedge.

Our results show that the presence of a level of detachment at the base of the sedimentary series leads to simulate compressive structure in the experiment at a similar scale than in the nature together. During the second half of the experiment, horst and graben start to open in the higher part of the wedge as observed in the Tell and Northern Atlas zones. Similarly, a free edge perpendicular to the direction of compression allows the formation of strike-slip and conjugate structures on the lowest part of the wedge. It is underlined in the surface by “en echelon” anticlines with variable hinge orientation. This pattern is very similar to the fault of Negrine-Tozeur and the N-S axis.

Thus, the best fit between model and nature has been obtained by taking into account both a ductile layer at the basement cover interface and a lateral free boundary. We performed direct observations bellow the ductile layer at the end of the experiments. Here we observed that:

- Basement slices bounded by thrusts may contribute to the uplift of the surface on the highest part of the wedge.
- Strike-slip and conjugate structures located on the lowest part of the wedge root into the basement.
- Folds located in the lower part of the wedge only affect the upper sand layer located suggesting a thin skin tectonic.
- Horst and graben located in the higher part of the wedge root into the ductile layer as already proposed by Belguith et al. [8].

4 Conclusions

Analog modelling carried into this study lead to the reproduction, in laboratory, of the complex Tunisian structural scheme that is characterized by the simultaneous association of compressive, strike-slip and extensive structures.

The results show that the presence of a level of detachment at the base of the sedimentary series is necessary to correctly represent the compressive and extensive structures in the central Atlas and north of Tunisia respectively. Similarly, a free edge perpendicular to the direction of compression allows the formation of strike-slip and conjugate structures similar to the fault of Negrine-Tozeur and the N-S axis. These experiments suggest that the main strike-slip structures and thrusts located north of Tunisia are rooted in the crust whereas normal faults to the north and reverse structures in Tunisia only affect the sedimentary cover.

References

1. European Geotraverse: EGT 1985 (1990)
2. Tunisian Company of Petroleum Activities: ETAP, 1990
3. Wildi (1984)
4. Rigane, A., and Gourmelen, C. (2011)
5. Belguith et al. (2011)
6. Booth-Rea et al. (accepted)
7. Davy and Cobbold (1988, 1991)
8. Belguith et al. (2013)

Compressional Tectonics Since Late Maastrichtian to Quaternary in Tunisian Atlas

Hedi Zouari, Achraf Zouari, and Fehmy Belghouthi

Abstract

Some authors consider that compressional tectonic to have begun in late Cretaceous and continued until Quaternary in the Atlas Mountains of Tunisia. Others propose relaxation periods characterized by distensional tectonic regime. In this study, we made use of detailed geological mapping, fault-slip field data and seismic data from the southern and northern regions of Tunisia, in order to highlight the variation of paleostress during tectonic inversion. Three directions of principal compressive stress σ 1 are deduced from fault-slip data: NW-SE, NE-SW and N-S. The compressive tectonics induce different tectonic regimes: strike slip, compressional strike slip, extensional strike-slip and compressive during Late Maastrichtian-Quaternary period in the Tunisian Atlas. Fold-axes present different directions: E-W, N-S, NW-SE and NE-SW. In some localities, major faults juxtapose folds which makes their axes are perpendicular. This structural complexity and the change of stress axis can mainly be related to two factors: (1) variation of geometry and kinematics of African plate boundary which can generate several shortening directions (2) heterogeneity of sedimentary cover due to detachment layers and to pre-existent structures as fault and diapirs which contributes to reorient the stress direction.

Keywords

Atlas • Tunisia • Compressive regime • Late Maastrichtian to quaternary • Paleostress

1 Introduction

Tunisian Atlas Mountains are located in the eastern part of northern Africa belt which trends roughly in an E-W direction. Tectonic deformation since late Maastrichtian to Quaternary interval is apprehended from a different point of view and remains a subject of much debate. Some authors [7] considered that the Mio-Pliocene was a period of relaxation, other authors [1, 5, 6] argued compressive tectonics during the same period. This study discusses and argues compressive tectonics since late Maastrichtian characterized by three directions of the principal compressive stress and several tectonic regimes.

2 Materials and Methods

Detailed geological mapping in several regions in the Tunisian Atlas has been performed using satellite data and field control. Seismic data was converted in depth and analyzed by Midland valley Move software. Fault-slip data (fault planes with slip line and slip sense) has been collected from several natural outcrops and analyzed for reconstructing paleostress tensor with the computer program Win Tensor [4] during Late Maastrichtian to Quaternary period (Fig. 1 and Table 1). The data was taken from different age layers. Tension gashes and stylolithes are also used to further support the determination of paleostress.

H. Zouari (✉) · A. Zouari · F. Belghouthi
Water Researches and Technologies Centre, Technopark of Borj Cédria, Natural Water Treatment Laboratory, PO Box 273 – touristic zone of Borj Cédria, 8020 Soliman, Tunisia
e-mail: hedi.zouari@certe.rnrt.tn; hedizouari3@gmail.com

A. Zouari
Faculty of Mathematical, Physical and Natural Sciences of Tunis, University of El Manar, Campus Universitaire El-Manar, 2092 El Manar, Tunis, Tunisia

F. Belghouthi
Faculty of Sciences of Bizerte, University of Carthage, 7021 Jarzouna, Tunisia

F. Rossetti et al. (eds.), *The Structural Geology Contribution to the Africa-Eurasia Geology: Basement and Reservoir Structure, Ore Mineralisation and Tectonic Modelling*,
Advances in Science, Technology & Innovation, https://doi.org/10.1007/978-3-030-01455-1_50

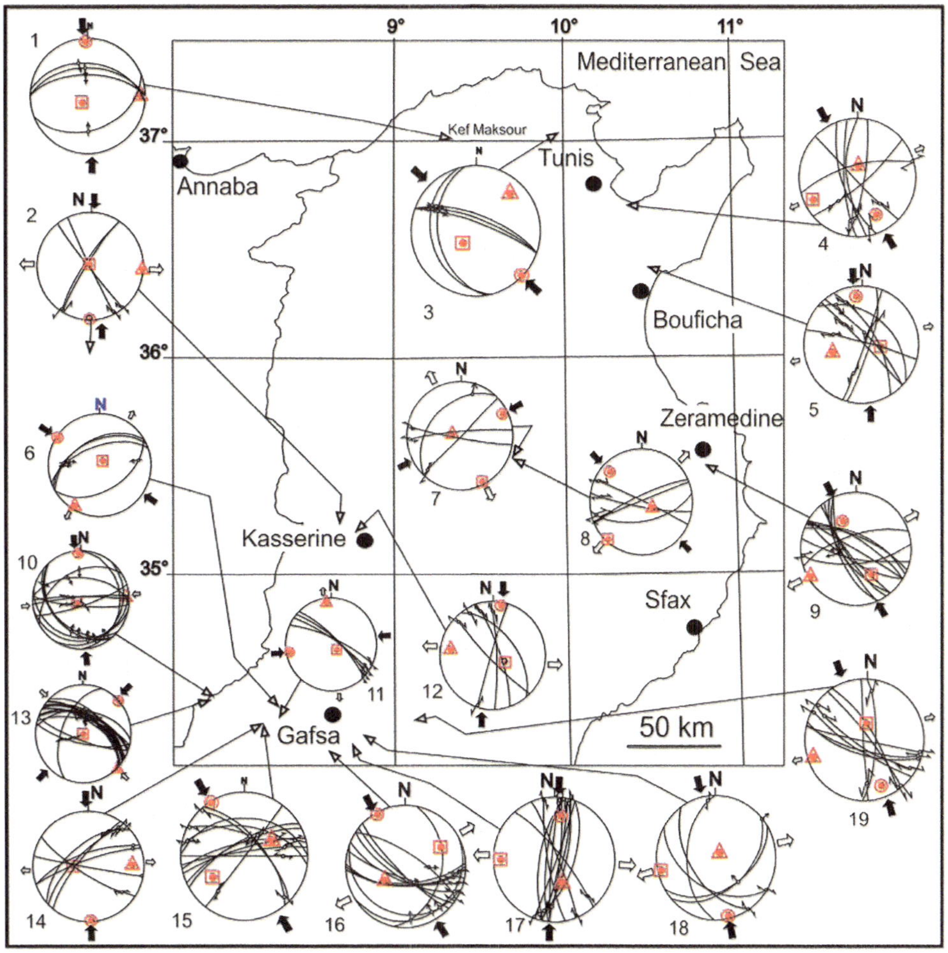

Fig. 1 Paleostress reconstruction for several localities in Tunisian Atlas

3 Results

Three compressive stress σ 1 axes are determined: NW-SE, N-S and NE-SW. Stress tensors reveal compressive and strike-slip tectonic regimes. In some area, folds coexist with strike-slip faults and normal faults indicating extensional strike slip regime. In other localities, folds are associated to reverse faults and strike-slip faults indicating compressional strike-slip regime. Growth synclines related to reverse faults took place during Maastrichtian-Quaternary period.

4 Discussion

Some authors consider coexisting compression and distension-related structures due to NW-SE compressive tectonics to have occurred for the whole compressive period [2, 3]. From this point of view, N120 faults are admitted as dextral slip during all compressive period. Withal, we demonstrated in this work that these faults are reactivated as reverse and sinistral strike-slip faults. Indeed, this study argues that N-S faults played as sinistral and dextral-slip faults. The evolution of compressional structure began in Late Maastrichtian and continued until Quaternary as the consequence of the convergence between Africa and Eurasia plates. The so-called grabens or troughs (of Foussana, Rohia, Grombalia, etc.) [2, 3] are in fact extensional structures of Cretaceous age. These structures have been folded during tectonic inversion as consequence of NE-SW shortening. In these structures, NW-SE Cretaceous faults are also reactivated with normal component due to NW-SE shortening that induce extensional strike-slip regime.

5 Conclusions

This study allowed the discussion and the demonstration of the following facts:

(1) Confirmation of continuity of shortening since late Maastrichtian to Quaternary period related to the convergence between Africa and Eurasia plate.
(2) Demonstrate that NW-SE extensional structures were folded during tectonic inversion due to N-S and NE-SW

Table 1 Parameters corresponding to the stress tensors drew in Fig. 1

Stereonet number	Locality	Layer	Azimuth of σ 1	Tectonic regime
1	Kef Maksour	Early Maastrichtian limestone	358	Pure compressive
2	Oum El Alia	Oligocene limestone	003	Strike-slip
3	Feja	Pliocene limestone	133	Pure compressive
4	Jbel Rsas	Eocene limestone	150	Strike-slip
5	Jbel Mounchar	Miocene sandstone	350	Strike-slip
6	Kef Ed Dour	Eocene marl and limestone	303	Compressional strike-slip
7	El Haouareb 1	Coniacian-Santonian limestone	067	Extensional strike-slip
8	El Haouareb 2	Coniacian-Santonian limestone	319	Pure strike-slip
9	Zeramedine	Miocene clay and sandstone	330	Strike-slip
10	Tamerza 1	Eocene limestone	356	Compressive
11	Oued Thelja	Early Maastrichtian	260	Compressional strike-slip
12	Kef Moungar	Santonian-Campanian	009	Strike-slip
13	Tamerza 2	Early Maastrichtian	046	Compressive
14	Ain Mtalga 1	Pliocene silts and conglomerate	177	Strike-slip
15	Ain Mtalga 2	Middle Quaternary conglomerate	328	Pure compressive
16	Jbel Sehib 1	Early Maastrichtian	330	Strike-slip
17	Jbel Sehib 2	Eocene limestone	007	Strike-slip
18	Jbel Sehib 3	Pliocene silts and conglomerate	168	Strike-slip
19	Jbel Ank	Pliocene silts and conglomerate	157	Strike-slip

σ 1 principal compressive stress direction

shortening. In the boundary of these structures, NW-SE faults are reactivated with normal component as a result of NW-SE shortening. This kinematic is related to extensional strike-slip regime, so it does not demonstrate the relaxation period proposed by some authors.

(3) The Different orientation of fold axis and the change of the principal compressive stress σ 1 direction can be related to the geometry and kinematic variation of Africa plate boundary and to the heterogeneity of sedimentary cover (e.g. Detachment layers, faults, diapirs).

References

1. Alyahyaoui, S., Zouari, H.: Synsedimentary folding process and transtensive tectonic during Late Miocene to Quaternary in northeastern Tunisia: case of Mateur-Menzel Bourguiba region. Arab. J. Geosci. **7**, 4957–4973 (2014). https://doi.org/10.1007/s12517-013-1111-2
2. Ben Ayed, N.: Evolution tectonique de l'avant pays de la chaîne alpine de Tunisie du début du Mesozoïque à l'actuel (Thèse de doctorat d'Etat). Paris Sud, Paris (1986)
3. Chihi, L.: Les fossés néogènes et quaternaires de la Tunisie et de la mer pélagienne, étude structurale et leur signification dans le cadre de la géodynamique de la Méditerranée (Thèse de doctorat d'Etat). Tunis el Manar II, Tunis (1995)
4. Delvaux, D., Sperner, B.: New aspects of tectonic stress inversion with reference to the TENSOR program. Geol. Soc. Lond. Spec. Publ. **212**, 75–100 (2003). https://doi.org/10.1144/GSL.SP.2003.212.01.06
5. Frizon de Lamotte, D., Saint Bezar, B., Bracène, R., Mercier, E.: The two main steps of the Atlas building and geodynamics of the western Mediterranean. Tectonics **19**, 740–761 (2000). https://doi.org/10.1029/2000TC900003
6. Khomsi, S., Ben Jemia, M.G., de Lamotte, D.F., Maherssi, C., Echihi, O., Mezni, R.: An overview of the Late Cretaceous-Eocene positive inversions and Oligo-Miocene subsidence events in the foreland of the Tunisian Atlas: structural style and implications for the tectonic agenda of the Maghrebian Atlas system. Tectonophysics **475**, 38–58 (2009). https://doi.org/10.1016/j.tecto.2009.02.027
7. Melki, F., Zouaghi, T., Chelbi, M.B., Bédir, M., Zargouni, F.: Tectono-sedimentary events and geodynamic evolution of the Mesozoic and Cenozoic basins of the Alpine Margin, Gulf of Tunis, north-eastern Tunisia offshore. Comptes Rendus Geosci. **342**, 741–753 (2010). https://doi.org/10.1016/j.crte.2010.04.005

The External Domain of Maghrebides Belt on the North–East of Algeria, Souk Ahras Segment: Definition of Structural Units, Block Structure and Timing of Thrusts Setting

Abdallah Chabbi, Asma Chermiti, Abdelmadjid Chouabbi, and Mohamed Benyoussef

Abstract

The detailed geological mapping supported by micropaleontological, tectonic and structural data of Souk Ahras region (North–East of Algeria) made it possible to define the main structural units of the external domain of the Maghrebides belt in the North–East of Algeria. The study area shows a block structure illustrating an allochthonous domain (Numidian unit, the northern Tellian unit and the southern Tellian) thrusting a folded para-autochthonous domain (the scaly unit of the Sellaoua type and the Dj. Graout unit). The current structure is edified into three phases at least: post Langhian phase, Tortonian phase and Pliocene phase.

Keywords

External domain • Allochthonous domain
Tellian nappes • Sellaoua unit • Souk Ahras
Algeria

1 Introduction

In the last century, the Maghrebides chain has commonly been divided into three distinctive domains, from north to south: the internal domain, the Flyschs domain and the external domain [1]. These domains have collided following the convergence of the African and European plates. This collision led to the scaling of the internal domain, the closing of the Flyschs and Tellian basin, and the thrusting of the series far to the south above the formations of the northern African plate margin [2].

The Souk Ahras zone is part of the external domain of the Maghrebides belt [3]. It offers an excellent structure built in the Neogene. Geological mapping, micropaleontological, tectonic and structural data allowed us to characterize the structural units and to suggest a scenario of units in the region's setting.

2 Materials and Methods

The study area is situated in the northeast of Algeria (Souk Ahras) which is spread into two domains:

A. Allochthonous domain: Dj Boubakhouch-M'Cid (BB-MC) Numidian nappe, Ouled Driss-Dj Boukebch (OD-BK) Tellian nappes),
B. Para-autochthonous domain: Dj Boubakhouch–Dj Bouallegue (BB-BL) of Sellaoua unit and Dj Graout–Merahna (GT-MR) unit.

This study is based on geological mapping supported by geological sections, using micropaleontological, tectonic and structure data. Over one hundred samples of marl, taken from different sections, have been dated. The sections are spread over four sectors of the region, from south to north: Graout–Merahna (GT-MR) sector (A-A' section), Boukebch (BK) sector (B-B section), Dj. Boubakhouch (BB) sector (C-C' section) and Ouled Driss (OD) sector (D-D' section). In this study, we will focus only on samples collected from different discontinuities contacts.

3 Results

3.1. Oued Medjerda–Lakhdara Section (A-A' Section) was investigated in the para- autochthonous of (GT-MR) sector (Fig. 1). It shows an NNE-SSW anticline structure

A. Chabbi (✉)
Mohamed Cherif Messaadia University, Souk Ahras, Algeria
e-mail: chabbiabdellah@yahoo.fr; a.chabbi@univ-soukahras.dz

A. Chabbi · A. Chouabbi
Geodynamics and Natural Resources Laboratory, Badji Mokhtar university, Sidi Amar, Annaba, Algeria

A. Chermiti · M. Benyoussef
Water Researches and Technologies Center, Ecopark of Borj Cedria, Soliman, Tunisia

F. Rossetti et al. (eds.), *The Structural Geology Contribution to the Africa-Eurasia Geology: Basement and Reservoir Structure, Ore Mineralisation and Tectonic Modelling*,
Advances in Science, Technology & Innovation, https://doi.org/10.1007/978-3-030-01455-1_51

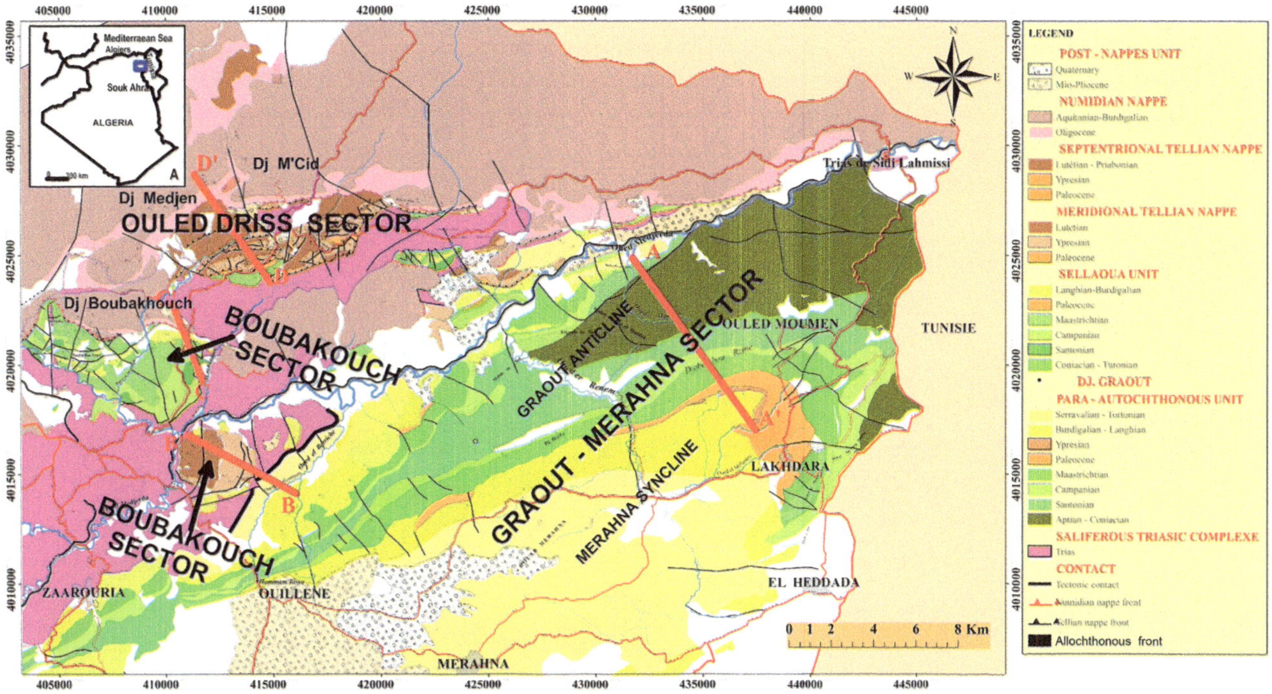

Fig. 1 Geological simplified map of the studied area and location of cross-sections

centered at GT succeeded by MR syncline. From North to South we can distinguish (Fig. 2a):

(a) Aptian–Coniacian: black marls, limestone and grey sandstone, (b) Santonian: grey marl and marl-limestone, (c) Campanian: a 160 m thick alternation of marl and limestone—marl with white limestone bar with a 100 m thick layer above, (d) Maastrichtian: 80 m of thick alternating grey marl with chalky limestone, and a bar of chalky limestones 100 m thick above. This Cretaceous series is overlaid by an alternating limestone and grey marl (120 m thick). Samples (S1–S3) which are collected from this level yielded *Globotruncanita falsostuarti, Rosita contusa, Gansserina gansseri* indicate the upper Maastrichtian.

However, Paleogene deposits are represented by (e) 260 m of thick schistose black marls rich in organic matter. The samples (S4–S7) which are collected from this level yielded *Morozovella trinidanensis, M. pseudobulloides and Globigerina triloculinoides,* (f) Ypresian–Lower Lutetian deposits are represented by a Nummulitic limestone bar 45 m in thickness. (g) Miocene is characterized by Glauconitic marls and sandstones, which overlay the previous term through an erosional surface marked by bioturbation. Samples S8–S10 and geological map [4] indicate Burdigalian–Tortonian age.

3. 2. Dj. Boukebch Section (B-B' Section): from SE to NO shows (Fig. 2b):

A. **Miocene** cover of **(GT-MR) unit**, thick series of brown marl and glauconitic sandstone. Samples (S11–S12) which are collected from the upper level are Serravallian age.
B. **Triassic salt complex**; clay-gypsum, dolomite, cargneule and anhydrite formations.
C. **The southern Tellian unit of BK**: to the south, this unit overlaps the Miocene (Serravallian) of O. El Berriche cover of the para-autochthonous domain. From bottom to top this unit is formed by: (a) **Paleocene**: approximately 5 m thick layer of black marl, (S13) is rich in benthic foraminifera with *Tritaxia midwayensis, Ammodiscus glabrata, Trochammina abrupta,* (b) **Ypresian**: 140 m thick, Nummulitic and lumachellic limestones. This level is overlaid by chalky limestones with *irregularis Nummulites, N. subirregularis, N. globulus, N. atacicus, N. subatacicus, N. gizehensis* [4] (c) **Lutetian deposits**; 200 m thick of alternating marls, brown clays and limestones. The marls yielded *Turritella carinifera,* and *Morozovella inaequispera, Globigerinatheka subconglobata* and Ostracodes (*Loculocyteretta*) (S14–S16). The series end with glauconitic sandstone.

3. 3. Dj. Bouallegue Section (C-C' Section) from the NW to the SE displayed in (Fig. 2c) shows:

(A) **The Sellaoua type unit of (BL)**: overlays the Triassic complex through a tectonic contact. This unit is made of: Santonian–Maastrichtian marly, marly-limestone

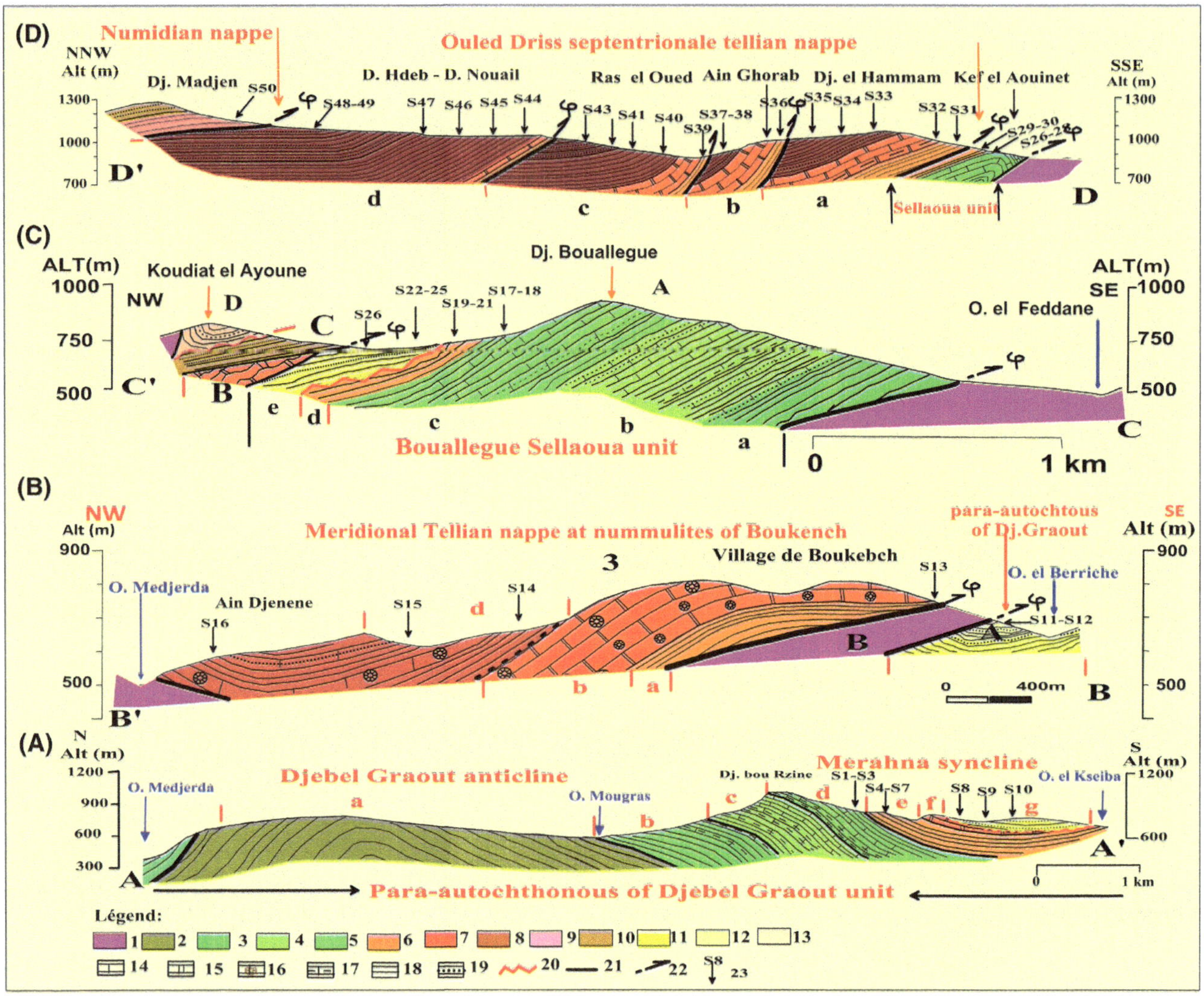

Fig. 2 Interpretative cross-sections: (1) Trias, (2) Aptian–Coniacian, (3) Santonian, (4) Campanian, (5) Maastrichtian, (6) Paleocene, (7) Ypresian, (8) Lutetian-Priabonian, (9) Oligocene, (10) Aquitanian, (11) Burdigalian–Langhian, (12) Serravalian–Tortonian, (13) Plio-Quaternary, (14) chalky limestone, (15) black limestone at silex, (16) Nummulites Limestone, (17) Marly–limestone, (18) Mudstone, (29) Sandstone, (20) Erosional surface, (21) Tectonic contact, (22) Thrust front, (23) Sample and its number

and chalky limestone 420 m thick (a, b, c). Samples (S17–18) collected from the uppermost part contains *Globotruncana stuartiformis, G. subspinosa, G. gansseri, G. Falsostuarti, G. havanensis*. (d) Paleocene, dark marl (nearly 20 m thick), samples (S19–S21) yielded Globigerina *triloculinoides, Planorotalites*. (e) Miocene: Glauconitic marls and sandstone with 70 m thick. Samples (S22–S26) collected from the uppermost part of these series provide *Praeorbulina glomerosa, P. transistoria, Globigerinoides trilobus, G. primorduis, G. altiapartura, G. bisphericus,* dating lower Langhian. The Miocene overlays Paleocene and Maastrichtian series, through an erosional surface.

(B) **BL Tellian unit**: shows black marl and dark limestones globigerina of Paleocene–Ypresian age.

(C) **BB Numidian unit** (Koudiet Ras El Ayoune): it is composed of marls and sandstone with 1000 m thick thrusting Tellian and Sellaoua units. It starts with a green clay 140-m-thick surmounted by sandstone layers separated by 100–150 m thick mudstone. We can attribute the Oligocene to lower Miocene this series [5–7].

(D) **Koudiet Ras el Ayoune post-nappe unit**: represented by polygenic conglomerates, sands, and red clays, of Mio-Pliocene. It is a continental molasse, folded to NW-SE syncline. This series is gulled by a Triassic formation overlapping the previous unit through sub-vertical tectonic contact.

3.4. Ouled Driss Section (D-D Section): from the SSE-NNW, it is represented by (Fig. 2d):

A. In the South, the Sellaoua type unit (OD scale) which is tectonically in contact on the Trias. The series starts with alternation of grey limestone and grey marl. Sample (S26–S28) collected from marls, yielded *Globotrucanita aegyptiaca, Gansserina gansseri and Rosita contusa* dating upper Maastrichtian. The series overlaid by 80-m-thick of black marls (S29-S30) including *Globigerina triloculinoides* which indicates lower Paleocene.
B. Farther north, the OD northern Tellian unit thrusts Sellaoua type unit. This unit contains four (04) thrusts, separated by SSW-NNE reverse faults; Dj. el Hammam thrust (a), Ain Ghorab thrust (b), Chaabet Ras el Oued thrust (c) and Douar Hdeb-Douar Nouail thrust (d). This unit consists of globigerina and flint carbonate material featuring a circalittoral deposit. Every thrust reveals upper Paleocene (P5) to Lutetian–Priabonian age [8].
C. **Dj. M'Cid Numidian unit:** is represented by 140-m-thick Oligocene green clay "sub-Numidian clays", surmounted by 900 m-thick, Oligocene–Burdigalien sandstone series, admitting clay levels.

3.1 Structure

The interpretative section (Fig. 3) shows a complex structure characterized by the stacking of thrusts allochthonous units on a folded para-autochthonous domain, organized in three blocks separated by major tectonic contacts involving the ductile formations of the Triassic. From south to north:

(a) **GT-MR block "Block A"**: represents a folded structure with the succession of the Merahna syncline to the south and the Dj Graout anticline to the north. This block characterizes platform sedimentation of Cretaceous age and shallow Paleocene–Ypresian sedimentation. These series are overlaid by unconformable marl, sandstone and glauconitic sandstone clays, of Burdigalian–Tortonian age, with an erosional surface at the base. This block represents a para-autochthonous structure.

(b) **BB–BL–BK block "Block B"**: the series of this block overlap the Miocene (Serravallian) of **"Block A"** through a major tectonic contact involving Triassic material. From bottom to top this block is formed by: BB–BL Sellaoua type unit, which presents a scaled structure implying Miocene series and overlapping by Northern Tellian unit with globigerina of Koudiet Ras el Ayoune from the North. After these units comes the BB Numidian unit that overlaps Tellian and Sellaoua units with a sub-horizontal tectonic contact. The block ends with Koudiet Ras el Ayoune post-nappe unit (Pliocene) which surmounts the Numidian nappe by a sedimentary contact.

(c) **MC–OD block "Block C"**: overlaps **"Block B"** by a major tectonic contact leading to the plastic formations of the Triassic which lead the folding of the Pliocene formations of the "block B". It is made of Sellaoua unit of OD which is thrusting by OD northern Tellian unit. The OD Tellian unit is characterized by a duplex structure of four thrusts (Upper Paleocene to late Priabonian). The delamination occurred in the Upper Paleocene marl. This block ends by MC Numidian unit.

4 Discussion and Conclusion

Calendar and setting of structural units: The superposition of units illustrated in (Fig. 3) allows us to distinguish 3 stages of emplacement of the structural units during Neogene.

(a) 1st **Phase** (The setting of Tellian and Numidia units): the geological map, sections (Figs. 1 and 2) and the interpretative section (Fig. 3) show that Tellian and Numidian units overlap the Burdigalien–Langhien series which covers the BB-BL Sellaoua units. On the one hand, the Numidian unit supports Pliocene deposits at of Koudiet Ras el Ayoune (D-D' Fig. 2). These arguments are in favor of laying nappes in Sellaoua basin, later to Langhian inferior and prior to Pliocene.

b) 2nd **Phase** the "C and D blocks" overlap the Burdigalian–Tortonian series cover of GT-MR block at O. el Berriche. The structure is in favor a compressional deformation which induced the setting "C and D blocks" in common motion. This displacement caused the scaling of Sellaoua unit (Dj. BB–BL unit) in the Langhian and moving the entire block to the south on

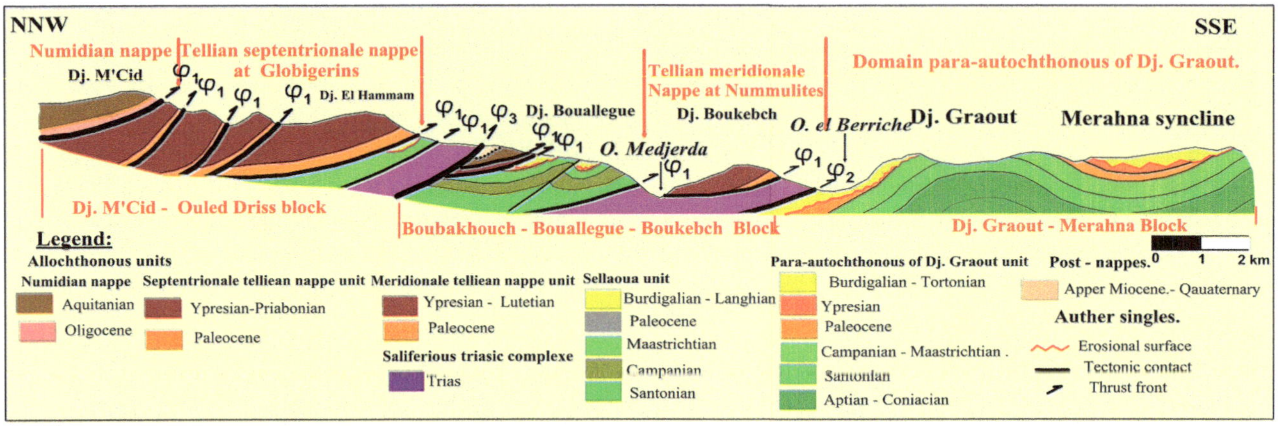

Fig. 3 Simplified interpretative cross-section of Souk Ahras structure

the para-autochthonous domain of GT-MR in the Upper Miocene. This level of delamination is major, leading the Triassic formations that should have been out-cropped by the diapirism phenomenon before the late Miocene period.

(c) **3rd Phase** (The setting of MC–OD block): the present structure shows that the united block (MC–OD and BB–BL–BK) flaking into two blocks separated by a plastic level consisting of Triassic formations (Fig. 3). The newly separated block (MC–OD) overlaps BB–BL–BK block to the south, inducing the folding of the Pliocene of Koudiet Ras el Ayoune unit. These facts are in favor of post-Pliocene displacement and illustrate a post- Pliocene compressional deformation.

This thrusting calendar is in line with that proposed to the west of investigated area by [6] in the region of Sidi Afif, Sedrata (near 40 km NW of study area) and that of [9] in the region of Hodna Mounts, central Algeria (near 500 km West of study area).

References

1. Bouillin, J.-P.: Le bassin maghrébin: une ancienne limite entre l'Europe et l'Afrique à l'Ouest des Alpes. Bull. Soc. geol. France **8** (4), 547–558 (1986)
2. Roure, F., Casero, P., Addoum, B.: Alpine inversion of the North African margin and delamination of its continental lithosphere. Tectonics **31**, TC3006 (2012). https://doi.org/10.1029/2011tc002989
3. Vila, J.M.: La chaine alpine d'Algérie orientale et des confins Algéro-Tunisien. Phd thesis, Univ. Pierre et Marie Curie (Paris VI), France, 665p (1980)
4. David, L.: Etude géologique de la haute Medjerda. Bulletin du Service de la carte géologique de l'Algérie, n° 11, Algérie, 304p (1956)
5. Rouvier, H.: Géologie de l'extrême Nord tunisien : tectoniques et paléogéographies superposées à l'extrémité orientale de la chaîne nord-maghrébine. Phd thesis, P et M. C. university, Paris, France, 898p (1977)
6. Vila, J.M., Feinberg, H., Lahondère, J.C., Gourinard, Y., Chouabbi, A., Magné, J., Durand-Delga, M.: The sandy uppermost Oligocene channel and the Miocene of Sidi- Affif area in their East Algerian structural setting -the Saharan Origin of the Numidian and the calendar of the Miocene overthrusts. Comptes Rendus De l'Academie Des Sciences Serie 2 **320**(10), 1001–1009 (1995)
7. Riahi, S., Soussi, M., Boukhalfa, K., Ben Ismail Lattrache, K., Dorrik, S., Khomsi, S., Bedir, M.: Stratigraphy, sedimentology and structure of the Numidian Flysch thrust belt in northern Tunisia. J. Afr. Earth Sci. **57**, 109–126 (2010)
8. Chabbi, A., Chouabbi, A., Chermiti, A., Ben Youssef, M., Kouadria, T., Ghanmi, M.: La mise en évidence d'une nappe de charriage en structure imbriquée: Cas de la nappe tellienne d'Ouled Driss, Souk-Ahras, Algérie. Courrier du Savoir, Algeria. N°21, pp. 149–156 (2016)
9. Bracene, R., Frizon de Lamotte, D.: The origin of intraplate deformation in the Atlas system of western and central Algeria: from Jurassic rifting to Cenozoic-Quaternary inversion. Tectonophysics **357**, 207–226 (2002)

Sedimentology and Petrographic Analyses of the Upper Cretaceous and Lower Eocene Carbonate Deposits in the Mateur-Beja Area (NW of Tunisia): The Repercussions of Tectonic Style on Sedimentation

Fehmy Belghouthi, Achraf Zouari, and Hedi Zouari

Abstract

The northwestern Tunisian Mateur-Beja area comprises a thick succession of carbonate rocks belonging to the Late Cretaceous (Abiod Formation) and Early Eocene (BouDabbous Formation and its lateral equivalent) times. Sedimentological and petrographic analyses carried out on these Formations with regard to their palaeogeographic reconstructions. Three sections from the upper Cretaceous deposits and twenty sections from the lower Eocene successions were selected and correlated for this study. The upper Cretaceous Abiod Formation is dominated by foraminiferal biomicrite deposited in upper slope to basin environments, usually influenced by gravity-driven currents. The lower Eocene carbonate successions are characterized by various and diverse facies associations deposited in inner platform to basinal environments. Both carbonate Formations are characterized by lateral variations in thickness and facies, although they were deposited under varying environmental circumstances. Close inspections show that changes in thickness and facies were mainly controlled by sea-floor topography and deposition rates. The conspicuous change in sea-floor topography is well expressed during drastic changes in tectonic style associated with active subsidence, while deposition rates significantly decrease in areas marked by active marine erosional currents. The upper Cretaceous and the lower Eocene carbonate deposits provide useful data about the relationships between tectonic and sedimentation.

Keywords

Eocene • Cretaceous • Carbonates • Sedimentology • NW Tunisia

1 Introduction

The upper Cretaceous and lower Eocene carbonate deposits in the north of Tunisia have been the focus of numerous studies dealing with their palaeogeographic implications [1, 3, 6, 7]. As a result of these studies, numerous descriptions and models were published, some of which focused on large areas while others concern a specific site. From a sedimentological point of view, the lack of unified terminology and classifications of these carbonate successions leads to numerous controversies about their depositional environments. For instances, the upper Cretaceous carbonate deposits were considered as monotonous over a large area [3]. However, recent sedimentological studies in the central and northern Tunisia [2, 7], showed frequent variations in facies and thickness even at small-scale areas, and that changes were particularly linked to irregular topography of the sea-floor [7]. In the same way, the subdivision of the early Eocene carbonates series into two lateral equivalent facies, "calcaires à nummulites" and "calcaires à globigérines" [5] does not faithfully reflect their depositional environment pattern. From our detailed observations in the Mateur-Beja area, there is many evidences showing that nummulite-rich deposits were affected by marine current activities, and therefore, they are frequently transported

F. Belghouthi · A. Zouari · H. Zouari (✉)
Water Researches and Technologies Centre, Technopark of Borj Cedria, Natural Water Treatment Laboratory, Touristic Zone of Borj Cedria, PO Box 273 8020 Soliman, Tunisia
e-mail: hedi.zouari@certe.rnrt.tn; hedizouari3@gmail.com

F. Belghouthi
Faculty of Sciences of Bizerte, University of Carthage, 7021 Jarzouna, Tunisia

A. Zouari
Faculty of Mathematical, Physical and Natural Sciences of Tunis, University of El Manar, Campus Universitaire El-Manar, 2092 El Manar Tunis, Tunisia

F. Rossetti et al. (eds.), *The Structural Geology Contribution to the Africa-Eurasia Geology: Basement and Reservoir Structure, Ore Mineralisation and Tectonic Modelling*,
Advances in Science, Technology & Innovation, https://doi.org/10.1007/978-3-030-01455-1_52

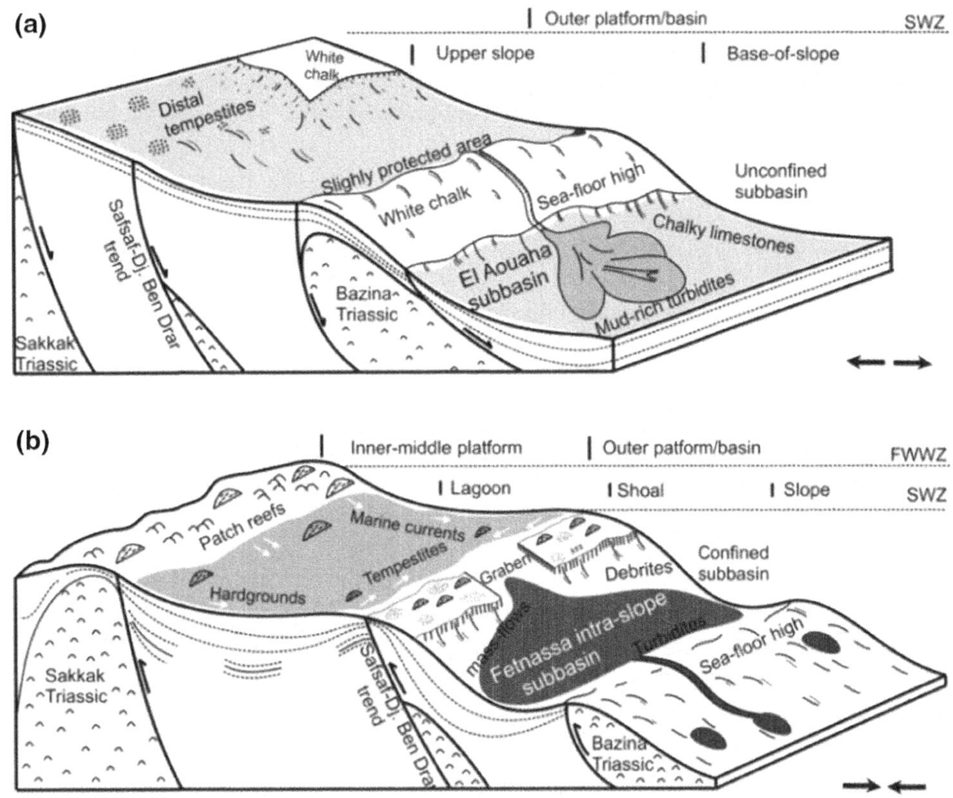

Fig. 1 Depositional environment interpretations of the Campanian-Maastrichtian (**a**) and lower Eocene carbonate successions (**b**) in the Mateur-Beja area. SWZ: storm-wave zone, FWWZ: Fair-weather wave zone

away from their original depositional sites and redeposited with globigerinid-rich deposits in many sites from the studied area. Such situations could pose many difficulties about the identification of facies using this classification. On the other hand, there are many additional facies (at least five) which can be observed in the lower Eocene succession that do not belong to this "bipartite" classification. Consequently, this begs many questions about their classification. In light of these discrepancies, the present study aims to (i) Update classifications of the upper Cretaceous and lower Eocene carbonate successions in the Mateur-Beja area (ii) Provide paleoenvironmental interpretations and palaeogeographic reconstructions of the studied area, and (iii) Highlighting the impact of tectonic styles on sedimentation.

2 Materials and Methods

Three sections from the upper Cretaceous outcrops and twenty sections from the lower Eocene succession were chosen, studied and correlated through the studied sector. The sedimentary succession was accurately studied bed-by-bed, described, sampled and sectioned for petrographic analysis. Microfacies classifications were defined and compared with standard microfacies types (SMF) [4].

3 Results

3.1 Facies Associations and Depositional Environments of the Upper Cretaceous Abiod Formation

The upper Campanian-lower Maastrichtian Abiod Formation consists of thick mixed pelagic and hemipelagic carbonate succession, with thicknesses ranging from 300 to 600 m, and are chiefly composed of planktonic foraminiferal biomicrite, associated with calcisphera, inoceramids, benthic foraminifera, brachiopods, in order of decreasing abundance. Depending on lithology, sedimentary characters and fossil contents, four main facies associations are recognized, deposited in three main sedimentary environments (Fig. 1a): (i) Partially protected outer platform environments, (ii) Base-of-slope environments and (iii) Basin.

3.2 Facies Associations and Depositional Environments of the Lower Eocene Carbonate Successions

The lower Eocene succession is composed of mixed deposits including, bioclastic limestones, glauconite and phosphate-rich marls, and organic-rich marlstones. Thicknesses of the carbonate succession range from 20 to 250 m-thick. Seven main facies associations are recognized, deposited in three main depositional environments (Fig. 1b): (i) Protected inner platform environments, (ii) mid-outer platform environments (iii) base-of-slope to basin environments.

4 Discussion

In the present work, we have provided detailed subdivisions and classifications of the Campanian-Maastrichtian and the lower Eocene carbonate deposits in the Mateur-Beja area. Unlike what was considered [3, 5, 6], both carbonate series are characterized by a wide spectrum of facies and significant lateral changes in thicknesses, and this is partly due to the irregular topography of the sea-floor. The impact of tectonic regime indirectly influenced on sediment types, compositions, and origins, especially when we compared both carbonate deposits. The lower Eocene carbonate series include more diverse facies belts derived from various local depositional environments taking place under an overall compressional tectonic regime. Many intra-slope confined “sub-basins” (Fig. 1b) such as the Fetnassa subbasins, appears and developed thanks to increased sedimentation rates and local subisidence. Sedimentation rates were enhanced by marine current activities (e.g. storm, turbidites) brought sediment from adjacent high-topographic regions of the inner and middle platform environments to the intra-slope depressions.

5 Conclusions

The main conclusions of the present study can be summarized as follows:

(1) An updated classification of the Campanian-Maastrichtian and the Lower Eocene carbonate successions in the Mateur-Beja area highlighting the presence of a wide spectrum of facies deposited in various sedimentary environments.
(2) Facies types, depositional environments and lateral variations of thicknesses and facies of both carbonate Formations give clues of unsteady tectonic styles. From a sedimentological point of view, we confirm the compressional tectonic regime during the Early Eocene times.

References

1. Belayouni, H., Guerrera, F., Martín Martín, M., Serrano, F.: Stratigraphic update of the Cenozoic Sub-Numidian formations of the Tunisian Tell (North Africa): Tectonic/sedimentary evolution and correlations along the Maghrebian chain. J. Afr. Earth Sc. **64**, 48–64 (2011)
2. Belghouthi, F.: Les facies des Formations Abiod et El Haria à Mateur: sédimentologie, environnements de dépôts et intérêt pétrolier. Mémoire de Master, Faculté des Sciences de Bizerte, 150 p (2014)
3. Burollet, P.F.: Contribution à l’étude stratigraphique de la Tunisie Centrale. Ann. Min. Géol., Tunis **18**, 350 p (1956)
4. Flügel, E.: Microfacies of Carbonate Rocks: Analysis, Interpretation and Applications, p. 983. Springer, Berlin (2004)
5. Fournié, D.: Nomenclature lithostratigraphique des séries du Crétacé supérieur au Tertiaire de Tunisie. Bull. Cent. Rech. Expl. Prod. Elf. Aquitaine **2**(1), 79–148 (1978)
6. Kujawski, H.: Contribution à l’étude géologique de la région des Hédil et du Béjaoua oriental. Ann. Min. Géol. Tunis **24**, 484 p (1969)
7. Negra, H.: Les dépôts de plate-forme à bassin du Crétacé supérieur en Tunisie centro-septentrionale (Formation Abiod et faciès associés). Stratigraphie, sédimentation, diagénèse et intérêt pétrolier. Thèse doctorat es-sciences, Université Tunis El Manar, Faculté des Sciences de Tunis, 649 p (1994)

Is the Mejerda Wedge-Top Basin of Northern Tunisia a Consequence of Basal Friction Change? a Theoretical and Experimental Approach

Mannoubi Khelil, Pauline Souloumiac, Dominique Frizon de Lamotte, Sami Khomsi, Bertrand Maillot, Seifeddine Gaidi, Mourad Bedir, and Fouad Zargouni

Abstract

In Northern Tunisia, the extensional Mejerda basin is developed over a major compressional boundary between the Tell fold-thrust belt to the north and the Atlas intra-continental system to the south. The abundance of Triassic salt structures in northern Atlas and under the Mejerda basin suggests a lateral change of the friction along the basal detachment. Such abrupt transition plays a significant role in the structural styles of the studied region and especially in the Mejerda basin. Therefore, we have developed prototypes with a basal frictional transition to study, through numerical and physical simulations, wedge deformation style. Accordingly, the one step results of all run numerical tests depict normal faults localized at this basal transition. In addition, the physical models in sandbox allow us to verify the mechanical results and to observe more deformation steps, in 2D and in 3D views. As a result, we observed via the sandbox experiments extensional features such as grabens. Finally, we highlight that compression can lead to build an extensional basin at the top of a thrust wedge as expression of basal friction change and we interpret the Mejerda basin accordingly.

Keywords

Basal friction transition • Numerical simulations
Analogue modelling • Sandbox experiments
Mejerda basin • Wedge-top basin

1 Introduction

The Cenozoic orogenic domain of the Maghreb is constituted by two different systems: the Rif-Tell collisional fold-thrust belt to the north and the Atlas intracontinental chain to the south [1]. These systems, which are not cylindrical are separated by areas of limited amounts of deformation (the Mesetas domains) in Morocco and western Algeria. In eastern Algeria and Tunisia, the Tell lies directly over the Atlas fold-thrust belt. This major structural boundary is underlined by the Mejerda basin (Fig. 1).

We present a set of numerical and analog models, which allow us to examine the parameters controlling the development of a sedimentary basin superimposed to a major compressional boundary. We will then discuss the structural conditions allowing the development of the Mejerda basin.

2 Modelling Data

2.1 Mechanical Simulation

For this study, we used the commercial software Optum G2 [2] that allows us to simulate through numerical tests, in terms of mechanics, the internal deformation at the onset of progressive deformation. All tests assume the same prototype where we changed only basal friction at the rear of the structure (Φd1) and at the front of the structure (Φd2) to evaluate the deformation as a function of the frictional properties on the main décollement. Φd1 representing the higher friction angle varies between 20° and 30° while Φd2

M. Khelil (✉) · S. Gaidi · F. Zargouni
Tunis El Manar University, 1068 Tunis, Tunisia
e-mail: khelilmannoubi@gmail.com

P. Souloumiac · D. F. de Lamotte · B. Maillot
Département Géosciences et Environnement, Université de Cergy-Pontoise, 95031 Cergy-Pontoise, France

S. Khomsi
Department of Petroleum Geology, Faculty of Earth Sciences, King Abdulaziz University, Jeddah, Saudi Arabia

M. Khelil · M. Bedir
Laboratoire Georessources, CERTE, Borj Cedria, 8020 Soliman, Tunisia

F. Rossetti et al. (eds.), *The Structural Geology Contribution to the Africa-Eurasia Geology: Basement and Reservoir Structure, Ore Mineralisation and Tectonic Modelling*,
Advances in Science, Technology & Innovation, https://doi.org/10.1007/978-3-030-01455-1_53

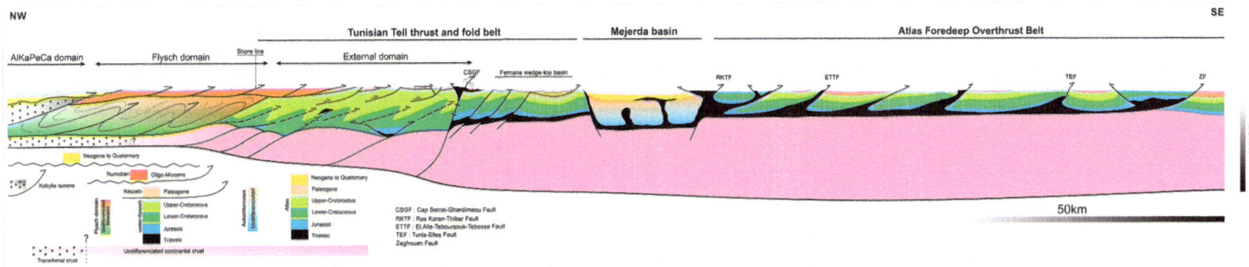

Fig. 1 Cross-section through Tell and Atlas belts of North Tunisia, Eastern Maghreb

varies between 5° and 15°. All calculated simulations show a failure localized near the basal friction transition and corresponding, mainly, to a normal fault. Mechanically, the material above the lower friction value is sliding faster than the one above the higher friction which creates a differential speed between the two blocks causing the collapse of the forwarding material. To illustrate this behavior, we selected the horizontal displacement (Ux). We identified three tectonic styles: curved normal fault, a listric low-angle fault (Fig. 2a), a straight normal fault (Fig. 2c) and an intermediate case (Fig. 2b). It depends on the speed difference, the more important it is, the more brutal the collapse is.

2.2 Sandbox Data

Several sandbox experiments are conducted using different materials on the décollement (glossy paper, sand paper, PVC, glass beads and silicon). Here, we present an experiment with glass paper (Φd1 = 43°) and Silicon (Φd2 = 5°) characterizing a high decrease of the basal friction. Theoretically, in case of high basal friction decrease we expect to observe a normal fault, as the numerical model shows [3].

The motor starts to push the system over the complex detachment. Few centimeters of shortening are enough to observe the collapse (S2 of Fig. 3a) situated near the friction

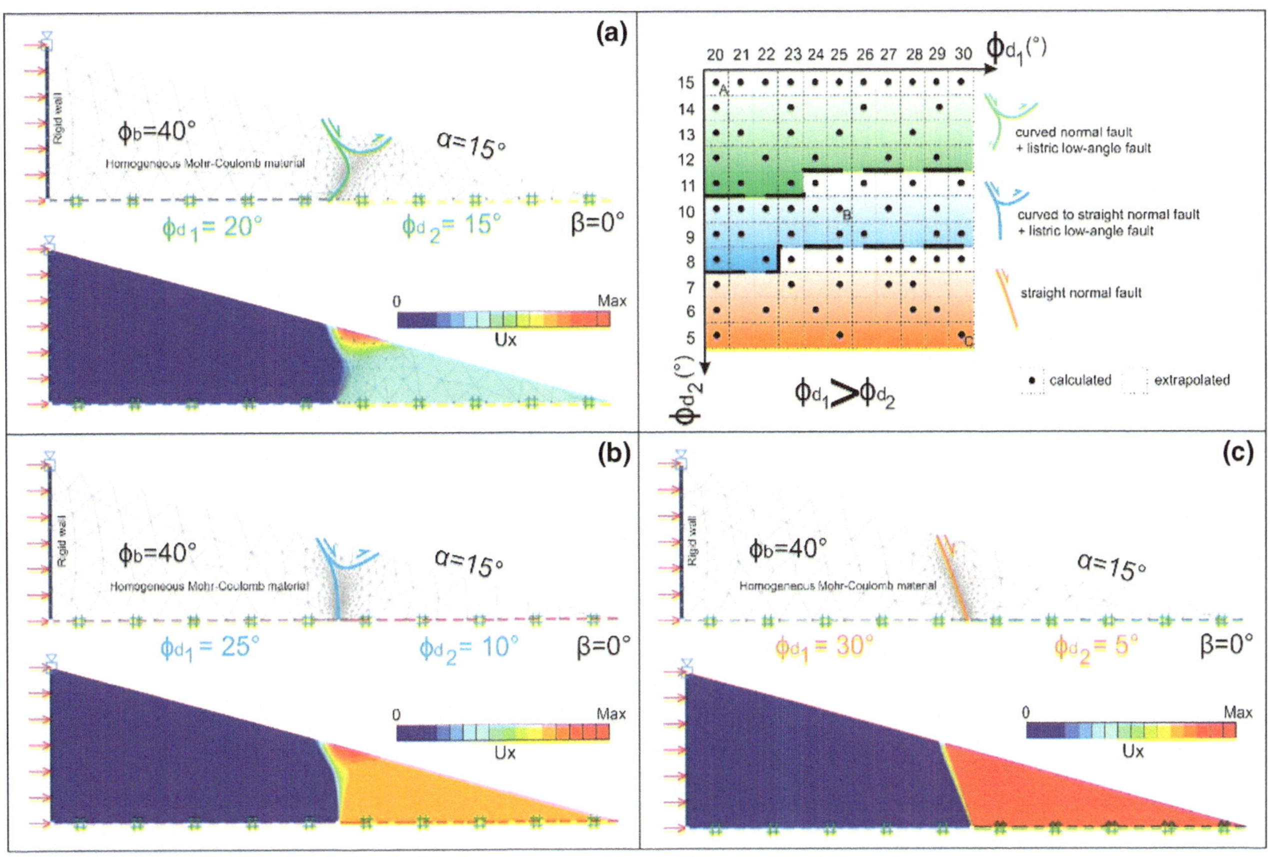

Fig. 2 Influence of basal décollement frictional proprieties change on faulting structural style. **a** Formation of a curved normal fault and a listric low-angle fault, **b** an intermediate case and **c** formation of a straight normal fault

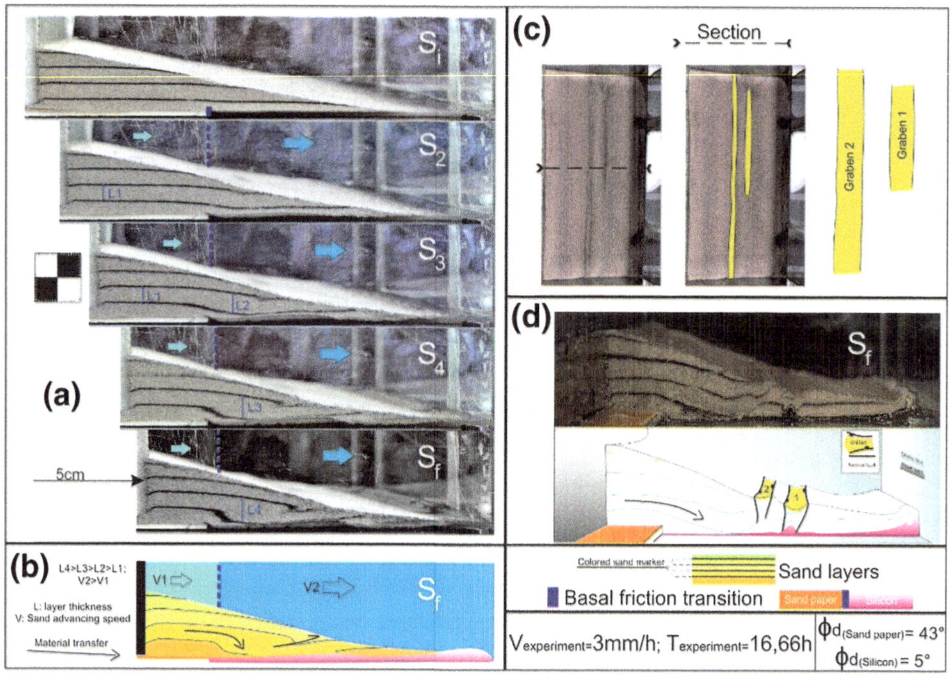

Fig. 3 Sandbox model: basal friction transition; deformation's different steps from side view (**a**), interpretation of the final step (**b**), interpretation of the final step from top view (**c**) and interpretation of the cross-section (**d**)

transition. By pushing the structure further, the same phenomena is accentuated (S2, S3 & Sf of Fig. 3a) as function of down-dip material transfer causes thickness variation (l4 > l3 > l2 > l1). The section (Fig. 3d), observed through the model, shows the youngest collapse near the transition, normal faults cutting clearly the sand markers and forming two grabens (Fig. 3c, d). Furthermore, the silicon intrudes vertically into sand layers under graben 1 (Fig. 3c, d) similar to diapiric movements in salt tectonics.

3 Discussion and Conclusions

As function of frictional transition, both approaches (numerical and physical), allow us to observe the development extensional features during compressional deformation. Besides, the Mejerda basin, separating the Tell and Atlas belts, is localized at the northern edge of the salt provinces. In other words, it is situated near the friction transition issued from the abundance of salt to the south of the advancing wedge.

So, we propose to interpret the Mejerda Basin as a wedge-top extensional basin developed in a compressional context as a consequence of a major change of the friction value along the basal décollement.

References

1. Wildi, W.: La chaîne tello-rifaine (Algérie, Maroc, Tunisie): structure, stratigraphie et évolution du Trias au Miocène. Rev. Geol. Dyn. Geog. Phys. Paris **24**(3), 201 (1983)
2. Krabbenhøft, K., Lyamin, A.V.: Optum G2. Optum Computational Engineering. www.optumce.com (2014)
3. Cubas, N., Avouac, J.P., Souloumiac, P., Leroy, Y.: Megathrust friction determined from mechanical analysis of the forearc in the Maule earthquake area. Earth Planet. Sci. Lett. **381**, 92–103 (2013)

Hinterland-Verging Thrusting in the Northern Sicily Continental Margin: Evidences for a Late Collisional Stage of the Sicilian Fold and Thrust Belt?

Attilio Sulli, Maurizio Gasparo Morticelli, Mauro Agate, and Elisabetta Zizzo

Abstract

The Sicilian Fold and Thrust Belt developed during Neogene-Quaternary times characterized by main African-wards tectonic transport direction. Recent investigations highlighted extensive hinterland-verging tectonic structures active during late Pliocene-Pleistocene time suggesting a late collisional stage of the Sicilian Fold and Thrust Belt that could be a precursor of a change in the subduction polarity in the central belt of Mediterranean.

Keywords

Backthrusting • Subduction polarity • Sicilian Fold and Thrust Belt • Central Mediterranean geodynamics

1 Introduction

Backthrusting, nappe refolding, and normal faulting frequently characterize late collisional stage of an orogen. Shortening driven by backthrusting is widely reported in the Alpine orogen, where it has been proposed to be responsible for the increase of subsidence [1]; moreover, delamination and backthrusting have been considered as related to subcritical condition of a Coulomb-type accretional wedge [2]. The southern Tyrrhenian is a key area where tectonic structures are accommodating the deformation in the transition belt between oceanic subduction and continental collision, both formed in the context of the convergence between Africa and Europe.

2 Geological Settings, Materials and Methods

We investigated the northern Sicily continental margin (the onshore as well as the offshore sectors), by using differently-penetrative seismic reflection data, including a deep crustal profile across central Sicily, calibrated with detailed stratigraphic and structural surveys and borehole data. Interpretation of the medium-to-high resolution seismic reflection profiles, combined with the deep seismic reflection data, led us to define both the deep and the shallower structural setting of the study area, revealing structural variations in the tectonic edifice architecture. Seismic reflection profiles available for the Northern Sicily and its offshore were used to unravel the deep structural setting of the submerged part of the Sicilian Maghrebian FTB. Sicily is a segment of the Apennine-Tyrrhenian System (Fig. 1) resulting from both the post-collisional convergence between Africa and Europe and the coeval roll-back of the subduction hinge of the Ionian lithosphere. During the last 15 My the building-up of the Sicilian Fold and Thrust Belt (SFTB) was characterized by a three-stage evolution: two main shortening events generated and developed at different structural levels (shallow- and deep-seated thrusts following a thin-skinned thrust-model) and at different time intervals, involving mainly the Meso-Cenozoic carbonate cover of the ancient African passive continental margin; these two events were followed by a more recent, thick-skinned thrust-model stage involving the crystalline basement in the internal sector of the chain and affecting the Plio-Pleistocene deposits in the frontal area as well. In northern Sicily (Cefalù seismic zone) the tectonic edifice is affected by extensional processes that were related to lateral extension on pre-existing faults and to an inferred mantle-upwelling beneath the Etna volcano [3].

A. Sulli (✉) · M. G. Morticelli · M. Agate · E. Zizzo
DISTEM - Palermo University, Via Archirafi 20-22, 90123 Palermo, Italy
e-mail: attilio.sulli@unipa.it

F. Rossetti et al. (eds.), *The Structural Geology Contribution to the Africa-Eurasia Geology: Basement and Reservoir Structure, Ore Mineralisation and Tectonic Modelling*,
Advances in Science, Technology & Innovation, https://doi.org/10.1007/978-3-030-01455-1_54

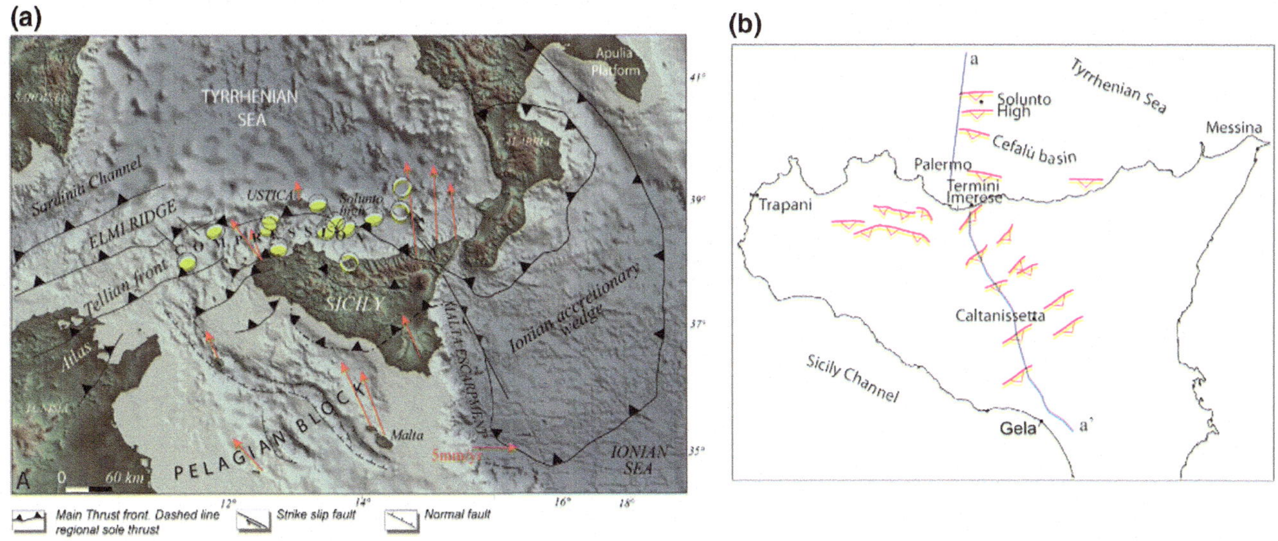

Fig. 1 **a** Simplified regional sketch of Sicily and its offshore in central Mediterranean Sea, showing the main tectonic features, focal mechanisms of earthquakes along the southern Tyrrhenian margin (from [9]) and GPS site measurements (red arrows; from [10]). **b** Distribution of the main recognized backthrust features (red line)

3 Results

The uppermost Neogene to Quaternary sedimentary successions, filling a complex wedge-top depozone developed on top of the Sicilian FTB, point out the prolonged syn-sedimentary tectonics, following the "step by step" growing and wedging of the Sicilian FTB during the progressive deepening of the deformational processes.

During the thin-skinned, shallow-seated Event 1 (Serravallian-early Tortonian) the emplacement of thin thrust sheets along low-angle 'décollement' planes is followed by the development of wide, wedge-top basins characterized by marine sedimentation, open towards the foredeep depozone. During the thin-skinned, deep-seated Event 2 (late Tortonian-Early Pliocene) thrusting of thick carbonate-platform units over the Iblean-Pelagian foreland was accompanied by internal deformation of the carbonate units, which were offset by high-angle, reverse, locally transpressional faults. This stage is associated with the development of small and confined wedge-top basins. During the thick-skinned Event 3 (Late Pliocene-Pleistocene) the deformation evolved at the depth of the crystalline basement by the activation of a forward sole thrust. This event evolved, in the inner sector, by activation of northward back-verging compressional faults that affected the already deformed tectonic wedge (Fig. 2).

4 Discussion

During the tectonic Event 3, the outward accretion of the FTB seems to have decreased in favor of back-verging thrusts and vertical growth, mostly in the inner sector of the orogen. Here a large number of N-verging thrusts controlled the tectonic evolution of the uppermost Miocene-Pleistocene deposits. In the offshore sector (southern Tyrrhenian border), some of these tectonic features seem to be responsible for the geometry of the main submerged structural highs, shaping the uneven physiography of the modern northern Sicily margin. The most recent of these features were locally reactivated by a compressional stress field with a dominant NW-SE orientation and responsible for the shallow (depth < 15 km) and low-to-medium magnitude seismicity recorded in the southern Tyrrhenian Sea. GPS site velocity up to 10 mm/y is consistent with right-transpressional deformation and convergence between Sardinia and Sicily. As a matter of fact widespread right-lateral strike-slip faults affect the main front of the Kabylian-Calabrian units in western-central northern Sicily margin [4], as well as the eastern sector along the Tindari fault system [3]. After having depicted the active deformational pattern, we analyzed the relationships with the regional (central Mediterranean) geodynamic setting. Along the northern Africa boundary (e.g. NW Algeria Neogene margin; [5, 6]) a late

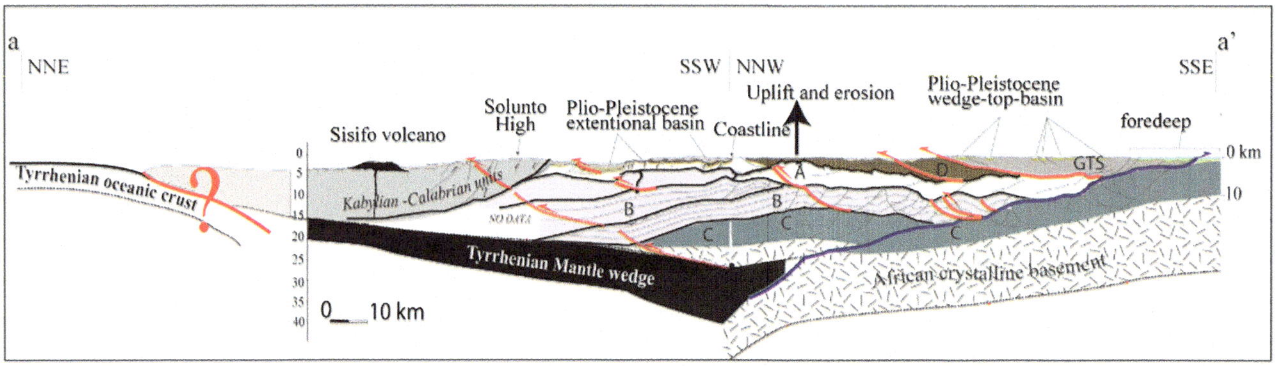

Fig. 2 Regional geological cross section showing the main tectonic units forming the Sicilian FTB. The main thrust related to the subsequent tectonic events have been distinguished: black line for thin skinned Event 1 and 2; blue line for thick skinned Event 3; red line for backthrust. **a** deep water succession; **b** and **c** carbonate platform succession; **d** Numidian Flysch deposits; GTS: Gela Thrust System. Section trace in Fig. 1b

Miocene-Quaternary northern migration of the plate margin producing opposite-verging structures is reported; moreover, a plate boundary reorganization during the latest 0.8–0.5 My with the development of backthrusts have been documented in the Mediterranean region [7], where shallow compressional structures and geodetic data [8] let to predict the development of a south-directed subduction of the Tyrrhenian lithosphere.

5 Conclusions

The results of this study suggest that the most recent tectonic processes in the study region are representative of a late collisional stage in the northern Sicily mountain building and, at a larger scale, they could be a precursor of a change in the subduction polarity in the central belt of Mediterranean Sea, as a consequence of the ongoing collision of the African promontory with the thinned continental to oceanic sectors (Algerian and Tyrrhenian basins) of the European plate.

References

1. Roure, F., Howell, D.G., Muller, C., Moretti, I.: Late Cenozoic subduction complex of Sicily. J. Struct. Geol. **12**, 259–266 (1990)
2. Torres Carbonell, P.J., Dimitri, L.V., Olivero, E.B.: Progressive deformation of a Coulomb thrust wedge: the eastern Fuegian Andes Thrust-Fold Belt. Geol. Soc. Lond. Spec. Publ. **349**, 123–147 (2011)
3. Billi, A., Presti, D., Orecchio, B., Faccenna, C., Neri, G.: Incipient extension along the active convergent margin of Nubia in Sicily, Italy: Cefalù-Etna seismic zone. Tectonics **29**, TC4026 (2010)
4. Sulli, A.: Structural framework and crustal characteristics of the Sardinia Channel Alpine transect in the central Mediterranean. Tectonophysics **324**(4), 321–336 (2000)
5. Yelles, K., Domzig, A., Deverchere, J., Bracene, R., De Lepinay, B.M., Strzerzynski, P., Bertrand, G., Boudiaf, A., Winter, T., Kherroubi, A., Le Roy, P., Djellit, H.: Plio-Quaternary reactivation of the Neogene margin off NW Algiers, Algeria: the Khayr-Al-Din bank. Tectonophysics **475**, 98–116 (2009)
6. Mauffret, A.: The Northwestern boundary of the Nubia (Africa) plate. Tectonophysics **429**, 21–44 (2007)
7. Goes, S., Giardini, D., Jenny, S., Hollenstein, C., Kahle, H.-G., Geiger, A.: A recent tectonic reorganization in the south-central Mediterranean. Earth Planet. Sci. Lett. **226**, 335–345 (2004)
8. Billi, A., Presti, D., Faccenna, C., Neri, G., Orecchio, B.: Seismo tectonics of the Nubia plate compressive margin in the south Tyrrhenian region, Italy: clues for subduction inception. J. Geophys. Res. 112 (2007)
9. Pondrelli, S., et al.: The Italian CMT dataset from 1977 to the present. Phys. Earth Planet. Inter. **159**, 286–303 (2006)
10. Devoti, R., Esposito, A., Pietrantoni, G., Pisani, A.R., Riguzzi, F.: Evidence of large scale deformation pattern from GPS data in the Italian subduction boundary. Earth Planet. Sci. Lett. **311**, 230–241 (2011)

Plio-Quaternary Shortening Structures in Northern Tunisia

Seifeddine Gaidi, Guillermo Booth-Rea, José Vicente Pérez, Fetheddine Melki, Wissem Marzougui, Mannoubi Khelil, Fouad Zargouni, and José Miguel Azañón

Abstract

Morphotectonic analysis shows the effects of local uplift related to active reverse and strike-slip faults in Northern Tunisia. Furthermore, we find that the main reverse-sinistral NE-SW oriented faults like the Alia-Teboursouk fault are segmented by conjugate WNW-ESE active dextral faults. The dextral faults, namely the Dhkila and Oued Zarga faults, have cut the drainage basins, uplifting and displacing the middle segment of the Tine River basin. The main anomalies of the normalized steepness index (Ksn) coincide with regions uplifted by the fault system or associated folds. Furthermore, some of the anomalies depict the footwall of late Miocene normal faults that exhumed competent Cretaceous sediments.

Keywords

Morphotectonic analysis • Plio-Quaternary • Shortening structures • North of Tunisia • Ksn index

1 Introduction

Northern Tunisia has evolved under NW-SE convergence between Africa and Eurasia during the Tertiary [1]. However, the main shortening structures presently active in the region have been forming and developing since the Pliocene, after an event of late Miocene orogenic collapse [2]. Here we describe two new dextral strike-slip faults that are linked to the Alia-Teboursouk reverse fault. Furthermore, we use morphotectonic tools to evaluate the relationship between active faulting and topographic uplift in Northern Tunisia.

S. Gaidi (✉) · F. Melki · M. Khelil · F. Zargouni
Department of Geology, Tunis El Manar University, 1060 Tunis, Tunisia
e-mail: gaidiseif@gmail.com

S. Gaidi · G. Booth-Rea · J. V. Pérez · J. M. Azañón
Geodinámica, University of Granada, Granada, Spain

W. Marzougui
Office National de Mines ONM, Tunis, Tunisia

2 Geological and Geomorphological Settings

The main structure in North Tunisia was built during the Early to Middle Miocene thrusting of the Numidian units over the autochthonous Atlas units [3, 4]. This South East directed thrusting event emplaced the sedimentary cover of the Paleotethys basin over the Mesozoic to tertiary North African passive margin sediments. However, during Middle to late Miocene, this orogenic wedge was extended by two orthogonal normal fault systems that were active until the Messinian, coeval to volcanism. Since the Plio-Quaternary, new shortening and strike-slip faults developed, producing the tectonic inversion of the late Miocene basins. Large SW-NE reverse-oblique faults accommodating NW-SE convergence characterize northern Tunisia at present [5, 6] (Fig. 1).

3 Methods

We have applied different morphometric analysis to characterize the tectonic activity. We use a 10 m resolution DEM to extract the main drainage lines and to delimitate two basins in the studied area: Joumine basin and Tine basin. Later, we have analyzed the drainage network and the topogragraphy using geomorphic indices. These indices detect anomalies that are possibly due to local changes in tectonic activity resulting in differential rock uplift. We use the normalized steepness index (Ksn) and Hypsometry curves.

Furthermore, we have carried out several field missions to map and characterize the active fault structures and folds in the region.

F. Rossetti et al. (eds.), *The Structural Geology Contribution to the Africa-Eurasia Geology: Basement and Reservoir Structure, Ore Mineralisation and Tectonic Modelling*,
Advances in Science, Technology & Innovation, https://doi.org/10.1007/978-3-030-01455-1_55

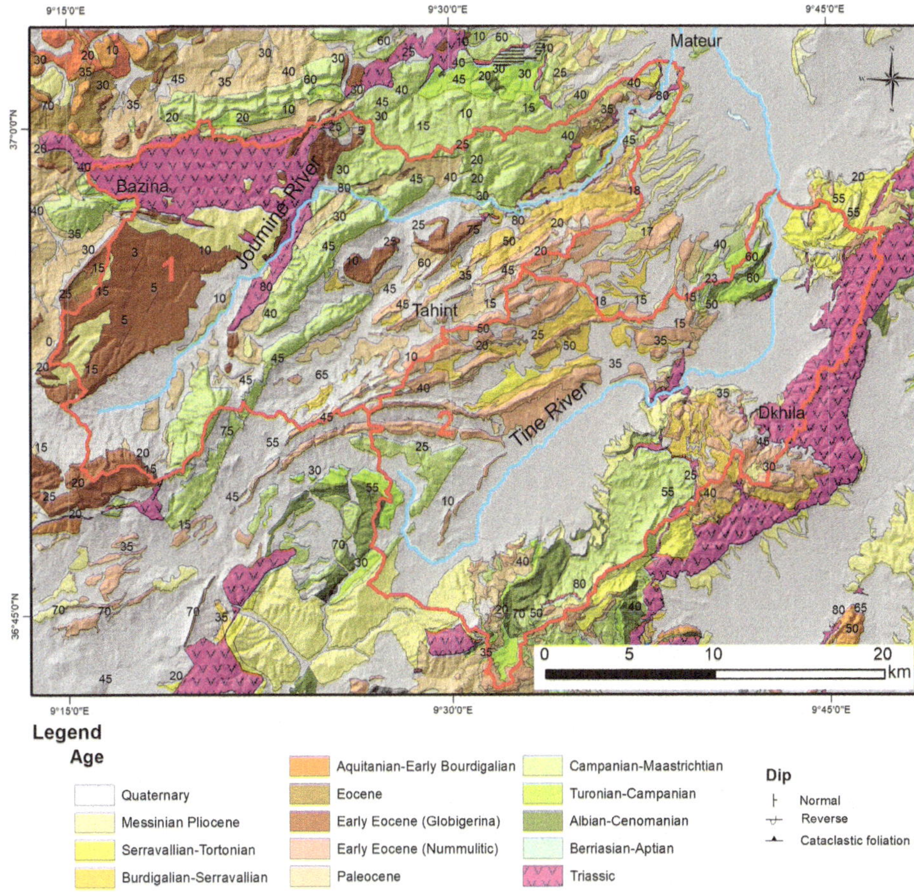

Fig. 1 Geological map of the study area (vectorised from Mateur, Hedhil, Tebourba and Beja geological maps: Edited by ONM Tunisia)

4 Results

Our field results show the existence of two large (>15 km) dextral strike-slip structures, namely the Dkhila (DK) and Oued Zarga (OZF) faults that affect Quaternary sediments (Fig. 2). In Dkhila village, we observe a large anomaly probably related to recent activity of the Dkhila Fault. This fault has N86°E strike and dips 75° towards the south with dextral-reverse kinematics. This fault produces a large nickpoint in the Tine River just southwards of the fault trace.

North of Dkhila fault, Plio-Quaternary terraces are folded and verticalized. We found that the Dkhila fault thrusted Triassic materials over Quaternary terrace sediments. The fault trace remains visible until the Joumine River valley, where different segments conform a horsetail configuration. In these segments, we identified 122°N striae in competent lower Eocene limestone blocks showing a dextral strike-slip movement (Fig. 3).

5 Discussion

The Ksn map of the studied region shows regions with differential rock uplift bounded by the newly mapped dextral Dhkila and Oued Zarga strike-slip faults. These faults cut the Tine River basin that has been uplifted in a plateau. Furthermore, other positive anomalies coincide with the recent Jebel Antra anticline in the Joumine region or with the footwall of late Miocene normal faults that exhumed Cretaceous limestones in the Jebel Aouana. Uplift related to recent shortening is favoring development of NW-SE

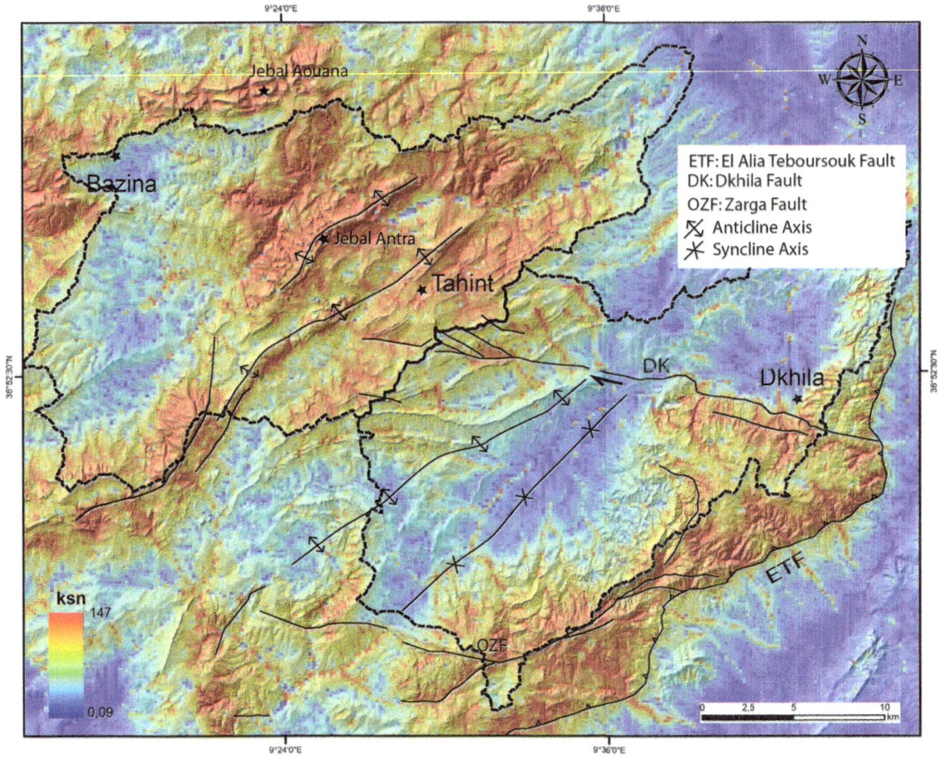

Fig. 2 Ksn heat map of the study area

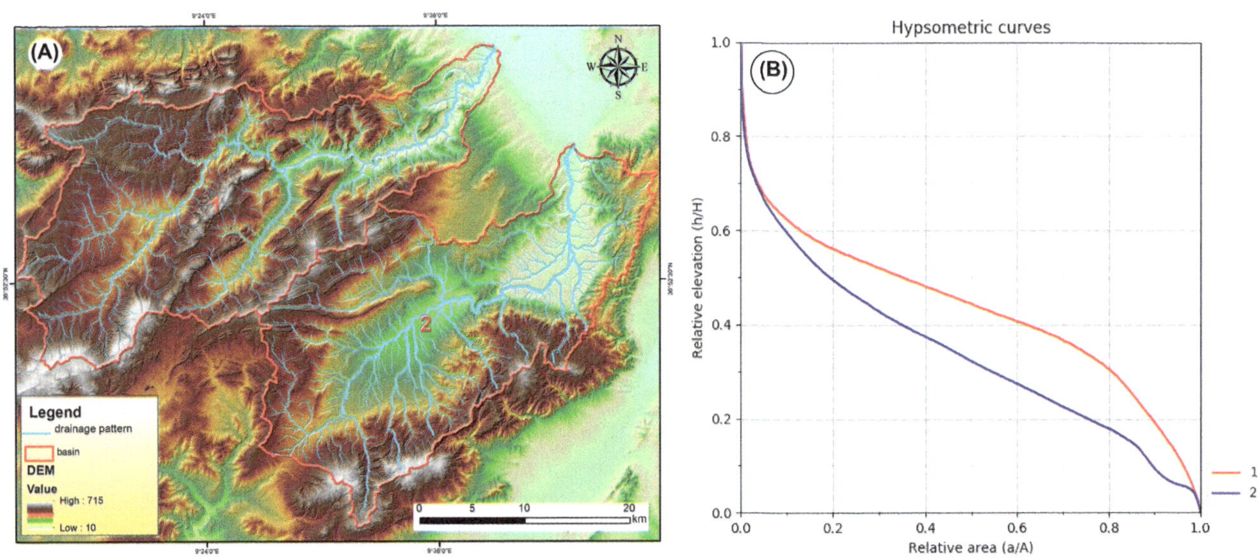

Fig. 3 **a** Digital elevation model with delimitation of the two basins and the drainage pattern. **b** Hypsometric curves for selected basins (Joumine: 1 and Tine: 2)

oriented transverse drainage that is capturing previous SW-NE directed basins formed in relation to the topographic gradient established by late Miocene extension.

References

1. Dewey, J.F., Helman, M.L., Turco, E., Hutton, D.H.W., Knott, S.D.: Kinematics of the western Mediterranean. In: Coward, M.P., Dietrich, D., Park, R.G. (eds.) Alpine Tectonics, pp. 265–283. Special Publication Geological Society of London, London (1989)
2. Booth-Rea, G., Gaidi, S., Melki, F., Marzougui, W., Azañón, J.M., Zargouni, F., Galvé, J.P., Pérez-Peña, J.V.: Late Miocene extensional collapse of Northern Tunisia. Tectonics (in press)
3. Rouvier, H.: Géologie de l'extrême Nord Tunisien: tectonique et paléogéographie superposées à l'extrémité orientale de la chaine maghrébine. Thèse Doc ès Sciences, Univ. Paris VI, 898p (1977)
4. Khomsi, S., Ben Jemia, M.G., de Lamotte, D.F., Maherssi, C., Echihi, O., Mezni, R.: An overview of the Late Cretaceous-Eocene positive inversions and Oligo-Miocene subsidence events in the foreland of the Tunisian Atlas: structural style and implications for the tectonic agenda of the Maghrebian Atlas system. Tectonophysics **475**(1), 38–58 (2009)
5. Melki, F., Zouaghi, T., Harrab, S., Sainz, A.C., Bedir, M., Zargouni, F.: Structuring and evolution of Neogene transcurrent basins in the Tellian foreland domain, north-eastern Tunisia. J. Geodyn. **52**(1), 57–69 (2011)
6. Khomsi, S., Frizon de Lamotte, D., Bédir, M., Echihi, O.: The Late Eocene and Late Miocene fronts of the Atlas Belt in eastern Maghreb: integration in the geodynamic evolution of the Mediterranean Domain. Arab. J. Geosci. **9**(15), 650 (2016)

The "Eovariscan Synmetamorphic Phase" of the Moroccan Meseta Domain Revisited; A Hint for Late Devonian Extensional Geodynamics Prior to the Variscan Orogenic Evolution

Hassan Ouanaimi, Abderrahmane Soulaimani, Christian Hoepffner, and André Michard

Abstract

In eastern parts of the Meseta Variscan domain of Morocco, Eo-variscan (Late Devonian) syn-metamorphic folding is described in the literature. In contrast, Late Devonian extension is clearly established in western parts of the Meseta domain and in its southern foreland (Anti-Atlas). We revisited three classical outcrops of the eastern Moroccan Central Massif (Zaian Mountains) where Eo-variscan folding has been described beneath the unconformable Tournaisian-Visean conglomerates and limestones. In fact, only one slaty cleavage occurs in the Ordovician and overlying Lower Carboniferous formations when their lithology is appropriate, and then results from the Variscan compression. The Lower Carboniferous unconformity can be ascribed to block tilting, likely coeval with the widely observed Late Devonian extension. The reality of a Late Devonian syn-metamorphic folding in easternmost Meseta (Midelt-Mekam zone) is poorly documented, either. At 330–320 Ma the Variscan Meseta orogen would include an eastern volcanic arc overlying early Variscan folds and a western forearc basin above the subducting Rheic lithosphere.

Keywords

Variscan Belt • Late Devonian • Volcanic arc
Forearc extension • Morocco

H. Ouanaimi (✉)
LGE-Ecole Normale Supérieure, Cadi Ayyad University, Hay Hassani, Marrakech, Morocco
e-mail: houanaimi@gmail.com

A. Soulaimani
Faculty of Sciences Semlalia, Cadi Ayyad University, Marrakech, Morocco

C. Hoepffner
Faculty of Sciences, Mohamed V University, Rabat, Morocco

A. Michard
Paris-Sud University, 91400 Orsay, France

1 Introduction

The Variscan Paleozoic orogen crops out as isolated massifs in the Atlas-Meseta domain of Morocco. The concept of an Eovariscan, likely Late Devonian syn-metamorphic folding phase has been introduced by Allary et al. [1] in their study of the Zaian Massif, in the eastern part of the Western Meseta domain (Fig. 1). This concept was popularized, sometimes under the name of "Phase bretonne", and applied to the Eastern Meseta that extends within (Tazekka), and east of (Midelt, Debdou-Mekkam, Jerada) the Mesozoic-Cenozoic Middle Atlas range [2, 3]. In contrast, extensional/transtensional or pull-apart geodynamics have been described in Western Meseta and Anti-Atlas from the 80s onward (see in [2, 4]), and even along the whole northern border of Africa [5]. Our aim is to question this surprising contrast between Western and Eastern Meseta, based on a direct revision of the Zaian unconformity.

2 Settings and Methods

The Zaian Mountains correspond to the Neoproterozoic-Upper Ordovician substratum of the Azrou-Khenifra Lower Carboniferous flysch basin. The latter is in turn overlain by a system of syn- to post-sedimentary NW verging thrust units (Nappes Zone). These nappes are rooted in the frontal part of Eastern Meseta. The Eastern and Western Meseta are arranged en echelon along the southern foreland of the Meseta orogen, i.e. the Anti-Atlas domain, which extends south of the South Meseta Fault. The available K-Ar ages of <2 micron white mica fractions from samples collected in the Azrou-Khenifra basin and its Ordovician substratum are all scattered between 335 ± 6 Ma and 283 ± 5 Ma [6]. In Eastern Meseta, a phyllite sample from the Midelt schists yielded a 366 ± 7 Ma date (Rb-Sr whole rock analysis) whereas samples from the Debdou-Mekkam slates yielded 368 ± 8 Ma and 372 ± 8 Ma (K-Ar fine-grained mica ages) [6].

F. Rossetti et al. (eds.), *The Structural Geology Contribution to the Africa-Eurasia Geology: Basement and Reservoir Structure, Ore Mineralisation and Tectonic Modelling*,
Advances in Science, Technology & Innovation, https://doi.org/10.1007/978-3-030-01455-1_56

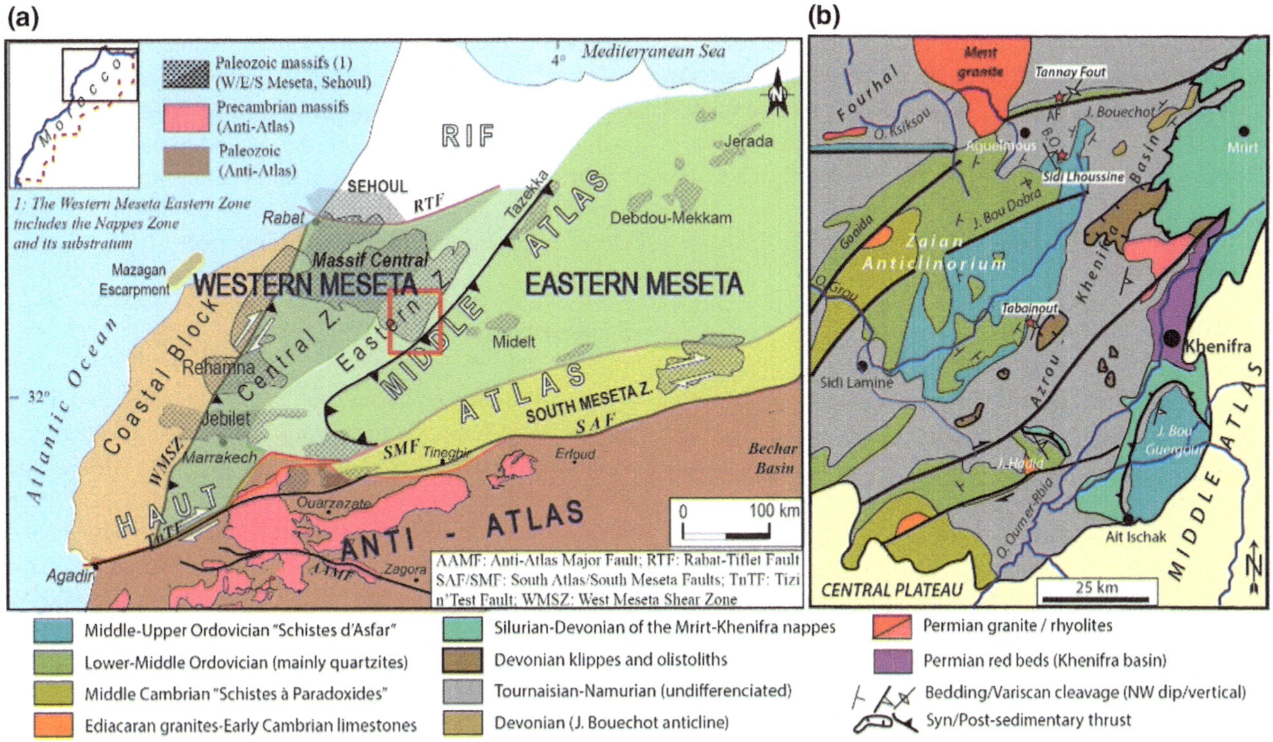

Fig. 1 **a** The Variscan Meseta domain of Morocco and its southern foreland (Anti-Atlas), modified after [3]. **b** (red box in **a**): Schematic structural map of the Zaian Mountains and surrounding areas after the Geologic map of Morocco, scale 1:1,000,000, modified. AF: Aguelmous fault. B.O: Bou Ouaouli crest

In this preliminary study we choose to study three key outcrops of the Zaian Mountains (red stars, Fig. 1b), and only used the classical methods of structural field geology. Thin sections of these outcrops are available in the literature, as well as all necessary paleontological data on the area.

3 Results

About 1 km west of Sidi Lhoussine, the SE dipping Bou Ouaouli crest consists of Lower Carboniferous limestones overlying unconformably Upper Ordovician siltstones. The Ordovician siltstones display a conspicuous, shallowly-dipping slaty cleavage that does not affect the meter-thick, lowermost sandy carbonate bed of the Bou Ouaouli crest [1]. However, (i) this cleavage transects minor folds that occur in the Ordovician slates and may have resulted from intraformational layer-parallel sliding; (ii) a foliation with similar shallow dip occurs in the uppermost, marly conglomeratic beds of the Lower Carboniferous slab, associated with plastic flattening of the calcareous elements, and (iii) no other tectonic cleavage is visible there in the Carboniferous formations, nor in the underlying Ordovician formations.

The Sidi Lhoussine outcrops are located at the northeastern pericline of the NNE-trending Bou Dobra anticlinorium. In contrast, the Tabainout unconformity is exposed along the southeastern flank of the same anticlinorium. In this area, the slaty cleavage is subvertical both in the Upper Ordovician silty formations and in the Visean-Namurian flysch that overlies the Tournaisian-Visean carbonates. We did not observe any cleavage specific to the Ordovician formations. The unconformity itself has been activated as a reverse fault during the Variscan orogeny.

In contrast, the Carboniferous unconformity remained virtually horizontal at Tannay Fout, in the axis of the Azrou-Khenifra folded flysch basin. There, the Middle-Upper Ordovician beds of the normal limb of a quartzite antiform dip broadly 45° to the SE and have been affected locally by open, NE-trending drag folds with subvertical axial planes and a pencil cleavage due to bedding-cleavage intersection. The cleavage does not affect the lowest Carboniferous layer, probably because of the mechanical properties of this coarse conglomeratic layer, in which the Variscan deformation is only recorded by numerous fractured and pitted pebbles. As at Sidi Lhoussine, the cleavage is, however, developed in the less competent beds of the upper part of the Carboniferous sandy-carbonaceous slab and in the overlying folded flysch series.

4 Discussion

The above observations show that no E-verging Eovariscan synmetamorphic folds exist in the Zaian Ordovician substratum of the Azrou-Fourhal Carboniferous basin. The Lower Carboniferous unconformity, which is paleontologically dated at J. Bouechot between the Famennian and Tournaisian, may indeed have resulted from block tilting in an extensional/transtensional setting (see Bouabdelli and Piqué (1996), in [2]). This is consistent with the observation that the large folds in Ordovician quartzite are parallel to the tighter folds in the surrounding flysch basins (Azrou-Khenifra and Fourhal basins), and with the Variscan K-Ar ages they both yielded [6].

We suggest that Eovariscan folding did not occur in the Eastern Meseta, either. The Ordovician low-grade "Schistes du Tazekka" have been affected by NW-verging folds at 330–329 Ma prior to the accumulation of latest Visean-Namurian subaerial conglomerates and calc-alkaline volcanics [2, 3]. Farther to the east, arc volcanism similar to that of the Tazekka occurred at Jerada, remaining under shallow marine to brackish conditions up to the upper Westphalian before being affected by pre-Triassic, E-trending open folds.

The weakly deformed Upper Carboniferous sequence contrasts with the underlying poorly dated metasedimentary series affected by tight NW-verging syn-metamorphic folds. The protoliths of the latter series includes Visean flysch with olistostromes in the Zekkara massif, Mid-Upper Devonian deep-sea fan rocks in the Debdou-Mekkam massifs, and Cambrian-Ordovician clastic formations in the Midelt massif [2, 3]. The *ca.* 370–360 Ma isotopic ages published from the area [6] cannot be regarded as robust as the dated samples were rich in detrital muscovite. Moreover, the granites that intrude the core of the dome-forming schists at Midelt are dated at 333 ± 2 and 319 ± 2 (U-Pb zircon; Oukemeni et al. (1995), in [2]), suggesting Late Visean-Namurian syn-metamorphic folding of the Midelt-Mekkam zone of Eastern Meseta. If the above interpretations are correct, the hypothesis that no convergence occurred during the Late Devonian in Eastern Meseta is also correct.

5 Conclusions

During the Late Devonian, the Zaian Mountains, and likely the Eastern Meseta domain would have been affected by a dominantly extensional regime as the Western Meseta and Anti-Atlas domains. The Eastern Meseta would have been affected by an early Variscan folding during the Visean, and then converted into a Visean-Namurian volcanic arc. This new interpretation of the Variscan evolution along NW Africa requires more structural and geochronological studies.

References

1. Allary, A., Lavenu, A., Ribeyrolles, M.: Etude tectonique et microtectonique d'un segment de chaîne hercynienne dans la partie sud-orientale du Maroc central. Notes Mém. Serv. Géol. Maroc. **261**, 1–170 (1976)
2. Michard, A., Soulaimani, A., Hoepffner, C., Ouanaimi, H., Baidder, L., Rjimati, E.C., Saddiqi, O.: The south-western branch of the Variscan Belt: evidence from Morocco. Tectonophysics **492**, 1–24 (2010)
3. Hoepffner, C., Ouanaimi, H., Michard, A.: La Meseta, un terrain vagabond ou la marge fragmentée de l'Anti-Atlas? Géologues (Soc. Géol. Fr.) **194**, 19–24 (2017)
4. Ouanaimi, H., Fekkak, A., Ettachfini, E.L., et al. (6 authors): Carte géologique du Maroc au 1/50.000, feuille Amizmiz. Notice explicative. Notes Mém. Serv. Géol. Maroc. **521** bis, 1–123 (2017)
5. de Lamotte, D.F., Shirazi-Tavakoli, S., Leturmy, P. et al. (6 authors): Evidence for Late Devonian vertical movements and extensional deformation in Northern Africa and Arabia. Integration in the geodynamics of the Devonian world. Tectonics **32**, 107–122 (2013)
6. Huon, S., Piqué, A., Clauer, N.: Etude de l'orogenèse hercynienne au Maroc par la datation K/Ar de l'évolution métamorphique de schistes ardoisiers. Sci. Géol. Bull. **40**, 273–284 (1987)

Paradigmatic Examples of Lateral Spreading Phenomena in the Betic-Rif and Maghrebian Chains

Seifeddine Gaidi, Cristina Reyes Carmona, Jorge Pedro Galve Arnedo, José Vicente Pérez, Fetheddine Melki, Booth Rea Guillermo, Antonio Jabaloy, Wissem Marzougui, and José Miguel Azañón

Abstract

Lateral spreading forms when fractured rocks or soil mass slide slowly over a softer underlying material. In this work, we studied two cases, one in the south of Spain and other in the north of Tunisia. Both are rock spreadings and occur along shear or tensile fractures. These mass movements involve in almost all cases rocks and occur along shear or tensile fractures.

Keywords

Rock spreading • Geomorphological analysis
GIS • Southwestern Mediterranean

1 Introduction

Lateral spreading is an extension of a cohesive and fracture rock or soil mass that slides slowly over a softer underlying material [1]. This phenomenon usually involves producing complex mass movements such as block sliding or flows.

One of the most important objectives in lateral spreading studies is the identification of their conditioning factors. In most of the cases, lithology, previous fractures and slope gradient are considered most important.

In this work, we characterized two examples of lateral spreadings in the southwestern Mediterranean region: the active lateral spreading of Albuñuelas in Granada (southern Spain) [2] and a huge lateral spreading in the Chgega mountain, in the region of Mateur (northern Tunisia) (Fig. 1).

2 Geological and Geomorphological Setting

In Tertiary basins of the southwestern Mediterranean region (Betic-Rif and the Maghrebian chains) are frequent stratigraphic configurations that involve silty marls beneath carbonate rocks. This particular stratigraphic setting is observed in the two examples analyzed in this work. The materials mobilized by the lateral spreading of Albuñuelas consist of two main geological formations: the Serravallian Albuñuelas Red Silts, and the Lower Tortonian bioclastic calcarenites and calcirudites [3]. The rocks affected by the Chgega lateral spreading are part of the Haria Formation (Paleocene) [4] that consists of dark clays and marls. The Garia Formation (early Eocene) formed by nummulitic limestones, and the Souar Formation (late Eocene) composed of clays and limestones [5].

3 Characterization of the Mass Movements

Lower Tortonian calcarenites and limestones slide over the Albuñuelas Red Silts and the Haria Formation marls, in the Albuñuelas and Chgega cases, respectively. The Garia limestones in the Chgega mountain show an intensive fracturation in two orthogonal families striking N140E and N45E (Figs. 2a and 3a, b).

In the two studied cases, the spreads take advantage of those discontinuities to fracture and separate large limestone blocks in the upper part of the mass movement. In the Albuñuelas area, the fractures (NW-SE) are perpendicular to the southeast downhill movement of the landslide (Fig. 2a). Therefore, the fractures of the rock favored the development of the spread. Moreover, the Albuñuelas case is a good example to appreciate the evolution of a lateral spreading (Fig. 4b) starting with open cracks (Fig. 3c) and grading to

S. Gaidi (✉) · F. Melki
Department of Geology, Tunis El Manar University, 1060 Tunis, Tunisia
e-mail: gaidiseif@gmail.com

S. Gaidi · C. R. Carmona · J. P. G. Arnedo · J. V. Pérez
B. R. Guillermo · A. Jabaloy · J. M. Azañón
Geodinámica, University of Granada, Granada, Spain

W. Marzougui
Office National de Mines ONM, Tunis, Tunisia

F. Rossetti et al. (eds.), *The Structural Geology Contribution to the Africa-Eurasia Geology: Basement and Reservoir Structure, Ore Mineralisation and Tectonic Modelling*,
Advances in Science, Technology & Innovation, https://doi.org/10.1007/978-3-030-01455-1_57

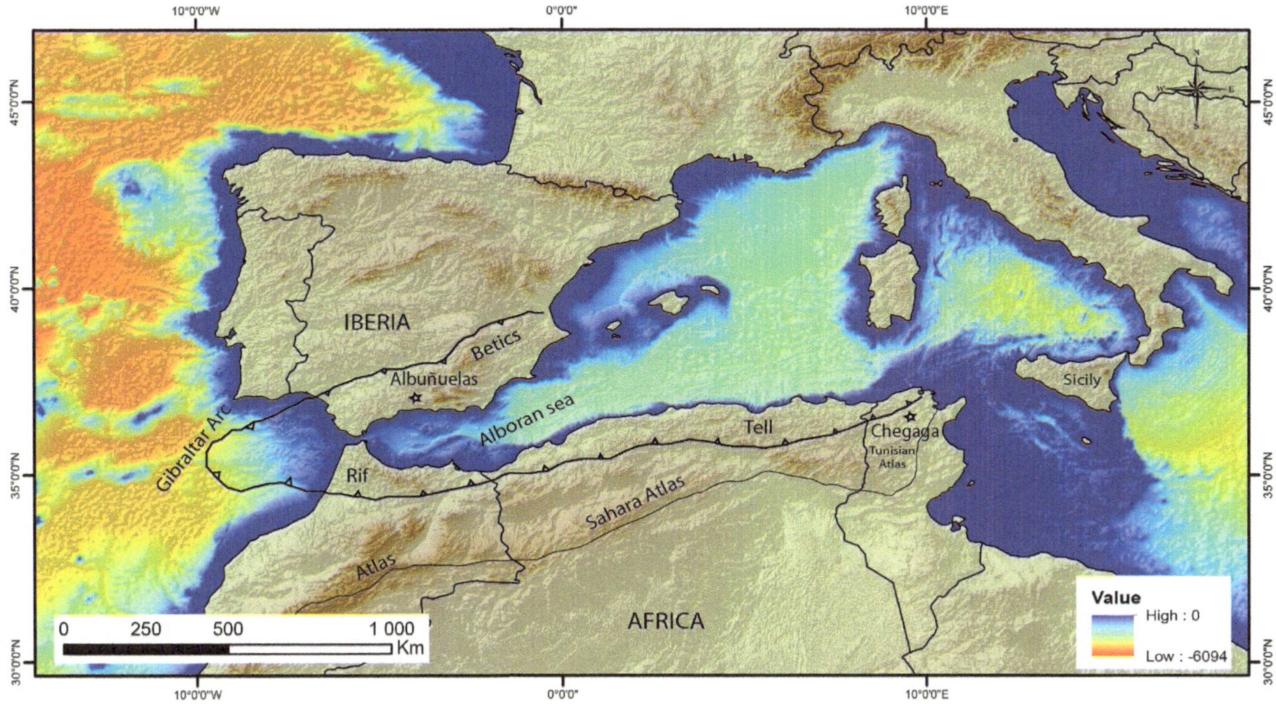

Fig. 1 Geographical and geological setting of the described lateral spreads

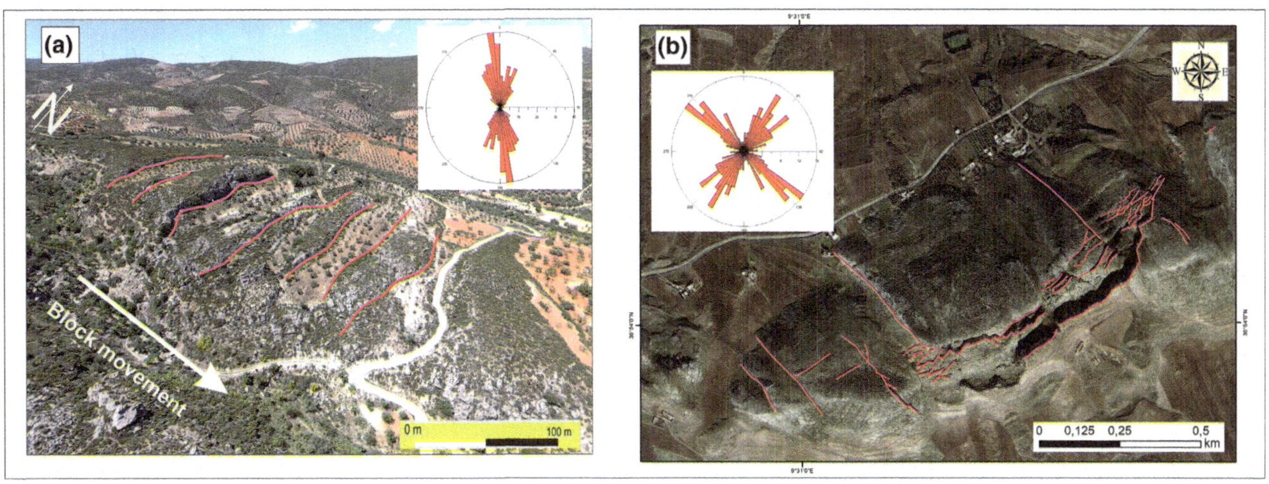

Fig. 2 a Aerial photograph of the Albuñuelas lateral spreading. b Google Earth® image of the Chgega mountain

an advanced stage of horst-and-graben structure at hillside scale. The subhorizontal disposition of the strata with a smooth inclination to the East provided an ideal configuration.

In the Chgega spread, the structural setting influences the geomorphic expression of the lateral spreading phenomenon. The mass movement is developed in a folded and faulted sedimentary succession and it has created a large trench more than 60 m deep in the water divide producing "double-ridge" (Figs. 2b and 3a, b).

With regard to the triggering factors of the described mass movements, the intense river incision due to active uplift of the Betic mountain chain may be the main responsible of the landslide in the Albuñuelas case. The longitudinal river profile of La Luna creek (Fig. 4a) shows the existence of a knick-point, indicating higher incision and unstable slopes. In the case of Chgega, seismic shaking may be one of triggering factors to initiate the landslide. The Chgega Mountain is located in a region affected by active tectonics, near of the Dkhilla fault, a N100 reverse dextral

Fig. 3 **a** Two orthogonal families striking N140E and N45E (Chgega Mountain); **b** open cracks (Chgega Mountain); **c** aerial photograph of the Albuñuelas lateral spreading

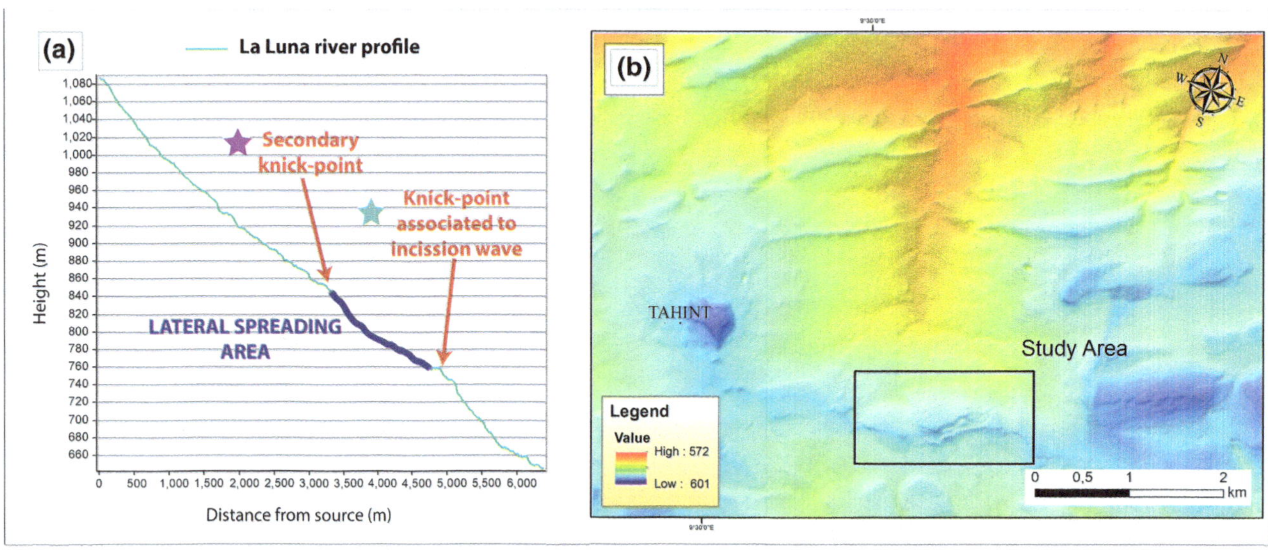

Fig. 4 **a** Longitudinal profile of the La Luna creek indicating the main knickpoints observed and the segment in correspondence to the described lateral spreading. **b** Map of the bulk erosion volume (Proxy for maximum Potential erosion) of the Chgega area. Note that the landslide was generated in the headwaters of the most eroded catchment

fault that show evidences of activity during Quaternary times. Due to the large size of the landslide scar, it is reasonable to think that a large earthquakes associated with the mentioned fault may have caused the opening of a large crack that subsequently formed the main scar of the mass movement. The latter opening of the trench and successive movements are probably related to periods of intensive precipitation and subsequent seismic activity. The evolution of this landslide could be also favored by the fluvial erosion that created the highest gradient in the area of Chgega just downhill where the lateral spreading occurred (Fig. 4b), thus creating the necessary space for the spread.

4 Final Remarks

The Albuñuelas and Chgega cases show two clear and remarkable examples which enable us to understand lateral spreading phenomena. They can also be references in identifying this type of mass movements in the Betic-Rif and Maghrebian chains because of their geological settings (i.e. structure and lithology) are common in these mountain belts. Aside from similar conditioning factors, the triggering factors are different in both cases. Whereas in the Albuñuelas case, torrential precipitations could be deemed as the most plausible triggering factor, in the Chgega, seismic activity may be necessary to initiate the movement of such huge rock mass.

References

1. Pasuto, A., Soldati, M.: Lateral spreading. In: Treatise of Geomorphology, vol. 7, pp. 239–248. Academic Press (2013)
2. Galve, J.P., et al.: Evaluation of the SBAS InSAR service of the European space Agency's Geohazard Exploitation Platform (GEP). Remote Sens. **9**, 1291 (2017)
3. Braga, J.C., Jimenez, A.P., Martín, J.M., Rivas, P.: SEPM Concepts in Sedimentology, vol. 5, pp. 131–139 (1996)
4. Rouvier, H.: Géologie de l'extrême Nord Tunisien: tectonique et paléogéographie superposées à l'extrémité orientale de la chaine maghrébine. Thèse Doc ès Sciences, Univ. Paris VI, 898p (1977)
5. Burollet, P.F.: Contribution à l'étude stratigraphique de la Tunisie centrale. Ana-les des Mines et de Géologie **18**, 352 (1956)

Remote Sensing Imagery Contribution to Structural Mapping and Analysis of Tin Tarabine Region (Central Hoggar, Algeria)

Kawther Araïbia, Kamel Amri, and Chakib Harouz

Abstract

Tin Tarabine is located in Central Hoggar, at the junction of three terranes of different geological and tectonic histories. The late Pan-African E-W compression produced in the Tuareg shield N-S trending folds linked to mega-systems of shear belts and strike-slip faults which dissect the Tuareg shield into N-S trending branches and blocks of many hundred kilometers lateral displacements. Remote sensing has been used to define the Pan-African deformation. Various treatments such as Contrast-enhanced Red-Green-Blue (RGB) composites, Principal Component Analysis, Band Ratios and Directional Filters have been applied on Landsat 8 OLI Satellite images of Tin Tarabine region. The aim of this study is to highlight the impact of major shear-zones on the region of study. The structural mapping has allowed us to identify and characterize their structural expression. Most of the structures found include faults, folds and lineaments with the highest lineaments densities obtained being mainly oriented NNE-SSW, NE-SW, NW-SE and N-S. We consider that during the final compressive phase the unquantified E-W shortening may have been absorbed by lateral movements along shear zones with a main sinistral component.

Keywords

Shear-zones • Pan-African • Central Hoggar
Tin Tarabine • Structural mapping

K. Araïbia (✉) · K. Amri
LGGIP/USTHB, BP N°32 El Alia, 16111 Bab Ezzouar, Algeria
e-mail: araibia.kawther@gmail.com

C. Harouz
LGOPH/USTHB, BP N°32 El Alia, 16111 Bab Ezzouar, Algeria

1 Introduction

Field mapping in complex terrain can be laborious such as in desertic areas and at times, inefficient, particularly over large areas and where access is made difficult or impossible due to high-relief topography such as in the Hoggar (Fig. 1). Fortunately, remote sensing has provided us with a means of acquiring data over large areas and at large mapping scales. In this work, we show the contribution of remote sensing in the study of deformation related to mega-systems of shear belts and strike-slip faults which dissect the Tuareg Shield into N-S trending branches and blocks [3] of many hundred kilometers lateral displacements which are related to the Pan-African Orogeny.

1.1 Geological Setting

The Tuareg shield, in the Trans-Saharan belt of Northern Africa was formed during the Pan-African orogeny and is characterized by major north-south oriented shear zones delimiting crustal blocks termed 'terranes' [1]. During the Tonian (1000–720 Ma) and Cryogenian (720–635 Ma) periods, they underwent a final and general E-W compression during which, major horizontal terrane movements along the plurikilometric submeridian shear zones occurred in this region wedged between the West African craton and the Saharan Metacraton [1, 2].

Tin Tarabine is a key region to study this final E-W compression period, because it is located in Central Hoggar, at the junction of two Archaeo-Eburnian terranes that belong to the LATEA superterrane: Azrou N'Fad and Egere-Aleksod, of polycyclic Precambrian history and the Serouenout terrane which is juvenile (Pan-African) [2]. The geodynamic scenario [1] considers a major east-dipping subduction in the west (680–620 Ma) and a west-dipping subduction (around 690–650 Ma). The major paleo-subductions divided the Tuareg Shield into three main

F. Rossetti et al. (eds.), *The Structural Geology Contribution to the Africa-Eurasia Geology: Basement and Reservoir Structure, Ore Mineralisation and Tectonic Modelling*,
Advances in Science, Technology & Innovation, https://doi.org/10.1007/978-3-030-01455-1_58

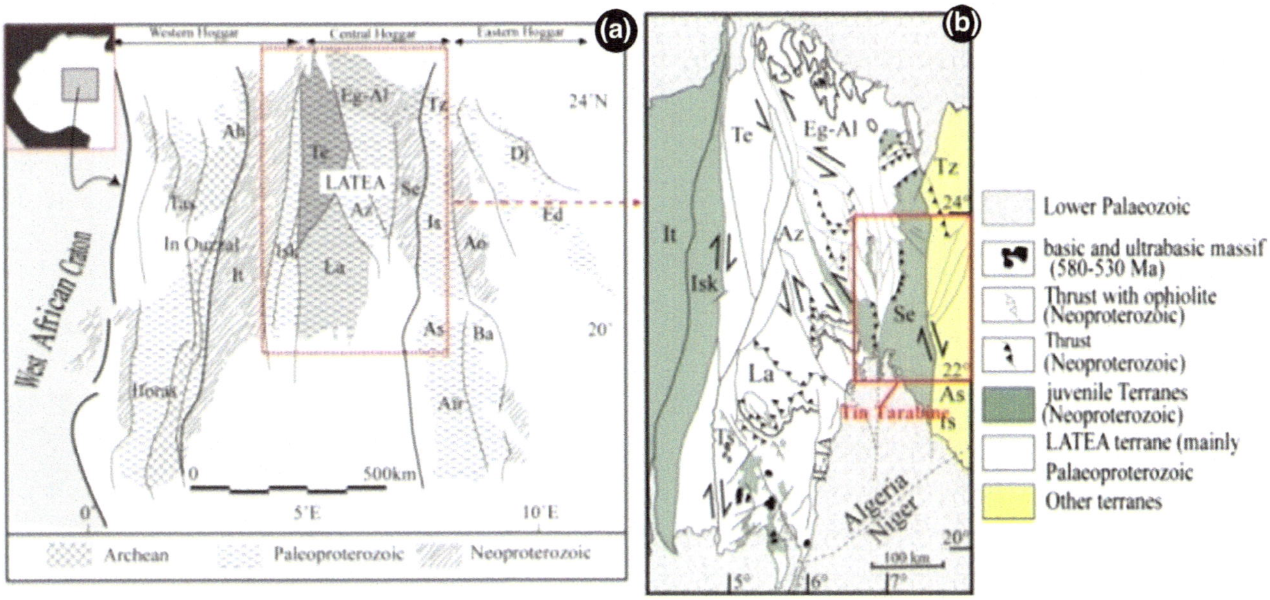

Fig. 1 **a** Schematic map of the Tuareg shield and the border areas by [1], showing the LATEA location. **b** Sketch map of terranes from the LATEA [2] showing the location of the study area

domains: Western terranes, Central terranes and Eastern terranes. In the Central Hoggar five terranes; Laouni, Azrou-n-Fad, Tefedest, Egéré–Aleksod form the LATEA superterrane. Therefore, the study of the contact between a part of this superterrane and another terrane within the Central Hoggar could be very interesting.

2 Materials and Methods

2.1 Remote Sensing Data and Data Processing

In this investigation Landsat 8 OLI data were obtained from the U.S. Geological Survey's Earth Resources Observation System (EROS) Data Center (EDC). They were acquired with low cloud cover and more rock exposures. The Landsat Scene used is the following: LC81910442017331LGN00. The datasets were processed using the ENVI (Environment for Visualizing Images) and Arc GIS software.

In order to map the different structures that affect the present terrain, the methodology used is to apply Contrast-enhanced Red-Green-Blue (RGB) composites (ETM4 ETM3 ETM2, ETM7 ETM5 ETM3 and ETM7 ETM6 ETM5), Principal Component Analysis (CP3762 CP3524 CP1362 and CP224 CP257 CP267), Band Ratios (4/3 7/6 5/4), (7/6 5/4 3/2) and (7/6 5/4 4/3) and directional filters (F045° F090° F000°) and (F045° F135° F000°) 3*3.

3 Results

The different techniques used include various enhancements of the image, directional filters being important ones for the study of the deformation, as they increase contrast in the image, allowed us to map a large number of lineaments. The highest densities obtained are oriented NNE-SSW, NE-SW, NW-SE and N-S, with the predominance of the first direction. Additionally, we have noticed, throughout the foliation trajectory map, that most of the fold axes are oriented NNE-SSW, NNW-SSE and N-S (Fig. 2).

Lineaments analysis was addressed to show the lineaments network geometry and to identify the dominant directions at a regional scale. The Landsat data were optimized using different standard spectral enhancements in order to put in evidence the main outcropping rocks based on textural characteristics linked to their spectral reflectance. This, allowed us to note numerous lateral displacements which we defined as strike-slip faults. Principal shear-zones are, then, associated with strike slip faults both sinistral and dextral but the sinistral ones are most dominant (Fig. 3).

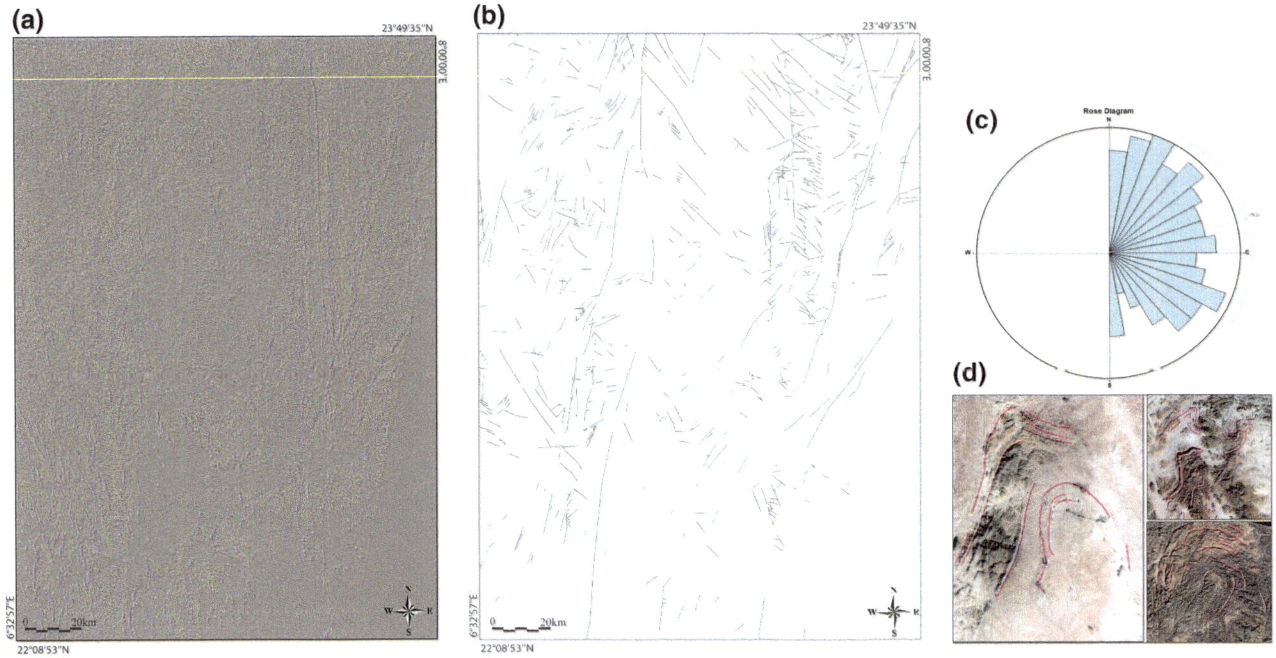

Fig. 2 a Directional filters (F045° F135° F000°); b Structural Lineaments Map; c Directional rose diagram from structural lineaments map; d Foliation trajectory drawn on ETM7 ETM6 ETM5 RGB composite

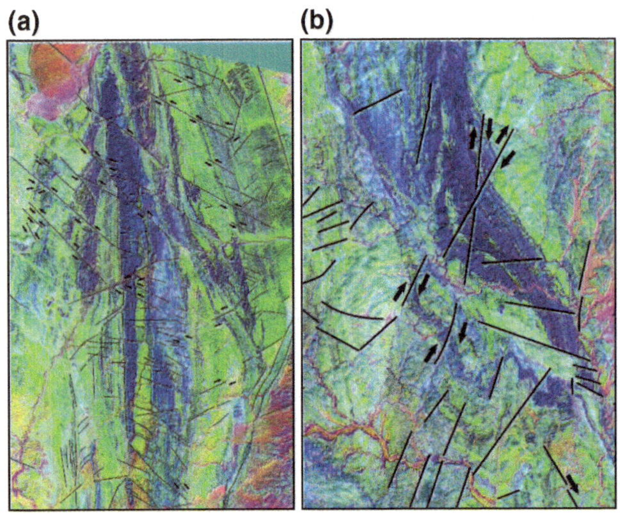

Fig. 3 Remote sensing allowed us to determine the different kinematics related to strike-slip faulting systems associated to mega shear-zones. a Predominance of sinistral strike-slip faults; b Predominance of dextral strike-slip faults

3.1 Lithological Mapping

The lithological analysis throughout PCA, Band ratios and RGB composites treatments, allows a clear separation into different rock types. Of course, the surface lithology interpretation should always be validated in the field. The geological map of the study area (Tazrouk; 1961) bears fewer information than the multi-spectral image, therefore, the use of Landsat data is of precious help for the study of litho-structural proprieties (Fig. 4) such as lithological units displacements, limits and organization.

4 Discussion and Conclusion

The various treatments applied in remote sensing on satellite image provide information on geological formations such as lithology and more detailed tectonic structures. It offers great information on the distribution, style, and kinematics of deformation associated with major tectonic events. The results show a significant deformation corresponding to the WNW-ESE to E-W overall direction of Pan-African compression. It is materialized by large submeridian shear-zones, strike-slip faults, folds and lineaments with the highest lineaments densities obtained mainly oriented NNE-SSW, NE-SW, NW-SE and N-S. It also led us to consider that the unquantified E-W shortening during the final compressive phase may have been absorbed by lateral movements along shear zones with a main sinistral component.

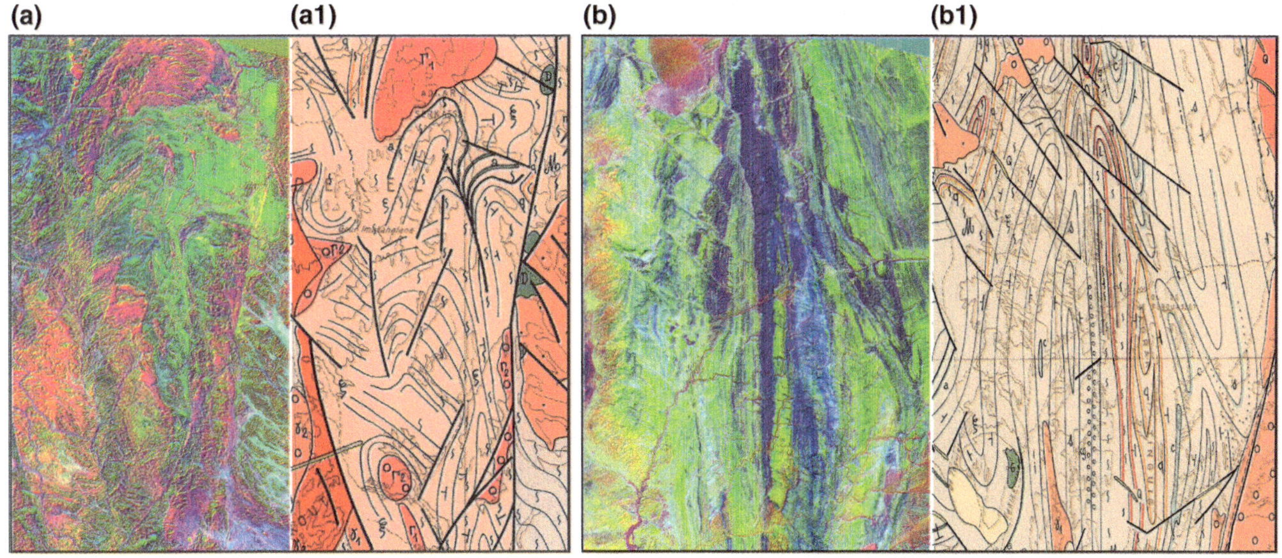

Fig. 4 Examples of contribution of remote sensing to geological mapping. By comparing parts of the geological map (**a**1; **b**1) of the study area (Tazrouk; 1961) and their equivalents on the Satellite Image (**a** principal component analysis: CP224 CP257 CP267; **b** Contrast-enhanced Red-Green-Blue (RGB) composites: ETM7 ETM6 ETM5) we can note that the use of satellite image provides us with more details

References

1. Caby, R.: Terrane assembly and geodynamic evolution of Central-Western Hoggar: a synthesis. J. Afr. Earth Sci. **37**, 133–159 (2003)
2. Liégeois, J.P., Latouche, L., Boughrara, M., Navez, J., Guiraud, M.: The LATEA metacraton (Central Hoggar, Tuareg shield, Algeria): behavior of an old passive margin during the pan-African orogeny. J. Afr. Earth Sci. **37**, 161–190 (2003)
3. Bertrand, J.M.L., Caby, R.: Geodynamic evolution of the Pan-African orogenic belt: a new interpretation of the Hoggar Shield (Algerian Sahara). Int. J. Earth Sci. **67**, 357–383 (1978)

New Evidence of Active Faulting Along the Ain Smara Fault in the Constantine Province (North East Algeria)

Yahia Mohammedi, Hamou Djellit, Abdlkrim Yelles-Chaouche, Mouloud Hamidatou, and Nassim Hallal

Abstract

The city of Constantine is one of the most populated cities in Algeria, located 300 km east of Algiers and represents the metropolis of eastern Algeria. In 1985, this city and its surroundings were shaken by an Ms = 6.0 earthquake generated by a pure left lateral strike-slip fault. This paper presents new results from multi-approach investigations carried out along the fault that generated this earthquake. The fault is composed of, at least, two main segments. The first segment responsible for the 1985 event displays clear geomorphic features of cumulative Quaternary displacement, which are manifested by a homogenous behavior of the hydrographic network in the epicentral area. However, the second segment was not reactivated during the earthquake. This segment shows evidence of recent Quaternary surface rupture with a left transpressional movement.

Keywords

Constantine • Neotectonics • Surface rupture
Transpression • Active faults

1 Introduction

The Tell Atlas (Algeria) is one of the Mediterranean regions that has to cope the most with seismic risk. Throughout its history, several moderate to strong earthquakes have been recorded (e.g. Algiers 1365 and 1716; Oran 1790; El Asnam 1980; Constantine 1985; Boumerdes 2003). The most known active faults were the subject of deep seismotectonic and paleosismological studies, mainly in the western and central part of Algeria (e.g. [1]). In the eastern part, with the exception of some studies that dealt with the problem at a regional scale [1, 2], few detailed studies on particular faults were carried out. This could be explained by the lack of important events and consequently a non-preservation of paleo seismic evidences.

During the last few centuries, several damaging earthquakes shook the Constantine province [2, 3]. At least three earthquakes were felt with a maximum intensity ranging from VIII to IX on the MSK scale (Fig. 1) [2, 3]; i.e., Constantine Ms = 5. 2 on 4 August 1908, Ms = 5.3 on 6 August 1947, and Ms = 6.0 on 27 October 1985 [4]. This last event was the strongest recorded event in the eastern Tell Atlas. The focal mechanism and the aftershock's distribution showed a pure vertical left-lateral strike slip fault trending NE-SW [2, 4]. Bounif et al. [2] reported that N55E coseismic surface ruptures were measured. These ruptures affected Quaternary deposits evidencing a left-lateral motion. In this paper, we present some results of geologic investigations carried out along the area shaken by the 1985 earthquake (i.e. Constantine and its surroundings). We have analyzed the neotectonic faults in this area with the aim of providing evidence of their recent activity and to select potential sites for future paleoseismological studies.

2 Methodology

The present study is based on two main approaches. The first approach used remote sensing techniques applied to a variety of spatial resources such as LandSat8, Eo Ali and ASTER GDEM from which a deep hydrographic analysis was performed in the epicentral area. The second approach is based on fieldwork observations and measurements along the Ain Smara fault zone (the western segment).

Y. Mohammedi (✉) · H. Djellit · A. Yelles-Chaouche
M. Hamidatou · N. Hallal
Centre de Recherche en Astronomie, Astrophysique et Géophysique, Bp 63 Route de l'observatoire, Bouzarèah, Alger, Algeria
e-mail: yahyouhe@yahoo.fr

F. Rossetti et al. (eds.), *The Structural Geology Contribution to the Africa-Eurasia Geology: Basement and Reservoir Structure, Ore Mineralisation and Tectonic Modelling*,
Advances in Science, Technology & Innovation, https://doi.org/10.1007/978-3-030-01455-1_59

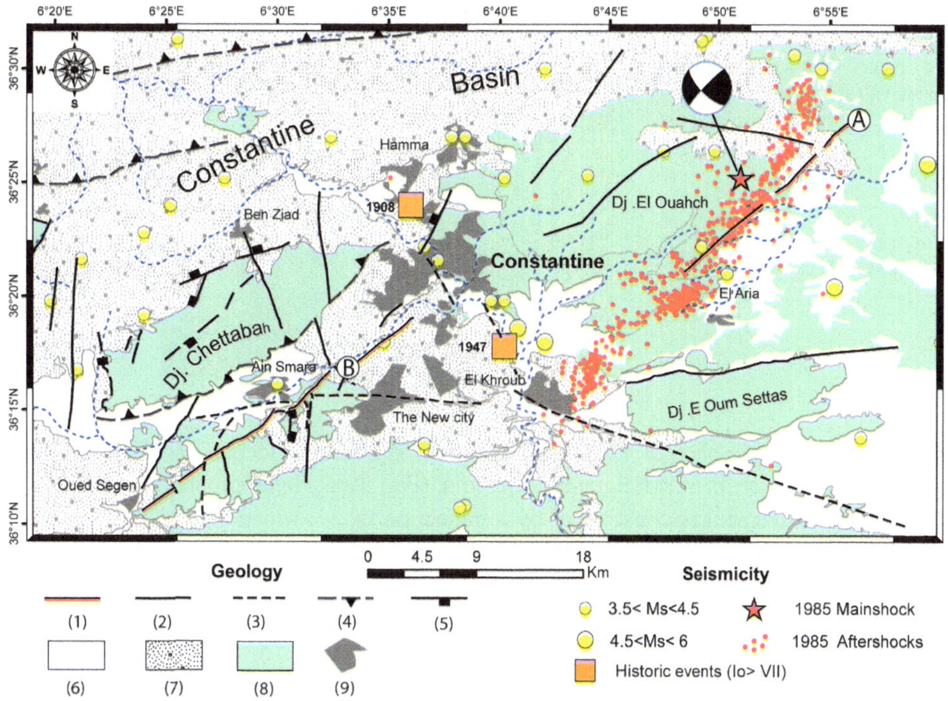

Fig. 1 Seismotectonic map of Constantine province showing the main neotectonic and active faults: (1) Active fault; (2) Neotectonic fault; (3) Inferred neotectonic fault; (4) Reverse fault; (5) Normal fault; (6) Quaternary deposits; (7) Mio-Pliocene deposits; (8) Ante-Miocene substratum; (9) Localities. Seismicity from [5] for the 1862-2014 period, aftershocks from [4], focal mechanism from [2]; geologic formations modified from [6]; tectonic features are from this study

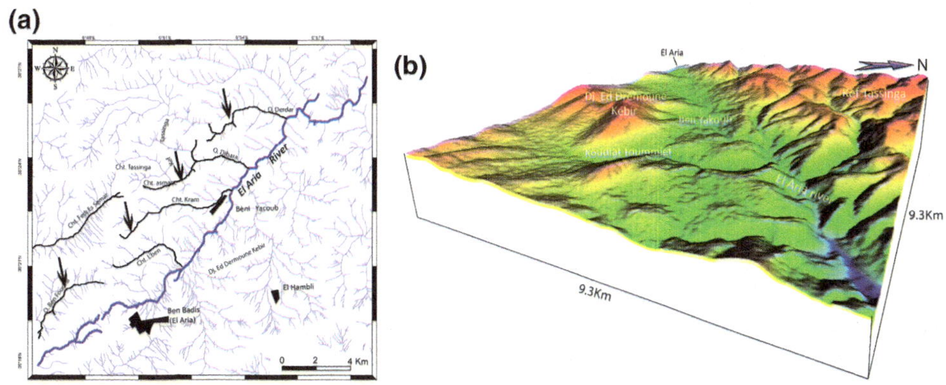

Fig. 2 Geomorphic evidences of Quaternary activity along El Aria fault: **a** Map showing the hydrographic network around the epicentral area of the 1985 earthquake, **b** 3d elevation model of the epicentral area showing the topographic lineament materialized by the El Aria river

3 Results

3.1 El Aria Fault

By using a medium resolution DEM (ASTER GDEM 30 m), we found tow geomorphic pieces of evidence along El Aria fault (fault (A) in Fig. 1): (A) Over a distance of 14 km long, the El Aria river has a direction close to the fault (Fig. 2b). This river marks the transition between northern high relief of Djebel El Ouahch and a relatively flat zone to the south. (B) The northern tributaries of the El Aria river are homogeneously anti-clockwise deflected (Fig. 2a), taking the direction of the N40E fault. The deflection occurs in the area where the secondary breaking surface are reported by [2].

3.2 Ain Smara Fault

This fault is materialized by a ten meters scarp, making it easily recognizable in the field or on the satellite imagery (fault (B) in Fig. 1). This fault seems to have experienced two types of Quaternary movements: (1) normal faulting (Fig. 3a) due to extensive tectonics, responsible of the present topographic contrast between high mountains and relatively low planes; (2) active transpressional tectonics,

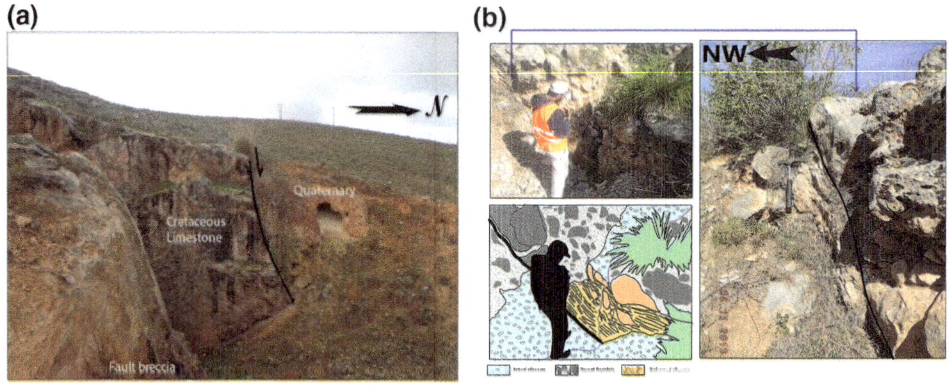

Fig. 3 Photos showing **a** recent Quaternary normal faulting along Ain Smara N040 fault **b** details of the recent rupture along the Ain Smara active fault

evidenced by the presence of 1 m scarp related to surface rupture affecting recent rockfalls and colluvium (Fig. 3b). The sinistral movement is constrained by a deflection of a small stream passing across the fault.

4 Discussion

The 1985 Constantine earthquake took place 20 km east of the city of Constantine. The epicenter and the aftershocks occurred in a zone with no clear evidence of active tectonics. However the geomorphic analysis in the 1985 epicentral area show evidences of Quaternary deformation in concordance with the ruptures mapped by [2].

The nearest neotectonic recognizable fault is located 30 km west of the epicentral area (i.e. Aïn Smara fault) which is considered as the western segment of the earthquake source. Although in several works the "Ain Smara fault" refers to the active structure that generated the earthquake (e.g. [7, 8]), we prefer to make a clear distinction between both segments. Hence we consider the El Aria fault as the active segment during the 1985 event.

5 Conclusion

The present study brings new evidences of the active tectonics along the Constantine zone. We highlighted a fault composed of two segments: (a) The El Aria fault which is the segment that generated the 1985 event and which shows evidences of Quaternary movements as revealed by the hydrographic network analysis; (b) The Ain Smara fault which is the second segment that experienced an upper Quaternary normal faulting followed by an actual transpression. This segment presents a recent surface rupture with a fault scarp.

References

1. Meghraoui, M.: Géologie des zones sismiques du N de l'Algérie paléosismologie, tectonique active et synthèse sismotectonique (1988)
2. Bounif, A., Haessler, H., Meghraoui, M.: The Constantine (northeast Algeria) earthquake of October 27, 1985: surface ruptures and aftershock study. Earth Planet. Sci. Lett. **85**(4), 451–460 (1987)
3. Harbi, A., Peresan, A., Panza, G.F.: Seismicity of Eastern Algeria: a revised and extended earthquake catalogue. Nat. Hazards **54**(3), 725–747 (2010)
4. Ousadou, F., Dorbath, L., Dorbath, C., Bounif, M.A., Benhallou, H.: The Constantine (Algeria) seismic sequence of 27 October 1985: a new rupture model from aftershock relocation, focal mechanisms, and stress tensors. J. Seismol. **17**(2), 207–222 (2013)
5. Mouloud, H., Badreddine, S.: Probabilistic seismic hazard assessment in the Constantine region. Northeast of Algeria. Arab. J. Geosci. **10**(6), 156 (2017)
6. Vila, J.-M.: Carte géologique de Constantine 1/200 000. Service de la carte géologique de l'Algérie (1977)
7. Harbi, A., Maouche, S., Ayadi, A.: Neotectonics and associate seismicity in the Eastern Tellian Atlas of Algeria. J. Seismol. **3**, 95–104 (1999)
8. Yelles-Chaouche, A., Boudiaf, A., Djellit, H., Bracene, R.: La tectonique active de la région nord-algérienne. C. R. Geosci. **338**(1–2), 126–139 (2006)

Gravimetric Study of the Khemis Miliana Plain (Upper Cheliff Basin): Structure of the Transition Zone Between the Cheliff and Mitidja Basin (North of Algeria)

Mohamed Bendali, Boualem Bouyahiaoui, Hassina Boukerbout, Billel Nèche, Salah-Eddine Bentridi, and Abdeslam Abtout

Abstract

Located in the North of Algeria, between two important seismogenic basins (the Cheliff and the Mitidja basin), the Khemis Miliana plain represents an important structure in understanding the connection between these two seismogenic basins. This Neogenic plain is known as a large depression covered with Quaternary alluvial deposits. It corresponds to an elongated East-West structure, is about 80 km long and 35 km wide. From September 2016 to April 2018 we achieved a new geophysical survey in this area. A total of 1536 gravity measurements points, spaced ~1 km apart were realized. This work leads to the establishment of a new structural image of the Khemis Miliana plain, using of different processing methods applied on gravity data. This allows the identification of the anomalies, causative structures and lineaments and the understanding of the connection between the Cheliff and Mitidja basins.

Keywords

Algeria • Upper Cheliff basin • Khemis Miliana plain
Neogenic basin • Gravimetric survey

M. Bendali (✉) · B. Bouyahiaoui · H. Boukerbout
B. Nèche · A. Abtout
Centre de Recherche en Astronomie, Astrophysique et Géophysique, Bp 63, route de l'observatoire, 16340 Bouzaréah, Alger, Algeria
e-mail: Bendalimohamed86@gmail.com; ; mohamed.bendali@univ-dbkm.dz

M. Bendali · S.-E. Bentridi
Laboratoire de l'Energie et des Systèmes Intelligents, University Djilali Bounaama, Khemis Miliana UDBKM, Route de Theniet El-Hed, 44225 Khemis-Miliana, Algeria

1 Introduction

The Neogenic Khemis Miliana plain is located in the North of Algeria and is known as being a vast depression, covered by alluvial quaternary deposits. It corresponds to an elongated East-West structure about 80 km long and 35 km wide (Fig. 1). Khemis Miliana plain is bordered at the North by the "Dahra" and "Zeccar" mountains, which are constituted of allochthonous Cretaceous and Upper Oligocene allochthonous formations [1]. To the South, it is limited by the "Ouarsenis" mountains, composed of Lower Jurassic and Cretaceous allochthonous formations [1]. According to a stratigraphical sequence of the Cheliff basin, this plain is constituted of Mio-Plio-Quaternary deposits overlaying a basement formed by thrust sheets with Mesozoic series [2].

2 Data Acquisition and Processing

In the present study, new gravimetric data was acquired throughout September 2016 to April 2018, carried out along the Khemis Miliana plain. A total of 1536 gravity stations, spaced ~1 km apart were measured using a terrestrial LaCoste & Romberg D220 gravimeter (black circles, Fig. 1). This data were tied to CRAAG-Bouzareah absolute gravity via reference points situated in the study area (blue circles, Fig. 1). To perform the gravimetric corrections we determined the topographic coordinates of each measuring gravimetric point by a bi-frequency differential GPS, Leica GS-10.

The values of the complete Bouguer gravity anomaly are calculated using the following equation:

$$\mathrm{gb} = \mathrm{gobs} - \mathrm{g}(\varphi) + (0.3086\,\mathrm{hs}) - (0.0419\,\mathrm{d.hs}) + \mathrm{gT}$$

where gobs is the station observed gravity (in mgals), $g(\varphi)$ represents the theoretical gravity calculated within the IUGG1967 system, hs is the station elevation (in meters), d

F. Rossetti et al. (eds.), *The Structural Geology Contribution to the Africa-Eurasia Geology: Basement and Reservoir Structure, Ore Mineralisation and Tectonic Modelling*, Advances in Science, Technology & Innovation, https://doi.org/10.1007/978-3-030-01455-1_60

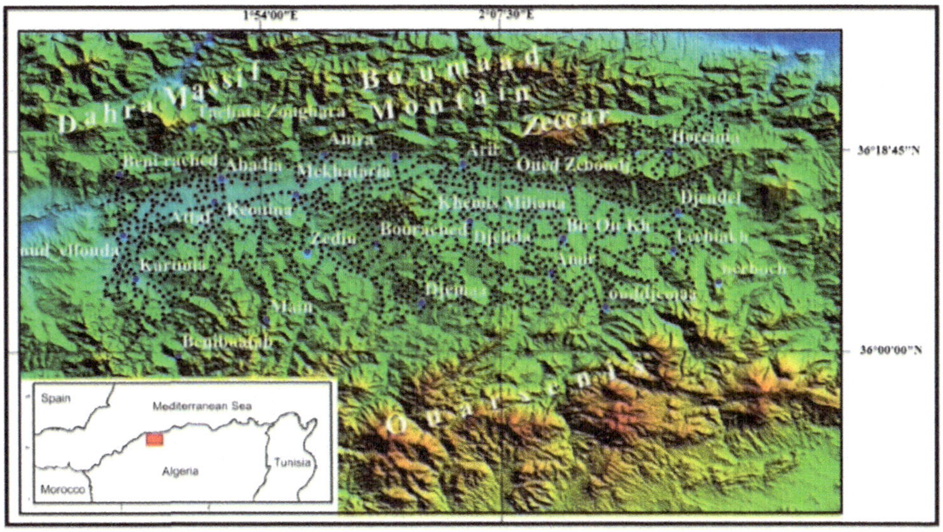

Fig. 1 Location map of the studied area superimposed on the topographic relief map from ETOPO1 1-min global relief (www.ngdc.noaa.gov). Black circles represent the gravimetric measurements and the blue circles represent the absolute gravity reference

represents the density correction and gT is the topographic corrections.

The Bouguer anomaly map shows values from −21.86 to +27.72 mGals (Fig. 2). The map shows the progression of the gravity field from south to north (Fig. 2). At the western part of this map, the Oued Fodda region is characterized by a positive anomaly, elongated E-W to NE-SW with values of about +25 mGals (red color, Fig. 2). To the east, South of Djendel, we notice a big circular negative anomaly with values less than −10 mGals (blue color, Fig. 2). The data was uniformly reduced with a density of 2500 kg/cm^3 for the Bouguer correction which is the mean density of the deposit formations ranging from Miocene to quaternary.

3 Results

We applied different processing methods (gradients, upward continuation and so on.) to the Bouguer map [3, 4]. We chose to show for instance, the results obtained by the use of gradient computation (Fig. 3) and the 3-D image obtained by the wavelet transform [5] (Fig. 4). The gradient map shows a set of NW-SE axes, to the West and a set of NE-SW axes, to the East. To the South of the area, some N-S axes appear at Bordj Amir Khaled and the Djemaa Ouled Cheick area.

In addition, the deepest structure is located in the NW of the area (South of Dahra Mounts); it reaches 11.5 km and is

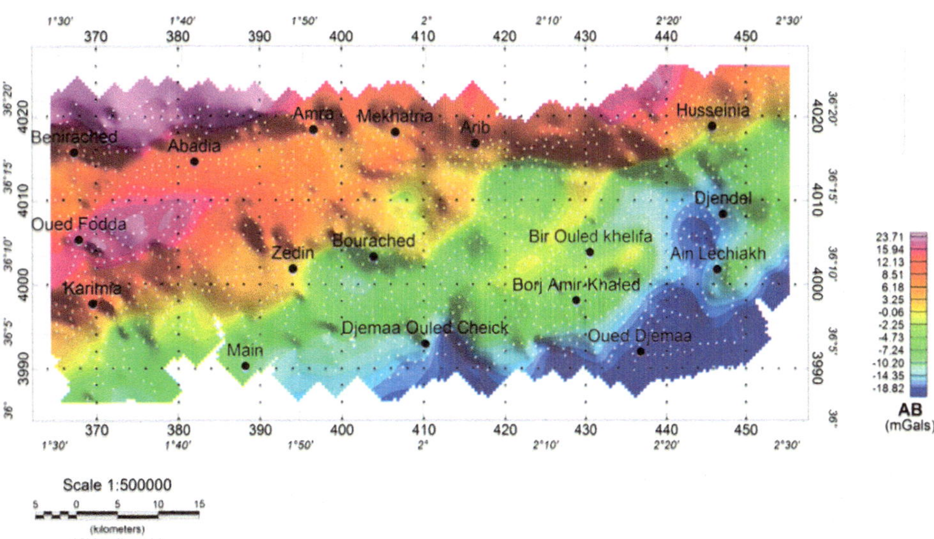

Fig. 2 The Bouguer anomaly of Khemis Miliana plain. White circles correspond to gravimetric measurements; black circles represent the GPS references stations

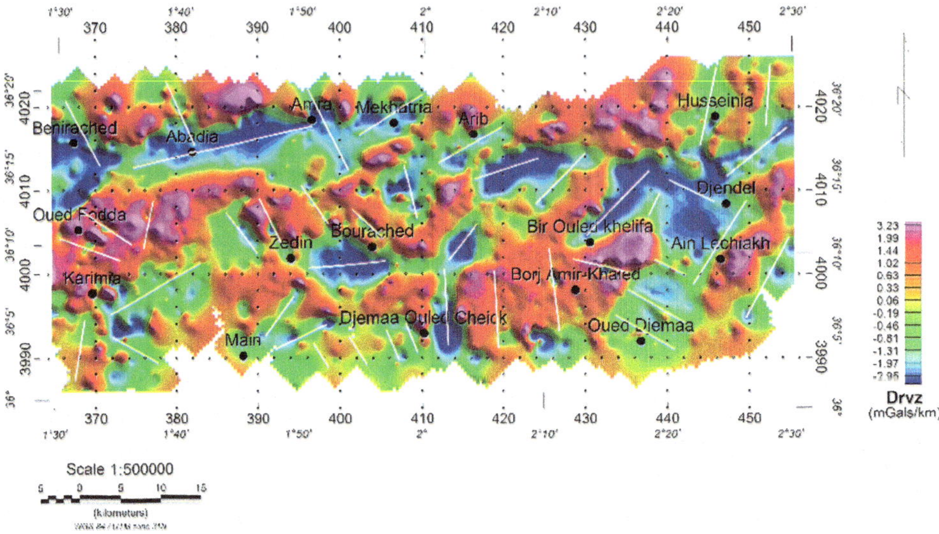

Fig. 3 Vertical gradient map of the Bouguer anomaly. The white lines correspond to faults and contacts

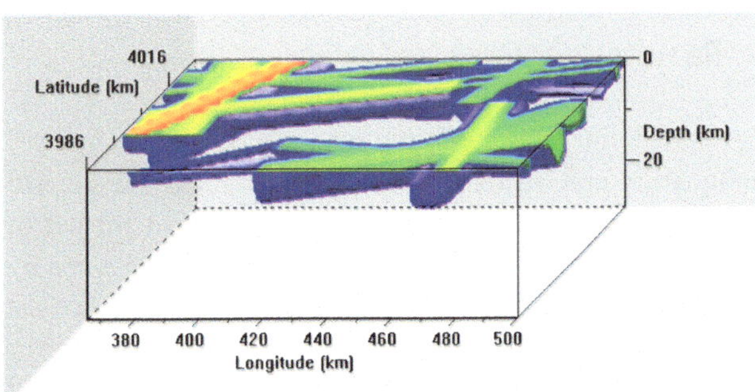

Fig. 4 Identified causative structures in Upper Chéliff basin

elongated in the NE-SW direction (Fig. 4). The depth of the northern limit of the basin is about 4.5 km while its southern boundary is at about 3.25 km. This suggests that the basin is not flat.

4 Conclusion

New gravimetric survey has allowed the imaging of deep structures in the upper Cheliff basin. The Bouguer anomaly map of the upper Cheliff basin shows two main distinguished domains. The first set of positive anomalies in the North, corresponding to Boumaad Mountain and Dahra massif. The second set of negative anomalies in the South correspond to upper Cheliff basin. The analysis and modeling of the gravimetric data illustrate the faults and lineaments both near the surface and in depth using different processing methods of gravity field interpretation (gradients, upward continuation, Euler deconvolution, wavelet transform ...). The identified axes are oriented NW-SE westernmost, and NE-SW to the east. The basin is not flat and its thickness is about 3.5 km at mean.

References

1. Belkebir, L., Bessedik, M., Ameur-Chehbeur, A., Anglada, R.: Le Miocène des bassins nord-occidentaux d'Algérie: biostratigraphie et eustatisme. Elf Aquitaine Editions, Pau, vol. 16, pp. 553–561 (1996)
2. Etude géologique des bassins néogènes sublittoraux de l'Algérie duNord occidental. Publications du service de la carte géologique de l'Algérie, vol. 12, p. 343
3. Blakey, R.J.: Potential Theory in Gravity and Magnetic Applications, 441 pp. Cambridge University Press, New York (1996) (Geophysics **35**, 293–302)
4. Abtout, A., Boukerbout, H., Bouyahiaoui, B., Gibert, D.: Gravimetric evidences of active faults and underground structure of the Cheliff seismogenic basin (Algeria). J. Afr. Earth Sci. **99**, 363–373 (2014). https://doi.org/10.1016/j.jafrearsci.2014.02.011
5. Boukerbout, H., Gibert, D.: Identification of sources of potential fields with the continuous wavelet transform: two-dimensional ridgelet analysis. J. Geophys. Res. **111**, 1–11 (2006)

Neotectonic Analysis of an Alpine Strike Slip Fault Zone, Constantine Region, Eastern Algeria

Sahra Aourari, Djamel Machane, Hamid Haddoum, Saber Sedrati, Nadia Sidi Saîd, and Djillali Bouziane

Abstract

The aim of this work is to describe the neotectonics indications observed near the Constantine region, located in the eastern part of the Algerian Maghrebides. The fault zone affected by the external domain, is oriented globally NE-SW. The fault zone reveals direct and indirect plausible tectonic indications visible in a number sites between Constantine and Mila. The microseismicity recorded during the last century in the vicinity of the area, is an argument in favor of the reactivation of an alpine tectonic structure during Quaternary.

Keywords

Alpine tectonic structure • Geomorphological indication • Neotectonic indications • Historical seismicity • Strike slip fault

1 Introduction

We would like to highlight sites revealing direct and indirect indications in relation with the tectonic activity during Pliocene and the Quaternary in the south part of the Constantine-Mila region, where recent deposits are very extensive. The neotectonic analysis will focus on morpho-tectonic interpretation and on structural analysis at different scales around the supposed faulted zone. In order to do so, we will discuss below, the escarpment of Djebel Felten which corresponds to the Ain Smara fault [1], the scarp of Djebel Maziout. The structural indicators associated with the fault zone, are well marked by folds, faults affecting the Neogene and by fractures affecting the Quaternary deposits in the southern part in post nappe basin. In addition, the uplift of the allochthonous massif of the High plain and the seismicity recorded particularly in Constantine are indications of recent tectonics.

2 Tectonic Setting

The Algerian Maghrebides [2] are complex and composed by meso- cenozoic sedimentary and metamorphic units, associated to the relative sliding and the convergence between the African-Eurasian plates [3–6]. The chain (Fig. 1) has been underground since Mesozoïc Era, several tectonic phases ago; followed by the Tertiary compressive phases, whose paroxysmal phase took place in low Miocene, then an extension phase induced a calc-alkaline magmatism during the middle-upper Miocene [7].

At regional scale, The Constantine region belongs to the external zone of Maghrebides (Fig. 2). The Constantine sea crossing—Ain M'lila, shows internal zones, flysch and external zones as well as Miocene post nappe basins [8–10]. Its structural context is characterized by regional fractures parallel to the general trend of the Tellian massifs. These fractures correspond to the large Miocene E-W overthrusting faults, which are shifted by conjugate NW-SE, NE-SW faults and N-S to NNE-SSW faults. These compressive structures were generated during the Late Cretaceous and Miocene [11]. In addition, Vila [1] specifies that the alpine tectonics in eastern Algeria, were produced in several phases: the Priabonian phase which generated the "front thrust" accident shown by an epizonal metamorphism, the flysch overthrust nappe, Tellian nappes and the folding of parautochthonous domain. The Burdigalian phase which is also thrusting generated

S. Aourari (✉) · D. Machane · S. Sedrati · N. Sidi Saîd
D. Bouziane
CGS, National Centre of Research Applied for Earthquake Engineering, 01 Kaddour Rahim Street, BO 255 Hussein Dey, Algiers, Algeria
e-mail: sahourari@hotmail.com

H. Haddoum
FSTGAT, USTHB University Sciences and Technology of Houari Boumedienne, Algiers, Algeria

F. Rossetti et al. (eds.), *The Structural Geology Contribution to the Africa-Eurasia Geology: Basement and Reservoir Structure, Ore Mineralisation and Tectonic Modelling*, Advances in Science, Technology & Innovation, https://doi.org/10.1007/978-3-030-01455-1_61

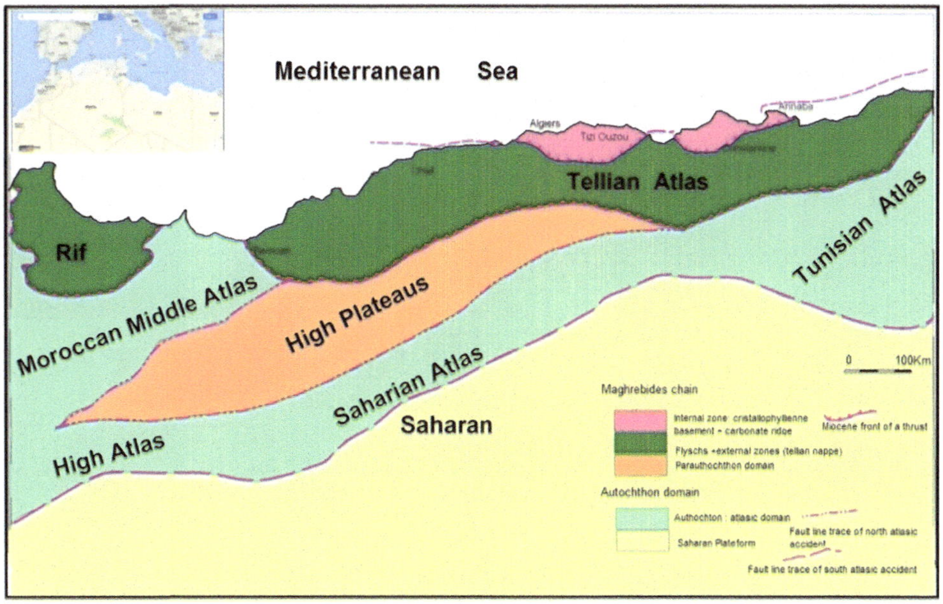

Fig. 1 The main geological domains of North of Algeria [redraw from Durand Delga et al. (1980)]

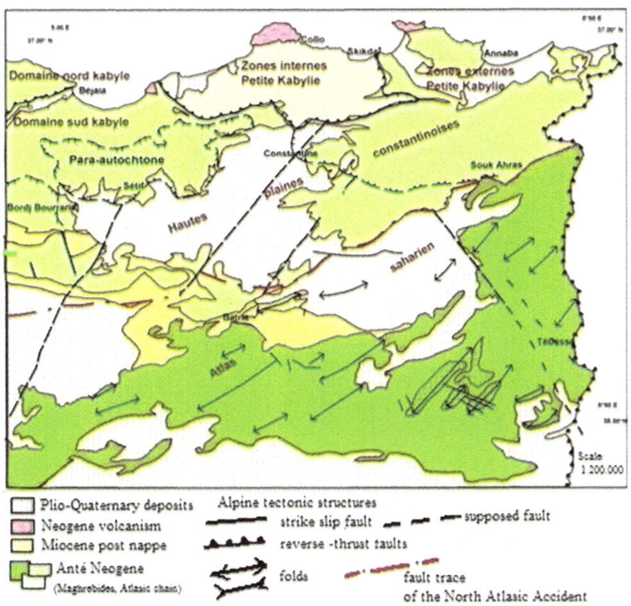

Fig. 2 Tectonic scheme of Eastern Maghrebides (from André Caire, 1961 modified) The differentunits: Socle kabyle (Paleozoïc terrains), kabylian dorsale (Jurassic Eocene deposits), flyschs units and tellian nappes (Jurassic–Cretaceous–Tertiary deposits), post nappes basins (Plio Quaternary deposit). Zone located inside the square: study areas

olistostromes, Numidian flysch and placative structures into internal zone. The Tortonian phase has generated a major extension creating post nappes basins. The active tectonics were a prelude to the southern margin of the western Mediterranean from the Pliocene, though a compression stress and strain oriented as follows NNW-SSE to N-S [12, 13]. Therefore, a number of alpine tectonic structures belonging to the network cited above seem active and currently induce earthquakes, in this part of the northern Algeria.

3 Historical Seismicity

The seismic activity that occurs in the Northern Algeria is the consequence of the geodynamics close to the plate boundary [14, 15]. Eastern Algeria seems less seismic compared to central and western Algeria [14]. The seismicity recorded is low to moderate and characterized by E-W, NE-SW distribution (Fig. 3). The seismic sources are localized in the Jijel area in coastal zones, Constantine-Guelma areas, the Mila area with a slight N-S concentration, and offshore where seismicity is mostly undetermined magnitudes. Far to the East, the Tebessa area which belongs to the Saharan Atlas chain is seismically active. These seismic sources are associated to an active NE-SW, E-W strike slip faults and E-W thrust faults [15–17] and also to a supposed active fault (Fig. 4).

In the Constantine region, a number of historical earthquakes induced earthquakes of maximum intensities VI MSK. The city of Constantine was shaken during the instrumental period by 03 earthquakes which occurred on August the 4th, 1908 [Ms 5.2], the earthquake of August 6th, 1947 magnitude [Ms 5.0] and that of October 27th, 1985 earthquake [Ms 6.0]. This last event is the strongest and was felt over 120 km from the epicenter and caused loss of life and extensive damage. Its macroseismic epicenter is located in El Aria locality where I_0 was assessed VIII MSK [18].

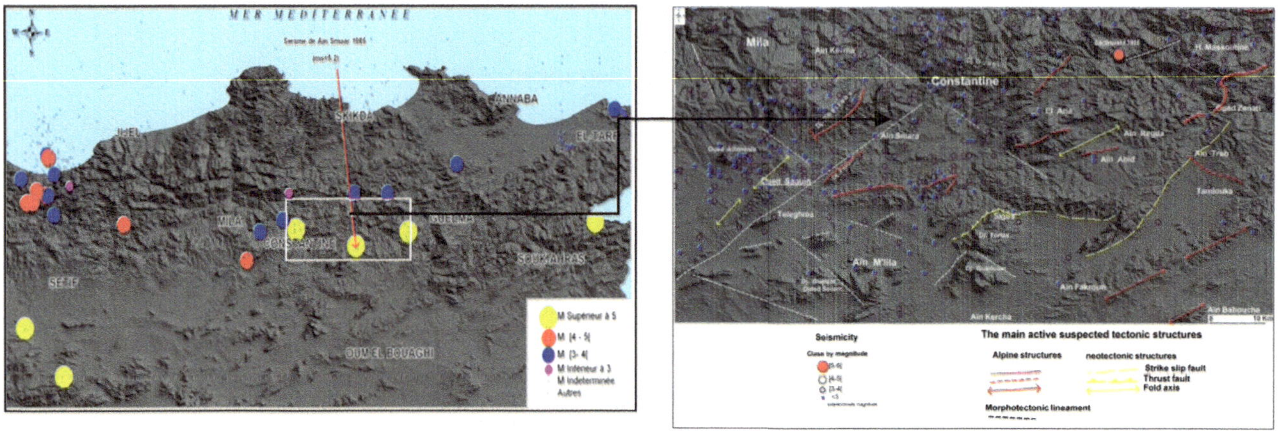

Fig. 3 **a** Seismicity map. Using Events taken from Harbi catalogue, 2010 (for 1860 to January 2008) and NEIC catalogue (data from 2008 to January 2012). **b** Seismicity associated to the main neotectonic structures

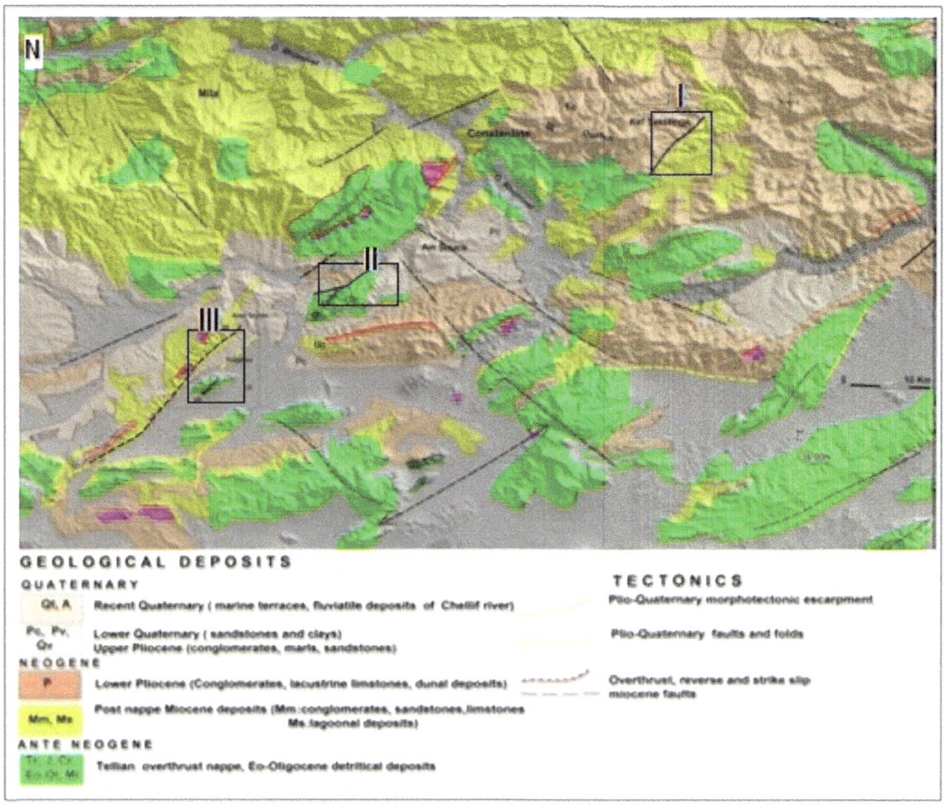

Fig. 4 Neotectonic map of Constantine region (on basis of geological map and Topography is from SRTM Digital Elevation Model with 90 m resolution)

4 Discussion: Quaternary Activity Along the Fault Zone

The study region presents geological indications concerning the Plio-Quaternary activity associated with a strike slip fault zone. It is oriented NE- SW and has a length of 60 km and it composed by three segments:

In the Constantine area, the northern segment corresponds to Kef Tassenga which reveals a fault that put in contact the Eocene-Oligocene flysch with the Miocene post nappes (geological map). This tectonic contact reveals a complex deformation zone, oriented ENE-WSW, where the C/S indicates a dextral transpressive movement of the post Eocene age. Also, the presence of fault planes oriented N060 sinistral affecting the shear zone (in I).

The second zone corresponds to the Ain Smara fault. This central segment is marked by abnormal contact between Cretaceous and Neogene, and crosses Oued El Rhummel to affect the Tellian Cretaceous massif. Its escarpment is clearly visible from the satellite images, it affects the northern flank of Djebel Felten oriented NE-SW (in II).

In the area of Mila, the southern segment put in contact the Plio-Quaternary of the Telaghma plain with Cretaceous massif (in III). In this area, many structural evidences are detected in the northern part of the plain, as Pliocene, Miocene post nappe inclined. Also, the asymmetry of the flanks of Djebel Maziout, is very remarkable. In fact, the northern flank presents minimum 04 scarps.

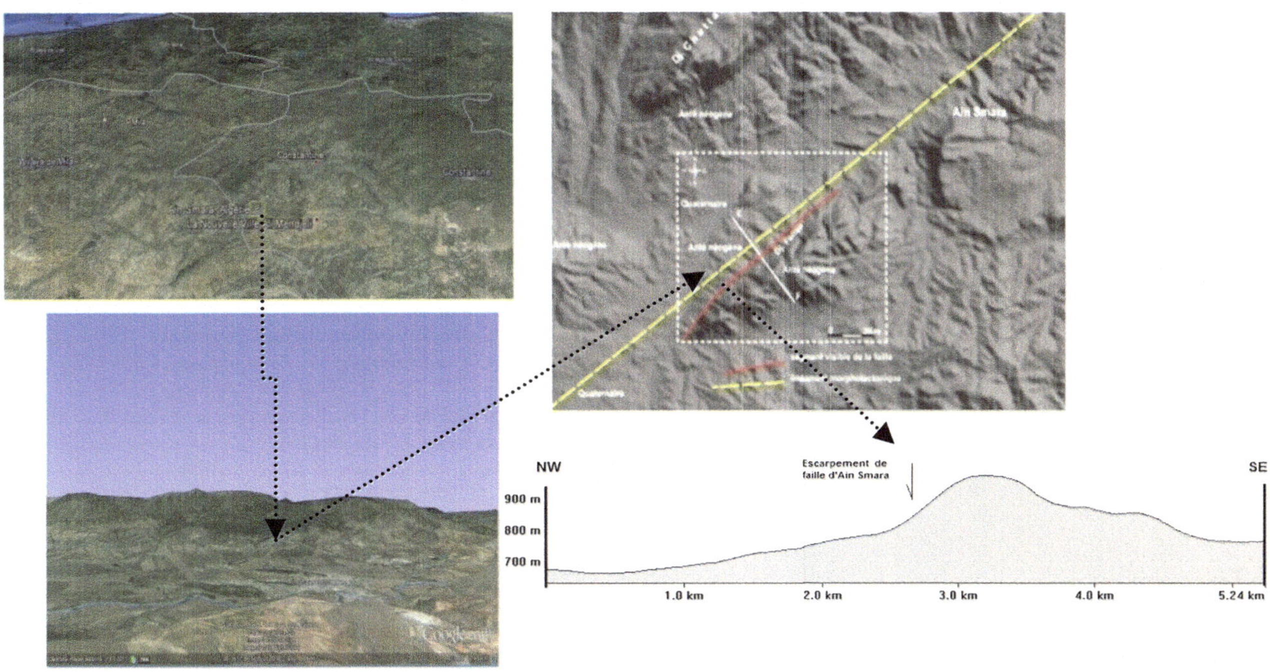

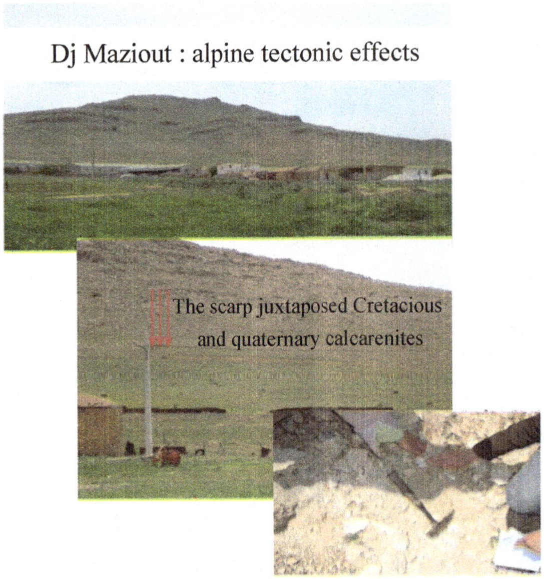

The scarp of Ain Smara fault is visible at the Jebel Felten NE-SW and is sub-vertical.

During the Constantine earthquake on the 27th of October, 1985, the epicenter was located at Ain Smara, specifically at the Kef Tassenga region [19]. The relocation of the main shock and aftershocks of the Ain Smara earthquake was estimated by [20]. The author confirms the reactivation of a strike slip fault oriented N210/215 about 14 km long and 10 km wide, with a depth ranging between 5 and 15 km.

5 Conclusions

We are interested in the analysis of the area between Constantine and Mila. From a structural point of view, the zone could be associated with a fault directed NE-SW and it connects the pre-Neogene with post nappe deposits (Mio-Plio-Quaternary). The activity of the supposed fault can be indicated by the concentration of instrumental epicenters in the vicinity of Constantine-Mila region. The map reveals three segments of significant lengths which can reach 25 km. These faults form the active domain in the Constantine region.

References

1. Vila, J.M.: La chaîne alpine d'Algérie orientale et des confins algéro-tunisiens. Thèse de Doctorat, université de Pierre et Marie Curie Paris-VI, vol. 3, 663 p (1980)
2. Durand Delga, M., Fontbote, J.M.: Le cadre structural de la Méditerranée occidentale. Mémoire B.R.G.M., 115, Orléans, pp. 57–85 (1980)
3. McKenzie, D.: Active tectonics of the Mediterranean region. Geophy. J. R. Astron. Soc. **30**, 109–185 (1972)
4. Tapponnier, P.: Évolution tectonique du système alpin en Méditerranée: poinçonnement et écrasement rigide plastique. Bull. Soc. Géol. France **19**, 437–460 (1977)
5. Bouillin, J.-P.: Le « bassin maghrébin » : une ancienne limite entre l'Europe et l'Afrique à l'Ouest des Alpes. Bull. Soc. Géol. France **II**(4), 547–558 (1986)
6. Kherroubi, A., Déverchère, J., Yelles, A., de Lépinay, B.M., Domzig, A., Cattaneo, A., Bracène, R., Gaullier, V., Graindorge, D.: Recent and active deformation pattern off the easternmost Algerian margin, Western Mediterranean Sea: new evidence for contractional tectonic reactivation. Mar. Geol. **261**(1–4), 17–32 (2009)
7. Wildi, W.: La chaine tello-rifaine (Algérie- Maroc- Tunisie): structure, stratigraphie et évolution du Trias au Miocène. Rev. Géol. Dynam. Géogr. Phys. Paris **24**(3), 201–297 (1983)
8. Raoult, J.F.: Géologie du centre de la chaîne numidique (Nord du Constantinois, Algérie). Mém. Soc. géol. France, N.S., t. III, n° 121, 164 p. (1974)
9. Bouillin, J.-P.: Géologie alpine de la Petite Kabylie dans la région de Collo et d'El Milia. Thèse de Doctorat, Paris - Toulouse, 511 p. (1977)
10. Aris, Y., Coiffait, P.E., Guiraud, M.: Characterisation of Mesozoic-Cenozoic deformations and palaeostress fields in the Central Constantinois, northeast Algeria. Tectonophysics **290**(1–2), 59–85 (1998)
11. Groupe de Recherche Néotectonique de l'Arc de Gibraltar L'histoire tectonique récente (Tortonien à Quaternaire) de l'Arc de Gibraltar et des bordures de la mer d'Alboran. Bull. Soc. Géol. France **19**(7), 575–614 (1977)
12. Philip, H.: Plio-Quaternary evolution of stress field in Mediterranean zones of subduction and collision. Ann. Geophy. **3**, 301–320 (1987)
13. Meghraoui, M., Morel, J.-L., Andrieux, J., Dahmani, M.: Tectonique plio-quaternaire de la chaîne tello-rifaine et de la mer d'Alboran - une zone complexe de convergence continent-continent. Bull. Soc. Géol. France (8), t. I, 167, 141–157 (1996)
14. Girardin, N., Hatzfeld, D., Guiraud, R.: La sismicité du Nord de l'Algérie. Bull. Soc. Géol. Fr. **2**, 95–100 (1977)
15. Harbi, A., Maouche, S., Ayadi, A.: Neotectonics and associate seismicity in the Eastern Tellian Atlas of Algeria. J. Seismol. **3**, 95–104 (1999)

16. Harbi, A.: Analyse de la sismicité et mise en évidence d'accidents actifs dans le Nord-Est Algérien, Thèse de Magistère, USTHB, Alger, 195 pp (2001)
17. International Seismological Centre (ISC): On-Line Bulletin. http://www.isc.ac.uk (2010)
18. Bounif, A., Haessler, H., Meghraoui, M.: The Constantine (Northeast Algeria) earthquake of October 27, 1985: surface ruptures and aftershock study. Earth Planet Sci. Lett. **85**, 451–460 (1987)
19. Bounif, A.: Etude sismotectonique en Algérie du Nord: contribution à l'étude d'un tronçon de la chaîne tellienne à partir des repliques du séisme de Constantine du 27 octobre 1985. Thèse de Magister, USTHB-Algiers (1990)
20. Oussadou, F.: Etude du champ de contrainte au nord de l'Algérie dans le contexte ibéro-maghrébin par inversion des solutions focales. Thèse de Doctorat, pp 375, USTHB-Algiers (2012)

Pan-African Transpression in the Tahifet Area (AzrouN'Fad Terrane, Central Hoggar, Algeria)

Kamel Amri and Yamina Mahdjoub

Abstract

The Tahifet area is located in Central Hoggar and belongs to the Azrou N'Fad terrane. The geological history of Tahifet is highly linked to two plurikilometric submeridian shear zones oriented NNW-SSE, which control the establishment of the Pan African batholite of Azir N'Fad. We have termed the West Azrou N'Fad Shear Zone (WASZ) and the East Azrou N'Fad Shear zone (EASZ). These sinistral ductile strike slip faults share the same deformation field as the overlaps determined in the studied area. This represents a structural argument in favor of a transpressive phase.

Keywords

First transpression • Gneiss • Exhumation
Tahifet

1 Introduction

The Tahifet area, which is part of the Azrou N' Fad terrane, is limited to the north by the first reliefs of the Atakor massif, to the west by Wadi Igharghar and to the east by the great wadi of Afara Héouine. To the south, the first outcrops of the Tassilian Ordovician sandstones limit it. This area consists of two units; the lower unit is mainly gneissic and is located in the depressions and the upper unit is mainly metasedimentary. The importance of this region is underlined by the existence of plurikilometric submeridian shear zones (NNW-SSE) [1, 2]. These shear zones separate the Tahifet region to the west from the Anfeg granite within the Laouni terrane which we have termed as the West Azrou N'Fad Shear Zone (WASZ) and to the east from the volcano-sedimentary formations of Aghefsa (Serouenout terrane) termed as the N'Fad Shear (EASZ) (Fig. 1) [2]. The geological history of Tahifet is highly linked to these submeridian shear zones which control the establishment of the pan-African batholiths.

The transpressive nature of these sinistral shear zones was demonstrated only by petrographic and geochemical arguments [3]. The main objective of this work is to demonstrate it by means of structural arguments. For that matter, the compatibility of strike-slip and overlapping structures is the best argument.

2 Methodology

In this work, we used the remote sensing techniques associated with field work data. For remote sensing a mosaic of two Landsat7 ETM+ scenes were used: p191r044_7t20010616 and p191r045_7t20000410 (taken respectively on June 16 2001 and April 10 2000; UTM projection, WSG84). Table 1 provides a summary of the different processings used in this study.

3 Structural Analysis

The structural analysis is mainly based on field observations. For this purpose, we studied separately the two shear zones. The sinistral movement is clearly defined by the reorientation of axial planes in the shear zone direction.

The foliations and lineation measurements (Fig. 2) specify and complete the trajectory's map of structures obtained by remote sensing.

K. Amri (✉) · Y. Mahdjoub
FSTGAT/USTHB, BP 32 El Alia Bab Ezzouar, 16111 Algiers, Algeria
e-mail: amrikamel@gmail.com

F. Rossetti et al. (eds.), *The Structural Geology Contribution to the Africa-Eurasia Geology: Basement and Reservoir Structure, Ore Mineralisation and Tectonic Modelling*,
Advances in Science, Technology & Innovation, https://doi.org/10.1007/978-3-030-01455-1_62

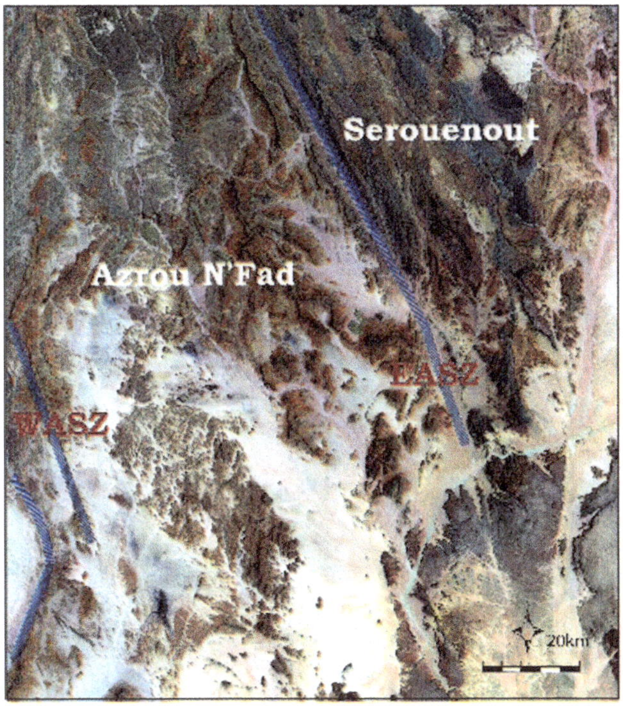

Fig. 1 Shear zones position in Azrou N'Fad terrane

3.1 The West Azrou N'Fad Shear Zone

The west Azrou N'Fad shear zone is expressed by two branches: (1) the wadi Igharghar shear zone (eastern limit of Laouni terrane) and (2) the Adrar Ayellum shear zone (see Fig. 1). The WASZ evolves progressively from NNW-SSE to NNE-SSE from the south to the north. Its affects the upper unit represented by a strike slip fault which is expressed by sub-vertical shear surfaces (C) parallel to sub-parallel to the foliation surfaces (S).

Overlaps observed in lithological units are related to these shears and show a maximum shortening axis compatible with the regional shortening axis which is globally east-west.

3.2 The East Azrou N'Fad Shear Zone

The EASZ is more than 80 km long and represents a NNW-SSE limit between the Azrou N'Fad and Serouenout terranes. Its affects the upper unit and volcano-sedimentary series of Aghefsa (Serouenout terrane) with subvertical planes showing a sinistral movement.

Table 1 Used methodology

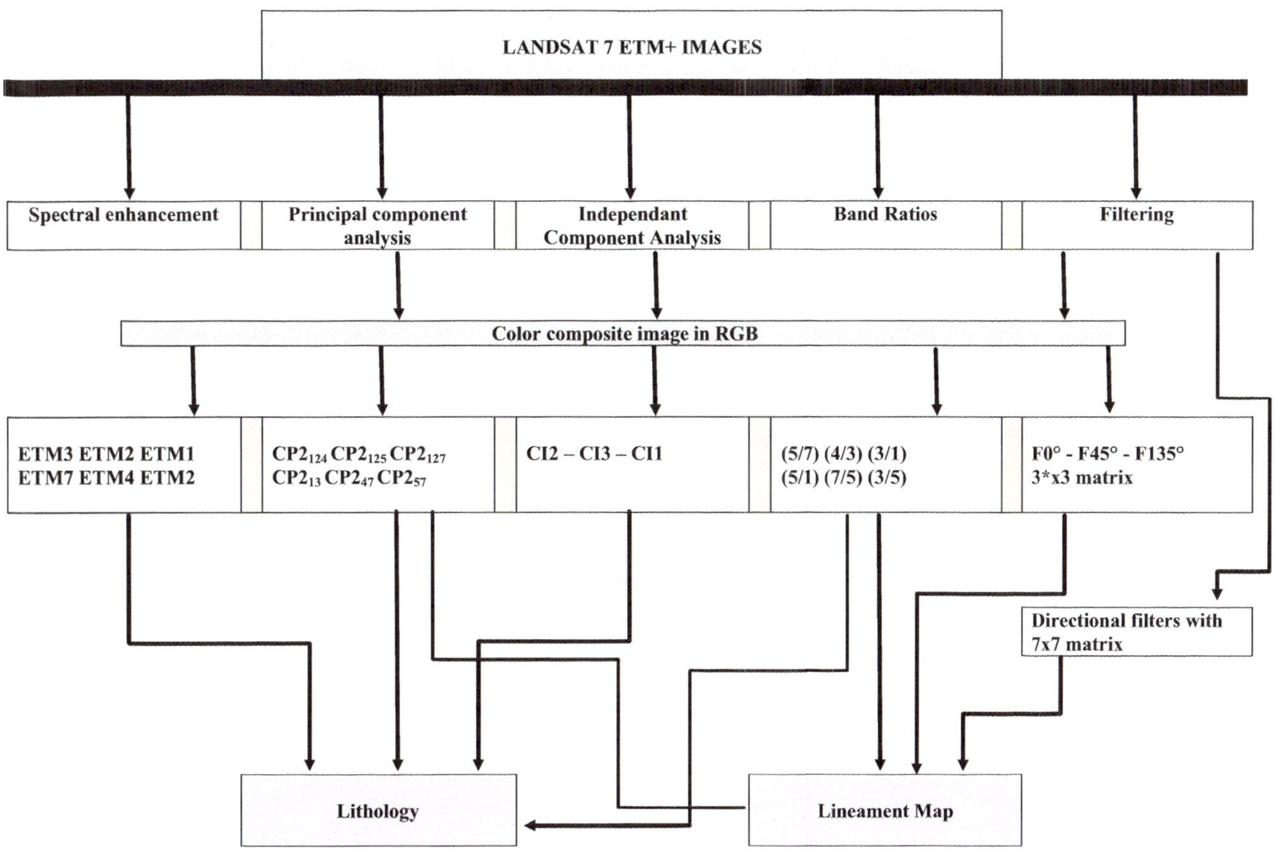

Fig. 2 Structural map of Tahifet area

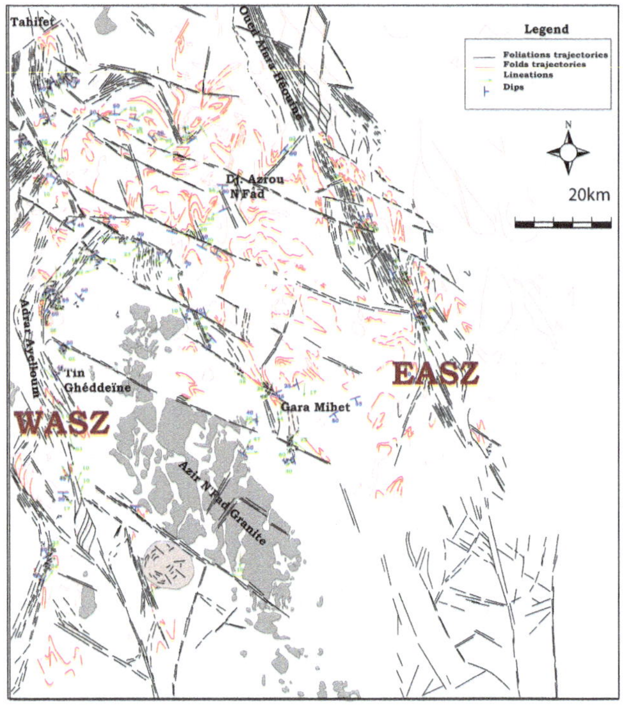

Fig. 3 Stereographic projections of WASZ

View of the EASZ

View of the WASZ

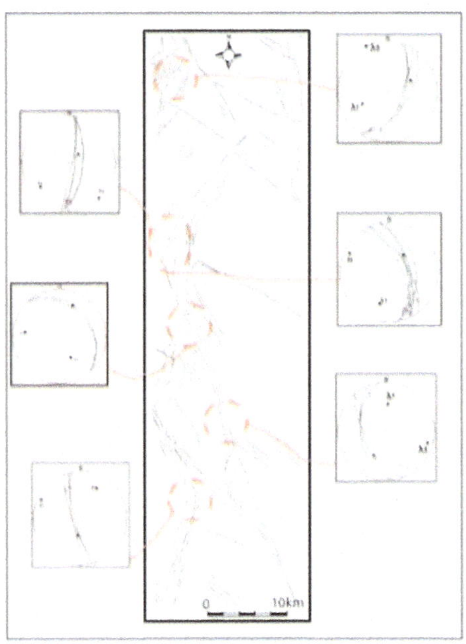

4 Discussion

The regular presence of oblique striations on shear planes (approximately 20–35°), allowed us to determine an oblique movement in sinistral shear planes. Component overlaps (inverse fault component) related to these shears shows their maximum shortening axis compatible with the regional shortening axis which is globally east-west. This scheme is always valid regardless of the shear dip.

The stereographic projections show a maximum shortening axis ($\lambda3$) oriented ENE-WSW. The NNW stretching lineation symbolize the $\lambda1$ elongation axis (Fig. 3). The local variation of the shortening axes is consistent with the non-coaxial character simple shear) of the deformation.

5 Conclusions

The WASZ and EASZ represent a sinistral transpressive deformation demonstrated by different criteria: oblique striations on shear planes, overlaps and associated folds. The E-W oriented shortening axis indicates a Pan-African phase which is linked to HP-BT exhumation rocks.

This transpressive phase made the Azrou N'Fad terrane overlap on the Laouni terrane to the west and on the Serouenout terrane to the East.

References

1. Briedj, M.: Etude géologique de la région de Tahifet (Hoggar central, Algérie). Implications géodynamiques. ThèsesUniversité Nancy (1993)
2. Amri, K.: Evolution thermomécanique de la region de Tahifet. Thèsedoctorat USTHB, 260p (2011)
3. Liégeois, J.P., Latouche, L., Boughrara, M., Navez, J., Guiraud, M.: The LATEA metacraton (Central Hoggar, Tuareg shield, Algeria): behaviour of an old passive margin during the Pan-African orogeny. J. Afr. Earth Sci. **37**, 161–190 (2003)

Aptian-Albian Diapirism and Compressional Tectonics Since Late Maastrichtian to Quaternary in Mateur-Tebourba Region (Northern Tunisian Atlas)

Achraf Zouari, Hedi Zouari, and Fehmy Belghouthi

Abstract

New outcrop and seismic based tectono-stratigraphic observations of the es Sakkak-el Baouala Range in northern Tunisia provide a more sophisticated model of its structural evolution than previously possible. This model envisages two phases of contrasting structural deformation to explain the linear salt structures now observed in outcrop. Early extension triggered reactive and then active diapirism with salt reaching the surface during the Aptian. Diapiric structures formed at this time were subsequently buried by Albian-Early Maastrichtian sediments before undergoing further intermittent growth in the late Maastrichtian and Cenozoic in response to compression from the north. Thrusts assisted salt flowage to the surface in the early Paleocene was followed by erosion and reburial in the later Paleocene and Eocene. Growth synclines developed along the thrust foot-walls of salt cored structures. Jbel el Hallouf provides an excellent example of a flanking syncline formed during the Aquitanain-Tortonian, accommodating thick Miocene deposits while the es Sakkak salt cored anticline continued to rise on the thrust hanging-wall, until burial by upper Tortonian sediments. A further phase of folding and thrusting occurred during the late Miocene. This created significant topographic relief which was subsequently eroded and redeposited as conglomerates in the adjacent valleys. Compressional deformation continues to influence sedimentation into the Pliocene and later.

Keywords

Tunisia • Mateur • Diapir • Albian-Aptian
Compressional deformation • Late Maastrichtian to quaternary

1 Introduction

This review presents a structural analysis of the es Sakkak-el Baouala Range in the northern Tunisian Atlas (Fig. 1). Triassic evaporites are exposed in this region along linear NNW-SSW trending outcrops, bounded by late Mesozoic and Cenozoic rocks. Three geological models have been proposed to explain its structural architecture:

(1) gravity spreading halokinesis [1–6],
(2) salt cored thrusting [7, 8] or
(3) salt glaciers [9–13].

More recent field mapping supported by seismic reflection data (Fig. 2) suggests the Triassic outcrops reflects two distinct phases of structural growth comprising diapirism associated with extensional faulting in the Aptian-early Albian followed by intermittent compression and inversion from late Maastrichtian to the present time. During the Cenozoic, sedimentation was controlled by folding, thrusting and strike slip faulting with rim-synclines developing along the flanks of thrusted diapirs during the Paleogene and early Neogene. Renewed thrusting and fault propagation folding in the late Miocene and Plio-quaternary resulted in steep or overturned south verging anticlines. This extended period of compression reflects the convergence of Africa and Eurasia and records the changing regional paleostress through time.

A. Zouari · H. Zouari (✉) · F. Belghouthi
Natural Water Treatment Laboratory, Water Researches and Technologies Centre, Technopark of Borj Cedria, PO Box 273 Touristic Zone of Borj Cedria, 8020 Soliman, Tunisia
e-mail: hedi.zouari@certe.mrt.tn; hedizouari3@gmail.com

A. Zouari
Faculty of Mathematical, Physical and Natural Sciences of Tunis, University of El Manar, Campus Universitaire El-Manar, 2092 El Manar, Tunis, Tunisia

F. Belghouthi
Faculty of Sciences of Bizerte, University of Carthage, 7021 Jarzouna, Tunisia

F. Rossetti et al. (eds.), *The Structural Geology Contribution to the Africa-Eurasia Geology: Basement and Reservoir Structure, Ore Mineralisation and Tectonic Modelling*, Advances in Science, Technology & Innovation, https://doi.org/10.1007/978-3-030-01455-1_63

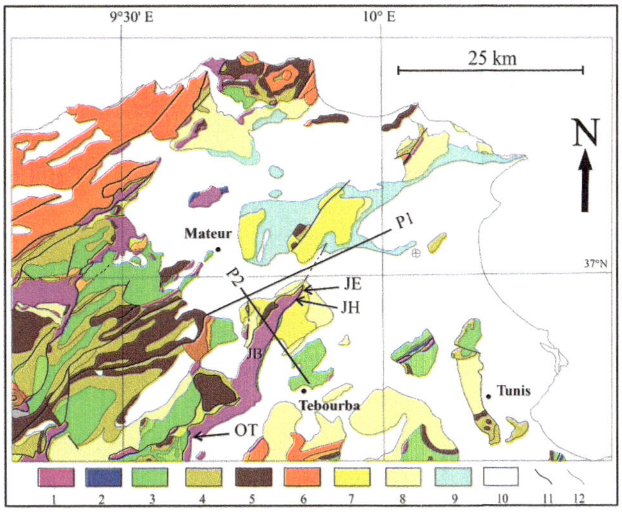

Fig. 1 Synthetic geological map adapted from previous work [14] including seismic profiles traces and localities cited in the text. (1) Triassic. (2) Jurassic. (3) Berriasian-Early Maastrichtian. (4) Late Maastrichtian-Paleocene. (5) Eocene. (6) Oligocene. (7) Aquitanian-Tortonian. (8) Messinian. (9) Pliocene. (10) Quaternary. (11) Faults. (12) Stratigraphic contacts. JE: Jbel es-Sakkak; JH: Jbel el Hallouf; OT: Oued Tazegga. P1 and P2: seismic profiles

Fig. 2 3D interpretation of Jebel el Baouala diapir

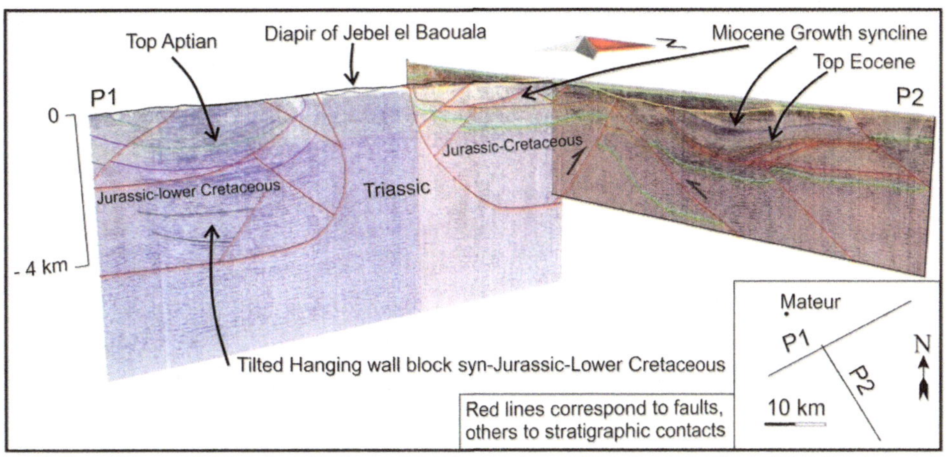

2 Materials and Methods

Satellite images were used to update earlier geological maps of the region, constrained by new field observations. Stratigraphic boundaries, fault intersections and structural orientations were all revised where necessary. The available seismic data (Fig. 2) was depth converted using Midland Valley Move software. This data provided a critical 3D framework to constrain the outcrop observations. Where seismic control was more limited, the structural architecture was reconstructed by extrapolating from neighboring seismic lines.

3 Discussion

Remapping suggests the salt canopy model proposed by Masrouhi et al. [13] and others is incorrect. They argue that Triassic evaporites were extruded onto the sea floor in response to a phase of Lower-Middle Albian extension and then flowed downslope to be covered by Upper Albian marls and limestones. However, this is not supported by field observation in the Oued Tazegga area where mobil Triassic rocks are covered by Lower Cretaceous clay and cherts. In other localities elsewhere, the Triassic is overlain by Aptian sediments. Geological cross sections prepared to illustrate the salt canopy model are contradicted by new tectono-stratigraphic evidences collected during this analysis. Instead this data points to a two-phase evolution of early diapirism followed by later compression/inversion.

4 Conclusions

Remapping of the es Sakkak-el Baouala Range suggests the following structural evolution:

- Jurassic to early Maastrichtian extension with fault-assisted Apto-Albian diapirism.
- Compressional deformation beginning in late Maastrichtian and continuing intermittently to the present time.
- Cenozoic sedimentation controlled by thrust propagating folding, strike slip faulting and salt flowage.
- Development of footwall growth synclines and salt cored structures during the Paleogene and early Neogene.
- Renewed thrust propagation folding in the late Miocene (Messinian) and Plio-Quaternary, responsible for south verging asymmetric anticlines.

References

1. Perthuisot, V.: Dynamique et pétrogenèse des extrusions triasiques en Tunisie septentrionale, Thèse de Doctorat és-Sciences. ed. Travaux du Laboratoire de Géologie, Ecole Normale de Paris (1978)
2. Perthuisot, V., Rouvier, H.: Les diapirs du Maghreb central et oriental: des appareils variés, résultats d'une évolution structurale et pétrogénétique complexe. Bull. Soc. Géol. Fr. **163**, 751–760 (1992)
3. Boukadi, N., Bedir, M.: L'halocinèse en Tunisie: contexte tectonique et chronologie des événements. C. R. Geosci. **231**, 480–482 (1996)
4. Jallouli, C., Chikhaoui, M., Braham, A., Turki, M.M., Mickus, K., Benassi, R.: Evidence for Triassic salt domes in the Tunisian Atlas from gravity and geological data. Tectonophysics **396**, 209–225 (2005). https://doi.org/10.1016/j.tecto.2004.12.003
5. Gharbi, R.A., Chihi, L., Hammami, M., Soumaya, A., Kadri, A.: Manifestations tectono-diapiriques synsédimentaires et polyphasées d'âge Crétacé supérieur–Quaternaire dans la région de Zag Et Tir (Tunisie centre-nord). C. R. Geosci. **337**, 1293–1300 (2005). https://doi.org/10.1016/j.crte.2005.07.006
6. Jallouli, C., Mickus, K., Turki, M.M., Rihane, C.: Gravity and aeromagnetic constraints on the extent of Cenozoic volcanic rocks within the Nefza-Tabarka region, northwestern Tunisia. J. Volcanol. Geotherm. Res. **122**, 51–68 (2003)
7. Truillet, R., Turki, M.M.: La tectonique tangentielle dans la zone des diapirs. l'exemple du Dj. Amar de l'Ariana (Tunisie septentrionale). C. R. Acad. Sci. Paris **291**, 325–327 (1980)
8. Truillet, R., Delteil, J.: Allochtonie Alpine de la "zone des diapirs" (Tunisie septentrionale). C. R. Acad. Sci. Paris 1143–1146 (1982)
9. Vila, J.-M.: Première étude de surface d'un grand "glacier de sel" sous-marin: l'est de la structure Ouenza-Ladjbel-Meridef (confins algéro-tunisiens). Proposition d'un scenario de mise en place et comparaisons. Bull. Soc. Géol. Fr. 149–167 (1995)
10. Vila, J.-M., Ben Youssef, M., Bouhlel, S., Ghanmi, M., Kassa, S., Miaadi, F.: Tectonique en radeaux au toit d'un "glacier de sel" sous-marin albien de Tunisie du Nord-Ouest: exemple du secteur minier de Gueurn Halfaya. C. R. Acad. Sci. Paris 563–570 (1998)
11. Ben Chelbi, M., Melki, F., Zargouni, F.: Mode de mise en place des corps salifères dans l'Atlas septentrional de Tunisie. Exemple de l'appareil de Bir Afou. C. R. Geosci. **338**, 349–358 (2006). https://doi.org/10.1016/j.crte.2006.02.009
12. Masrouhi, A.: Les appareils salifères des régions de Mateur, Tébourba et de Medjez-el-Bab (Tunisie du Nord) (Ph.D. thesis). Tunis-el-Manar University, Tunis (2006)
13. Masrouhi, A., Bellier, O., Koyi, H., Vila, J.-M., Ghanmi, M.: The evolution of the Lansarine-Baouala salt canopy in the North African Cretaceous passive margin in Tunisia. Geol. Mag. **150**, 835–861 (2013). https://doi.org/10.1017/S0016756812000763
14. Ben Haj Ali, M., Jedoui, Y., Dali, T., Ben Salem, H., Memmi, L.: Carte Géologique de la Tunisie au 1/50 000. Ed. Serv. Géol. ONM, Tunis (1985)

USE of Landsat 8 OLI Images to the Characterization of Hercynian Deformation of the Ougarta (South-West Algeria)

Chakib Harouz, Kamel Amri, Rachid Hamdidouche, and Kawther Araibia

Abstract

The range of Ougarta is located in the suture zone between the Panafrican domain of West Africa and the West African craton. It is a strongly folded region; these folds were guided by large pan-African shear zone during the Hercynian and later compression phases. The specific treatments applied to the Landsat 8 OLI images allowed for the determination the lithologies recognized in the field and the structures associated with the large pan-African shear-zones.

Keywords

Ougarta • Panafrican • Hercynian • Remote sensing Structural mapping

1 Introduction

The interpretation of landsat 8 images in the arid areas has been very successful lately [9]. The aim of this study is to demonstrate the contribution of remote-sensing imagery in the characterization of the Hercynian deformation in the region of interest. This is achieved through the combination of field measurements, previous studies and the processing of multispectral data collected using Landsat 8 OLI.

C. Harouz (✉)
LGEOPH/USTHB, BP N°32 El Alia, 16111 Bab Ezzouar, Algeria
e-mail: chakib-harouz@live.fr

K. Amri · K. Araibia
LGGIP/USTHB, BP N°32 El Alia, 16111 Bab Ezzouar, Algeria

R. Hamdidouche
LGBSO/USTHB, BP N°32 El Alia, 16111 Bab Ezzouar, Algeria

2 Materials and Methods

2.1 Geological Setting

The Ougarta range is located in the South-Western region of Algeria (Fig. 1). It also belongs to the North-Western part of the Saharan platform. It is made of two beams: the Saoura in the East and the Daoura in the west. These are presented as elongated and straight ridges NW-SE. Moreover, they are separated by the Earroui erg. The Ougarta range appears as a succession of anticlinal structures oriented NW-SE. The Ougarta is located at the junction between the West African Craton and the Pan-African area of the PFS. It is regarded as the consequence of the converges of these two domains through the panafrican orogeny. However, it is considerably related to previous orogenies, more specifically it is linked to the Hercynian. Fabre et al. [8] suggest a hypothesis about the region of Ougarta. It was located on the passive margin of the West African craton and the suture line would be towards the East in the region of Bled El Mass and major accidents NW-SE would be the result of fracturing this margin.

2.2 Remote Sensing Data and Data Processing

The diversity of the structure and the lithology of the region has induced us to use remote-sensing images so to extract the relevant geological information. We will process the satellite images drawn from the Landstat 8 OLI so to highlight as many structural indicators as possible. Moreover, this will allow for a good discrimination of the lithology in regards to the region of the Ougarta where the fracturation has well evolved.

We have used an extract from a mosaic of two Landstat 8 OLI images, these include the majority of the range, in addition to using an extract from a geological map (Kerzaz 1951).

F. Rossetti et al. (eds.), *The Structural Geology Contribution to the Africa-Eurasia Geology: Basement and Reservoir Structure, Ore Mineralisation and Tectonic Modelling*,
Advances in Science, Technology & Innovation, https://doi.org/10.1007/978-3-030-01455-1_64

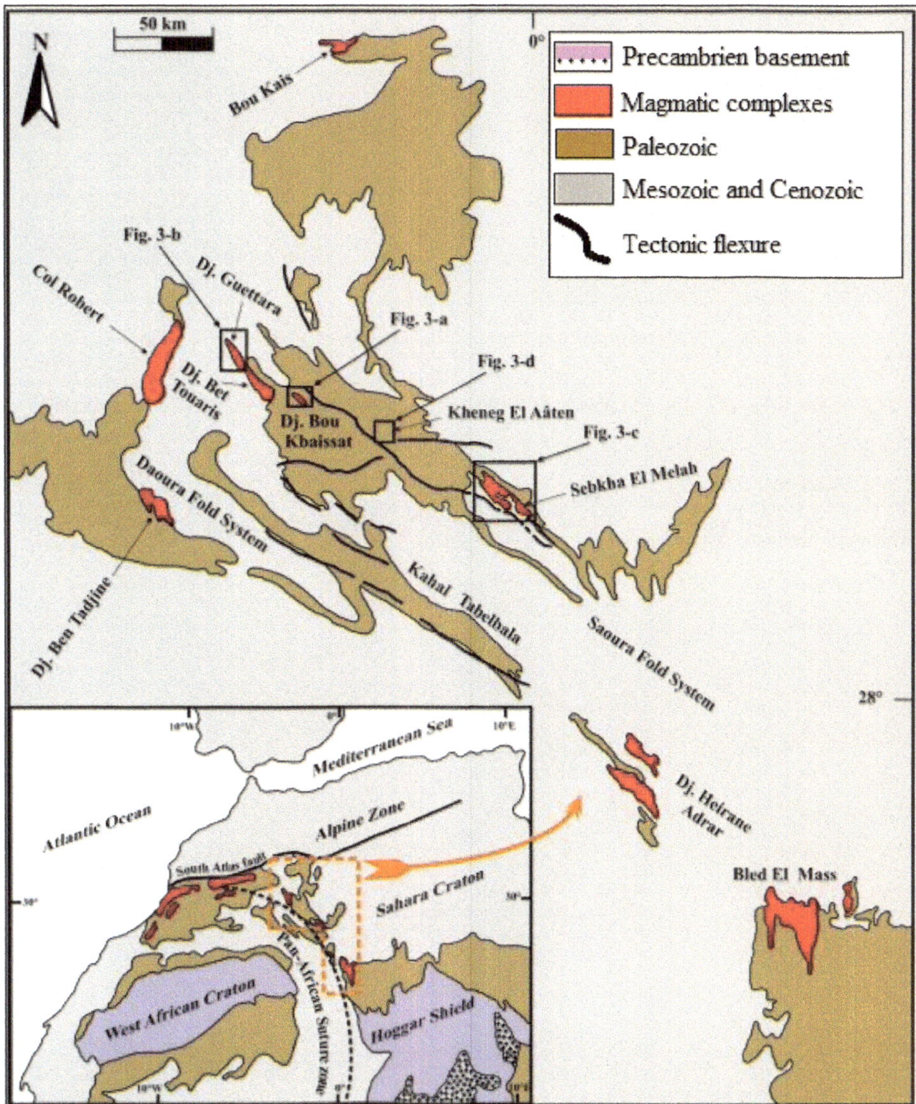

Fig. 1 Geological map of the Ougarta belt (after ([3, 4, 7, 13, 11, 6, 5, 10, 12])

2.3 Different Treatments and Methodology

The ENVI software provides remote-sensing image processing techniques for the design of lithological discrimination and structural analysis. To have an overview of the facies of the region of Ougarta, we have established different treatments namely, those related to the colour compositions, the principal component analyzes and the band ratios.

The band ratios are based on the notion of reflectance, it consists in dividing the digital number (DN) [2] in one band by the DN of another band for the same pixel. The use of these indices makes it possible to reduce the effects of the topography and to enhance the contrast of reflectivity between the mineral surfaces.

The Principal Component Analysis allows to obtain nebands that lose their spectral character and remain in linear combination with the original images. Principal Component Analysis is used to reduce the number of output bands, in order to gather the most relevant information.

The choice of bands for band ratios and PCA is dictated either by the correlation matrix or by the covariance matrix [1]. The selection of the bands was established by the correlation matrix.

This discrimination allowed us to highlight the Precambrian substratum that is outcropping at the anticlines core. It consists of volcanic rocks (Fig. 2a, b).

The determination of faults' networks involves the representation of the different directions. Linear maps and treatments with several 3×3 matrix directional filters (0°, 45°, 90°, 135°) and (R: 0°, G: 45°, B: 135°) (Fig. 3a) were necessary to highlight the different linear directions. The lineaments were extracted automatically on this map (Fig. 3b) for that we used the software "Geomatica".

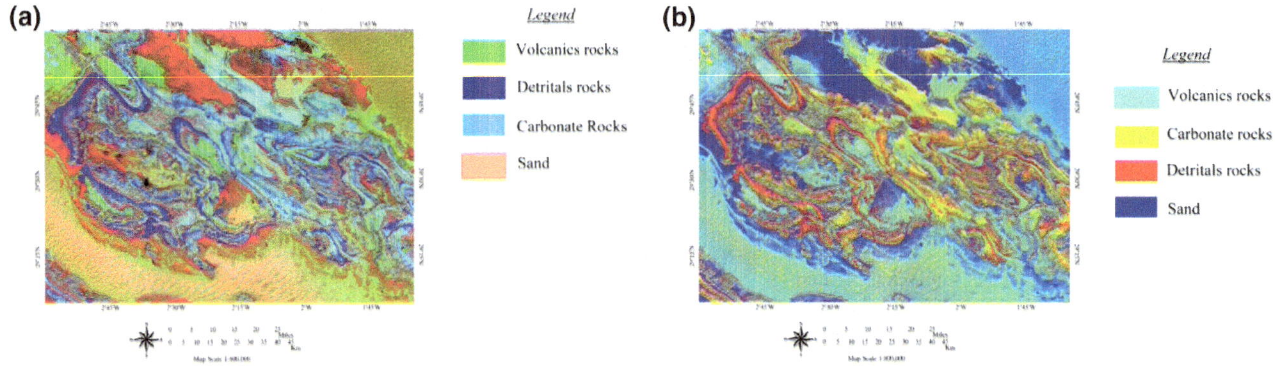

Fig. 2 **a** Color composite of band ratios in R (7/5) G (7/6) B (5/7). **b** Color composite of band ratios in R (7/6) G (7/5) B (5/4)

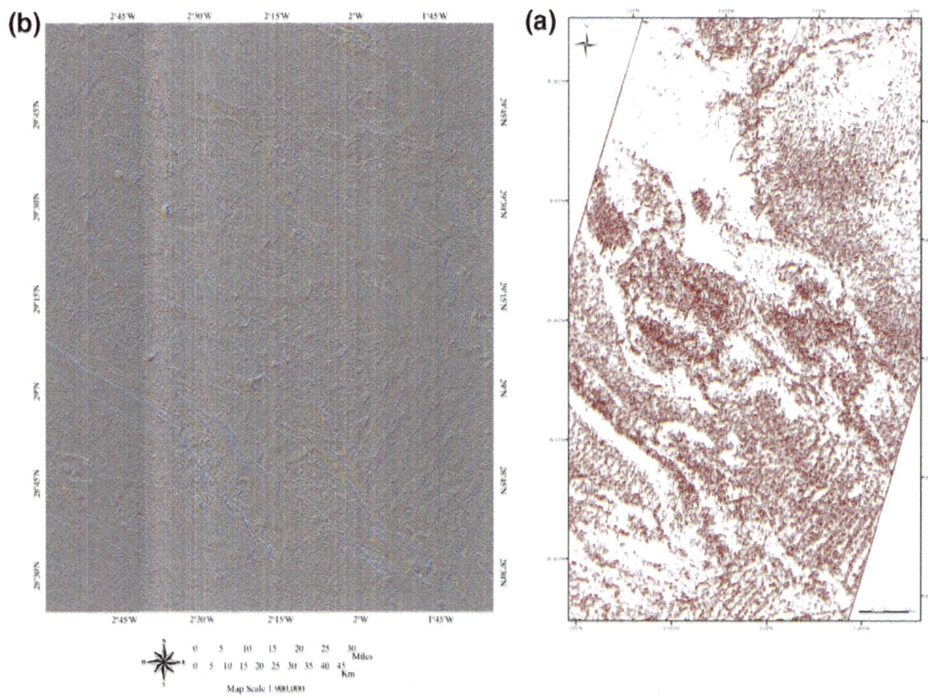

Fig. 3 **a** Directional filters R (0°) G (45°) B (90°); **b** Structural lineaments map

2.4 Result

The processing and analysis of the data collected by the Landsat 8 OLI has provided us with information about the lithology of the region. This has permitted us to locate the structures associated to the hercynian deformation. The contribution of treatments aimed at the structural aspect allowed consistent interpretations. Indeed, the results obtained made it possible to highlight three major directions NW-SE, E-W and NE-SW.

3 Conclusions

These treatments enabled us to delineate the lithologies recognized in the field and the structures associated with the large pan-African shear-zones as well as the folded structures associated with them. It appears that during the Hercynian and later compression phases, the folds were guided by large pan-African shear-zones which are predominantly oriented NW-SE, E-W and NE-SW.

References

1. Amri, K., Mahdjoub, Y., Guergour, L.: Apport des images Landsat 7 ETM+ pour la discrimination lithologique et l'étude structurale de la région d'AfaraHéouine, Tahifer, Hoggar Central. Journée d'Animation Scientifique (JAS09) de l'AUE Alger Novembre 2009, 7p (2009)
2. Brahimi, B.: Cartographie et caractérisation de la déformation de la zone de cisaillement intraIn Ouzzal (Hoggar occidental, Algérie). Thèse du Magister, USTHB, Alger, 97p (2011)
3. Donzeau, M.: Etude structurale dans le Paléozoïque des Monts d'Ougarta. 3 ème cycle thesis, Orsay (1971)
4. Donzeau, M., Zellouf, K.: La jonction Ougarta–Anti–Atlas: structure de la région de Tadaout El Berhil (Saha occidental algérien; Soc. Hist. Nat. Afrique S.N.E.D. Tome 64: Fascicule, 3, 4, pp. 171–184 (1973)
5. Dostal, J., Caby, R., Keppie, J.D., Maza., Neoproterozoic magmatisme in South-Western Algeria (Sebkha el Melah inlier): a northerly extension of the Trans-Saharan orogen. J. Afr. Earth Sci. **35**, 213–225 (2002)
6. Ennih, N., Liégeois, J.P.: The Moroccan Anti-Atlas: the West African passive margin with limited Pan-African activity. Precambr. Res. **112**, 289–302 (2001)
7. Fabre, J.: Introduction à la géologie du Sahara algérien et des régions voisines. Soc. nat. Ed. Diff, ed., Alger, 422p (1976)
8. Fabre, J., Ait-kaci-Ahmed, A., Bouima, T., Moussinie-Pouchkine, A.: Le cycle molassique dans le rameau trans-saharien de la chaine pana-fricaine. J. Afr. Earth Sci. **7**(A), 41–55 (1988)
9. Gani, N.D.S., Abdessalam, M.: Remote sensing analysis of the Gorge of the Nile, Ethiopia with emphasis on Dejen-Gohatsion region. J. Afr. Earth Sci. **44**, 135–150 (2006)
10. Ghienne, J.F., Boumendjel, K., Paris, F., Videt, B., Racheboueuf, P., Salem, H.A.: The Cambrian-Ordovician succession in the Ougarta Range (Western Algeria, North Africa) and interference of the Late Ordovician glaciations on the development of the Lower Palaeozoic transgression on northern Gondwana. Bull. Geosci. **82**, 183–214 (2007). http://dx.doi.org/10.3140/bull.geosci.2007.03.183
11. Seddiki, S., Remaci-Benaouda, N., Cottin, J.Y., Moine, B.N., Ménot, R.P., Perrache, C.: The volcano-sedimentary Boukais inlier (south-western Algeria): evidence for lithospheric thinning during the late Neoproterozoic. J. Afr. Earth Sci. **39**, 257–266 (2004)
12. Villeneuve, M.: Paleozoic basins in West Africa and the Mauritanide thrust belt. J. Afr. Earth Sci. **43**, 166–195 (2005)
13. Villeneuve, M., Cornée, J.J.: Structure, evolution and palaeogeography of the West African craton and bordering belts during the Neoproterozoic. Precambr. Res. **69**, 307–326 (1994)

Part VI

The Alpine Himalayan Convergence Zone

The Role of Izeh Transverse Fault Zone in Zagros, Iran

A. Yassaghi

Abstract

In this work, kinematics of the Izeh fault zone as one of the major transverse faults in Zagros and its influence as younger deformation on the belt structure are analyzed. Izeh fault zone with 165 trend in Azimuth and right-lateral strike-slip kinematics, cross-cuts the High Zagros and the Folded Belt zones of Zagros fold-thrust belt. Surface deformation of Izeh fault zone on the cover sediments includes changes on the trend of the belt major structures and development of minor folds and faults that are developed independently or overprinted on the belt major structures. Detailed structural mapping showed the presence of five major restraining step-over zones between the mapped right lateral strike-slip en-echelon faults along the Izeh fault zone. Oblique convergence of the Arabian Plate with the Central Iran is the case for the reactivation of the fault zone and the surface deformations along it.

Keywords

Izeh fault zone • Zagros • Basement transverse fault
En-echelon fault pattern • Restraining zone

1 Introduction

Zagros fold-thrust belt (ZFTB) is transversely cross-cut by several strike-slip faults. These faults are identified as basement reactivated faults [1, 2]; Burberry [3], lateral ramp of the main Zagros reverse faults [4] and strike-slip faults due to oblique convergence of Arabian Plate and Central Iran (e.g., [5]).

The Zagros transverse faults and their influence on the deformation of the belt were mainly studied based on the interpretation of satellite images [6], integrated remote sensing images with geomagnetic data [2], variety of thickness and sedimentary facies [4], earthquake data [7] and young deformation measurements taken by GPS [8, 9]. These studies are mainly concentrated on the Kazerun and Sabzpushan faults and were not supported by detailed structural evidences.

The Izeh fault zone is defined as an N-trending lineament on the tectonic maps of the Zagros [10]. The main purpose of this work is to document the post collisional role of the Izeh transverse fault zone during the Zagros evolution and to confirm its relation to the basement fault using a combination of remote sensing imagery interpretation, detailed structural mapping and analysis.

2 Methods

We used Landsat ETM+, IRS PAN-1C with resolution of 5.8 m, DEMs obtained from 1:50,000 topographical maps and 90 m SRTM as well as detailed field data to map structures related to the Izeh Fault Zone (Fig. 1).

3 Results

Detailed interpretation of the satellite images as well as field investigation and mapping of the Izeh fault zone structures revealed two main groups of major and minor structures (Fig. 1). Folds that are clearly recognized on satellite images are the major structures and their trends gradually change from the NW to N-trending where they pass the Izeh fault zone. Reverse faults that are the other major structures are also bent where they pass the Izeh fault zone. For example, the trace of the MFF changes from the NW to NNW where it crosses the Izeh fault zone (Fig. 1).

Minor structures mapped for the first time in six selected areas during this study refer to those structures formed within the Izeh fault zone. The orientations of these minor

A. Yassaghi (✉)
Department of Geology, Tarbiat Modares University,
P. O. Box: 14115-175 Tehran, Iran
e-mail: yassaghi@modares.ac.ir

F. Rossetti et al. (eds.), *The Structural Geology Contribution to the Africa-Eurasia Geology: Basement and Reservoir Structure, Ore Mineralisation and Tectonic Modelling*,
Advances in Science, Technology & Innovation, https://doi.org/10.1007/978-3-030-01455-1_65

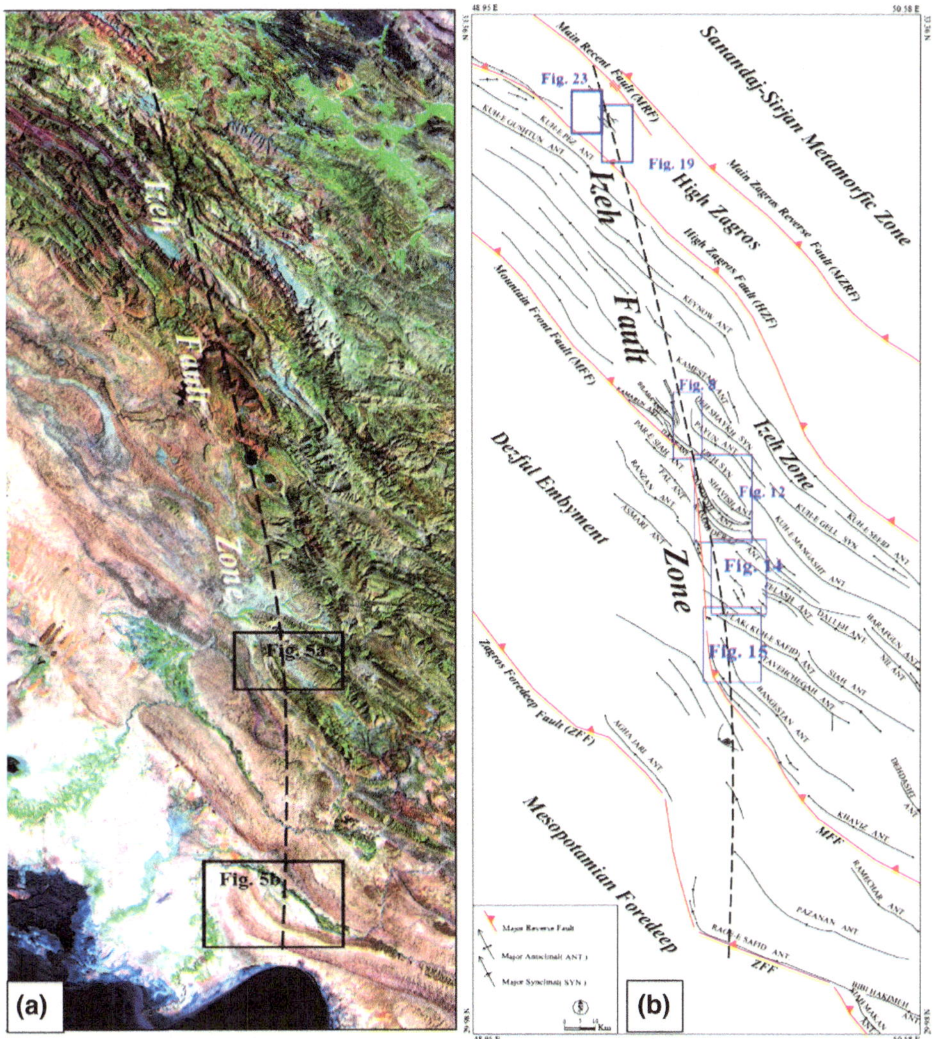

Fig. 1 Satellite image (**a**) and a detailed structural mapping (**b**) of the Izeh fault zone

folds and faults are generally different from that of the Zagros major structures but are printed over them. These minor folds and faults are mainly developed in competent limestone rocks of both Mezosoic and Cenozoic. The distribution pattern of the minor structures between these eight en-echelon strike-slip faults indicates five restraining step-over zones along Izeh fault zone. These restraining zones are the result of the Izeh subsurface fault reactivation.

For a detailed geometric and kinematic analysis of the folds, their aspect ratio (AR) (hinge length/fold width) as geometrical parameter was determined using geological maps, satellite images, field data and underground contour maps. The folds' aspect ratios are divided into two sets. The first set with higher AR > 10 and longer hinge length is considered as forced folds. The second set with AR between 2 and 10 is considered as buckle folds with en-echelon arrangement. It is proposed that the second set is generated within the Izeh fault zone, while the first set is the result of the Zagros shortening along major reverse faults that their axial trends have rotated when they pass the Izeh fault zone.

4 Discussion

The right-lateral strike-slip kinematics of the Izeh fault zone were interpreted using the curvelinear geometry of the major structures affected by the fault as well as the en-echelon pattern of minor structures developed within the fault zone. The minor faults within the Izeh fault zone have en-echelon arrangements and are offset to develop step-over geometry. This geometry in a strike-slip fault system, generally produces either zones of extension at releasing step-over regions or compression at restraining step-over regions [11]. Detailed analyses of the minor folds and thrusts mapped mostly in the six selected areas along the Izeh fault zone indicate that they are spatially developed in step-over zones in between these faults. Since these faults have right-stepping en-echelon patterns, these minor structures have developed in restraining zones between a set of right-lateral strike-slip faults. Therefore, the Izeh fault zone along which these restraining zones have formed themselves is a deep-seated dextral Principal Deformation Zone.

The thickness variations along the Izeh fault zone in the isopach maps drawn for the time interval between the Jurassic till Tertiary imply that the fault is a deep-seated basin bound and therefore provide evidence that the fault is an inherited structure active since at least Jurassic. The presence of a thick viscous Hormuz salt covers of the Zagros basement as it constraint by physical modelling [12] caused formation of a series en-echelon structures on cover sediments during the fault activity.

5 Conclusions

The Izeh basement fault is reactivated as an NE dipping oblique-slip fault with normal dip-slip component since Permian-Triassic (during which the Neotethys is initiated) until the upper Jurassic and its displacement when it has changed to reverse oblique-slip since the beginning of oblique conversion of the Arabian Plate and Central Iran in the upper Cretaceous until the last stage of the continent-continent collision in the upper Miocene. The development of the Izeh fault zone structures on Late Neogene rocks confirms the activity of the Izeh fault zone even after continent-continent collision of the Arabian Plate and Central Iran and based on earthquake data, this reactivation is continuing until the present time.

References

1. Hessami, K., Koyi, H.A., Talbot, C.J.: The significance of strike slip faulting in the basement of the Zagros fold and thrust belt. J. Petrol. Geol. **24**, 5–28 (2001)
2. Yassaghi, A.: Integration of landsat imagery interpretation and geomagnetic data on verification of deep-seated transverse fault lineaments in SE Zagros, Iran. Int. J. Remote Sens. **27**(18–20), 4529–4544 (2006)
3. Burberry, C.M.: The effect of basement fault reactivation on the Triassic—recent geology of Kurdistan, North Iraq. In: Earth and Atmospheric Sciences, vol. 432 (2015)
4. Sepehr, M., Cosgrove, W.: Structural framework of the Zagros fold- thrust belt, Iran. Mar. Petrol. Geol. **21**, 829–843 (2004)
5. Allen, M.: Roles of strike-slip faults during continental deformation: examples from the active Arabia-Eurasia collision. Geol. Soc. London Spec. Publ. **38**, 329–344 (2010)
6. Fürst, M.: Strike-slip faults and diapirism of the South-Eastern Zagros ranges. In: Proceeding of Symposium on Diapirism, vol. 2, pp. 149–181. Bander Abbas, Hormozgan, Iran (1990)
7. Baker, C., Jackson, J., Priestley, K.: Earthquakes on the Kazerun line in the Zagros Mountains of Iran: strike- slip faulting within a fold and thrust belt. Geophys. J. Int. **115**, 41–61 (1993)
8. Palano, M., Imprescia, P., Agnon, A., Gresta, S.: An improved evaluation of the seismic/geodetic deformation-rate ratio for the Zagros Fold-and-Thrust collisional belt. Geophys. J. Int. **213**(1), 194–209 (2017)
9. Vernant, P., et al.: Present-day crustal deformation and plate kinematics in the Middle East constrained by GPS measurements in Iran and Northern Oman. Geophys. J. Int. **157**, 381–398 (2004)
10. NIOC: Tectonic map of south-west Iran. Scale 1:2,500,000. Natl. Iran. Oil. Co., Explor. prod., Tehran, Iran (1976)
11. Sylvester, G.: Strike slip faults. Geol. Soc. Am. Bull. **100**, 1666–1703 (1988)
12. Darrell, S., Ferrill, D.A., Stamatakos, J.A.: Role of ductile decollement in the development of pull-apart basins: Experimental results and natural examples. J. Struct. Geol. **21**, 533–554 (1999)

The Eastern Black Sea and Caucasus Domain Origin and Its Tectonic Evolution: New Insights from Results of a Decade of Field Works and of Geophysical Research

Marc Sosson, Randell Stephenson, Shota Adamia, Ara Avagyan, Talat Kangarli, Vitaly Starostenko, Tamara Yegorova, Yevgeniya Sheremet, Eric Barrier, Yann Rolland, Marc Hassig, Zoé Candaux, Victor Alania, Onice Enukidze, Nino Sadradze, Lilit Sahakyan, Ghazar Galoyan, and Sargis Vardanyan

Abstract

The current work is focused on our main results concerning the tectonic evolution of the Black Sea-Greater Caucasus domain obtained during decades of detailed studies in the framework of DARIUS programme and GDRI South Caucasus Geosciences project of the CNRS/INSU. Results of this analytical work allowed us to establish a paleo-reconstruction of the whole of the southern margin of the European craton, during much of the Phanerozoic (Barrier E et al in Paleotectonic reconstruction of the Central Tethyan Realm. CCGM/CGMW, Paris, 2018 [3]). Focusing on Black Sea-Greater Caucasus domain we present a state of the art about analysis of this area, highlighting the perspectives of new investigations at question not solved by the tectonic and geodynamic evolution of these basins.

Keywords

Black sea basins • Greater caucasus • Lesser caucasus • Neotethys • Paleotectonic reconstruction

1 Introduction

The Black Sea-Caucasus (BS-CA) domain (Fig. 1) is the complex result of the Meso-Cenozoic tectonic history of the Tethys Ocean (Paleotethys and Neotethys) closure. Despite scenarios proposed to explain the BS-CA basins opening, the underlying processes controlling their formations remain unclear (e.g. [4, 21–23, 36]).

One of the reasons is the unknown nature of the lithosphere (oceanic? or highly extended continental lithosphere?) of the BS and CA basins. The other questions are: How has the BS basin preserved its existence when this region sustained shortening due to two collisions since the Latest Cretaceous? How do structural heterogeneities within a strong continental lithosphere control the opening of the BS-CA basins, as back-arc basins (BABs), behind a long-lived subduction zone? Was it a slab roll-back in the upper mantle that influenced on the stress field, driving the opening of the BS-CA BABs within the thick Eurasian continental lithosphere? This short paper outlines the main results and hypothesis, worked out by our working group (DARIUS programme and the International Research Group (GDRI) South Caucasus Geosciences project of the CNRS/INSU France, Armenia, Azerbaijan, Georgia, Ukraine) and points out to the scientific objectives for this optimal tectonic domain [30, 32].

M. Sosson (✉) · Y. Sheremet · Y. Rolland · Z. Candaux
S. Vardanyan
University Côte d'Azur, CNRS, UNS, Obs. Côte d'Azur, IRD, UMR Géoazur, 06560 Valbonne, France
e-mail: sosson@geoazur.unice.fr

R. Stephenson · S. Vardanyan
University of Aberdeen, Aberdeen, UK

S. Adamia · V. Alania · O. Enukidze · N. Sadradze · S. Vardanyan
I. Javakhishvili Tbilisi State University, Tbilisi, Georgia

A. Avagyan · L. Sahakyan · G. Galoyan · S. Vardanyan
Institute of Geological Sciences, Yerevan, Armenia

T. Kangarli
Institute of Geology, Baku, Azerbaijan

V. Starostenko · T. Yegorova
Subbotin Institute of Geophysics, Kiev, Ukraine

E. Barrier
Université Paris VI P&M Curie, CNRS, ISTEP, Paris, France

M. Hässig
Department of Earth Sciences, University of Geneva, Geneva, Switzerland

F. Rossetti et al. (eds.), *The Structural Geology Contribution to the Africa-Eurasia Geology: Basement and Reservoir Structure, Ore Mineralisation and Tectonic Modelling*, Advances in Science, Technology & Innovation, https://doi.org/10.1007/978-3-030-01455-1_66

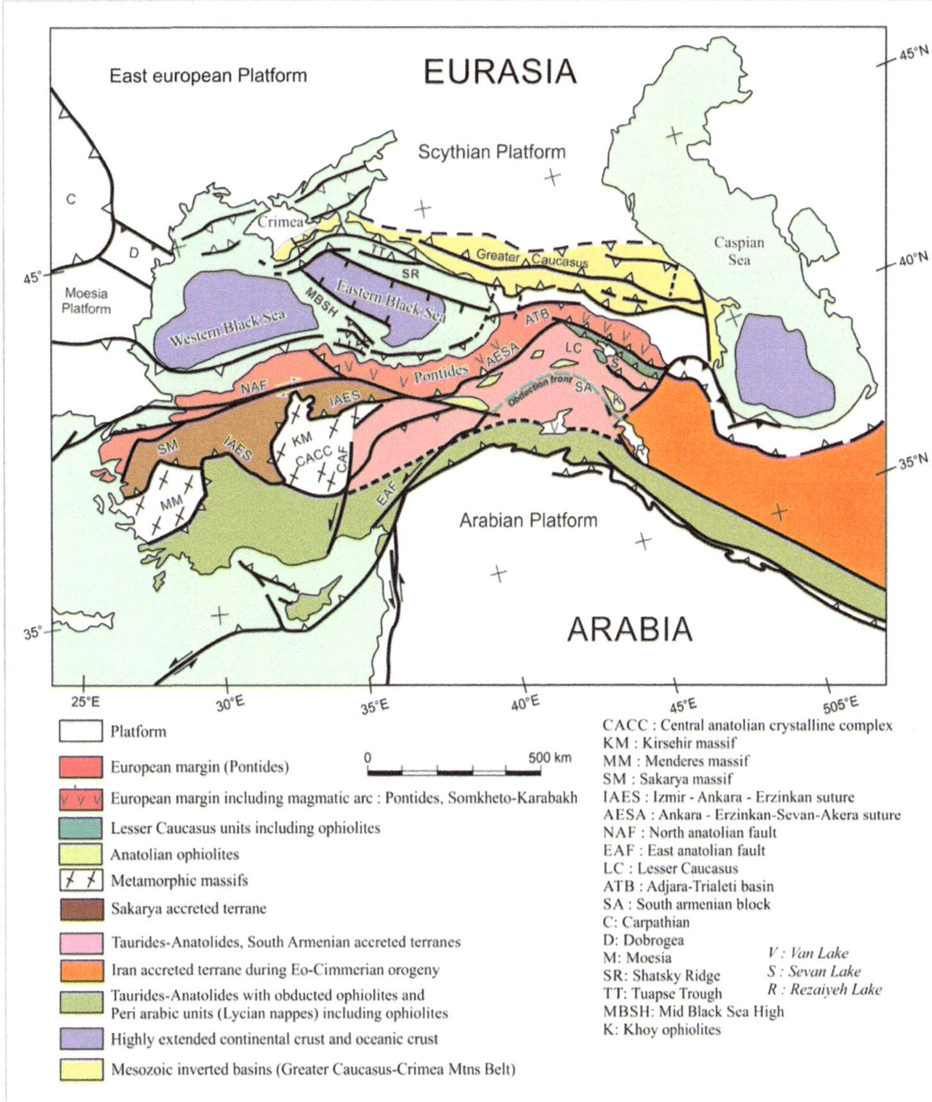

Fig. 1 Structural sketch map of the Black Sea—Caucasus domain, based on our results, pre-Quaternary structures are shown in the BS (modified from [31]

2 Geodynamic and Tectonic Settings

The Meso-Cenozoic BABs in the BS-CA domain formed within the strong European continental lithosphere lying over a long-lived subduction zone (100–120 Ma of duration) ([1–3, 5, 7–10, 14, 16–20, 24–27, 35–37]). Long-lived subduction process is evinced by tomographic images obtained beneath Eurasia and Anatolia. These images reveal the presence of what can be interpreted as cold subducted lithosphere, remnant of the Neotethys ocean, in the upper (from 500 to 660 km depth) and lower mantle [6, 15, 34, 38].

The process of subduction produced the GC basin (a BAB) opening in the Early-Middle Jurassic (the oceanic crust was not formed), and then the Eastern BS basin which opened during the Cretaceous (see paleo-tectonic maps of Barrier et al. [3], e.g. [28, 29] and ref. inside). The Eurasian margin, however, was inverted: First, in the Late Triassic-Early Jurassic (Cimmerian orogeny) and then, during Late-Cretaceous-Paleocene (Alpine orogeny). The paleo-tectonic reconstruction suggests that in the late Early Jurassic (Toarcian) the estimation length of the oceanic plate between Gondwana and Laurasia was approximately 3500 km. Thus, pre-existing geodynamic settings, obviously influenced on the following opening of the BS basins, as well as on the obduction process within the Neotethys plate, which will conduct the ophiolitic units to be over thrusted on top of the Taurides-Anatolides-South Armenia microplate (TASAM) [11–13, 31].

3 Results

The detailed field observations (mapping, structural data), analyses (micropaleontological and geochronological datings, petrological data, geochemistry) and geophysical data (seismic profiles) allow reconsidering tectonic evolution of the BS-CA domain. Our main results are as following:

- Significantly upgraded the knowledge about the tectonic evolution of the Crimean Mountains (CM) and performed the tectonic reconstruction of the northern BS margin that includes three main tectonic provinces: North Dobrogea, CM and the GC. It allowed the precising of the timing of the BS basins' rifting stage during the Cretaceous.
- Based on the aforementioned, we suppose that the reactivation of inherited structures within the Eurasia lithosphere is a main parameter controlling the location of the extended crust in the BABs of BS and GC basins.
- Established modalities and timing of the late Cretaceous obduction of the Sevan-Akera back-arc basin on the TASAM.
- In the Lesser Caucasus (LC) there is only one suture zone, the Sevan-Akera, which is the eastward continuity of the Izmir-Ankara-Erzincan suture zone. It corresponds to the suture between Eurasia and the TASAM from the latest Cretaceous to the Late Eocene.
- Defined structures and their propagation (thin versus thick skin tectonic) of the LC and GC foreland basins from Paleocene to Miocene and structural patterns of inverted Paleocene-Eocene Adjara-Trialeti basin.
- Performed paleotectonic and palinsplastic reconstruction (since the Jurassic) of the convergence (till Miocene) between the TASAM and Eurasia margin.

4 Conclusion

Analysis of our results lead to the following hypotheses: (1) structural heterogeneities (mainly anisotropic lithospheric-scale shear zones) within the Eurasian upper plate could be responsible for the geometry and the spatio-temporal evolution of the GC-BS BABs (during rifting and during inversion), (Permo-Triassic basins, BS and GC); (2) the forces driving the opening of BABs are directly related to subducted slab motion and its behaviour in the mantle (roll back, break-off, slab window?).

References

1. Adamia, S.A., et al.: J. Struct. Geol. **3**, 437–447 (1981)
2. Barrier, E., Vrielynck, B.: Commission for the Geologic Map of the World (CCMW, CCGM), Paris (2008)
3. Barrier, E., et al.: Paleotectonic Reconstruction of the Central Tethyan Realm. CCGM/CGMW, Paris, http://www.ccgm.org. Atlas of 20 maps (scale: 1/15,000,000) (2018)
4. Cloetingh, S., et al.: Sed. Geol. **156**, 169–184 (2003)
5. Dercourt, J., et al.: Tectonophysics **123**, 241–315 (1986)
6. Faccenna, C., et al.: Earth Planet. Sci. Lett. **242**, 85–97 (2006)
7. Finetti, I., et al.: Bolletino di Geofisica ed Applicata **30**, 197–324 (1988)
8. Forte, A.M., et al.: GSA Bull. **122**(3–4), 465–486 (2010). https://doi.org/10.1130/b26464.1
9. Forte, A.M., et al.: Tectonics **32** (2013). https://doi.org/10.1002/tect.20032
10. Görür, N.: Tectonophysics **147**, 247–262 (1988)
11. Hässig, M., et al.: Terra Nova **28**, 76–82 (2016a)
12. Hassig, M., et al.: J. Geodyn. **96**, 35–49 (2016b)
13. Hässig, M., Rolland, Y., Sosson, M.: Geol. Soc. London Spec. Publ. **428** (2017). https://doi.org/10.1144/sp428.10
14. Khain, V.Y.: American J. Sci. **274**, 16–23 (1974)
15. Lei, J., Zhao, D.: Earth Planet. Sci. Lett. **257**, 14–28 (2007)
16. Letouzey, J., et al.: In: International Symposium on the Structural History of the Mediterranean Basins, Technip, Paris, pp. 363–376 (1977)
17. Mosar, J., et al.: Geol. Soc. London Spec. Publ. **340**, 261–280 (2010)
18. Nikishin, A., et al.: (2012) http://dx.doi.org/ https://doi.org/10.3906/yer-1005-22
19. Nikishin, A., et al.: (2013). https://doi.org/10.3103/S0145875213030058
20. Nikishin, A., et al.: (2015). https://doi.org/10.1016/j.marpetgeo.2014.08.017
21. Okay, A.I., et al.: Geology **22**, 267–270 (1994)
22. Okay, A.I., et al.: Tectonics **32**, 1247–1271 (2013)
23. Okay, A.I., Nikishin, A.M.: Int. Geol. Rev. **57**, 1051–1076 (2015). https://doi.org/10.1080/00206814.2015.1010609
24. Robinson, A., et al.: Mar. Pet. Geol. **13**, 195–223 (1996)
25. Rolland, Y.: Gondwana Res. **49**, 130–146 (2017)
26. Saintot, A., Angelier, J.: Tectonophysics **357**, 1–31 (2002)
27. Saintot et al.: Geol. Soc., London Mem. 481–505 (2006)
28. Sheremet, Y., et al.: Geol. Soc. London Spec. Publ. **428**, 265 (2016). https://doi.org/10.1144/SP428.14
29. Sheremet, Y., et al.: Tectonophysics **688**, 84–100 (2016). https://doi.org/10.1016/j.tecto.2016.09.015
30. Sosson, et al.: Geol. Soc. London Spec. Publ. **340**, 329–352 (2010)
31. Sosson, M., et al.: C. R. Geosci. (2016). https://doi.org/10.1016/j.crte.2015.11.002
32. Sosson, M., et al.: Geol. Soc. London Spec. Publ. **428** (2017). https://doi.org/10.1144/SP428.16
33. Spadini, G., et al.: Tectonophysics **266**, 139–154 (1996)
34. Spakman, W.: Geophys. J. Int. **107**, 309–332 (1991)
35. Stampfli, et al.: Episodes **24**, 222–228 (2001)
36. Stephenson, R.A., Schellart, W.: Geol. Soc. London Spec. Publ. **340**, 11–21 (2010)
37. Zonenshain, L.P., Pichon, X.: Tectonophysics **123**, 181–211 (1986)
38. Zor, E.: Geophys. J. Int. **175**, 1273–1282 (2008). https://doi.org/10.1111/j.1365-246X.2008

Northern Dobrogea (Romania) and the Crimean Mountains (Russian Federation): The Key Areas in the Tectonic Evolution of the Black Sea Basin

Yevgeniya Sheremet, Marc Sosson, Antoneta Seghedi, and Mihaela C. Melinte-Dobrinescu

Abstract

The work poses the question about the impact of inherited structures in the Black Sea back-arc basin (BAB) tectonic evolution. The new structural analysis of the Northern Dobrogea (ND) and the Crimean Mountains (CM) shows that the origins of structural patterns of both regions are in close relationship with deep faults/or fault zones. The comparative analyses of structures, of tectonic stages and their duration allow us to better understand the connections in time and space between the ND and the CM, against the back-ground of the long-living subduction. In particular during: (1) the Cimmerian orogeny; (2) the opening of the BS and (3) the inversion of the BS during the Cenozoic shortening.

Keywords

Crimean mountains • Northern dobrogea
Tectonic evolution • Structural analysis
Black sea basins

1 Introduction

The underestimation of the impact of ancient crustal structures in the formation and configuration of the back-arc basin (BAB) in the geodynamic context of subduction can be the cause of why non-of the proposed scenarios concerning the opening of the BS basin e.g. [1, 2] can explain all the geological facts and, especially, the new data, observed and described around the BS in recent decades.

The Crimean Mountains (CM) and Northern Dobrogea (ND) represent key areas where this question can be studied: the CM exposes inverted ancient structures of the Triassic (or earlier) rift e.g. [1], while the ND lies in the vicinity of the longest lineament of Europe—the Tornquist-Teisseire fault zone, extending up to the Black Sea e.g. [3] (Fig. 1). Reconsidering the geology of the northern part of the Black Sea basin in the context of new investigation will bring us closer to understanding the pre- and post-tectonic evolution of the Black Sea basin.

In the current work we present comparative results of structural analysis in the ND and the CM based on field observation, new sampling for dating (micropale-ontology), collecting tectonic data (paleo stress field) and seismic line interpretation from the Dobrogea foredeep (Ukraine) that in its turn adds new puzzles into the fol-lowing tectonic stages of the BS tectonic history as the Cimmerian orogeny, the opening of the BS and its inversion during the Cenozoic.

2 Tectono-Stratigraphic Settings of the Crimean Mountains and of the Northern Dobrogea

The Mesozoic stratigraphy of both regions, in view of recent geological and geophysical data obtained about the CM [4, 5, 6, 7] has certain similarities allowing to trace out a space-time correlation within the main tectonic stages of the Black Sea evolution: (1) the Triassic carbonated facies of the ND and flysch basin deposits of the CM link to (a) a formation of the Triassic fore-/back-arc Trough within the southern Laurasian margin due to northward subduction of Paleotethys ocean plate [8–14], and (b) the inversion of this basin during the Cimmerian orogeny (Latest Triassic-Earliest Jurassic time-span) e.g. [10]; (2) the Cretaceous

Y. Sheremet (✉) · M. Sosson
University Côte d'Azur, UNS, CNRS, Obs. de La Côte d'Azur, IRD, UMR Géoazur, 06560 Valbonne, France
e-mail: sheremet@geoazur.unice.fr

A. Seghedi
GeoEcoMar, Str. Dimitrie Onciul, Nr. 23-25, 024053 Bucharest, Romania

M. C. Melinte-Dobrinescu
National Institute of Marine Geology and Geo-Ecology, 23-25 Dimitrei Onciul Street, 024053 Buharest, Romania

F. Rossetti et al. (eds.), *The Structural Geology Contribution to the Africa-Eurasia Geology: Basement and Reservoir Structure, Ore Mineralisation and Tectonic Modelling*,
Advances in Science, Technology & Innovation, https://doi.org/10.1007/978-3-030-01455-1_67

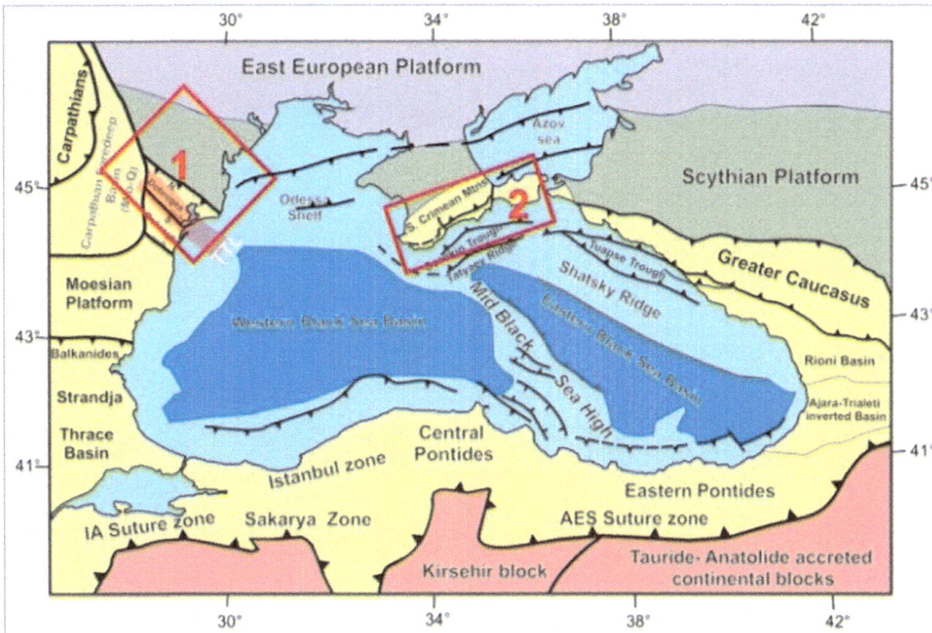

Fig. 1 General structural sketch map of the Black Sea area, modified from Sheremet et al. [4]. Red squares correspond to the ND (1) and CM (2); TTL—Tornquist-Teisseire fault zone

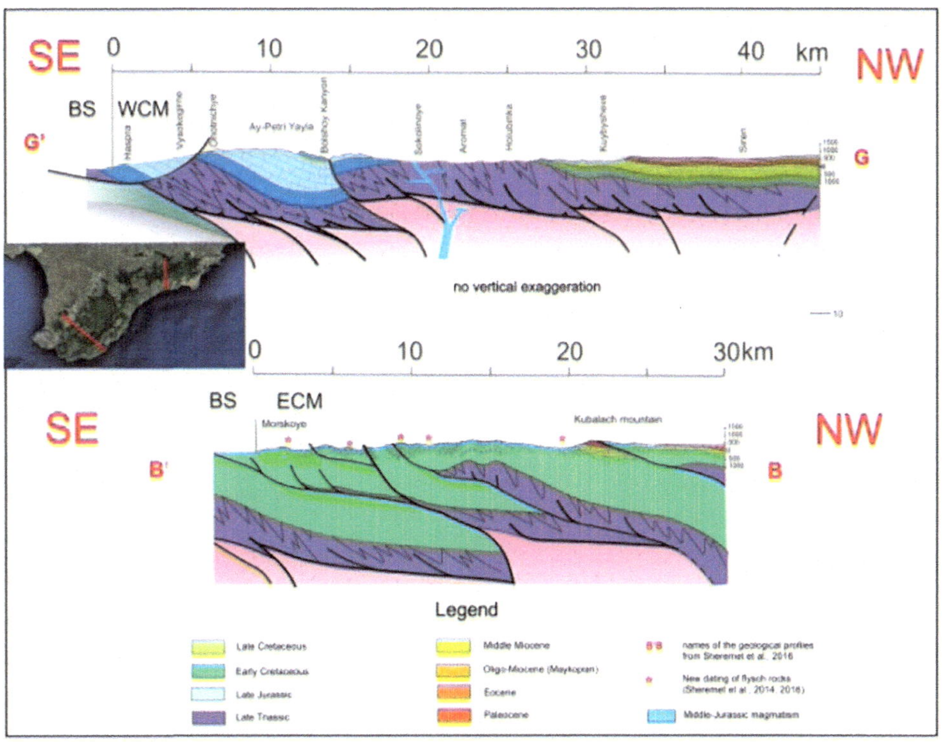

Fig. 2 Geological cross-sections via WCM (upper) and ECM (lower)

terrigenous deposits and carbonates of both areas [4, 6, 7, 13, 15, 14, 16] reveal main features of the Black Sea rifting.

Due to the absence of the Cenozoic rocks in the Northern Dobrogea, the tectonic stage of the compression, initiated by the collisional processes e.g. [7], in the ND and the CM could be compared using the structural analysis, the preliminary results of which are shown below.

3 Results

Results of the structural analysis of the CM are shown on the following Figs. (2, 3 and 4). There is a significant difference in the age of sediments involved in the deformations within the Western and Eastern CM (WCM and ECM respectively) (Fig. 2).

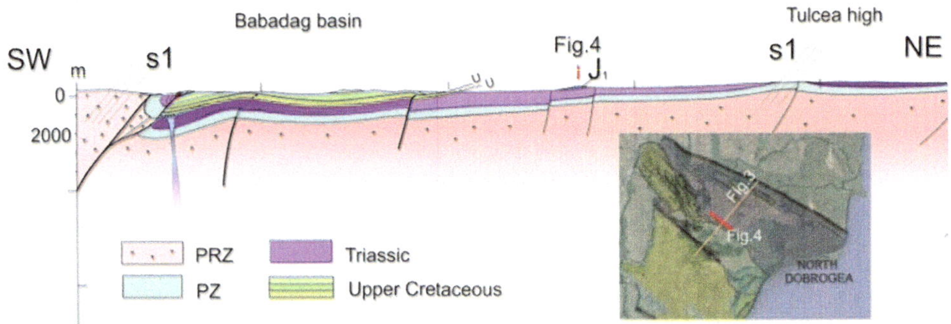

Fig. 3 Geological cross-section of the ND. PRZ—proterozoic basement; PZ—paleozoic rocks

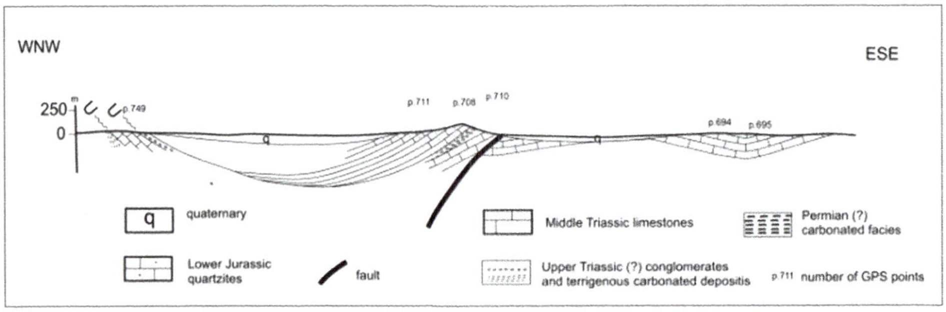

Fig. 4 Geological cross-section, showing the Cimmerian structures

The field data from the ND (Fig. 3) show post Cretaceous compression, while that of Fig. 4, built perpendicular to the SW-NE section indicates the Cimmerian deformations (Latest Triassic-Earliest Jurassic).

4 Discussion

Preliminary results of structural analysis from the ND (Fig. 4) allow us to precise the northerly vergence e.g. [13] of the compressional trend of the Cimmerian orogeny as NW-SE (same as in the Western CM) and not NE, as it was earlier suggested e.g. [17].

The opening of the Cretaceous Babadag basin as well as its Cenozoic inversion occurred immediately near the NW-SE deep fault zones, the TTL, presented with one of its faults, the Pecenega-Camena fault. In the CM, the NW-SE fault zone predetermined the different geological compositions of the Western and of the Eastern CM. The results obtained from the CM e.g. [4, 6] together with the Albian rift-related unconformity of the Western CM, proved that opening of the Eastern BS preceded the one of the Western BS basin.

5 Conclusions

The ancient crustal structures predetermined the tectonic evolution of the CM and ND, and, in turn, impacted the BS BAB's tectonic evolution.

References

1. Sydorenko, G., Stephenson, R., Yegorova, T., Starostenko, V., Tolkunov, A., Janik, T., Omelchenko, V.: Geological structure of the northern part of the Eastern Black Sea from regional seismic reflection data including the DOBRE-2 CDP profile. Geol. Soc. London Spec. Publ. **428**, SP428-15 (2016)
2. Stephenson, R.A., Schellart, W.: The Black Sea back-arc basin: insights to its origin from geodynamic models of modern analogues. Geol. Soc. London Spec. Publ. **340**, 11–21 (2010)
3. Seghedi, A.: Palaeozoic formations from Dobrogea and Pre-Dobrogea–an overview. Turkish J. Earth Sci. **21**(5) (2012)
4. Sheremet, Y., Sosson, M., Gintov, O., Muller, C., Yegorova, T., Murovskaya, A.: Key problems of stratigraphy in the Eastern Crimea Peninsula, some insights from new dating and structural data. Geol. Soc. London Spec. Publ. **428** (2016a)
5. Meijers, M.J.M., Vrouwe, B., van Hinsbergen, D.J.J., Kuiper, K. F., Wijbrans, J., Davies, G.R., Stephenson, R.A., Kaymakci, N., Matenco, L., Saintot, A.: Jurassic arc volcanism on Crimea (Ukraine): implications for the paleo-subduction zone configuration of the Black Sea region. Lithos **119**, 412–426 (2010). https://doi.org/10.1016/j.lithos.2010.07.017
6. Sheremet, Y., Sosson, M., Ratzov, G., Sydorenko, G., Voitsitskiy, Z., Yegorova, T., Gintov, O., Murovskaya, A.: An offshore-onland transect across the north-eastern Black Sea basin (Crimean margin): evidence of Paleocene to Pliocene two-stage compression. Tectonophysics **688**, 84–100 (2016)
7. Sosson, M., Stephenson, R., Sheremet, Y., Rolland, Y., Adamia, Sh, Melkonian, R., Kangarli, T., Yegorova, T., Avagyan, A., Galoyan, Gh., Danelian, T., Hässig, M., Meijers, M., Müller, C., Sahakyan, L., Sadradze, N., Alania, V., Enukidze, O., Mosar, J.: The eastern Black Sea-Caucasus region during the Cretaceous: New evidence to constrain its tectonic evolution. C. R. Geoscience **348** (2016), 23–32 (2016). https://doi.org/10.1016/j.crte.2015.11.002
8. Barrier, E., Vrielynck, B.: Palaeotectonic Maps of the Middle East. Tectono-sedimentary-palinspastic Maps from Late Norian to

Piacenzia. Commission for the Geological Map of the World (CGMW/CCGM)/UNESCO. (http://www.ccgm.org) Atlas of 14 maps, scale 1/18 500 000 (2008)

9. Nikishin, A.M., Ziegler, P.A., Bolotov, S.N., Fokin, P.A.: Late Palaeozoic to Cenozoic evolution of the Black Sea-Southern Eastern Europe region: a view from the Russian platform. Turkish J. Earth Sci. **20**, 571e634 (2012). http://dx.doi.org/ https://doi.org/10.3906/yer-1005-22
10. Okay, A.I., Altiner, D., Kilic, A.M.: Triassic limestone, turbidites and serpentinite–the Cimmeride orogeny in the Central Pontides. Geol. Mag. **152**(03), 460–479 (2015)
11. Sengör, A.M.C.: The Cimmeride Orogenic System and the Tectonics of Eurasia. Geol. Soc. Am. Spec. Pap. **195**, 81 (1984)
12. Ustaömer, T., Robertson, A.H.F.: Late Palaeozoic marginal basin and subduction-accretion: the Palaeotethyan Küre complex, central Pontides, northern Turkey. J. Geol. Soc. **151**(2), 291–305 (1994)
13. Tari, G., Dicea, O., Faulkerson, J., Georgiev, G., Popov, S., Stefanescu, M., Weir, G.: Cimmerian and Alpine stratigraphy and structural evolution of the Moesian Platform (Romania/Bulgaria). AAPG Memoir **68**, 63–90 (1997)
14. Okay, A.I., Sunal, G., Sherlock, S., Altıner, D., Tüysüz, O., Kylander-Clark, A.R.C., Aygül, M.: Early Cretaceous sedimentation and orogeny on the active margin of Eurasia: southern Central Pontides, Turkey. Tectonics **32**, 1247–1271 (2013)
15. Krezsek, C., Bercea, R.-I., Tari, G., Ionescu, G.: Cretaceous sedimentation along the Romanian margin of the Black Sea: inferences from onshore to offshore correlations. In: Simmons, M. D., Tari, G.C., Okay, A.I. (eds.) Petroleum Geology of the Black Sea. Geological Society, London, Special Publications, 464. https://doi.org/10.1144/SP464.10
16. Okay, A.I., Nikishin, A.M.: Tectonic evolution of the southern margin of Laurasia in the Black Sea region. Int. Geol. Rev. **57**(5–8), 1051–1076 (2015). https://doi.org/10.1080/00206814.2015.1010609
17. Hippolyte, J.-C.: Geodynamics of Dobrogea (Romania): new constraints on the evolution of the Tornquist-Teisseyre Line, the Black Sea and the Carpathians. Tectonophysics **357**, 33–53 (2002)

Caucasian-Arabian Syntaxis, The Alpine-Himalayan Continental Collisional Zone

Evgenii V. Sharkov

Abstract

The Caucasian-Arabian segment of the huge Late Cenozoic Alpine-Himalayan orogenic belt marks a zone of the indentation of the Arabian plate into the Eurasian plate. The segment is formed by two tectonic domains: (1) the EW-striking Greater Caucasus (GC) at the north, which represents the southern margin of East European Craton, tectonically uplifted along the Main Caucasian Fault (MCF), and (2) Caucasian-Arabian Syntaxis (CAS) at the south; it includes arc-like tectonic domains of the Lesser Caucasus and Eastern Anatolia, vast thrust-folded belt between Eurasian and Arabian plates. A large NS-trending positive isostatic anomaly is characterized for CAS which suggests the presence of a mantle plume head beneath it. CAS is characterized by intense Neogene-Quaternary volcanism where two different types of magmatism occur: (1) extensive basaltic plateau with geochemical features of plume-related rocks, as well as (2) calc-alkaline volcanics, which are typical for supra subduction settings. However, according to seismic data, there is no subduction zone beneath the GC, but some north-directed seismic zones up to 50–70 km depth occurred beneath the CAS which, probably, can generate calc-alkaline magmas. So, we are suggesting that "mixed" magmatism in Caucasian-Arabian convergence considered here two independent magma sources: mantle plume and "embryonic" subduction zones.

Keywords

Greater caucasus • Caucasian-Arabian syntaxis Continental collision • Mantle plume • Foreland's tectonomagmatic processes • "Mixed" magmatism

E. V. Sharkov (✉)
Institute of Geology of Ore Deposits, Petrography, Mineralogy and Geochemistry RAS, Moscow, Russia
e-mail: sharkov@igem.ru

1 Introduction

The Caucasian-Arabian segment of the Alpine-Himalayan belt consists of two domains: the EW-striking Greater Caucasus (GC) at the north and the Caucasian-Arabian Syntaxis (CAS) at the south. The Greater Caucasus is represented by the south margin of the East European Crston, uplifted along the Main Caucasian Fault [11]. The latter is a part of modern super-regional mega-fault which stretches out from Kopetdag through the Caspian Sea, Caucasus and Crimea; very likely, that its further continuation is the Tornquist-Teisseyre Fault. This mega-fault separates areas of Alpine convergence from Eurasian plate sensu stricto [19]. The CAS includes arc-like tectonic domains of the Lesser Caucasus and Eastern Anatolia which are represented by thrusted and folded belt between Eurasian and Arabian plates, i.e. between the MCF in the north and Bitlis-Zagros Fault (BZF) in the south. A large NS-trending positive isostatic anomaly is characterized by CAS, which assumes the presence of a mantle plume head beneath it, which conforms to the existence of modern plume-related volcanics here [19]. So, the discussed below case represents an example of continental convergence above a mantle plume head, and this work is devoted to identifying the geological-petrological features of such convergence.

2 Geological, Petrological and Geophysical Data

2.1 Geological Features

The modern (Alpine) structure of the Caucasus is formed by NS horizontal compression generated by the interaction of two plates: the Arabian indenter and the East European Craton. In the late Cenozoic, this plate interaction resulted in the transverse shortening of the CAS up to 400 km, mainly at the expense of the south territory of the Main Caucasian

F. Rossetti et al. (eds.), *The Structural Geology Contribution to the Africa-Eurasia Geology: Basement and Reservoir Structure, Ore Mineralisation and Tectonic Modelling*,
Advances in Science, Technology & Innovation, https://doi.org/10.1007/978-3-030-01455-1_68

Fault (MCF) [1, 11]. Many scientists believe that the MCF is due to the over-thrust or under-thrust of the Transcaucasian massif under the Greater Caucasus, e.g. [7, 13, 18–21]. However, direct geological observations suggest the linear shape of the fault, as well as rather steeply dipping imbricate structures and/or reverse faults. In addition, geophysical data suggest a steep or vertical angle of the MCF up to the depths of 70–80 km [11, 17, 20]. Consequently, according to Leonov [11], the fact that the MCF is a reverse fault with a large rate of vertical displacement and minimal horizontal displacement should be accepted.

In this case, the shortening south of the MCF could result from a lateral "diffluence" or "spreading" of lithospheric material under the push of the Arabian indenter in two opposite directions away from the GC in front of the rigid East European Craton (Fig. 1). More evidence for this comes from geological [8] and GPS data from the zone of the Africa-Arabia-Eurasia continental collision [16]. We suggest that the convergence of Arabia and Eurasia resulted in lithosphere shortening and bi-lateral transportation of excessive lithosphere material, i.e., its "diffluence" both to the south-west and south-east of the collision zone which formed the tectonic sheets of the eastern Asia Minor Peninsula and Zagros Mountains, respectively. Obviously, the removal of a significant part of the basement in front of the MCF can be responsible for the difference between the structures of the northern and southern Caucasian domains.

2.2 Deep-Seated Structure of the Region

Arabian-Caucasus syntaxis is characterized by high seismicity. Shallow focuses sharply predominate and very rare relatively deep earthquake (up to 120–140 km) found mainly on the north-east of the region, beside the area of modern magmatism [19]. The seismic data available do not reveal the subduction zone beneath the Greater Caucasus, although some earthquake's hypo-central depths can reach 175 km [6]. However, beneath the transitional fold and thrust zones between the north margin of Arabian plate (BZF) and the Greater Caucasus, north-directed seismic zones occurred reaching up to 50–70 km of depth. Judging by the seismic data, descending tongue-like leakages of crustal material pushed into the mantle plume head. They look like possibly peculiar "embryonic" subduction zones, where crustal material is involved in descending currents (Fig. 1).

The absence of subduction raises the question about the reasons of aforementioned substantial (probably ~400 km) shortening of the south territory of MCF. As mentioned above, this could result from a lateral "diffluence" or "spreading" of a lithospheric material under pressure of the Arabian indenter in two opposite directions away from the GC in front of the rigid East European Craton (Fig. 1). Obviously, the removal of a significant part of the basement in front of the MCF can be responsible for the difference between the structures of the northern and southern Caucasian domains. Moreover, part of crustal material is involved in the north-directed descending flows beneath the folded-thrusted belt.

2.3 Modern Volcanism

The CAS includes Neogene-Quaternary volcanic belt, which is extended from Eastern Anatolia via the Lesser Caucasus

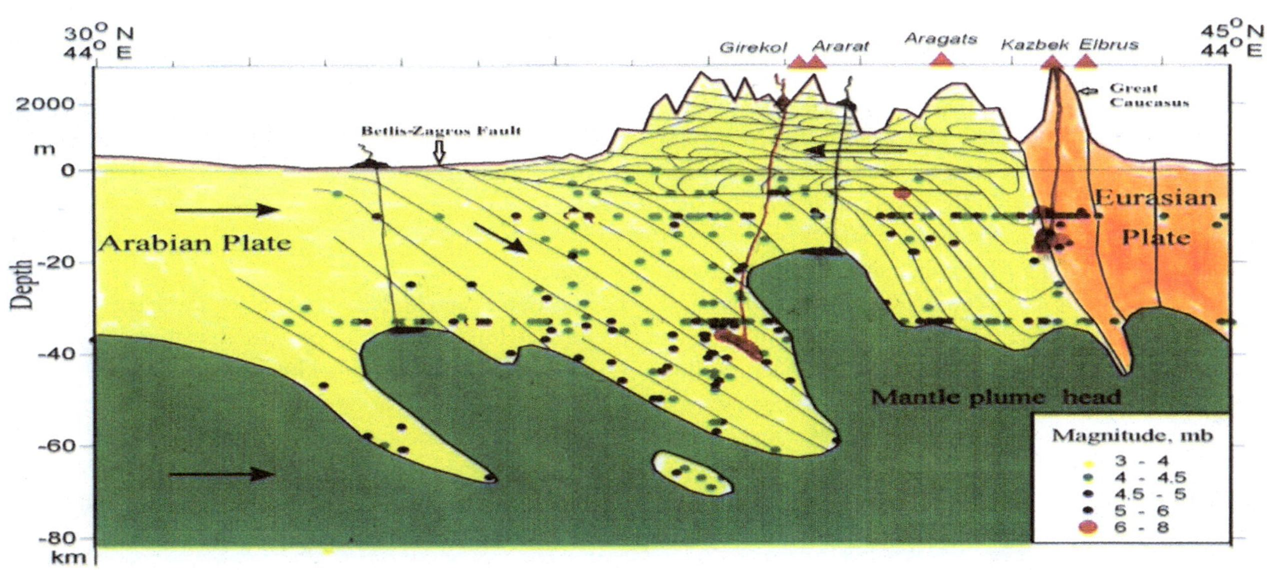

Fig. 1 Proposed deep-seated structure beneath the CAS's fold-thrust zone

up to the Great Caucasus. The belt is composed of two types of volcanic rocks: (1) extensive plateau basalts possessing geochemical characteristics of intra-plate (plume-related) rocks, and (2) calc-alkaline volcanics, which are petrologically and geochemically close to those formed in a supra-subduction setting. Isotopic-geochemical data proved that both types of magmatism contain some admixture of each other's material [9, 10, 14, 15]. Therefore, two independent types of magmatism have developed in the zone of Arabian-Eurasian collision, i.e. specific type of "mixed" magmatism occurred here.

We mentioned above the positive isostatic anomaly over fold-thrusted zone gives evidence of the presence of a mantle plume beneath the CAS. It is in a good agreement with the existence of plateau basalts here, which are exposed to adiabatic melting of local ledges on the plume head. In other words, we suggest that calc-alkaline volcanic rocks are apparently derived from "embryonic" subduction zones. It is also very likely that fragments of these rocks, represented by heavy garnet granulites and eclogites, can sink into hot plume head material and dissolve in it. As a result, newly-formed plume-derived magmas of basaltic plateaus were contaminated by crustal material. Calc-alkaline melts can be partly contaminated by mantle-derived magmas in such dynamic occurrence.

3 Discussion

Zone of Arabian-Eurasian convergence is located on the north termination of the huge belt of Cenozoic basaltic plateaus which develop along all west margin of Arabian Plate (Afro-Arabian LIP [4, 5]), and stretch out like an "arm" from Afar Hot Spot. Because such type of magmatism is considered with adiabatic melting of the mantle plume matter, we suggest that this belt reflect the morphology of the crest of a large mantle plume head and the basaltic volcanism outlines its contour in projection to the surface. As a result, the plume head has an elongated shape here and goes on for about 3500 km from the Gulf of Aden up to the Greater Caucasus, at 300–400 km wide. It looks like sub-lithospheric flow in the mantle which gradually moves from Afar via Arabia to the north [22]. Judging on aforementioned geological, petrological, geochronological and geophysical data, magmatism CAS, in essence, is a part of the Afro-Arabian LIP, developed above the northern termination of the same Afro-Arabian mantle plume but already evolved under specific conditions of collision zone. We suggest that northern end of plume-related sublithospheric mantle flow from Afar finally "rests" against East European Craton, partly "dives" under it and lifts its margin, forming the Greater Caucasus Range.

The beginning of igneous activity in this belt is in concordance with the first stages of the spreading of the Red Sea ($\sim$30 Ma; [2]) and its activity has continued since historical time. On the basis of incompatible trace element content, the volcanic activity in Syria can be divided into two stages: the first lasting from $\sim$26 to $\sim$6 Ma and the second from $\sim$6 to 5 Ma to recent times [12, 22]. This temporal shift in composition of magmas is related to a major tectonic re-organization of the region which occurred during upper Miocene and dates back from the beginning of Arabia-Eurasia collision.

However, the situation of the first-stage basaltic magmatism is more complicated. It stretches out only up to the Syrian region. Owing to this, it follows that the "arm" was rather shorter during lower-middle Miocene and began to extend to the north since the late Miocene. So far, the eruptions of basaltic magma have continued at all its length.

From these data, we can conclude that the Afro-Arabian mantle plume has already existed for 30 million years and periodically was supplied with pulses of fresh plume material which provided the stable lava composition all over the distance of the "arm". Most probably, the "arm" represents something like a tube-like occurrence in which mobile plume material extends to the north from Afar Hot Spot. Basaltic plateaus, which trace a crest of this "tube", very likely, mark location of "fingers" (local plumes) at its surface.

The CAS is located at the north termination of this belt. Judging by the geophysical and petrological data, the north continuation of the Afro-Arabian mantle plume "dive" under the Main Caucasus Fault [19]. It led to lifting the Greater Caucasus and providing an appearance of the Caucasian Mineral Vody igneous complex as well as Elbrus and Kazbek volcanoes. U-Pb isotopic data proves the presence of the material of the Afar Hot Spot in the Elbrus's volcanic rocks [3]. The deepest earthquakes beneath the GC are considered with dynamic processes below Afar-Caucasus mantle flow.

4 Conclusions

1. The Caucasian-Arabian segment of the huge Cenozoic Alpine-Himalayan collision belt is formed by two domains: the EW-striking Greater Caucasus (GC) at the north and the Caucasian-Arabian Syntaxis (CAS) at the south. GC, in essence, represents uplifted along the Main Caucasian Fault southern margin of the East-European Craton, whereas CAS includes arc-like tectonic domains of the Lesser Caucasus and Eastern Anatolia. The features of the CAS consist of large NS-trending positive

isostatic anomaly, which assumes presence of a mantle plume head beneath it. Existence of plume-related magmatism in CAS supports this idea.
2. The Alpine (late Cenozoic) structure of the Caucasus is determined by NS horizontal compression generated by the interaction of two plates: the Arabian indenter and the Eurasian Plate, which resulted in the appearance of transitional vast fold-thrust zone and in transverse shortening of the territory up to 400 km, mainly at the expense of the area south of the MCF.
3. There is no subduction zone beneath the GC. However, some north-directed seismic zones occurred beneath transitional fold-thrust zones between the Arabian plate (Bitlis-Zagros Fault) and the MCF. They are traced up to 50–70 km depth and look like peculiar "embryonic" subduction zones. We assume that the mentioned above shortening occurred due to: (1) the tectonic "diffluence" of crustal material apart from the Arabian indenter in front of the East European Craton, and (2) involving another part of crustal material in the descending flows.
4. The Neogene-Quaternary volcanic belt is extended from Eastern Anatolia to the Lesser Caucasus and farther to the Greater Caucasus. Two types of volcanic rocks are predominant here: (1) the extensive plateau basalts possessing geochemical characteristics of plume-related rocks, and (2) calc-alkaline volcanic rocks originated from "embryonic" subduction zones.
5. The CAS is located at the north termination of the Afro-Arabian LIP and developed in specific conditions of continental collision above mantle plume.

References

1. Bazhenov, M.L., Burtman, V.S.: Structural Arcs of Alpine Belt (Carpatian, Caucasus, Pamir), p 167. Nauka Publ., Moscow (in Russian) (1990)
2. Camp, V.E., Roobol, M.J.: The Arabian continental alkali basalt province. Geol. Soc. Am. Bull. **101**, 71–95 (1989)
3. Chugaev, F.V., Chernyshev, I.V., Lebedev, V.A., Eremina, A.V.: Isotopic composition of plumbum and origin of Quaternaly lavas of Elbrus volcano (Greater Caucasus, Russia): data of high-precisions MC-ICP-MS method. Petrology **21**, 16–27 (2013)
4. Eppelbaum, L.V., Katz, Y.: A new regard on the Tectonic Map of the Arabian-African region inferred from the satellite gravity analysis. Acta Geophys. **65**, 607–626 (2017)
5. Ernst, R.E.: Large Igneous Provinces. Cambridge University Press, Cambridge (2014)
6. Godoladze, T., Sandvol, E.A., Mackey, K.G., et al.: The caucasian seismic network (CNET): seismic structure of the Caucasus. Geophys. Res. Abstracts **20**, EGU2018-19744 (2018) (EGU General Assembly 2018)
7. Khain, V.E.: Regional Geotectonics. Alpian-Mediterranean Belt, p 344. Moscow, Nedra (1984) (in Russian with English abstract)
8. Kopp, M.L.: Late Alpine collisional structure of Caucasus region. In: Leonov, Y.G. (ed.) Alpine History of the Great Caucasus, pp. 285–316. GEOS, Moscow (2007) (in Russian)
9. Lebedev, V.A., Chernyshev, I.V., Sharkov, E.V.: Geochronological scale and evolution of Late Cenozoic magmatism within Caucasian segment of the Alpine Belt. Dokl. Earth Sci. **441**(2), 1656–1660 (2011)
10. Lebedev, V.A., Sharkov, E.V., Unal, E., Keskin, M.: Late Pleistocene Tendurek Volcano (Eastern Anatolia, Turkey): II. Geochemistry and petrogenesis of the rocks. Petrology **24**(3), 234–270 (2016)
11. Leonov, Y.G.: Cimmerian and late Alpine tectonics of the Greater Caucasus. In: Leonov, Y.G. (ed.) Alpine History of the Great Caucasus, pp. 317–340. GEOS, Moscow (2007) (in Russian)
12. Lustrino, M., Sharkov, E.: Neogene volcanic activity of western Syria and its relationship with Arabian plate kinematics. J. Geodyn. **42**, 115–139 (2006)
13. Mosar, J., Mauvilly, J., Enna, N., et al.: The Greater Caucasus: a new tectonic map! Geophys. Res. Abstracts **20**, EGU2018-7934 (2018) (EGU General Assembly 2018)
14. Oyan, V., Keskin, M., Lebedev, V.A., et al.: Petrology and geochemistry of the Quaternary Mafic Volcanism in the north-eastern branch of Lake Van, Eastern Anatolian Collision Zone. Turkey. J. Petrology **58**(9), 1701–1728 (2017)
15. Pearce, J.A., Bender, J.F., De Long, S.E., et al.: Genesis of collision volcanism in Eastern Anatoly. Turkey. J. Volcan. Geotherm. Res. **44**, 189–229 (1990)
16. Reilinger, R., McClusky, S.: Nubia-Arabia-Eurasia plate motions and the dynamics of Mediterranean and Middle East tectonics. Geophys. J. Int. **186**, 971–979 (2011)
17. Rogozhin, E.A., Gorbatikov, A.V., Stepanova, Yu., M., et al.: Structure and modern geodynamics of the Greater Caucasus in the light of new data on its deep-seated structure. Geotectonics **2**, 36–49 (2015)
18. Saintot, A., Brunet, M.-F., Yakovlev, F., et al.: The Mesozoic-Cenozoic tectonic evolution of the Great Caucasus. In: Gee, D.G., Stephenson, R.A. (eds.) European Lithosphere Dynamics. Geological Society of London Memoir, vol. 32, pp. 129–145 (2006)
19. Sharkov, E., Lebedev, V., Chugaev, A., et al.: The Caucasian-Arabian segment of the Alpine-Himalayan collisional belt: geology, volcanism and neotectonics. Geosci. Front. **6**, 513–522 (2015)
20. Shempelev, A.G., Prutsky, N.I., Kukhmazov, S.U.: Materials of geophysical investigation along Near-Elbrus profile (volcano Elbrus-Caucasian Mineral Waters). In: Tectonics of the Earth's Crust and Mantle. Proceedings of XXXVIII Tectonic Meeting, pp. 361–365. GEOS, Moscow (2005) (in Russian)
21. Trifonov, V.G., Dodonov, A.E., Sharkov, E.V., et al.: New data on the Late Cenozoic basaltic volcanism in Syria, applied to its origin. J. Volcan. Geotherm. Res. **199**, 177–192 (2011)
22. Trifonov, V.G., Sokolov, S.Y.: Sublithospheric flows in the mantle. Geotectonics **51**(6), 535–548 (2017)

Tectonic and Eustatic Sea-Level Fluctuations During the Early Cretaceous: A Case Study from the Berriasian–Valanginian Sequences in Saudi Arabia

Abdullah O. Bamousa

Abstract

Reconstruction of sea-level changes during the Early Cretaceous from eastern part of Saudi Arabia provided an amazing picture for mixed eustatic and tectonic influences on the sea-level fluctuations during that time. The studied successions consist mainly of carbonates with minor interbedded siliciclastics, forming the Silayy, Yamamah and Buwaib formations in north of Ar Riyad city. Based on integrated sedimentary facies analysis, a relative sea-level curve for the Berriasian–Valanginian interval has been established and represented by two transgressive–regressive facies cycles. Correlation of the recognized sea-level events with those recorded from Europe provided useful information about the origin of these events. Absence of the Berriasian sea-level fall events in the study area attributed to tectonic subsidence, which created accommodation spaces for sediments to be deposited during the late Berriasian time. On the other hand, the Valanginian sea-level events of the study area are well-matched with those recorded globally, therefore supporting an isochronous formation of Valanginian facies cycles of eustatic origin.

Keywords

Early cretaceous • Facies cycles • Relative sea-level • Central Arabia

1 Introduction

In the North of Arriyad city, the measured successions crop out on the eastern and western limbs of a NW-plunging major anticline, known as Irq Banban anticline that formed during the Alpine-Himalayan orogeny [1]. These successions are thick and well-exposed, represented by the lowermost Cretaceous carbonate with minor siliciclastic rocks of the Silayy, Yamamah and Buwaib formations [2, 3]. From the Middle Jurassic to the Early Cretaceous times, most of the Eastern parts of the Arabian Plate underwent subsidence, leading to rising of relative sea-level and wide distribution of marine sedimentation [4, 5]. Valanginian transgression covered a large part of eastern Arabia and the studied sediments were deposited in a proximal position compared to the distal positions of Oman and the United Arab Emirates [6, 7], giving it a great importance in tracking fluctuations of relative sea-level during those times. The detailed facies analysis of the studied sequences [7] provided useful data about the sea-level changes. This work offers a discussion about the origin of these sea-level fluctuations.

The present study aims at introducing new information about the origin of forces responsible for the Early Cretaceous sea-level changes imprint at that time in the eastern part of Saudi Arabia. Consequently, global correlation will be enhanced and the geological history of eastern Arabia will be broadened.

2 Geologic Setting

The study area is located to the north of Ar Riyad city, where the Berriasian–Valanginian successions are thick, reaching to about 100 m, and well-exposed (Fig. 1). During the Early Cretaceous, the study area was situated under a shallow marine environment at the southern margin of the Neotethys at a close position from the equator [8]. This resulted in a deposition of thick shallow marine carbonate rock with limited siliciclastic input during that time. The Basal part of the exposed successions started with interbedded bioclastic limestones of the Silayy Formation (Berriasian), followed by argillaceous coarse-grained limestones of the Yamamah Formation (lowermost Valanginian) and ended with intercalated sandy limestones and sandstone beds of the upper lower Valanginian Buwaib Formation (Fig. 1).

A. O. Bamousa (✉)
Department of Geology, Faculty of Science, Taibah University, Madinah, 41411, Saudi Arabia
e-mail: abamousa@taibahu.edu.sa; aobamousa@gmail.com

F. Rossetti et al. (eds.), *The Structural Geology Contribution to the Africa-Eurasia Geology: Basement and Reservoir Structure, Ore Mineralisation and Tectonic Modelling*,
Advances in Science, Technology & Innovation, https://doi.org/10.1007/978-3-030-01455-1_69

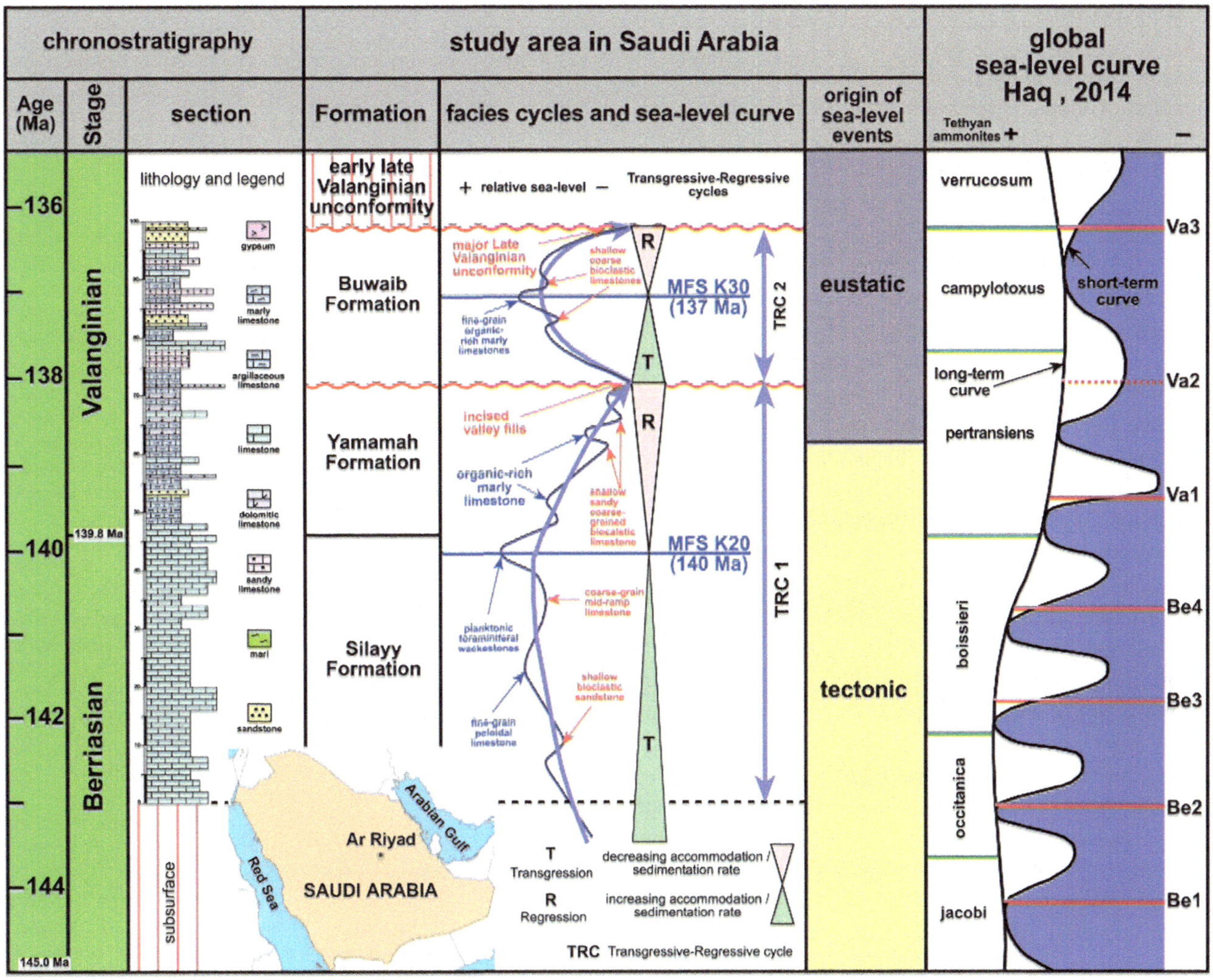

Fig. 1 Correlation of the reconstructed sea-level curve of the study area with the global curve of Haq [10] during Berriasian–Valanginian times

3 Results

Facies of the Berriasian–Valanginian Silayy, Yamamah, and Buwaib formations are showing a deposition in a homoclinal carbonate ramp setting, with a wide range from lagoonal back-ramp to outer ramp settings. A relative sea-level curve for the lowermost Cretaceous successions of the study area can be established based on the available detailed stratigraphic and sedimentologic data. According to the shallow marine proximal position of the study area during Early Cretaceous times and the depositional environments of the different facies associations, this curve can be subdivided into two transgressive–regressive cycles (Fig. 1). The first transgressive-regressive cycle comprises the upper Berriasian–lowermost Valanginian Silayy and Yamamah formations, while the second transgressive-regressive cycle spans the time from the lower Valanginian to lower upper Valanginian. This facies cycle is represented by the entire Buwaib Formation. In addition, each transgressive-regressive cycle has been divided into different cycle sets and high-frequency cycles, based on the stacking of the different facies (increasing/decreasing accommodation space and sediment supply). It provided a detailed tracing of the sea-level changes during the deposition of the studied successions. Correlation of the reconstructed sea-level fluctuations of the studied sequences with that of Europe [9] and with the global curve [10] is showing absence of the sea-level fall events (Be 3, Be 4, and Va 1 sequence boundaries) from the study area during the late Berriasian time. On the other hand, the sea-level fall events which are recorded globally during the Valanginian time and are well documented are in Saudi Arabia (Fig. 1).

4 Discussion

The studied successions are considered to be the lowermost sequences of the Cretaceous rocks in Saudi Arabia; represented by, from older to younger, Silayy, Yamamah and Buwaib formations. Based on integrated facies, a homoclinal carbonate ramp is suggested to be the suitable depositional model for the investigated successions in the study area. A sea-level curve has been constructed based on the facies cyclicity detected. This curve is divided into two transgressive–regressive cycles, representing the Berriasian–Valanginian interval in the study area. The sea-level curve of the studied successions is showing continuous sedimentation during the Berriasian time, in contrast to the interruption of deposition with four sequence boundaries globally at this time. This is clearly indicative of a tectonic effect in the study area during the late Berriasian. This tectonic event was probably due to a subsidence movement of the sea floor, responsible for the creation of accommodation spaces for sediments to be deposited without any unconformities. The subsidence is a part of the geodynamic episode which started with uplift leading to partial erosion of the Jurassic rocks on the eastern part of the Arabian plate, followed by significant thermal subsidence [5]. The studied sedimentary basin became quieter during the Valanginian and the studied successions start showing unconformity surfaces, which are globally recorded. This supports an isochronous formation of Valanginian facies cycles by eustatic origin. Therefore, the studied sequences are showing a mix of late Berriasian tectonic and early Valanginian eustatic effects during the deposition of the studied sequences.

5 Conclusions

A sea-level curve for the Berriasian–Valanginian successions of the study area has been reconstructed based on facies cyclicity. Correlation of the recognized sea-level curve with the global curve [10], is showing an absence of the late Berriasian sea-level falls, while the Valanginian unconformities are well-represented. this could be the result of local tectonic subsidence culminating during the Berriasian time in the study area. On the other hand, the good matching of the recognized sequence stratigraphic surfaces during Valanginian times clearly indicates eustatic origin for these surfaces.

References

1. Bamousa, A.O., Memesh, A.M., Dini, S.M.: Morphostructural evolution of Ath Thumamah depression, North Riyadh, Saudi Arabia. Carbonates Evaporites **29**, 65–72 (2014)
2. Powers, R.W., Ramirez, L.F., Redmond, C.D., Elberg, E.L.: Geology of the Arabian Peninsula: Sedimentary geology of Saudi Arabia. United States Geological Survey Professional Paper 560-D, p. 147 (1966)
3. Hughes, G.W.: Saudi Arabian Late Jurassic and Early Cretaceous agglutinated foraminiferal associations and their application for age, palaeoenvironmental interpretation, sequence stratigraphy, and carbonate reservoir architecture. In: Hart, M.B., Kaminski, M. A., Smart, C.W. (eds.) Proceedings of the Fifth International Workshop on Agglutinated Foraminifera, Grzybowski Foundation, vol. 7, pp. 149–165 (2000)
4. Shebl H.T., Alsharhan, A.S.: Sedimentary facies and hydrocarbon potential of Berriasian-Hauterivian carbonates in Central Arabia. In: Simmons, M.D. (ed.) Micropalaeontology and Hydrocarbon Exploration in the Middle East, Chapman and Hall, London, ISBN 0412427702 (1994)
5. Sharland, P.R., Archer, R., Casey, D.M., Davies, R.B., Hall, S.H., Heward, A.P., Horbury, A.D., Simmons, M.D.: Arabian plate sequence stratigraphy. GeoArabia Special Publication 2, Gulf PetroLink, Bahrain, p. 371, with 3 charts (2001)
6. Pratt, B.R., Smewing, J.D.: Early Cretaceous Platform-Margin configuration and evolution in the Central Oman Mountains, Arabian Peninsula. Am. Assoc. Petrol. Geol. Bull. **77**, 225–244 (1993)
7. Nagm, E., Bamousa, A.O., Memesh, A., Babikir, I., Dini, S.: Relative sea-level changes and sedimentary facies development of the lowermost Cretaceous (Berriasian–Valanginian) cycles in the north of Ar Riyad city, Saudi Arabia. J. Asian Earth Sci. **163**, 163–176 (2018)
8. Scotese, C.R.: Map Foilo 31, Early Cretaceous (Berriasian, 143.0 Ma), PALEOMAP PaleoAtlas for ArcGIS, volume 2, Cretaceous Paleogeographic, Paleoclimatic and Plat Tectonic Reconstructions, PALEOMAP Project, Evanston, IL (2013)
9. Hardenbol, J., Thierry, J., Farley, M.B., Jacquin, T., de Graciansky, P.-C., Vail, P.R.: Cretaceous sequence stratigraphy, Chart 4. In: de Graciansky, P.-C., Hardenbol, J., Jacquin, T. and Vail, P.R. (eds.) Mesozoic and Cenozoic Sequence Stratigraphy of European Basins. SEPM (Society for Sedimentary Geology) Special Publication 60 (1988)
10. Haq, B.U.: Cretaceous eustasy revisited. Glob. Planet. Change **113**, 44–58 (2014)

Obduction and Collision Tectonics in Oman: Constraints from Structural and Thermal Analyses

Eugenio Carminati, Luca Aldega, Luca Smeraglia, Andreas Scharf, and Frank Mattern

Abstract

In Oman, Permian-Mesozoic sediments belonging to the Arabian passive margin were overthrust by Hawasina deep-water sediments and Semail Ophiolite in late Cretaceous. Passive margin sequences and Hawasina units are now exposed in the Hawasina, Jabal Akhdar and Saih Hatat domes and at the Oman Mountains thrust front. Structural analyses were performed in the Jabal Akhdar Dome and in the Musandam area (where the Oman Mountains are characterized by a syntaxis), and samples from sub-ophiolite sediments were collected in the Hawasina, Saih Hatat and Jabal Akhdar domes as well as at the Jabal Salakh and the Musandam thrust front. The thermal evolution of sub-ophiolite sediments was investigated by clay-size-fraction X-ray diffraction (analyses for Musandam, Hawasina and Saih Hatat samples are in progress) and 1D thermal modeling allowed us to infer that sub-ophiolite units of central Oman were thrust over by a 4.5 km-thick pile of Ophiolite and Hawasina rocks during the Coniacian age, and exhumed in late Campanian. The sub-ophiolite rocks at the Jabal Salakh front suffered a burial of 3.35 km of clastic and allochthonous units. The Musandam thrusts are characterized by a first phase of dip slip and a second phase of dextral strike slip kinematics.

Keywords

Ophiolite • Obduction • Oman • Collision
Thermal maturity

E. Carminati (✉) · L. Aldega · L. Smeraglia
Sapienza University, 00185 Rome, Italy
e-mail: eugenio.carminati@uniroma1.it

A. Scharf · F. Mattern
Sultan Qaboos University, Muscat, Oman

1 Introduction

The Semail Ophiolite constitutes an excellently exposed and complete ophiolite complex [1]. However, ophiolite thickness evaluation was achieved via measurements of stratigraphic sections, whose correlation is difficult due to internal deformation of the ophiolites and the lack of regional markers [2]. From the Hawasina and Jabal Akhdar tectonic windows, from a NNE-SSW fold of Hawasina sediments located SW of the Saih Hatat window and at the Jabal Salakh Range and Musandam fronts (Fig. 1), we collected samples of sub-ophiolite sedimentary rocks (Permian-Mesozoic sediments of Arabia passive margin, the Hawasina pelagic sediments) to constrain their thermal evolution via 1D thermal modeling and to reconstruct the thickness of the allochthonous units. In addition, structural analyses were carried out in the Jabal Akhdar Dome and in the Musandam area, where the Oman Mountains are characterized by a syntaxis (the thrust changes its orientation from NW-SE to NE-SW). Using these original multidisciplinary data, we constrained the Cretaceous-Cenozoic Oman Mountains evolution.

2 Methods

We performed X-ray diffraction (XRD) analyses to determine whole-rock composition and illite content in mixed layers I-S by using a Scintag X1 X-ray system (CuKα radiation) at 40 kV and 45 mA. Mixed layers of illite-smectite (I-S) and the transformation sequence diosmectite-random-ordered mixed layers (R0)-ordered mixed layers (R1 and R3)-illite-di-octahedral K-mica (muscovite) were used as indicators of the thermal evolution of sub-ophiolite sediments [1, 3]. These inorganic thermal indicators were used to constrain 1D thermal models, performed using Basin Mod® 1-D software.

F. Rossetti et al. (eds.), *The Structural Geology Contribution to the Africa-Eurasia Geology: Basement and Reservoir Structure, Ore Mineralisation and Tectonic Modelling*, Advances in Science, Technology & Innovation, https://doi.org/10.1007/978-3-030-01455-1_70

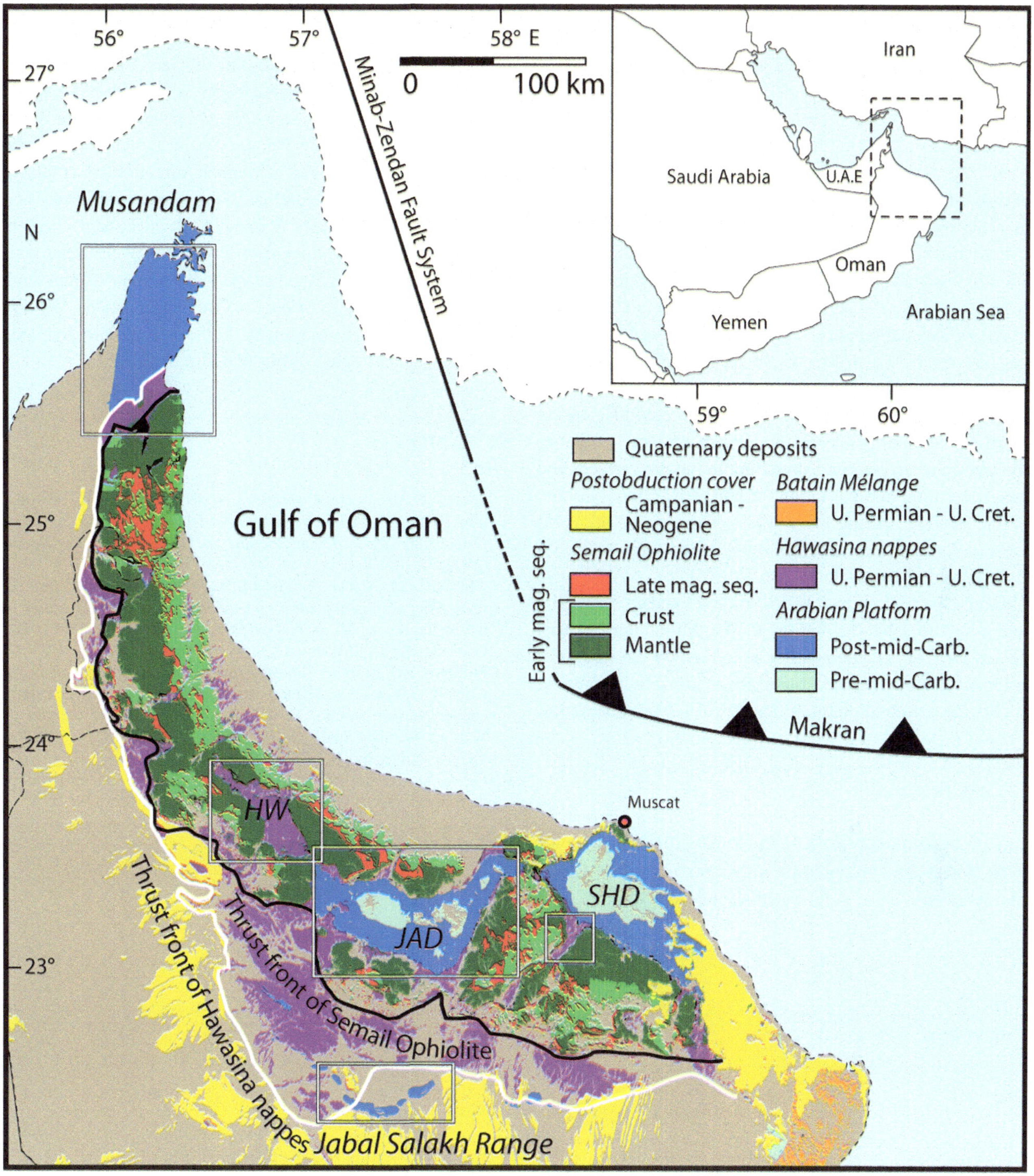

Fig. 1 Geological map of Oman, showing the study areas (black rectangles; modified and redrawn after [1]). Haybi thrust sheets, not described in the text, comprise the most distal (oceanward) Hawasina deposits. JAD: Jabal Akhdar Dome. SHD: Saih Hatat Dome. HW: Hawasina window

Structural analyses consisted in measuring kinematic indicators along faults and strain analyses on oolite-rich carbonates using the Flinn [4] and the Fry [5] methods.

3 Results

Thermal analyses for the Musandam, Hawasina and Saih Hatat samples are still in progress. The samples from the northern boundary of the Jabal Akhdar window show an illite content between 85 and 92%, long-range ordered mixed layer I-S of clay minerals and presence of pyrophyllite or/and paragonite, suggesting peak burial temperatures between 150 and 200 °C. In the southern flank of the window, peak paleotemperatures were 120–150 °C, indicating a decrease in the thickness of ophiolite/Hawasina toward the south. These results are consistent with plastic deformation (flattening and stretching in the x-z surface) and pressure solution of ooids along the northern boundary of the tectonic window. Along the southern flank, ooids show no ductile deformation, and brittle rheology is consistent with reduced overburden.

Mixed layers I-S show R1–R3 structures and ca. 80% illite content in the Jabal Salakh area. The Musandam thrusts, oriented NE-SW, i.e., perpendicular to the NW-SE thrusts of the Oman Mountains, are characterized by a first phase of dip-slip and a second phase of dextral strike-slip kinematics.

4 Conclusions

From combined analysis of thermal and structural data, we conclude that the sub-ophiolite sediments of central Oman were overthrust by a ~4.5 km-thick pile of Ophiolite and Hawasina rocks during Coniacian and exhumed later in Campanian [2]. In the Jabal Salakh Range, sub-ophiolite rocks were buried under 1.35 km of clastic sediments and thrust by Hawasina sediments with a thickness of about 2 km. These results suggest a decrease in the thickness of the allochthonous units from NE to SW, in agreement with the strain analysis results and with their direction of emplacement. The peculiar orientation of the Musandam thrust and the change of kinematics (from dip- to strike-slip) suggest that they were possibly developed with a NW-SE trend (i.e., parallel to the remaining thrusts of the Oman Mountains), and were subsequently rotated to their present-day NE-SW strike during the development of the Musandam Syntaxis and then sheared again by dextral slip.

References

1. Searle, M.P.: Structural geometry, style and timing of deformation in the Hawasina Window, Al Jabal al Akhdar and Saih Hatat culminations, Oman Mountains. GeoArabia **12**(2), 99–130 (2007)
2. Aldega, L., Carminati, E., Scharf, A., Mattern, F., Al-Wardi, M.: Estimating original thickness and extent of the Semail Ophiolite in the eastern Oman Mountains by paleothermal indicators. Mar. Pet. Geol. **84**, 18–33 (2017)
3. Pollastro, R.M.: The illite/smectite geothermometer: concepts, methodology and application to basin history and hydrocarbon generation. In: Nuccio, F., Barker, C.E., (eds.) Application of Thermal Maturity Studies to Energy Exploration, pp. 1–18. SEPM Special Publication (1990)
4. Flinn, D.: On folding during three dimensional progressive deformation. Q. J. Geol. Soc. London **118**, 385–428 (1962)
5. Fry, N.: Random point distribution and strain measurement in rocks. Tectonophysics **60**, 89–105 (1979)

Evidences of Neo-Tectonic Landforms Between Srinagar and Bagwan Area in Lower Alaknanda Valley (Garhwal Himalaya), India

Devi Datt Chauniyal

Abstract

In this paper an attempt has been made to identify the places of recent displacement and upliftment along the North Almora Thrust (NAT) in the Lower Alaknanda Valley of Lesser Garhwal Himalaya. The basic aim of this paper is the identification of the first record sites of recent neo-tectonic activities and associated landforms along N-S trending transverse faults. The study indicates that active movements along thrust plane and transverse fault were active. The evidences of neo-tectonic activity along strike slip transverse faults have been observed in the Quaternary sediments exposed 2nd and 3rd level river terraces. Structurally, quaternary sediments were gently folded and at places show warping and gentle dips with prominent thrust planes. It may be concluded that neo-tectonic movements along NAT were prominent during 15–9 ka and show the recurrence of tectonic activity. The minor structural features on Quaternary sediments are folding, faulting, thrusting, warping and tilting. The final results of the study indicate that the neo-tectonic movements along NAT were more active during 15–9 ka. It is concluded that NAT in Srinagar section has been reactivated from time to time.

Keywords

Neo-tectonics • North Almora Thrust • Transverse fault Quaternary period • River terraces

1 Introduction

Neo-tectonic implies all kinds of vertical and horizontal crustal movements during the Quaternary period. Most of evidence of recent neo-tectonic activities is attributed to the reactivation of the WNW-ESE trending faults and lineaments in the recent and sub recent time in Himalaya [6]. In the study area, NW-SE trending NAT is a major tectonic thrust in the lesser Himalaya, which passes between Srinagar and Bagwan trending EW in Garhwal Lesser Himalaya. The NW to SE direction trending the North Almora Thrust (NAT) is one of the major tectonic thrusts in Lesser Himalaya which passes through an NW-SE direction near Srinagar (Garhwal). The area under investigation lies in the lower Alaknanda valley between Srinagar and Bagwan (E to W) of Garhwal Lesser Himalaya (Fig. 1). This paper describes the results of a continuous investigation of neo-tectonic activities and associated features in the Quaternary sediments exposed river terraces of Alaknanda River in Lesser Garhwal Himalaya.

2 Methods and Material

The present research work is essentially based on the interpretation of topographic maps, CARTOSET 1 image, Digital Elevation Model (DEM) and later detailed field investigations. The DEM data reveals the mapping of NAT zone, its lineaments, drainage analysis and geomorphic features. Four significant sites have been identified in the present thrust zone. In order to ascertain the date of samples that were taken from sand layers present in terrace scarps. The sand samples have been taken from the sand layer of terrace scarps in order to calculate age (Table 1). One of the samples was obtained from Chauras (Site 4) and was dated by Juyal et al. [1]. The initial approximated age of 3 other sites has been calculated by the Institute of the Seismological Research, Gandhinagar, Gujarat, India.

D. D. Chauniyal (✉)
Department of Geography, HNB Garhwal University, Srinagar (Garhwal), Uttarakhand, India
e-mail: devidattchauniyal@gmail.com

F. Rossetti et al. (eds.), *The Structural Geology Contribution to the Africa-Eurasia Geology: Basement and Reservoir Structure, Ore Mineralisation and Tectonic Modelling*,
Advances in Science, Technology & Innovation, https://doi.org/10.1007/978-3-030-01455-1_71

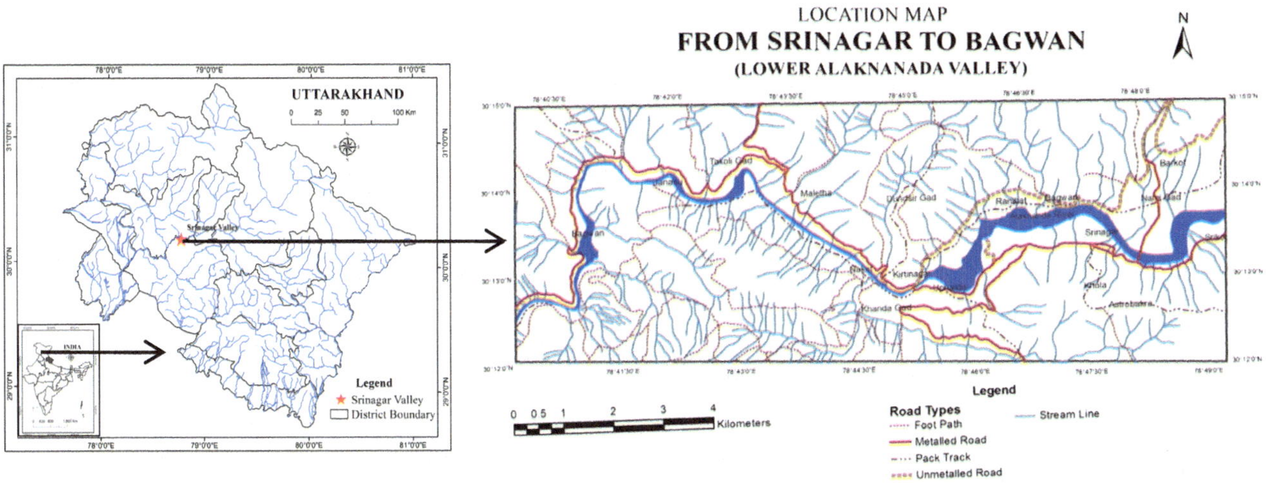

Fig. 1 Location maps the study area

Table 1 Neo-tectonic sites, their geographical locations and approximate age

Sample site	Latitude	Longitude	Depth (m)	Height (m)	Age (ka)
1. Bagwan	30° 13′ 22″N	78° 40′ 46″E	7	527	10
2. Chopriya village	30° 14′ 01″N	78° 42′ 28″E	12	526	9
3. Kirtingar	30° 12′ 59″N	78° 44′ 40″E	15	546	9
4. Chauras	30° 13′ 17″N	78° 48′ 21″E	10	562	10–15

3 Results

3.1 Tectonic Setting

NAT is a major tectonic unit of Lesser Himalaya [2]. The northern flank of Almora Nappe is marked by NW-SE to WNW-ESE trending tectonic planes in the Srinagar area known as NAT. It is a high-angle reverse fault dipping southwards all along its extension. It separates the Quartzite of Garhwali Group in the north from Chandpur Group (Phyllite) in the south [4]. It is estimated that the tectonic plane along NAT was reactivated in the Quaternary era [6]. The NAT and associated strike slip transverse faults gave rise to an imbricate stack of highly sheared rocks. The continuous movements around this zone has resulted in the development of numerous NW-SE trending transverse faults/lineaments [2]. Major and minor lineaments indicate that the area was passing through differential upliftment [5]. The higher level of neo-tectonic activities and uplift factor along this tectonic plane has been combined with geomorphic observations. The development of episodic 6 level terraces at Chauras (Fig. 2), paleo-channel at Supana, Chauras, Ufalda and Bhilkedar, meander bends, faulted scarp, triangular facet, knick points, landslide and epigenetic gorge are witness of vertical movement along the transverse fault. Four neo-tectonic sites are shown on the GIS based 3D model along the N-S trending faults near NAT. Deformity is developed wherever the lineament is seen across the River.

3.2 Geomorphology

Out of the tectonic features, six levels of unpaired river terraces are identified at Chauras (site 4) but their sequence is not found throughout the river course (Fig. 2). Only 4 levels of terraces are identified at sites 1, 2 and 3. The other

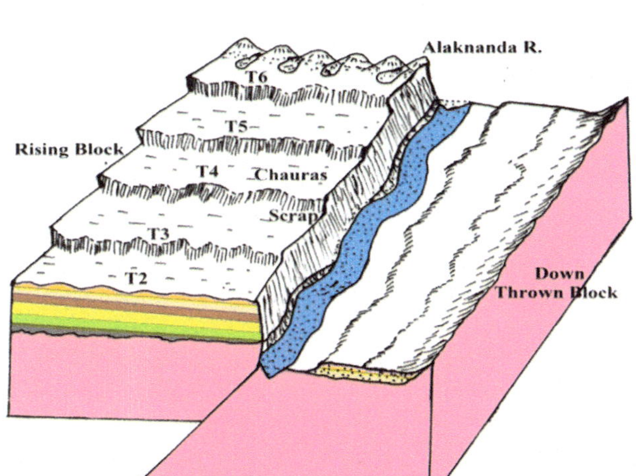

Fig. 2 Six level of terraces on uplifted block of Chauras

geomorphic features are river meanders, paleo channel, colluvial fans, talus cone and scree slopes along the thrusted zone and valley wall slopes. The Quaternary deposits consist of heterogonous materials along with boulders, gravels, pebbles sand, silt etc. Juyal et al. [1] has suggested that aggradations in the valley took place during Late Pleistocene (25 ka) and 15–9 ka (Table 1) in two phases.

3.3 Evidences of Tectonics

Site 1 Bagwan. At Bagwan, a north to south trending transverse strike slip fault along the Chandrabhaga stream, displacement on the sedimentary layer of 3rd level river terrace is found which is clearly exposed at a road cut section. At this place folded and faulted structures in sandy and boulder layers clearly indicate the neo-tectonic activities of past 10 ka.

Site 2 Chopriya Village: At the Chopriya village the neo-tectonic features are observed and identified on the sedimentary bed of the 3rd level terrace scarp along the Takoli Gad transverse fault. Sedimentary beds clearly show tilted and open folded structure. Deformation process has been shown in a hypothetical model with bed rock (Fig. 3). Movement along the strike slip fault has resulted in the development of unpaired terraces and narrow gorge of Takoli Gad before joining the Alaknanda. The vertical movements on NE facing slope has also made the slope extremely vulnerable to landslides and mass movement around this site.

Site 3 Kirtinagar: The Kirtinagar site lies near the confluence of Dhundsir Gad along the Badrinath road. The area is characterized by narrow Dhundsir Gad to a gorge of a nearly vertical wall and meandering channel course (Fig. 4). The neo–tectonic movement along strike-slip fault is manifested in the pronounced geomorphic rejuvenation of the landscape. The geomorphological evidences of the tectonic activities in this site have been documented in the field photographs. There are numerous neo-tectonic features on the 3rd level terrace scarp i.e. warping, faulting, thrusting, open folding, irregular folding, pinching, swelling, tilting structures and displacement of layers. Among these are landslide, abrupt change in channel course, uphill facing scarp, fault displacement and gorge are the prominent features around it (Fig. 4).

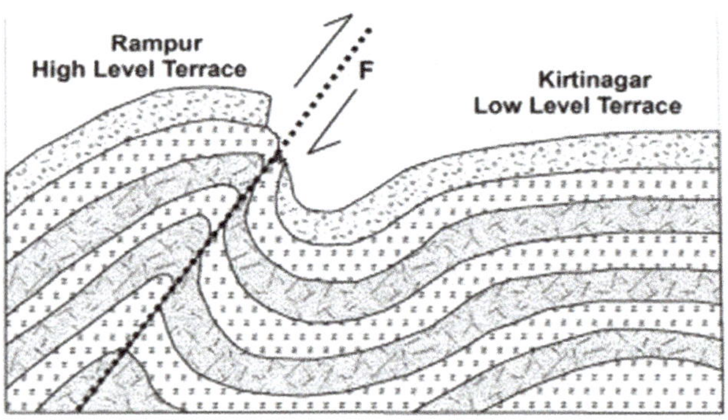

Fig. 3 A model of deformation along fault line

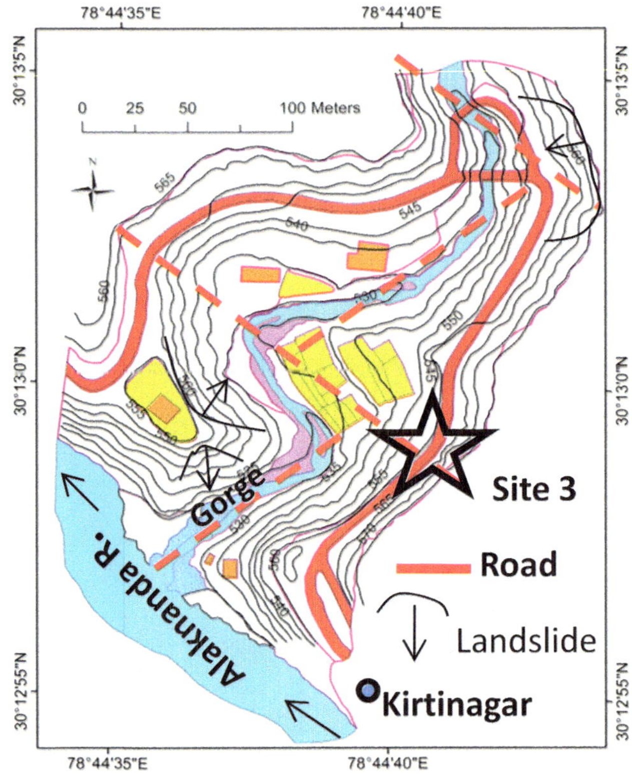

Fig. 4 Neo-tectonic features at Kirtinagar (site 3)

Site 4 Chauras: This site is one of the active segments of the NAT zone of Grahwal Lesser Himalaya. Numerous evidences of active tectonic features are documented on sedimentary deformation structures. The neo-tectonic features were well exposed between the second and third level terrace scarp during the 2013 flood. Deformation structures are in the form of open folding, faulting, and tilting structures. The continuous vertical movement along NAT has resulted in multi levels (six) of unpaired terraces (Fig. 2).

4 Discussions

It is a well-known fact that there is a continued convergence of Indian Plate with the Eurasian plate. Thrusts and faults in the Himalayan region are the result of this convergence. When thrust and faults in this region are reactivated, it causes the development of tectonic features and geomorphic landforms. Hence it reflects that the structural layout of the associated faults is neo-tectonically and very seismo-tectonically active [3]. Geomorphic evidences of neo-tectonic features and displacements with in Quaternary sediments deposited by rivers and glaciers are observed along the N-S trending associated transverse faults.

5 Conclusions

After **investigating** 4 neo-tectonic sites, it can be concluded that the development of NW-SE trending transverse faults/lineaments and displacement within Quaternary sediments **reveal** that the NAT is active in the Srinagar section. The sediments deposited on the terraces are estimated to be about 9–15 ka. It implies that the NAT zone has been active prior to <9 ka and it was more active during 15–9 ka. **A** close relationship **exists** between the major and minor structural elements along the river Alaknanda in Srinagar valley. The neo-tectonic features have been reported for the first time preliminary in nature which requires further investigation.

References

1. Juyal, N., et al.: Late quaternary fluvial aggradations and incision in the monsoon-dominated Alaknanda valley, Central Himalaya, Uttarakhand, India. J. Quat. Sci. **25**(8), 1293–1304 (2010)
2. Kothyari, G.C.: Quaternary reactivation of NAT in Central Kumaun: implication to neotectonic rejuvenation, Lesser Himalaya, Uttarakhand. Unpublished Ph.D. thesis, submitted to Kumaun University, Nainital in Geology, pp. 75–91 (2007)
3. Rajendra, C.P., Rajendra, K.: The status of central seismic gap: a perspective gap on the spatial and temporal aspects of the large Himalayan earthquakes. Tectonophysics **395**, 29–39 (2005)
4. Sati, S.P., et al.: Geomorphic indicators of neotectonic activity around Srinagar (Alaknandada Basin),Uttarakhand. Curr. Sci. **92**(6), 824–829 (2007)
5. Tyagi, A.K., et al.: Identifying areas of differential uplift using steepness index in the Alaknanda Basin, Garhwal Himalaya, Uttarakhand. Curr. Sci. **97**(25), 1473 (2009)
6. Valdiya, K.S.: Reactivation of Terrane defining boundary thrust in central sector of the Himalaya: implications. Curr. Sci. **81**, 1418–1430 (2001)

Part VII

Tectonic Modelling

From Rifting to Spreading: The Proto-Oceanic Crust

Philippe Schnürle, Maryline Moulin, Alexandra Afilhado, Mikael Evain, Afonso Loureiro, Nuno Dias, and Daniel Aslanian

Abstract

The sequence of processes that lead from continental rifting to oceanic drifting during the genesis of passive margins remains debated. Wide-angle seismics has, over the past decades, proven an efficient tool to image the architecture and to quantify the acoustic velocity of continental margins. These data, combined with structural interpretation and stratigraphic/kinematic/mechanical numerical modeling, have led to better characterization of the crustal and mantellic processes active from rifting to drifting. The Protoc project is focused on the accretion of proto-oceanic crust that occurs before the more homogeneous and stable oceanic drift.

Keywords

Proto-oceanic crust • Passive margins • Wide-angle seismics

1 Introduction

The transition between the so-called intermediate domain of the passive continental margins and the first true oceanic crust is still a matter of debate and the exact boundary of the first true oceanic crust can be hard to define in many if not all passive margins. The nature of the intermediate domain, after a more or less sharp continental necking zone, is also a subject of discussion. In some places, as in the western and eastern Mediterranean sea, systems which may have produced this first oceanic crust are proposed with opposite spreading directions [1–4]. In other places, as in the Central segment of the South Atlantic [5–8] or in the Provencal Basin [9–11] the crucial role of the lower continental crust is highlighted. Moreover, upper mantle may be exhumed as in the Galician margin [12, 13] or intruded in the lower continental or oceanic crust [7, 14]. Last, in many places, a proto-oceanic crust is observed before the true «normal» oceanic crust [6, 7, 9, 10] as on the Ceara-Maranhão margin, where the 4–5 km thick crust of the intermediate domain is interpreted as an exhumed lower continental crust, followed by an about 50 km-wide band of proto-oceanic crust before reaching a more normal but only 5 km thick oceanic crust. McKenzie et al. [15] show that a thick continental crust (equal or greater than about 30 km) and magmatic activity can reduce the lower crustal viscosity to a level at which flow is possible. Therefore, one proposition may be that lower crust can either "flow" into the first accreting process, as earlier suggested by Bott [16] for the volcanic margin, and/or flow laterally along different margin segments. The lower crust can also be involved in the exhumation processes to create the first "proto-oceanic crust" or even be mixed with the upper mantle underneath [6]. These points are far from being inconceivable if we keep in mind that continental break-up nearly always occurs inside previous orogenic belts and that the material involved in the process is therefore highly heterogeneous.

2 Methods

For twenty years Ifremer with its academic and industrial partners have imaged this transition zone (see Fig. 1) in the Mediterranean sea (Provencal Basin), the Central Atlantic (Morocco margins), the Equatorial Atlantic (Berrenhinhas-Maranão-Ceara Margins), the central segment of the South Atlantic (Angola Margin, Santos Basin, Jequitinhonha-Almada-Camamu-Sergipe-Alagoas Margins) and the Indian (Mozambique Margins) oceans. Forward modeling of the wide-angle seismic profiles acquired at sea and on land during

P. Schnürle (✉) · M. Moulin · M. Evain · D. Aslanian
IFREMER, REM/GM/LGS, Centre de Brest, 29280 Plouzané, France
e-mail: Philippe.Schnurle@ifremer.fr

A. Afilhado
Instituto Superior de Engenharia de Lisboa, Lisbon, Portugal

A. Loureiro · N. Dias
Faculdade das Ciencias da Universidade de Lisboa, Instituto Dom Luis, Lisbon, Portugal

F. Rossetti et al. (eds.), *The Structural Geology Contribution to the Africa-Eurasia Geology: Basement and Reservoir Structure, Ore Mineralisation and Tectonic Modelling*,
Advances in Science, Technology & Innovation, https://doi.org/10.1007/978-3-030-01455-1_72

Fig. 1 Locations of the wide-angle and reflection seismic exploration conducted by Ifremer and its Collaborators since 1994, marked in orange, and experiments proposed in the next future are marked in green

these experiments reveals an evolution from the continental necking zone to the true (but usually thin) oceanic crust, with a domain of exhumed material, most generally exhumed lower continental crust, and the existence of a proto-oceanic crust. For instance, in the Santos Basin, the two 700 km long parallel wide-angle profiles exhibit in the middle of the basin very different velocity structure, although they are only spaced by 50 km. The southernmost profile presents a structure of thin exhumed continental crust while the northern profile shows an anomalous and probably heterogeneous crust, which can be inferred as proto-oceanic crust on a failed rift. On the Ceara-Maranhão margin, the 4–5 km thick crust of the intermediate domain is interpreted as exhumed lower continental crust. Seawards and parallel to the coast, an about 50 km-wide band of proto-oceanic crust is described before reaching a more normal but of only 5 km thick oceanic crust.

Three new combined wide-angle and reflexion seismic experiments are proposed for the next years.

The CAMPOSEIS project, on the Campos Margins, in collaboration with MARUM, GEOMAR and Univ. of Brasilia on a German research vessel (AWI-Marum-Ifremer funding) and two projects proposed this year at the French National Commission of the Offshore Fleet: the BasAlg project, in the Algerian Basin, and the ArcMal project, in the Levant Basin.

Thanks to these new cruises, this rather unique data set will cover most of geological and geodynamical settings. The aim of the Protoc project is, with our French and international partners, to use different processing methods, to compile and compare these data in order to characterize the transition domain with a focus on the role of lower continental crust in the formation of proto-oceanic crust.

3 Conclusions

Characterizing the sequence of processes that lead from continental rifting to oceanic drifting during the genesis of passive margins is key to our fundamental knowledge of the structure and history of the earth. Clarifying the nature and behavior of the first oceanic crust contributes to understanding the processes presently active at mid-oceanic ridges and transform faults. The respective role and behavior of the lower crust and upper mantle between rift and drift remain to be established. Furthermore, the dynamic of the upper mantle is responsible for topography building, leading to consecutive erosion, subsidence and isostatic rebound, and therefore connected to the quantification of global sedimentary mass transfers through time (Source to Sink). Finally, continental margins are the most populated and developed area on earth. Deep earth dynamics (topography, erosion, tectonics) are strongly related to naturals hazards such as earthquakes, slope instabilities, tsunamis and mass transfers have important consequences on the geo-resources and geothermal energy. The ability to read and understand the link between deep Earth dynamics and surface processes has therefore important societal impacts.

References

1. Van Hinsbergen, D.J.J., et al.: Origin and consequences of western Mediterranean subduction, rollback, and slab segmentation. Tectonics **33**, 393–419 (2014). https://doi.org/10.1002/2013TC003349
2. Leroux, et al.: Atlas of the stratigraphic markers in the western Mediterranean sea with focus on the Gulf of Lion. Commission de la Carte Géologique du Monde (in press)
3. Gardosh, M.A., et al.: Tethyan rifting in the Levant region and its role in Early Mesozoic crustal evolution. Geol. Soc. London Spec. Publ. **341**, 9–36 (2010). https://doi.org/10.1144/SP341.2
4. Montadert, L., et al.: Petroleum systems offshore Cyprus. AAPG Memoir **106**, 301–334 (2014). https://doi.org/10.1036/13431860M1063611
5. Moulin, M., et al.: Geological constraints on the evolution of the Angolan margin based on reflection and refraction seismic data (ZaiAngo project). Geophys. J. Int. **162**, 793–810 (2005). https://doi.org/10.1111/j.1365-246X.2005.02668.x
6. Aslanian, D., et al.: Brasilian and Angolan Passive Margins: the kinematic constraints. Tectonophysics Spec. Issue Role Magmatism **468**, 98–112 (2009). https://doi.org/10.1016/j.tecto.2008.12.016
7. Evain, M., et al.: Deep structure of the Santos Basin-São Paulo Plateau System (SSPS). Geophys. J. Int. **120**(8), 5401–5431 (2015). https://doi.org/10.1002/2014jb011561
8. Loureiro, A., et al.: Imaging exhumed lower continental crust in the distal Jequitinhonha basin, Brazil. J. S. Am. Earth Sci. **84**, 351–372 (2018). https://doi.org/10.1016/j.jsames.2018.01.009
9. Moulin, M., et al.: Deep crustal structure across an young passive margin from wide-angle and reflection seismic data (The SARDINIA Experiment)—I. Gulf of Lion's margin. BSGF **186**, 309–330 (2015). https://doi.org/10.2113/gssgfbull.186.4-5.309
10. Afilhado, A., et al.: Deep crustal structure across an young passive margin from wide-angle and reflection seismic data (The SARDINIA Experiment)—II. Sardinia's margin. BSGF, ILP Special volume **186**, 331–351 (2015). https://doi.org/10.2113/gssgfbull.186.4-5.331
11. Jolivet, L., et al.: Continental breakup and the dynamics of rifting in back-arc basins: the Gulf of Lion margin. Tectonics **34**, 662–679 (2015). https://doi.org/10.1002/2014tc003570
12. Boillot, G., Froitzheim, N.: Non-volcanic rifted margins, continental break-up and the onset of sea-floor spreading: some outstanding questions. Geol. Soc. London Spec. Publ. **187**, 9–30 (2001). https://doi.org/10.1144/gsl.sp.2001.187.01.02
13. Lavier, L., Manatschal, G.: A mechanism to thin the continental lithosphere at magma-poor margins. Nature **440**, 324–328 (2006). https://doi.org/10.1038/nature04608
14. Klingelhoefer, F., et al.: Imaging proto-oceanic crust off the Brazilian Continental Margin. Geophys. J. Int. **200**(1), 471–488 (2014). https://doi.org/10.1093/gji/ggu387
15. McKenzie, D., et al.: Characteristics and consequences of flow in the lower crust. J. Geophys. Res. **105**(B5), 11029–11046 (2000). https://doi.org/10.1029/1999JB900446
16. Bott, M.H.P.: Evolution of young continental margins and formation of shelf basin. Tectonophysics **11**, 319–337 (1971). https://doi.org/10.1016/0040-1951(71)90024-2

Passive Margin and Continental Basin: Towards a New Paradigm

Daniel Aslanian, Maryline Moulin, Philippe Schnürle, Mikael Evain, Alexandra Afilhado, and Marina Rabineau

Abstract

The existence of an unique thinning process generating passive continental margins must be considered for discussion: the diversity of their structural morphology is indeed related to tectonic heritage, geodynamic context and probably mantle heat segmentation. However, the recurrence of certain general features pleads in favour of common rules of the first-order. The commonly used models, like simple shear, pure shear or polyphase models, to explain the lithospheric stretching and consequent crustal thinning of passive continental margins, exclude exchanges between the lower continental crust and upper mantle, and are thus referred as conservational models. They imply large amount of horizontal movement and are not able to explain the observations collected in the most conjugate margins neither fit the kinematic constraints. Based on wide-angle seismic images on passive margins in the Mediterranean sea, the Central Atlantic, the Equatorial Atlantic, the Central segment of the South Atlantic and the Indian oceans, we are now able to propose a new paradigm.

Keywords

Lower continental crust • Thinning process
Passive margin • Continental basin • Aborted rift

1 Introduction

Since the well-known models of McKenzie [1] and Wernicke [2], understanding the formation of passive continental margins, that is to say mainly the way that continental lithosphere is thinned leading to subsidence, remains one of the main challenges in Earth Sciences. As quoted already by Aslanian and Moulin [3], in the last decades many recent observations and discoveries have modified our basic views of margin formation: the emerged or shallow marine position of margins until the break-up, leading to subaerial emplacement of basalts or salt, followed by basin subsidence during the plate divergent phases, the absence of extensional faults in many continental margins, the anomalous heat flow in the transitional domain, the presence of a strong reflector in the lower crust (the S reflector in the Iberian margin [4] or the T reflector in the Gulf of Lion Margin [5, 6]), the discovery in few places of serpentinized mantle [7], and the detail that some part of lower crust may be missing in the process [3, 8], have all modified our basic concepts of margin formation. Physical and numerical models have attempted to explain the presence of exhumed mantle observed in few places or the wide transitional domain and/or the formation of sag-basins [9–13], but these models have not been placed in precise paleogeographic reconstruction and in global paleo-geographical evolution. In fact, whatever conservational model is applied, it implies considerable amount of horizontal movement that is not been observed in the field. Putting together the observation of kinematic constraints, passive margin, continental basins and aborted rift geometries allow us proposing a new paradigm.

2 Materials and Methods

Since more than twenty years Ifremer (Institut français de recherche pour l'exploitation de la mer), with its academic and industrial partners, have conducted simultaneously and

D. Aslanian (✉) · M. Moulin · P. Schnürle · M. Evain
IFREMER, REM/GM/LGS, Centre de Brest,
29280 Plouzané, France
e-mail: aslanian@ifremer.fr

A. Afilhado
Instituto Superior de Engenharia de Lisboa, Lisbon, Portugal

M. Rabineau
CNRS, UMR6538, LGO (CNRS/UBO/UBS), IUEM,
Plouzané, France

F. Rossetti et al. (eds.), *The Structural Geology Contribution to the Africa-Eurasia Geology: Basement and Reservoir Structure, Ore Mineralisation and Tectonic Modelling*,
Advances in Science, Technology & Innovation, https://doi.org/10.1007/978-3-030-01455-1_73

independently kinematic studies and wide-angle experiments (see Fig. 1) in the Mediterranean sea (Provencal Basin), the Central Atlantic (Morocco margins), the Equatorial Atlantic (Berrenhinhas-Maranão-Ceara Margins), the Central segment of the South Atlantic (Angola Margin, Santos Basin, Jequitinhonha-Almada-Camamu-Sergipe-Alagoas Margins) and the Indian (Mozambique margins) oceans. Each experiment was dedicated to specific a priori scientific questions, which was thought to be solvable at this specific location.

The link between kinematic models and passive margin geometries is so strong that their studies should not be conducted separately: conceptual models of continental thinning must be situated on precise paleogeographic maps in order to test their consequences on global view and to be validated. Furthermore palaeogeographic reconstructions must take into account the detailed observations collected on the passive margins. Indeed, the understanding of the genesis and evolution of a continental passive margin implies (1) to define the exact conjugate margins system, its tightest initial position and to constrain the horizontal motions (2) to define its current and initial geometries and its crustal segmentation and nature, which can be done by combining very good multi-channel seismic (MCS) and wide-angle seismic data (3) and to constrain its vertical motions (subsidence) by studying the above sedimentary sequences with a focus on paleo-environments and stratigraphy sequences.

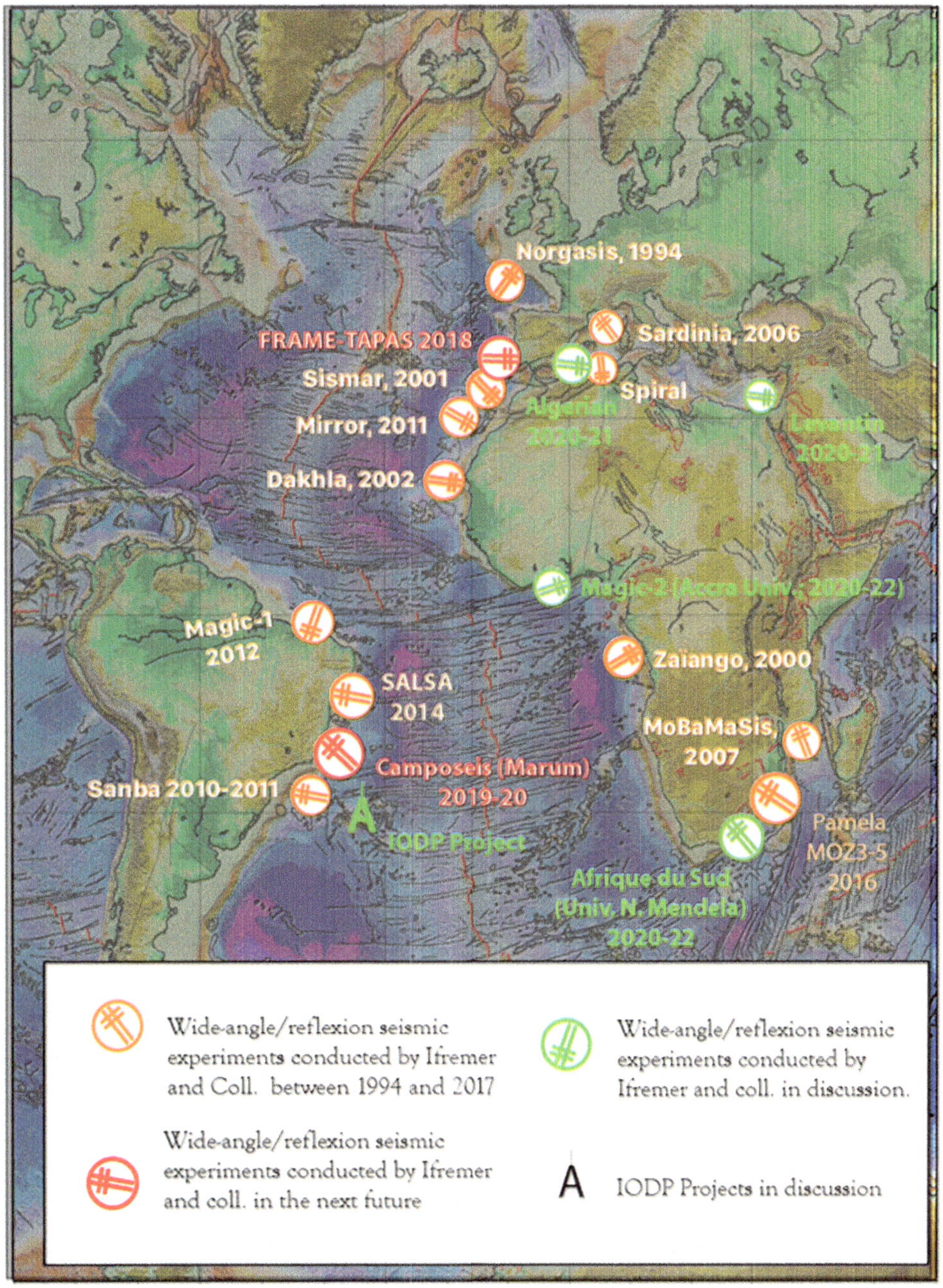

Fig. 1 Localisation of the wide angle/reflexion seismic experiments conducted since 1994 and scheduled for next years by Ifremer and coll. These experiments were accompanied systematically by independent kinematic studies. The new paradigm is based on their combined results

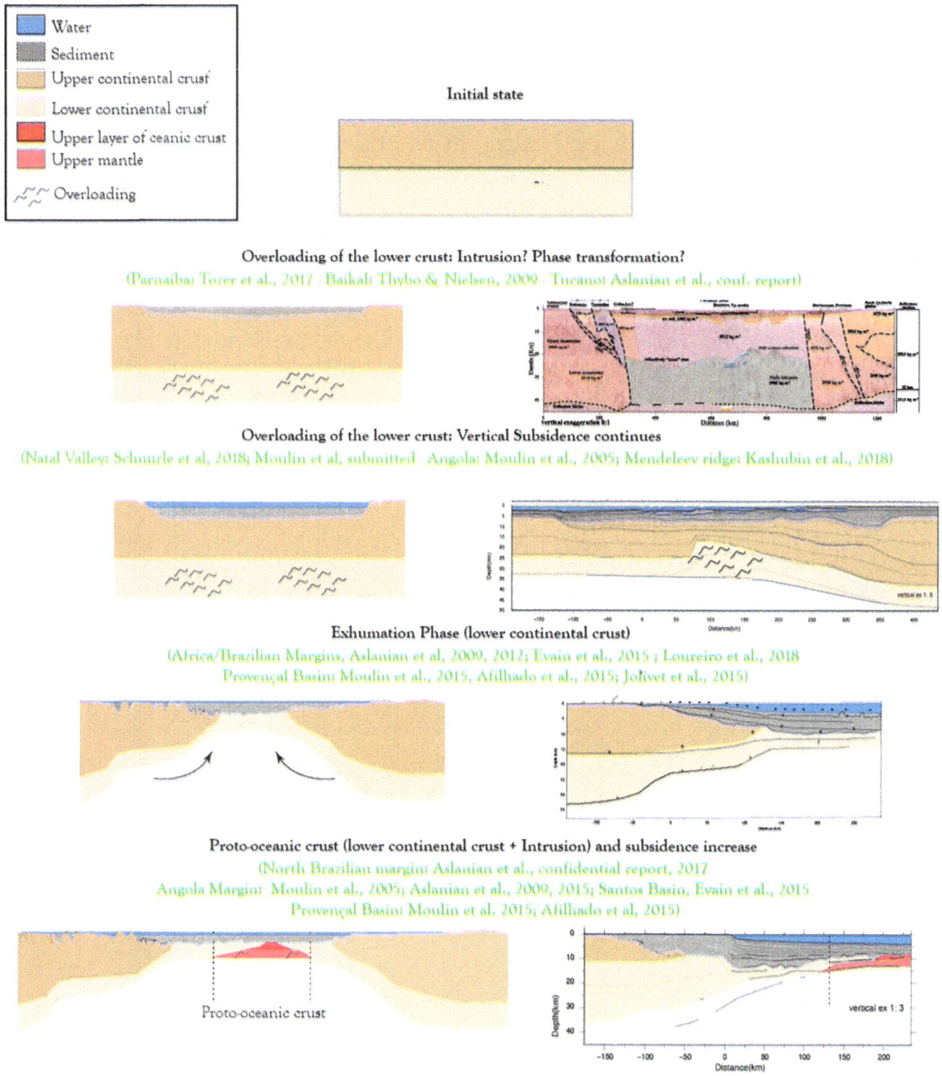

Fig. 2 Left: Model of sedimentary basin and passive margin genesis emphasizing the crucial role of the lower continental crust; Right: some examples for each phase with real data, from the top to bottom after [15, 16, 28, 31]

Determining the initial position of the main tectonic plates, on which are located the passive margins, is thus of a crucial importance: all propositions, all interpretations, all conceptual models must be based on precise paleogeographic maps in order to test their consequences on global view and to be validated.

3 Towards a New Paradigm

We propose here a virtual walk all along the Ifremer dedicated cruises together with a re-evaluation of the paleogeographies before breakup on the Mediterranean Sea [14–18] Central [19, 20], Equatorial and South Atlantic [8, 21–26] and West Indian Oceans [27, 28]. Together with the recent information given by continental basins and aborted rifts [29–31] we propose a new paradigm for the thinning process which confirms the crucial role of the lower continental crust [3, 8] and its relation with the upper mantle, with a phase of overloading of the lower continental crust, a phase à exhumation the lower continental crust and a phase of proto-oceanic crust which involved the lower continental crust and the upper mantle, before a more typical oceanic crust (Fig. 2).

References

1. McKenzie, D.: Some remarks on the development of sedimentary basins. Earth Planet. Sci. Lett. **40**(1), 25–32 (1978)
2. Wernicke, B.: Uniform sense normal simple shear of continental lithosphere. Can. J. Earth Sci. **22**, 108–125 (1985)
3. Aslanian, D., Moulin, M.: Paleogeographic consequences of conservational models in the South Atlantic Ocean. In: Mohriak, W.U., Danforth, A., Post, P.J., Brown, D.E., Tari, G.C., Nemcok, M., Sinha, S.T. (eds.) Conjugate Divergent Margins. Geological Society, London, Special Publications, vol. 369 (2012). http://dx.doi.org/10.1144/SP369.5

4. Hoffmann, H.J., Reston, T.J.: The nature of the S reflector beneath the Galicia Bank rifted margin: preliminary results from pre-stack depth migration. Geology **20**(12), 1091–1094 (1992)
5. De Voogd, B., et al.: First deep seismic reflection transect from the Gulf of Lions to Sardinia (ECORS-Crop profiles in Western Mediterranean). Geodynamics **22**, 265–274 (1991)
6. Pascal, G., et al.: The ocean continent boundary in the Gulf of Lions from analysis of expanding spread profiles and gravity modelling. Geophys. J. Int. **113**, 701–726 (1993)
7. Boillot, et al.: Ocean–continent boundary off the Iberia margin: a serpentinite diapir west of the Galicia Bank. Earth Planet. Sci. Lett. **48**, 23–34 (1980)
8. Aslanian, D., et al.: Brasilian and Angolan passive margins: the kinematic constraints. Tectonophysics Spec. Issue Role Magmatism **468**, 98–112 (2009)
9. Brun, J.P., Beslier, M.O.: Mantle exhumation at passive margins. Earth Planet. Sci. Lett. **142**, 161–173 (1996)
10. Lavier, L., Manatschal, G.: A mechanism to thin the continental lithosphere at magma-poor margins. Nature **440**, 324–328 (2006). https://doi.org/10.1038/nature04608
11. Kusznir, N.J., Karner, G.D.: Continental lithospheric thinning and breakup in response to upwelling divergent mantle flow: application to the Woodlark, Newfoundland and Iberia margins. In: Karner, G.D., et al. (eds.) Geological Society, London, Special Publications, vol. 282, pp. 389–419 (2007)
12. Reston, T.J.: The opening of the central segment of the South Atlantic: symmetry and the extension discrepancy. Pet. Geosci. **16**, 199–206 (2010). https://doi.org/10.1144/1354-079309-907
13. Huismans, R.S., Beaumont, C.: Depth-dependent extension, two-stage breakup and cratonic underplating at rifted margins. Nature **473**, 74–78 (2011). https://doi.org/10.1038/nature09988
14. Afilhado, A., et al.: Deep crustal structure across an young passive margin from wide-angle and reflection seismic data (The SARDINIA Experiment)—II. Sardinia's margin. BSGF, ILP Special **186**, 331–351 (2015). https://doi.org/10.2113/gssgfbull.186.4-5.331
15. Moulin, M., et al.: Deep crustal structure across an young passive margin from wide-angle and reflection seismic data (The SARDINIA Experiment)—I. Gulf of Lion's margin. BSGF **186**, 309–330 (2015). https://doi.org/10.2113/gssgfbull.186.4-5.309
16. Aslanian, D., et al.: Structure and evolution of the Gulf of Lions: the Sardinia seismic experiment and the GOLD (Gulf of Lions Drilling) project. Lead. Edge **31**(7), 786–792 (2012)
17. Leroux, E., et al.: Sedimentary markers: a window into deep geodynamic processes. Terra Nova **27**(2), 122–129 (2015). https://doi.org/10.1111/ter.12139
18. Pellen, R., et al.: The Minorca Basin: a buffer zone between Valencia and Provençal Basins. Terra Nova **28**(4), 245–256 (2016)
19. Sahabi, M., et al.: Un nouveau point de départ pour l'histoire de l'Atlantique Central. C. R. A. S. (2004)
20. Klingelhoefer, F., et al.: Crustal structure variations along the NW-African Continental margin: a comparison of new and existing models from wide-angle and reflection seismic data. Tectonophysics **674**, 227–252 (2016)
21. Moulin, M., et al.: Geological constraints on the evolution of the Angolan margin based on reflection and refraction seismic data (ZaiAngo project). Geophys. J. Int. **162**, 793–810 (2005)
22. Moulin, M., et al.: A new starting point for the history of the South Atlantic Ocean. Earth Sci. Rev. **98**(1–2), 1–37 (2010). https://doi.org/10.1016/j.earscirev.2009.08.001
23. Moulin, M., et al.: Kinematic keys of the Santos—Namibe Basins. Geological Society, London, Special Publications, vol. 369 (2012) http://dx.doi.org/10.1144/SP369.3
24. Evain, M., et al.: Deep structure of the Santos Basin-São Paulo Plateau System (SSPS). Geophys. J. Int. **120**(8), 5401–5431 (2015). https://doi.org/10.1002/2014jb011561
25. Loureiro, A., et al.: Imaging exhumed lower continental crust in the distal Jequitinhonha basin. Brazil. J. South Am. Earth Sci. **84**, 351–372 (2018). https://doi.org/10.1016/j.jsames.2018.01.009
26. Pinheiro, J., et al.: Modeling onshore-offshore wide-angle seismic data across the Alagoas-Sergipe passive margins, NW Brazil. J. South Am. Earth Sci. (2018)
27. Leinweber, V., Klingelhöfer, F., Neben, S., Reichert, C., Aslanian, D., Matias, L., Heyde, I., Schreckenberge, B., Jokat, W.: Deep crustal structure of the Mozambic margin from wide-angle and reflection seismic data (The MoBaMaSis experiment). Tectonophysics **599**, 170–196 (2013)
28. Moulin, M., et al.: Gondwana breakup, passive margin genesis and contourite: insigths from the Natal Valley. submitted to Geology
29. Thybo, Nielsen: Magma-compensated crustal thinning in continental rift zones. Nature **457** (2009). https://doi.org/10.1038/nature07688
30. Kashubin, et al.: Crustal structure of the Mendeleev Rise and the Chukchi Plateau (Arctic Ocean) along the Russian wide-angle and multichannel seismic reflection experiment "Arctic-2012". J. Geodyn. (2018). https://doi.org/10.1016/j.jog.2018.03.006
31. Tozer, B., Watts, A. B., Daly, M. C.: Crustal structure, gravity anomalies, and subsidence history of the Parnaíba cratonic basin, Northeast Brazil. J. Geophys. Res. Solid Earth **122**, 5591–5621 (2017). https://doi.org/10.1002/2017JB014348
32. Schnürle, P., Leprêtre, A., Verrier, F., Evain, M., Aslanian, D., Leroy, S., de Clarens, P., Dias, N., Afilhado, A., Moulin, M.: Crustal structure of the Natal Valley from combined wide-angle and reflection seismic data (MOZ3/5 cruise), South Mozambique Margin, SEISMIX conference (2018)
33. Jolivet, L., Gorini, C., Smit, J., Leroy, S.: Continental break-up and the dynamics of rifting in back-arc basins: the Gulf of Lion margin. Tectonophysic (2015)

Interpretation of Global GPS-Vectors Map

Muhammad Hassan Asadiyan Falahiyeh

Abstract

Distribution of GPS-vectors denotes that the Earth's crust is twisted from North Pole and South Pole spirally around two hinges: Aleutian and Scotia. This torsion field originates from the Mecca-spiral with two arms: Dahw (rolling) and Tahw (spreading). D and T are two main components of geodynamics. In this paper, I constrain the general vorticity of the Earth's crust within a sigmoid called global sigmoid (GS). Head and tail of GS is approximated by two wings: A and B, which are rolled toward the central bar (Mecca). A-wing bounds divergent velocity vectors (source) and B-wing bounds convergent velocity vectors (sink). The highest divergent velocity starts from Nazca (inflection line of GS) and ends (converges) toward the Mariana Trench. The minor divergent along the Mid Atlantic Ridge occurred in the excess part (0.618–0.382). The torsion field partitionizes into two orthogonal belts: in the east hemisphere, it is tuned along the E/W-direction, and in the west hemisphere, it is tuned along the N/S-direction. Orthogonal tuning applies inverse polarization in the GPS-vectors as well as geomorphology, gravitational anomaly and magnetic anomaly. Creation of a mountainous spot (Himalaya) in opposite direction of abysmal spot (Bermuda) is a typical sample of morph-polarization. The Bermuda is located between two mountainous spots (Rocky and Ands) but the Himalaya is between two abysmal spots (Gulf of Indus and Gulf of Bengal).

Keywords

Dahw (rolling) • Tahw (spreading) • GPS-vectors Sigmoid • Spiral • Geodynamic pole • Catastrophic zone

M. H. A. Falahiyeh (✉)
Payam-e Noor University, Ahwaz, Iran
e-mail: asadiyan@pnu.ac.ir

1 Introduction

Interpretation of GPS-vector's distribution in small scale and big scale is a serious challenge for the Plate Tectonics (PT). I chose Self Interpretation Method (SIM) for understanding Earth-dynamics. SIM means the study of the Earth's face, which appears in the GPS-vectors map, morphologic map, gravity map and magnetic map. For resolving this problem here, I highlight just few of the lineaments in the pictures, and the rest can be guested them in such a way to have a harmonic shape such as spiral, Mobius, Ƨ-curve, S-curve, and Euler Elastic Curve (Ω-like curve). These topological elements could be traces of folding, faulting, trench or ridge, as actually the alphabets of Earth-symphony. This paper attempts to explain this nice symphony just within four pages.

2 Materials and Methods

The distribution pattern of GPS-vectors shows general deformation on the Earth's crust. According to observational study, the best model that fits with the global GPS-map is sigmoid. It seems that the Earth's crust like a spring follows cycloidal path i.e. as it extends from two ends at the same time; the cycloid's circum is shrinking. Continental Decay Point (CDP) is the extreme point of Dahw, and North-Pole (NP) or Oceanic Decay Point (ODP) is the extreme point of Tahw (Fig. 1). The distance between CDP and ODP represent the amount of friction between Dahw-Field and Tahw-Field. The typical sample of DT in the surface is Arabia-Cylinder, which is located in the center of Global-Sigmoid; it spreads out along t-axis and rolls out along the d-axis. The GS showed in this picture bounds the Global Mobius, the GM forms from the forward segment shown by S-curve and retard segment shown by Ƨ-curve in the sketch. For better demonstration of the subject, in Fig. 1, we draw two sketches: the left one shows vorticity of the

F. Rossetti et al. (eds.), *The Structural Geology Contribution to the Africa-Eurasia Geology: Basement and Reservoir Structure, Ore Mineralisation and Tectonic Modelling*,
Advances in Science, Technology & Innovation, https://doi.org/10.1007/978-3-030-01455-1_74

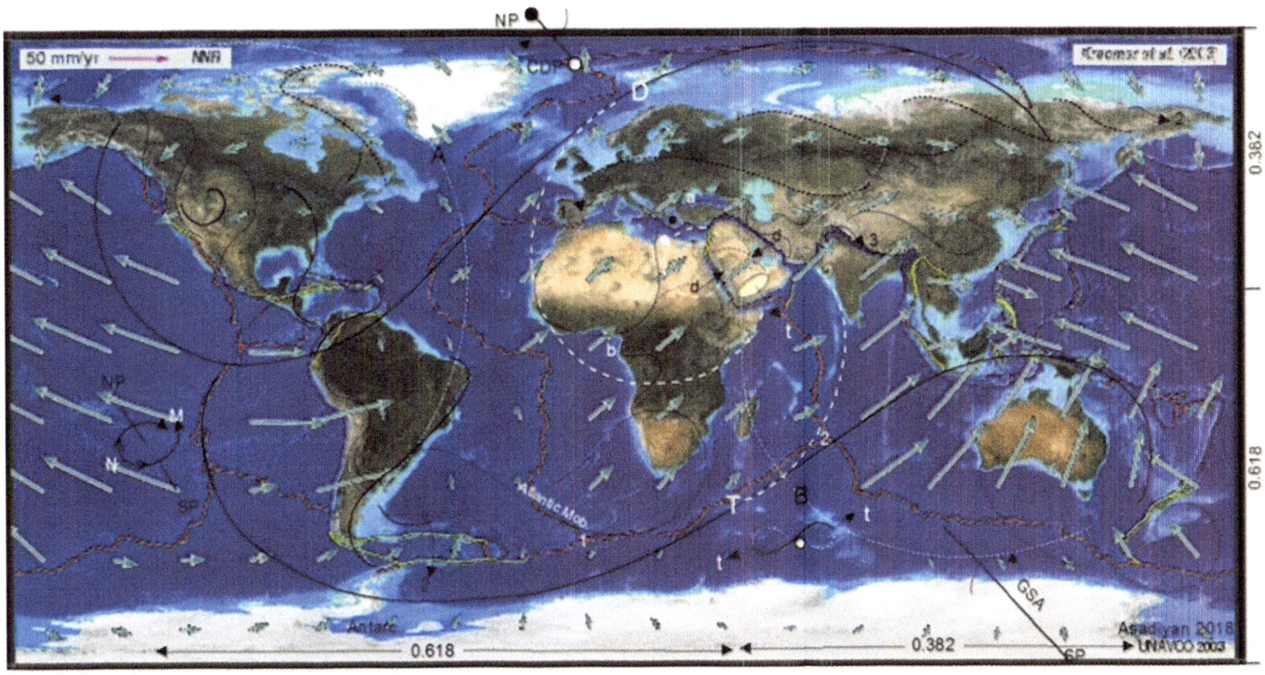

Fig. 1 Global gps-vectors map. Left sketch is global cycloid. gps-vectors diverge from inflection point N (Nazca) and terminate in the M (Mariana). Right sketch is plan form of global cycloid which named global mobius

Earth in the glob and the right one shows vorticity of the Earth in the plan. In the glob-model, the vorticity of the Earth is symbolled by a cycloid but in the plan-model, it is symbolled by a Mobius. Actually, GS is a cycloid in which the GPS-vector diverges from its inflection point (Nazca) and converges toward its terminal point (Mariana). White circle in the small sketch is cross section of the Arabian-Cylinder. Cycloid defamation in the surface converts to sigmoidal deformation; hence, we see best sigmoidal structures in the shear zones.

Figure 1, shows GPS-vectors in global scale in match with global sigmoid. D/T-spiral originates from the Mecca and ends to the Aleutian and Scotia; this is a mother spiral, which all structures are born out of its womb. Global sigmoid represents the Earth-Vortex, a/b-wing represents primary dahwing (a trusted over b) and A/B-wing represents secondary dahwing. A (convex share) and B (concave shore) wings bound the biggest divergent and convergent vectors, respectively. Geomagnetic-axis passes through A-C. Global Sigmoid Axis (GSA) rotates reversely from NP and SP around the Aleutian and Scotia. The inverse tuning of Eurasia in the E-W direction between 1 and 2 produces big orogeny in the middle part but the inverse tuning of America in the N-S direction between 1 and 2 produces big sea in the middle part (mass-polarization). Between 1 and 3 is an extending region; hence, many lakes and seas have been created here. GS with extensional t-axis along the direction of propagation and comprational d-axis in the orthogonal direction (retard path passes through the Red Sea). A/B-wings roll toward Geodynamic Pole. GPS-vectors diverge from the Arctic and converge in the Antarctica (opposite to the geomagnetic field). Horizontal Aleutian-hinge bounds the seas but vertical scotia-hinge bounds the lands. The sense of vectors changes along the MAR. The highest divergent starts from Nazca (inflection line of GS) and ends (converge) toward Mariana. Minor curvature occurs in the excess-part (0.618–0.382), showing slower divergent vectors.

Figure 2a, show the GPS-vectors in local scale. GPS-vectors diverge with respect to the cross-point of bended-mobius. Bended-mobius is an envelope of the African-Apex. Two arrows show the divergent folding in Arabian-Shield and Nubian-Shield. Figure 2b, is a geoid. δ-type cyclone bounds Africa, the east-trail connected to Himalaya and the west-trail connected to South-America. The highest and lowest anomalies have form in the extreme region between the two trails. The Sharpest topographic point has formed in the cross point of African-Mobius. Low anomaly in Central Africa has formed in the south of the cross point. The corner pictures are from the Persian Gulf: left one shows symmetry in morphology with respect to the central-axis and right one show Mobius-form of gravity anomaly. African-Mobius and Persian Gulf's Mobius sang one sing. Conclusion from this small sample denotes that: *boudin-structures formed due to torsion not simply due to extending.*

During the spreading of D-arm from the central bar to the outer bar, the decay length [1] extends from d to D. The

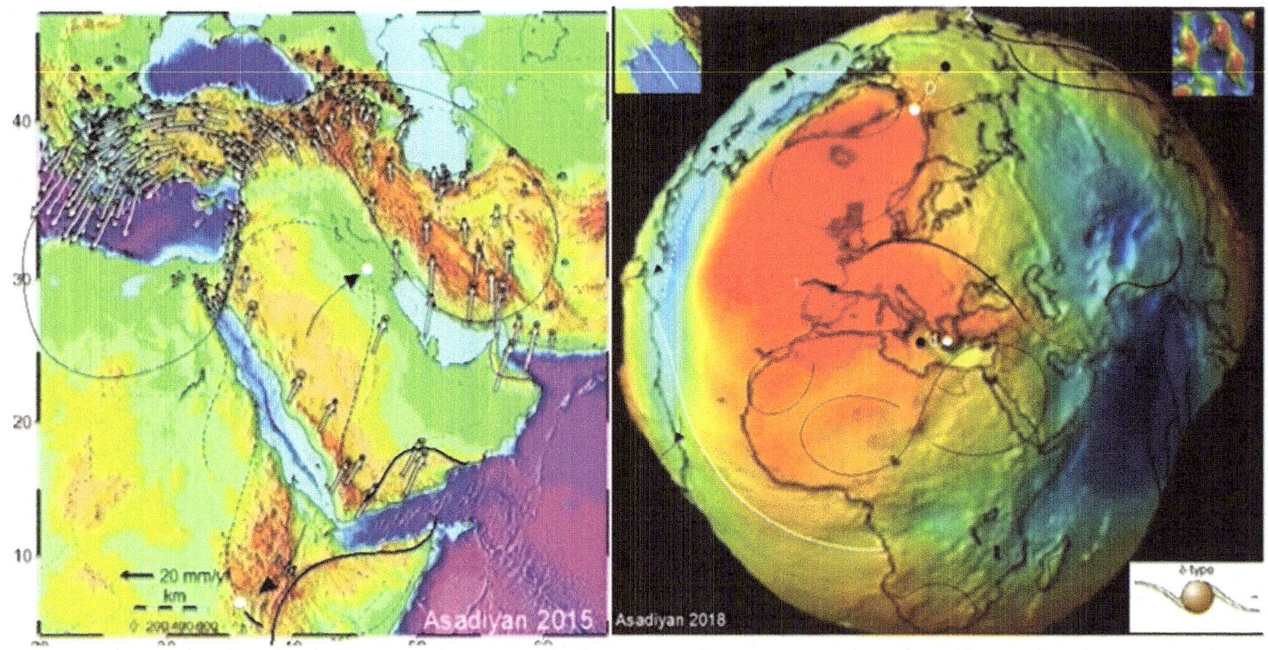

Fig. 2 a distribution of gps-vectors in small scale which completely match with form of mobius. b distribution of gravity anomaly which completely match with δ-type of tectonic cyclone

black and white-points are inflection point and CDP, respectively. Notice that these fundamental points formed from the same dynamics (combination of reciprocal mobiuses). Notice also that high anomaly has formed due to the interferences of these mobiuses. Another important point in this picture is the inverse tuning of Eurasia between 1 and 2 which produce Euler Elastic Curve. The Ural Mountain (Mid Continental Ridge) has formed in the middle part of EEC, and the Mid Atlantic Ridge has formed between the African-Cyclone and the American-Cyclone (Fig. 3).

This idea not only interprets the distribution of GPS-vectors and gravity anomaly but also the distribution of magnetic anomalies. Figure 3 shows magnetic anomalies along Atlantic Ocean. Zooming in inverse anomalies in the white Mobius (∞-like curve) could dissolve the mystery of symmetrical strip along the Atlantic. It seems that magnetic reversal or reciprocal anomalies (+ and −) created along the Orthogonal Saddles, Vertical Saddle (VS) and Horizontal Saddle (HS). This Mobius (Etendeka) is exactly located in the *inflection region* of the African-block (couple of this Mobius on the other side is Santos). In the inflection point, anomaly suddenly changes from green to red. This scenario is repeated in bigger scale between the African-Cyclone (black curve) and the American-Cyclone (white curve). Look carefully to the arrangement of strips along the catastrophic-zones (cz) in the middle part of the loops of Atlantic-Mobius (blue/red ∞-like curve); this inverse arrangement could be created due to the *inverse tuning of tectonic-cyclones*. These strips have been created along the VS and HS of Atlantic-Mobius. Each loop (red and blue) of Atlantic-Mobius fractaly has been partitionized into two divergent loops: a and b. Due to progressive tuning, the Atlantic-floor has been partitionized to 64 strips. The left lateral strike slip fault confirms this divergence. Disc-like anomaly inside the ellipsoid is a good evidence for Spiral Tectonic [2]. Anomaly in Santos Basin Brazil follows Mobius-strip (white dotted curve). Symmetry with respect to *inflection line* or *catastrophic zone* is also seen in geomorphology (refer to the left corner picture in Fig. 2b). *When the polarity of spiral changes (after 180° rotation), inflection occurs.*

3 Discussion

It seems that after creation of the core and mantel by mother-spiral, the first nucleation of tectonic emerge from Mecca. After billions of years of spiraling (dahwing and thawing) the spiral-arms have reached current state. In the primary period, GP was coinciding with NP. According to $f_c = h/2d$ formula, tidal friction (f_c) in the primary period was maximum (because d was minimum) but in the secondary period when d increased to D, the friction decreased to 2.08×10^{-3}, h is the average of elevation differences between the continents and oceans (h is assumed to be about 5 km, d about 500 km and D about 1200 km). D is a distance between NP and CDP (Fig. 2b). It seems that when first period of spiral interferes with the second period, we

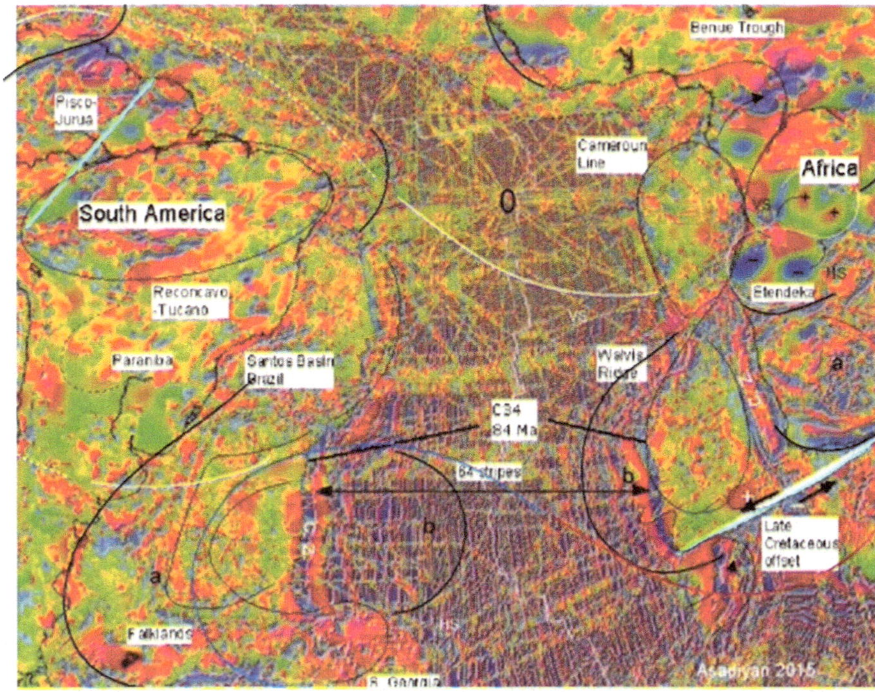

Fig. 3 Distribution of magnetic strips along Mid Atlantic Ridge is resultant of cyclone dynamics between African-cyclone and American-cyclone

have seismicity (also when second period terminates around the fire ring). This why the safest point (Mecca) is exactly located in the opposite direction of the most active region (Fire Ring). Unlike PT understanding of sea floor spreading in the Atlantic is most difficult part in this idea.

4 Conclusions

Due to spiraling, GPS-vectors diverge from the convex-shore and converge to concave shore (like a river). Global Sigmoid is the best model matching with Earth-spiraling and the distribution of GPS-vectors. Small GPS-vectors along MAR occurred in excess part (0.618–0.382) of Fibonacci-Proportion. Sea floor spreading and symmetrical strips along the Atlantic occurred due to the inverse tuning of the tectonic cyclones of Africa and South America. Magnetic polarization in the A/B-wing (Global-Mobius) exactly occurred as in the Blue/Red-wing (Atlantic-Mobius).

References

1. Edwards, B.F., Smith, D.H.: River meandering dynamics. Phys. Rev. E **65**, R046303 (2002)
2. Asadiyan, M.H., Zamani, A.: Could Dahw/Tahw dissolve problems of plate tectonics? Asian J. Earth Sci. **3**(4), 190–212 (2010)
3. The Holy Quran, 79/30 and 91/6

Printed by Printforce, the Netherlands